Animal
Play
Behavior

418

P.482
- well-known
patterns

P 64

Testable
about

Cooperd
P.394

well
Effects of play
Weather on play
P.303

P.52

Moment Mistake
P.338

P.492 Symons, V, M.B

P.429 P.7 - Inter x-sex + x-age RLT

Sea. analysis P.413

1 year old not
play/fight with
2 year old females
P.435

Effects + function P.316

A context P.308

3 general classes of function P.357

Dark side of play
P.438

Self-handicapping Sequently
inconsistent with R-F hypothesis
P.396

Physical Training
P.310

Social Relations
P.373

Play → Agg
P.435

£11.95N

Animal Play Behavior

Robert Fagen

University of Pennsylvania

New York   Oxford
OXFORD UNIVERSITY PRESS
1981

Library of Congress Cataloging in Publication Data

Fagen, Robert.
  Animal play behavior.

Bibliography: p.
  Includes index.
  1. Play behavior in animals. I. Title.
QL763.5.F33   599.05'1   80-11396
ISBN 0-19-502760-4
ISBN 0-19-502761-2 pbk.

The author gratefully acknowledges permission to reprint excerpts from the following sources:

Bateson, G. (1956). A. Hasler, quoted by G. Bateson, The message "This is Play," in *Group Processes* (Trans. 2nd Conf.) (ed. B. Schaffner). Macy Foundation, N.Y., p. 235. (p. 470)

David, J. H. M. (1975). Observations on mating behaviour, parturition, suckling and the mother-young bond in Bontebok (*Demaliscus dorcas dorcas*). *J. Zool., Lond.* 177: 203–223. By permission The Zoological Society of London. (pp. 197–198)

Fagen, R. (1976). Exercise, play, and physical training in animals in *Perspectives in Ethology, Vol. 2* (eds. P. P. G. Bateson and P. H. Klopfer). Plenum, N.Y. pp. 189–219.

Fagen, R. M. (1977). Selection for optimal age-dependent schedules of play behavior. *Am. Nat.* 111: 395–414. By permission The University of Chicago Press, © 1977 by The University of Chicago. All rights reserved. (pp. 374–376)

Fagen, R. M. and T. K. George (1977). Play behavior and exercise in young ponies (*Equus caballus L.*). *Behav. Ecol. Sociobiol.* 2: 267–269. (p. 313, Table 5–6)

Galdikas, B. (1978). Orangutans and hominid evolution in *Spectrum: Essays Presented to Sultan Takdir Alisjahbana on his Seventieth Birthday* (ed. S. Udin). Dian Rakyat, Jakarta, pp. 304–305. By permission. (pp. 323–324)

Hadamard, J. (1945). "A Testimonial from Professor Einstein" in *Psychology of Invention in the Mathematical Field*. Princeton University Press, Princeton, N. J., pp. 142–143. Reprinted by permission of Princeton University Press, copyright 1945 © 1973 by Princeton University Press. (pp. 469–470)

Hall, K. R. L. (1965). Behaviour and ecology of the wild Patas monkey, *Erythrocebus patas*, in Uganda. *J. Zool., Lond.* 148: 15–87. By permission The Zoological Society of London. (pp. 100–101)

Henry, J. D. and S. M. Herrero (1974). Social play in the American black bear. *Am. Zool.* 14: 371–389. By permission. (p. 141)

Printed in the United States of America

Koford, C. B. (1957). The Vicuna and the puna. *Ecol. Monogr.* 27: 153–219. Copyright 1957 by The Ecological Society of America. (p. 188)

Kuhn, T. S. (1962). *The Structure of the Scientific Revolution*. By permission The University of Chicago Press, p. 45, © 1962 by the University of Chicago. All rights reserved. Published 1962. (p. 40)

McGrew, W. C. (1977). Socialization and object manipulation of wild chimpanzees in *Primate Bio-Social Development* (eds. Suzanne Chevalier and Frank E. Poirier). By permission Garland STPM Press, N.Y., pp. 276–277. Copyright © 1977 by Garland Publishing, Inc.

Milne, L. (1924). *The Home of an Eastern Clan: A Study of the Palaungs of the Shan States*. Clarendon Press, Oxford, pp. 8–9. By permission of Oxford University Press. (p. 292)

Murray, J. A. H., H. Bradley, W. A. Craigie, and C. T. Onions (eds.) (1933). *The Oxford English Dictionary*. Clarendon Press, Oxford. (pp. 509–512)

Nabokov, V. (1974). From *Look at the Harlequins* by V. Nabokov. McGraw-Hill Book Company, N.Y., pp. 8–9. Copyright © 1974 by McGraw-Hill International, Inc. Used with permission of McGraw-Hill Book Company. (p. 449)

Onions, C. T. (ed.) (1973). *The Shorter Oxford English Dictionary*. Clarendon Press, Oxford.

Schaller, G. B. (1972). *The Serengeti Lion*. Reprinted by permission The University of Chicago Press, p. 159, © 1972 by the University of Chicago. All rights reserved. Published 1972. (pp. 165, 167)

Sidor, E. (1962). "Being," © 1962 by The New York Times Company. Reprinted by permission. (p. 493)

Simpson, M. J. A. (1976). The study of animal play reprinted from *Growing Points in Ethology*, edited by P. P. G. Bateson and R. A. Hinde, by permission of Cambridge University Press, p. 386, © Cambridge University Press 1976. (p. 98)

Stefansson, V. (1944). *The Friendly Arctic*. Macmillan Publishing Co., N.Y., pp. 53–54. Copyright © 1944 by Vilhjamur Stefansson. Reprinted by permission of McIntosh and Otis, Inc., N.Y. (pp. 191, 194)

van Lawick-Goodall, J. (1968). The behaviour of free-living chimps in Gombe Stream Reserve. *Animal Behav. Monogrs.* 1: 161–311. By permission Baillière Tindall, London. (p. 109)

Preface

Animal Play Behavior addresses a major biological paradox. Why do young and old animals of many species spend time and energy, and even risk physical injury, performing the apparently unproductive behaviors colloquially called play? What makes this "useless" activity so important that animals literally risk their lives for it? And, even more curiously, why are humans both enchanted and enraged by play? My principal purpose in writing the book was to better define these and other biological questions about play by formulating an evolutionary approach to the development of behavior based on recent advances in theoretical biology. This approach led me to challenge the traditional separation of evolutionary and developmental approaches to behavior. I prefer to address this historical dichotomy by distinguishing between ultimate and proximate questions about behavioral development.

Animals that play would seem to be at an evolutionary disadvantage. By sacrificing time, energy, and safety for play, they negatively influence their chances of surviving to reproduce, all in order to perform behavior that lacks an obvious beneficial product. But play is important, both to animals and to scientists, for the following reasons:

1. Play experience may develop adaptive behavioral flexibility (Bruner 1976, Einon, Morgan, & Kibbler 1978, Sutton-Smith 1979).

2. Play experience apparently accelerates development (Clark

et al. 1977), stimulates brain growth (Ferchmin et al. 1980), and aids recovery from the effects of malnutrition (Grantham-McGregor et al. 1979) and of social deprivation (Cummins & Suomi 1976); it has been found to help hospitalized children (Jolly 1968, Morris 1968, Susser & Watson 1969).

3. Play is central to behavioral epigenesis because it is a crucial mediator of environmental effects on the phenotype. Both as a set of epigenetic rules and as a mechanism for modifying epigenetic rules in response to individual play experience, play is an essential element in gene-culture interaction, for the key to the biological study of culture is precisely behavioral epigenesis (Lumsden & Wilson 1981). Long recognized as the crux of the anthropological (Huizinga 1950) and psychological (Erikson 1977) study of culture, play may soon become a focal element of cultural biology as well.

4. Play seems to provide a foundation for cognitive and motor skills used in interacting with the inanimate and social environment (Stern 1974b, Symons 1978a) and can be a significant source of physical exercise (Fagen & George 1977).

5. Play may facilitate recognition of kin, social attachment formation, and social bond maintenance and reinforcement (Bekoff 1978b, Bekoff & Byers 1979, Wilson 1973, Wilson & Kleiman 1974).

These claims may prove to be exaggerated, but they are grossly understated in comparison to many public statements about the importance of play. For example, the announcement of a First National Conference on the Vital Role of Play in Learning, Development and Survival (Washington, D.C., May 2/3, 1979) stated, "Play is vital to the healthy development of all of the so-called higher animals. Play is a biological imperative. Play heals, naturally and as a tool in psychotherapy with children. Play for your life: the stakes are survival."

If even one of the statements, inferences, and claims listed above proves true, then aggression and sex can no longer be considered the most important problems for behavioral research. I believe that general recognition of play as a central issue in the study of behavior is imminent as a result of the investigations cited above. Furthermore, play is scientifically important for additional reasons. Play is virtually the only nonhuman behavior to furnish analogs of human language (Marler 1977) and human deception (Thorpe 1966). Play provides possible evidence for human uniqueness as well as for biological continuity between human and nonhuman behavior. *Homo sapiens*—man the toymaker—is the only species known to fashion objects for use in play by its offspring. This form of tool behavior is uniquely human. In contrast, E. O. Wilson (1977a, b) views play as one of five phenomena most suited to sociobiological investigation of possible genetic foundations of human nature. To psychologists and clinicians, play is an important clinical sign of well-being (Cummins & Suomi 1976, Mason & Berkson 1975). Scientists concerned with parents, children, and society use play as one indicator of the "health" of a parent-child relationship, as suggested by occurrence of play only in the closest and most mutually responsive relationships between confined rhesus macaque mothers and their infants (Hinde & Simpson 1975). Play is important in the study of hormones and behavior (Goy 1970), in mammalian systematics (Griffiths 1978), in defining essential differences between insect and mammalian societies (E. O. Wilson 1971b, pp. 218, 460), in assessing the well-being of farm animals under intensive management (Marx, Schrenk, & Schmidtborn 1977), and in other areas of biology, of anthropology, of psychology, of animal science, and of veterinary medicine. For all these reasons, there is a need for a scientific monograph on play behavior that covers, as objectively as possible, what is known and what is unknown about play, in order to assess its importance.

Animal Play Behavior stresses the functional and evolutionary biology of play in animals. Although students of behavior recognize that an understanding of animal play would offer important insights on development and on social behavior, progress has been slow. One difficulty is that past research on these topics has lacked an adequate theoretical basis. Current biological understanding of the evolution of behavior, although still incomplete, represents a significant advance and may offer new insights into the behavior of developing organisms.

The book represents a contribution to the evolutionary biological study of behavioral development, a relatively new, but rapidly growing area of biological science. Sociobiological, evolutionary psychobiological, and developmental behavioral ecological approaches are stressed. This perspective serves to identify, to characterize, and in part, to resolve theoretical and conceptual issues that have confused the study of play. The analysis relates new theoretical ideas on development and social behavior to current information on animal play. In so doing it serves to broaden the scope of play research by suggesting new hypotheses and new opportunities for study.

I view play as a behavioral tactic available to immature and mature individuals and expressed in accordance with strategies dictated by natural selection. Evolutionary consideration of the behavior of sexually immature organisms and its development is the basis for this approach. The entire life history reflects biological adaptation, and biological constraints on behavior at each age include trade-offs between maintenance, survivorship, growth, fat deposition, and reproduction within and between different ages. These constraints almost always involve behavior of other organisms, including parents, siblings, additional kin, and unrelated peer and non-peer conspecifics as well as members of other species. But play can also be studied from a developmental perspective in individuals. It is customary to separate the analysis of behavioral development from the analysis of behavioral evolution. This distinction can clarify and simplify the study of many behaviors. Unfortunately, play does not appear to be an easily separable problem. It is precisely the interaction between development and evolution that has made the analysis of play behavior so difficult.

Concepts of social evolutionary biology and evolutionary developmental ethology are central to the work. Thus evolutionary bases of behavioral development and current theory on social behavior of young animals are outlined in order to place the natural history of play in a biological context that facilitates further analysis and insight. Structural and functional ethological definitions of play and the serious definitional and research strategic problems associated with play are discussed. The range of the behaviors covered is wide, from animal social playfighting to the solo capers and gambols of hoofed mammals, from the leaps and dashes of house mice to the rhythmic, mutually stimulating informal games human mothers and their children play. Formal games with rules, human sports, and a miscellany of animal behaviors having no apparent biological goal will be excluded for reasons to be outlined in the text. Probable effects or consequences of play (both beneficial and harmful) are discussed. Immediate causation and motivation of play as well as age, sex, habitat, and species differences will be discussed in the light of biological theory.

The book should be of interest to researchers and students in animal behavior and in ethology, particularly to those familiar with modern evolutionary views of behavior, and to sociobiologists and behavioral ecologists. Comparative and developmental psychologists and traditional ethologists will find that the work will challenge many of their beliefs about the study of behavior. General readers in biology, anthropology, psychology, sociology, and child development should be interested in the book, especially in the chapters on natural history, in which no technical background in modern biology is assumed. In theoretical chapters, I assume biological background at the level of Dawkins (1976b) and Wilson & Bossert (1971), but these chapters are not highly mathematical.

A reader of this book can learn about the natural history of animal play behavior in depth. My review of this diverse phenomenon includes many important accounts overlooked by or unavailable to previous commentators. The reader can also expect an introduction to current biological ideas about play.

The study of animal play deepens our understanding of the lives of young mammals and birds. Young animals have chasing and wrestling matches. They play elaborate games resembling human tag, hide-and-seek, "King of the castle," and blind man's buff. Young primates even enjoy being tickled. Just as these amusing antics may once have spurred human efforts at domestication, undeniable parallels between animal play and play of human children give young animals honorary membership in human society. Our intellectual concepts of young animals' worlds, as well as the growing human commitment to other species' welfare and to protection of these species' environments, are enriched by the study of animal play.

I am indebted to Stuart Altmann, P. P. G. Bateson, Sarah Blaffer Hrdy, Ruth Chodrow, Stephen Jenkins, Roger Lewis,

John Pfeiffer, and Donald Symons for reading and commenting on one or more drafts of this book. Their advice strengthened the manuscript in many ways. I did not act on every suggestion received, but their insights and responses were consistently rewarding.

I thank J. Altmann, M. Bekoff, and D. McDonald for reading early drafts of Chapters 1, 5, 6, 7, and 8.

Particular sections of Chapter 3 were read by specialists familiar with the animal species discussed in those sections. These biologists and the sections they read are as follows: D. B. Croft (marsupials), P. Moehlman (*Canis mesomelas*), D. Müller-Schwarze (rodents, *Antilocapra americana*, cervids, birds), N. Owen-Smith (*Ceratotherium simum*), and S. Wilson (*Phoca vitulina*). I am grateful for their readings and for the comments supplied.

For discussions, permission to cite unpublished material, and responses to queries I thank J. and S. Altmann, J. and J. Baldwin, E. M. Banks, D. Barash, G. Barlow, P. P. G. Bateson, B. Beck, M. Bekoff, J. Berger, C. Berman, I. Bernstein, K. Bildstein, N. Bishop, J. T. Bonner, W. H. Bossert, A. Brownlee, I. Chase, D. Cheney, T. Clutton-Brock, D. B. Croft, I. DeVore, K. Dolgin, D. Einon, S. Ellis, V. Eterović, J. Fagen, P. Ferchmin, B. M. F. Galdikas, V. Geist, S. J. Gould, W. Greenough, Z. Havkin, D. Kay, G. Kuester, V. Langman, C. Lumsden, N. Mankovich, J. Maynard Smith, M. McClintock, D. McDonald, G. Michener, P. Moehlman, G. Moran, D. Müller-Schwarze, N. Owen-Smith, N. Owens, P. Parker, S. Pellis, J. Pfeiffer, R. Powell, A. C. Purton, S. Ralston, A. Rasa, G. Rathbun, M. Rutzmoser, G. Salanga, T. Schoener, D. Singer, R. Small, D. Sweeney, I. Taylor, J. Terborgh, C. Wemmer, E. O. Wilson, S. Wilson, A. Zahavi, and E. Zimen. As this long list indicates, I have received information from many colleagues; my apologies to anyone whose name has inadvertently been omitted.

I am especially grateful to Janet Levy for her generous permission to cite results and discussions in her doctoral dissertation and to Helena Rivero for a detailed summary of field notes on baboon play. My special thanks also to Ellen Sidor for her kindness in allowing me to employ her poem "Becoming" in the Epilogue. On a sadder note, I acknowledge unpublished material on play in dasyurid marsupials, a uniquely valuable contribution to play research by the late Graham Settle.

I was privileged to observe and photograph behavior of carnivores and other mammals in the Chicago Zoological Society (Brookfield Zoo) collection. For this opportunity, and for the superb assistance and cooperation that facilitated my research at Brookfield Zoo, I thank Director George B. Rabb, Curators Benjamin Beck and Pamela Parker, Lead Keepers James Rowell and Kenneth Lang, and Keepers Tony Blueman, Eric Delbecq, Johanna Torpey Fagen, Dennis Grimm, Craig Kitchen, Karen Nesetril, Rodger Phillips, and Roger Reason.

For the unusual opportunity to observe and photograph play of young Trakehner horses, I thank Dr. and Mrs. David Goodman, owners of Wonderland Farms, West Chester, Pa., and Ms. Sheila Whaley, Farm Manager.

Publishable-quality photographs of animal play are rare. For this reason, I am very grateful to Owen Aldis, Stuart Altmann, Joel Berger, John M. Bishop, Tony Blueman, Dave Dorn, Maitland A. Edey, Johanna Fagen, Dian Fossey, Stephen J. Krasemann, Hugo van Lawick, Phillip Parker, Rodger Phillips, the late Graham Settle, Philip Teuscher, and Susan Wilson, who generously permited me to reproduce their work. A particularly fine selection of original material on play was made available to me by John M. Bishop, a gifted photographer whose contributions to anthropology, primatology, human ethology and animal behavior research significantly enriched this volume.

For typing and clerical assistance I thank R. Chodrow, I. Fagen, J. Fagen, M. Fagen, M. Kruse, D. Rofini, W. Ross, K. Whitmore, C. Woolley, and most of all Gitta Salanga. Gitta typed most of the final manuscript and carried out a wide range of clerical and administrative tasks. Her help made the final stages of the project particularly easy.

I am also grateful to Joyce Berry, Michael Cook, Carol Miller, and Robert Tilley, my editors at Oxford University Press, for their help.

I thank the Geraldine R. Dodge Foundation for sponsoring me as the first Geraldine R. Dodge Scholar in Animal Behavior during my final year of work on this book. I first observed wild animals at play in the Great Swamp Wilderness (Morris County, New Jersey)—a unique area in which Mrs. Geraldine Rockefeller Dodge took personal interest and on whose behalf the Geraldine R. Dodge Foundation has since been active. The association between my studies of play and the Geraldine R. Dodge Foundation may therefore be traced to the late 1960's, when the Great Swamp first received protection that enabled me, and many other New Jersey ecologists and ethologists, to take field courses in the Great Swamp and to conduct field studies there.

Many older natural history accounts cited in the text were located by four students: Ruth Chodrow, Anita Hayworth, Andrea Morden, and Mark Siragusa. Their bibliographic research contributed significantly to my quest for accounts of play in an enormous and inadequately indexed body of natural history literature. One person could not have accomplished this task.

No scholarly work is complete without obligatory apologies for dinners and deadlines missed. Some busy scientists offer public thanks for grants and private thanks to ghostwriters. It was my pleasure to finish this book in a country house, aided by three cats, by one dog, and especially by Johanna Fagen, to whom this book is dedicated.

Gradyville, Pennsylvania   R. F.
September, 1980

Contents

. . . pinguesque in gramine laeto
inter se adversis luctantur cornibus haedi.
Vergil, *Georgics* II:525–526

Animal
Play
Behavior

1

Introduction
and Synopsis

INTRODUCTION

The night before his execution, Padraic Pearse, Irish revolutionary, wrote not of love and revolution, but of "beauty that will pass": a leaping squirrel, small rabbits in a field at evening,

> Or children with bare feet upon the sands
> Of some ebbed sea, or playing on the streets
> Of little towns in Connacht,
> Things young and happy.
>
> *Last Lines,* 1916

   A biologist might interpret Padraic Pearse's heroic self-sacrifice in the Easter Week Rebellion as kin altruism, stemming from Pearse's genetic relatedness to the people of Ireland, but who would dare hazard an explanation of his last lines? It seems audacious to view literature, drama, and the fine arts through the glass of contemporary biology. Yet the first modern commentator on animal play, the aesthetic philosopher Karl Groos (1898), did just that. Groos sought a biological basis for the study of play in order to obtain new scientific insights into human artistic activity. To Groos, play represented a simplified form of artistic creation, one more amenable to scientific analysis than the far more complex expressions found in human cultures. Indeed, the relationship of aesthetic philosophy to Groos's investigations is comparable to the relationship of ethical philosophy to sociobiology, as outlined by E. O. Wilson (1975, 1978).

Animal play behavior, as Karl Groos fully recognized, poses a problem in aesthetic philosophy. Comparing animal play to human artistic activity, Groos found in both an "as if" quality of conscious illusion and self-deception. Human dance seemed akin to the dashes, leaps, capers, and gambols of young ungulates and rodents. Drama, sports, and games with rules represented human forms of nonagonistic playfighting and play-chasing. Sculpture, architecture, tool manufacture and tool use, and painting corresponded to the repetitive and variable forms of object manipulation that often follow initial sensory inspection. Naive though it now appears, this theory had considerable influence in its day. Behavior previously viewed as un-biological, as distinctively human, as exclusively culture dependent, now became part of a biological continuum stretching from dance and painting to the play of little animals. Surely it was a triumph of evolutionary thought.

Groos's audacious synthesis of biology and art rested on the best evolutionary theory and on the best field observations of his time. Nevertheless, despite the presence of several profound and original insights, it failed, at least in part. Groos's theory was seriously flawed, his observations incomplete (Beach 1945, Loizos 1966). Today his works are seldom read except by specialists, who find him charming and historically interesting, but seriously misguided. Yet his influence lives. Indeed, the study of animal play has never quite overcome its embarrassment at Groos's attempt to unite evolutionary animal psychology with aesthetics. As E. O. Wilson (1975 p. 164) has observed, "no behavioral concept has proved more ill-defined, elusive, controversial, and even unfashionable than play." What do biologists mean by "play," and how do they approach the phenomenon?

To biologists, the term "play" denotes three distinct, but intersecting sets of behavior patterns. Whether play is taken to represent one type of behavior or more than one depends on the mode of categorization used and on underlying assumptions about behavioral classification, as will be argued in Chapter 2. Play behaviors represent structural transformations and functional rehearsals or generalizations of behaviors or behavioral sequences. In other contexts, these behaviors yield relatively specific and immediate beneficial effects. Winning a disputed resource, obtaining a food item, escaping a predator, or using a tool are examples of such effects.

The most familiar form of animal play, playfighting (Figs. 1-1,1-2,1-3) and play-chasing (Figs. 1-4,1-5) consists of cooperative nonagonistic chasing, wrestling, and hitting, as exemplified by the friendly tussles and pursuits of puppies and kittens. Unlike agonistic fighting, such social play does not settle conflict over access to a disputed resource, nor does it result in enduring dispersal of participants or a change in their relative dominance status. Animal playfighting and play-chasing have their own set of distinct communication signals and social conventions. For example, certain facial expressions and postures elicit play, but are absent from agonistic encounters.

In a second form of play, locomotor-rotational exercise, a

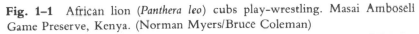

Fig. 1-1 African lion (*Panthera leo*) cubs play-wrestling. Masai Amboseli Game Preserve, Kenya. (Norman Myers/Bruce Coleman)

Fig. 1–2 Playfighting between juvenile and infant rhesus macaques (*Macaca mulatta*). The juvenile, in an advantageous (on-top) wrestling position, has rolled the infant onto its side. Infant with play-face. La Parguera, Puerto Rico. (John M. Bishop)

Fig. 1–3 Himalayan langur (*Presbytis entellus*) juveniles playfighting. Melemchi, Nepal. (John M. Bishop)

Fig. 1–4 Domestic dogs (*Canis familiaris*) play-chasing through shallow water. Massachusetts. (John M. Bishop)

Fig. 1–5 Himalayan langurs (*Presbytis entellus*) play-chasing (note play-signal). The one-year-old infant swings upside down and looks back at the six-month-old infant chasing it. Melemchi, Nepal. (John M. Bishop)

Fig. 1–6 Domestic horse (*Equus caballus*) foal cavorting in solo play. Anglo-Trakehner colt, aged 18 days. An adult horse watches. Wonderland Farms, Pennsylvania. (Robert Fagen)

lamb, foal, or other animal may run around, leap vertically in the air again and again, gamboling and capering, twisting its body and kicking out with its hindfeet (Figs. 1-6, 1-7, 1-8, 1-9). Free-living young primates swing (Fig. 1-10), roll, slide, and jump repeatedly. They may even turn back flips or perform somersaults. Such play is distinguished from avoidance of a predator, a conspecific, or a parasite and from captivity artifacts by its non-threatening context, by the player's loose body tone, by its sensory non-focus or inattentiveness, and by the repetition and ease of interruption of its activity.

Young animals, especially infant animals, often repeat the same developing locomotor or manipulative behaviors with slight variation at a given stage of mastery. These post-mastery behaviors are widespread in animals. They are especially well known in infant primates (including humans) and include jumping to a vertical support, running away from the mother and back to her, and various repetitive manipulations of objects, of

Fig. 1–7 Domestic horse (*Equus caballus*) foal rear-kicking in solo play. Anglo-Trakehner colt, aged 11 days. Wonderland Farms, Pennsylvania. (Robert Fagen)

the player's own body, or of its mother's body (Fig. 1-11). Repetitive locomotor-rotational and manipulative post-mastery behaviors merge with a third category of play behavior called diversive play or diversive exploration (Hutt 1966, 1970). These varied and vigorous effector interactions with an inanimate object follow initial sensory inspection. They may be distinguished from interactions with the same object leading to information about the properties of the object that do not depend on the behavior of the observer. For example, after human children three to five years old had visually inspected, felt, and touched a novel object whose levers rang bells, their facial expressions and postures changed from "intent" to "relaxed," and they performed "repetitive motor movements, manipulations of long duration accompanied by visual inspection of other objects, and a sequence of activities incorporating both the novel object and other toys" (Hutt 1966). These children might pat the lever re-

Fig. 1–8 Domestic horse (*Equus caballus*) foal propping in solo (or companion-oriented) play as second foal watches closely. Anglo-Trakehner colt, aged 18 days. Wonderland Farms, Pennsylvania. (Robert Fagen)

Fig. 1–9 Domestic horse (*Equus caballus*) foal head-twisting in solo play as he returns at a gallop to his mother's side. Anglo-Trakehner colt, aged 18 days. Wonderland Farms, Pennsylvania. (Robert Fagen)

Fig. 1-10 Himalayan langur (*Presbytis entellus*) swinging from vegetation. Melemchi, Nepal. (John M. Bishop)

peatedly, lean on the lever making the bell ring continuously while looking around the room, and run around the room with a toy truck, ringing the bell each time the object was passed. Hutt (1966) concludes that as inspection gives way to play, the emphasis changes from the question of "what does this *object* do?" to "what can *I* do with this object?" Analogously, as a basic skill is mastered, practice gives way to diversive play. For example (Bruner 1973a p. 302):

The six-month-old infant, having learned to hold on to an object and get it easily to his mouth, now begins a program of variation; when he takes the object now, he holds it to look at, he shakes it, he bangs it on his high chair, he drops it over the edge, and he manages shortly to fit the object into every activity into which it can be put.

Another familiar example of diversive play following mastery is that of a human child who has recently learned to ride a

Fig. 1–11 Cleo, an infant mountain gorilla (*Gorilla gorilla*), plays in contact with Flossie, her mother. With evident delight (note Cleo's facial expression), the six-month-old infant grasps and pulls her mother's body hair. Parc des Volcans, Rwanda. (Dian Fossey, © National Geographic Society)

bicycle. The child may weave, sway back and forth, and shout while riding with both hands off the handlebars.

Biologists approach the phenomenon of play in diverse ways that reflect prevailing assumptions about the study of behavior. We can identify two previous stages of inquiry into play. The first tradition is one of subjective description (often inadequate by present standards), followed by inductive, post facto interpretation. Beach (1945) sharply criticized this approach and called for better observations, better definitions, and objective tests of existing hypotheses. The second stage, coinciding with the growth of field research on animal behavior, yielded detailed, often quantitative and less subjective accounts of play in many animal species. These investigations resulted in two major symposia on animal play, one in Britain (Jewell & Loizos

1966) and another in the United States (Bekoff 1974b). By defining a style of research on play that was long on observation, but short on theory, these symposia answered Beach's (1945) call for controlled observation of activities customarily regarded as play. In the absence of testable theory, each contributor observed different characteristics of behavior. It proved almost impossible to correlate the data that resulted, much less to relate them to fundamental theory on development or on social interaction. However, no one appeared to object strenuously to this situation. These investigators, all trained ethologists, shared sufficient unstated assumptions and biological common sense to permit some preliminary structural comparisons to be made between the form of play in different families or genera of mammals and in different environments (Müller-Schwarze 1978a).

A single fallacy undercuts these previous approaches to play. It is fallacious to assume that any behavior is studied in a theoretical vacuum. The quality of observations is only as good as the quality of the observer's theory, however implicit and unconscious such theory may be. For precisely this reason Beach (1945) favored a hypothetico-deductive approach to animal play. He failed because existing theory was inadequate and because adequate theory simply did not exist. Hypotheses about play need a sound biological foundation. It is such a foundation that previous studies of play appear to have lacked.

The purpose of this book is to define a biological approach to animal play using newer basic scientific knowledge about social behavior and about development. This basic biology is presented by Barash (1977), Dawkins (1976b), Geist (1978a), and E. O. Wilson (1975) for social behavior and by P. P. G. Bateson (1976a,b, 1978a,b, 1979), Gould (1977), Konner (1977a), and Stearns (1976, 1977) for development, based on formal theoretical breakthroughs by Cole (1954), Feldman & Cavalli-Sforza (1976, 1977), W. D. Hamilton (1964), Levins (1968), Maynard Smith & Price (1973), and others. The book presents a novel framework for studying play. Previous traditions of play research bypassed evolutionary theory and failed to exploit biological ideas about development, skill, and social behavior, whereas I propose to rely on these modes of analysis. This is not to say that the evolutionary developmental approach is original with me. Independent evolutionary analyses of play by at least three investigators (Fagen 1974a, 1976b, 1977, 1978a, Konner 1975, 1977a, Symons 1974, 1978a,b) indicate that the time is propitious for this new perspective. Müller-Schwarze (1978a p. 384) appears to agree that the major unsolved questions in the study of play are evolutionary questions; in his words, "we are far from understanding 'the evolution of play behavior' as a process. . . . The evolutionary stages of play behavior are there, but how they evolved remains a question that is still to be answered."

A severe, if unintentional indictment of the situation that still prevails in some areas of animal behavior research, including most studies of play and behavioral development, was offered by a prominent ethologist who introduced Müller-Schwarze's (1978a) anthology of pre-1975 play research as follows:

The present-day students of animal behavior have all they can do to keep abreast of developements [sic] within their own area of special interest, let alone in the field as a whole—and of course we have long since given up attempts to maintain more than a superficial awareness of what is happening "in biology," "in psychology," "in sociology," or in any of the broad fields touching upon or encompassing the behavioral sciences.

These comments reveal an important philosophical difference between traditional ethologists and comparative psychologists, on one hand, and sociobiologists, behavioral ecologists, and evolutionary morphologists, on the other. Investigators belonging to the former group are content with superficial awareness of cognate fields. They continue to view animal behavior in terms of the home-grown, informal, but increasingly sophisticated mechanistic ideas of ethology and comparative psychology. Excellent if limited studies of play by Baldwin & Baldwin (1977), Bekoff (1972, 1974c, 1976a), Müller-Schwarze (1966, 1968, 1978b), Poole (1978), and others exemplify this approach. On the other hand, behavioral ecologists and their colleagues in evolutionary morphology look to cognate fields for insight and theory (Gould 1977, E. O. Wilson 1975). It has become increasingly evident that the study of animal play urgently needs rejuvenation by the same biological ideas that have already transformed so many other aspects of animal behavior research.

The approach to behavioral development presented here is also evolutionary. It is controversial because evolutionary and developmental questions about behavior have long been viewed as intellectually separable and conceptually distinct. Tinbergen himself (1951 p. 187) stated that "the study of ontogeny of behavior cannot be expected to contribute much to the study of evolution."

Other contemporary students of animal behavior have identified the study of development as central to all of current behavioral biology. "Of the problems confronting the student of behaviour that demand attention in the next decade or so, none is more pressing than the understanding of behavioural development" (Marler 1975). Prior solution of problems of ontogeny is necessary if the principal goal of sociobiology is to be achieved (Hinde 1976).

The topic of behavioral development seems sufficiently broad to accommodate evolutionary as well as mechanistic approaches. Indeed, some biologists have already called for evolutionary studies of behavioral development. Citing earlier discussions by C. H. Waddington and others, Ernst Mayr (1963 p. 148) stated the issue succinctly as follows:

It is evident that phenotypic modifiability, phenotypic stability (owing to developmental canalization) and behavioral flexibility are biological phenomena of considerable evolutionary importance. In the past these phenomena were erroneously considered the exclusive domain of the developmental physiologist. They deserve much greater attention by evolutionists than they have so far received.

An evolutionary approach to development has profound consequences. For example, Lumsden & Wilson (1981) identify behavioral epigenesis as the key to the study of gene-culture interaction. Play is central to behavioral epigenesis, both as a set of epigenetic rules and as a mechanism for modifying epigenetic rules in response to individual play experience (see pp. 350–355). Long recognized as a pivotal anthropological topic, the phenomenon of play may soon become a focal element of cultural biology as well.

Not one full-length, single-author work on animal play as a whole has yet appeared in this century—a century during which the neo–Darwinian synthesis, the growth of field biology, the increasing sophistication of behavioral science, and the parallel development of computer science and systems analysis have made biological analysis of behavior intellectually legitimate. The time is propitious for formulating a comprehensive body of improved biological theory on play and for reexamining existing information in the light of new ideas. This volume was written to fill the need for a book-length review of play behavior as a whole in the light of modern evolutionary theory. No previous author has applied biological theory of animal conflict, development, altruism, manipulation, or deception to the study of play. Dynamic mathematical models of the flow of a play bout and of the processes generating play partner distributions have not previously been formulated, but such models can serve to unify separately reported information. They are needed in order to clarify, to quantify, and to test ideas about initiation, maintenance, possible breakdown, and termination of play (including effectiveness of play-signals), about conflicts of interest over play, about mechanisms for resolving such conflicts, and about individual responsibilities for maintaining play relationships. They also serve as a springboard for sociobiological analyses, including theoretical predictions about relative magnitudes of the models' parameters in different situations. The study of play also needs to recognize recent biological research on mechanisms mediating environmental enrichment effects, on enrichment effects themselves, and on development of skill.

Previous volumes on play may be divided into three categories: anthologies with editorial comments, proceedings of symposia, and analyses of particular aspects of play in selected species. Important works in these three categories are Bruner, Jolly, & Sylva (1976), Bekoff (1974b), and Symons (1978a), respectively. These volumes defined the study of play for the late 1970's by applying ethological, comparative psychological and biological modes of analysis.

I discuss old and new issues in the ever-controversial field of play research. Symons (1978a) joins battle with Miller (1973), with Sutton-Smith (1975, 1979), and with me over a central issue in play research: whether play ever represents a biological adaptation for producing novel behavior. This question cannot be considered without addressing the general issue of possible functions of play. If Symons is correct and if we are wrong, sociobiologists seeking a biological basis for the study of art and creative ability must turn to behavior other than play for their comparative data. I wish to consider this and other controversial topics. For example, is social play purely cooperative behavior that benefits both partners (Bekoff 1978b) or is it a covert competitive tactic by which one partner benefits and the other suffers (Geist 1978b)? As is traditional in behavioral science, Bekoff and Geist argue this question typologically, without regard to intermediate degrees of competition or cooperation, without distinguishing between individual and inclusive fitness, without recognizing possible scaling of effects with environmental or genetic variables, and without considering possible variation across species. Phrased in absolutist, either/or terms, their debate seems unlikely to yield productive results. However, it raises important issues that I will discuss.

Probably, the most controversial aspect of this book is its approach to human behavior via the biology of play. It seems a long way from the dashes, leaps, capers, and gambols of young mammals to playground games, dance, competitive gymnastics, and sculpture. Extrapolation from animal play to human culture seems facile, speculative, and even superfluous. Surely, known and hypothesized biological relationships of play to development and social interaction are of sufficient interest and controversy in their own right. But I suspect that Karl Groos was at least partly correct. Some of Groos's thought has a surprisingly modern flavor. His emphasis on the biocultural importance of conscious illusion and self-deception anticipates current sociobiological concerns. He offers one of the first evolutionary discussions of the biology of interaction among heredity, learning, and environment. Admittedly, his comments on these topics strike us as antique, but that is hardly surprising for a book written in the last decade of the nineteenth century. Nearly every commentator on play, from Groos to Huizinga (1950), Morris (1962b), Thorpe (1962, 1974), E. O. Wilson (1975), and Bruner (1976) raises the issue of possible connections between animal play and various "unbiological" or seemingly non-utilitarian aspects of human culture, from sports and games to drama, fine arts, and literature. In planning and writing this book, I encountered considerable pressure to join this tradition. Speaking to biological audiences, I often cited examples of animal play that to me seemed fascinating in their own right. These examples included young zebras, growing up without playmates, who leave their family forever in order to find someone their own age to play with; chimpanzee brothers and sisters who tickle and chase together; and kittens of the rare Asian Pallas cat that tumble over and ambush each other in madcap games of hide-and-seek. Afterwards, to my discomfort and embarrassment, I would chiefly be asked "people questions." Did human language originate in hominid play? Might a physiological basis for art, music, and dance be found by examining brain structures homologous to those that mediate the play of nonhuman primates? How might the biological study of animal play aid in bringing human culture and creativity within the domain of evolutionary analysis?

Man's insatiable curiosity about himself and biologists' desire for a fair hearing, their impatience with the social sciences and the humanities, a pragmatic sense of concern, and perhaps even good old-fashioned avarice increase the pressure to comment on human implications of animal play. To do so would significantly extend the scope of this book, would allow justifiable accusations of intellectual and political naivete, and would place me in the position of discussing broad areas in which I have no expertise. Yet none of these considerations kept earlier authors from rushing blithely in. Therefore, not entirely reluctantly, I restrict detailed speculation on play and culture to my penultimate chapter. There are four reasons for discussing this topic at all.

(1) As Konner (1977a) points out, rigorous and humane biological investigation of phenomena previously considered nonbiological can make it more difficult for self-styled experts to misapply biology to human behavior for irresponsible ends. Of course, serious investigation of a topic by responsible scientists can always open the door to irresponsible analysis of that topic. It has even been claimed that any attempt to analyze human behavior biologically is irresponsible (Hirsch 1976).

(2) There is need, both within and especially beyond biol-

ogy, for sound, up-to-date basic biological research on play. The biological study of play can have clear relevancy to education (e.g., Hall 1906), child rearing, and psychotherapy (e.g., Erikson 1977), as well as to aesthetics and other areas of the humanities.

(3) It can be great fun to leap into an established, but alien field of knowledge and to try to recreate it around original ideas based on a novel point of view. This leap is a form of "deep play" (Bentham 1840) in which "the stakes are so high that it is, from (Bentham's) utilitarian standpoint, irrational for men to engage in it at all. . . . In genuinely deep play, this is the case for both parties. They are both in over their heads" (Geertz 1972).

(4) The relationship of biology to the arts is not merely one of subversion. By viewing play behavior from a sufficiently great variety of formal theoretical perspectives, including those furnished by sociobiological models, we significantly expand our perception and understanding of animal play. The resulting descriptive information and conceptual insight may directly stimulate dancers, filmmakers, musicians, writers, and other artists. Indirectly, it may also support efforts at preserving animals and their environments, those perennial sources of artistic and cultural material and inspiration.

The central argument of this biological book on play behavior may be summarized, briefly and simply, as follows. Animals risk time, energy, and injury to play. They do so in certain environments and because play is important for development in those environments. Play is a behavioral mediator between the environment and the phenotype. Through play, the cerebral cortex is stimulated to grow, to develop, and therefore to take a larger role in control of behavior, making that behavior more flexible. Simultaneously, play experience produces adaptive modifications of effector structures, such as muscle, bone and connective tissue, used in that behavior. Through play, animals acquire physical ability and develop social relationships. Such plasticity evolved because of economic trade offs in brain development: the optimal balance between cortical and subcortical control of behavior depends on environmental information, and the experience of play reliably serves to indicate that the animal is in an appropriate environment for such responses to develop, as well as to develop the responses themselves, and the anatomical structures that support them, through repetition and variation. How can such behavior evolve when the plasticity underlying the effects involved is subject to manipulation by others in the service of their own selfish interests, which may not coincide with those of the subject? The probable answer to this question of evolutionary stability is that play occurs under circumstances in which current or future resources cannot be competed for in this way; that such damaging behavior would be costly to the actor as well as to the recipient and therefore generally evolutionarily unstable against a strategy of not playing at all; that social play is designed to include only messages that cannot be faked or only stimuli that cannot be successfully manipulated to damage another; that developmental resilience, including the ability to detect and resist attempts at damaging manipulation through play, holds such damage to an evolutionarily tolerable (though not necessarily negligible) level; and that social play evolved through kin selection, with close kin favored as playmates, making cheating even less likely given the conditions for play already described above.

SYNOPSIS

1. *The problem.* Although biologists generally agree that the study of play has lagged far behind other areas of animal behavior research, they disagree about the reasons for this lack of progress. Some commentators feel that the problem resides in the definition, or lack of definition, of play behavior. Others feel that the study of play suffers from excessive speculation and that further descriptions, quantitative data, or experiments are needed. I argue that because previous theory was simply inadequate by contemporary standards in theoretical biology, descriptions, observations, and experiments based on this theory were necessarily inadequate as well, no matter how comprehensive and well designed such empirical work might have been. Therefore, the chief need of play research at this time is for more, newer, and better theory.

2. *Defining play*. Like other coarse-grained behavioral categories, including aggression and even sex, play is difficult to define. There are different ways of defining behavior and different senses in which behavior is defined, as well as different conceptual schemes for definition. "Play" is a convenient, but scientifically inexact term used to denote certain locomotor, manipulative, and social behaviors characteristic of young (and of some adult) mammals and birds under certain conditions in certain environments. The term "play" may be defined structurally or functionally. It may refer to a category of behavior, to a behavioral or social relationship, or even to a mental state. Play can be a theoretical extreme point of an abstract continuum whose opposite extreme is agonistic, predatory, or predator avoidance behavior. The scientific study of play consists of a set of behavioral descriptions and a set of theoretical, observational, and experimental investigations of the behaviors described, including causation, results, and correlates. For the purposes of this book, play includes nonagonistic fighting and chasing maintained by social cooperation; solo locomotor and rotational movements performed in the absence of threatening predators, parasites, and conspecifics; developing locomotor or manipulative behavior repeated with slight variation at a previously established level of mastery; and diversive effector interactions with an inanimate object subsequent to the termination of an initial phase of sensory and mastery activity, including exploratory manipulation, directed toward the object. These behaviors are the outward expressions of a set of distinct, but functionally related developmental pathways that may run parallel to each other, merge, or branch. Therefore, the distinctness (by any arbitrary criterion) of these categories of play may depend on the species, sex, and developmental state of the individual studied.

3. *Natural history of animal play*. A significantly enlarged body of empirical information on animal play, in part the result of the recent resurgence of field and laboratory ethological research, rewarded my perusal of existing literature as summarized in this section. Among the most interesting of these observations are accounts published many years ago and overlooked in subsequent reviews of animal play. Ethologists and sociobiologists now recognize that biologically important field observations

frequently occur in the older natural history literature and should not be dismissed as mere anecdotes.

Professional field naturalists have documented remarkable and spectacular instances of animal play. The value of these accounts is assured on aesthetic grounds alone. My review is intended to give these accounts the broad visibility they deserve. But the everyday play of familiar and well-studied animals, including Norway rats, domestic cats, horses, and rhesus monkeys, is of compelling scientific interest as well. Play of these and other well-known species is described at length. Other major features of my discussion are a significantly enlarged body of evidence for interspecific play and object play in free-ranging wild animals, detailed information on play in birds, including convincing descriptions of complex social play in parrots, and evidence that social play may frequently escalate into agonistic fighting under certain conditions.

4. *Continuity*. The natural history of play in mammals and birds is remarkably uniform. When compared across species, behavioral content and structure of play appear to vary little if at all around norms of nonagonistic contact, including wrestling or sparring, or nonpredatory locomotion and body rotation, including pursuit, capture, handling, and evasion. Most differences in play behavior between species mirror behavioral, demographic, anatomical, or environmental constraints. However, complexity or elaborateness of play appear to increase with relative brain size across a wide phylogenetic spectrum. Especially interesting from this standpoint is the phenomenon of play in birds, behavior that may have evolved independently from mammalian play. The early study of avian play was flawed by misinterpretation of courtship behavior or of maturational precursors of locomotor behavior. Recent studies of parrot behavior demonstrate surprisingly elaborate social and object play, including play-signals and cooperative social conventions. However, the scale of complexity of avian play appears to be shifted relative to that of mammals. When we compare birds and mammals that have comparable ecological niches and that form equivalent societies, we find similarities in the structure and content of play, but differences in its elabo-

rateness. The play of birds is less complex than that of sociobiologically equivalent mammals and seems to reflect corresponding differences in relative brain size or complexity.

5. *Presong.* A playlike behavior of young birds is presong, a developmental precursor of full song. Presong consists of fragments of song that may be repeated and varied again and again by a young bird learning to sing or by an adult after the end of breeding season. Its metaphorical resemblance to play has impressed many observers. As a result of experimental studies of song learning, the developmental history of bird song is well understood, and extrapolation of principles derived from this research to other developmental processes seems possible.

6. *Discontinuity.* Play is a well-documented phenomenon in mammals and birds, but we lack convincing reports of play in other vertebrates and in invertebrates. Possible reasons for this discontinuity are insufficient study, lack of objective criteria for play in organisms anatomically unlike mammals, or true biological differences. Vertebrate and invertebrate species believed not to play may exhibit parental care, agonistic fighting among juveniles, complex social cooperation, tool use, learning, environmentally induced brain plasticity, capacity for elevated aerobic metabolism in vigorous exercise, and other phenomena sometimes viewed as necessary conditions for the evolution of the capacity to play. However, these conditions are clearly not sufficient. In mammals and birds, endothermy, relatively large brains, capacity for generalized learning, and ability for long-term individual recognition are concomitants of play that suggest reasons for this apparent sharp behavioral discontinuity in animals, although they may fail to explain the apparent absence of play in certain species of birds and mammals.

7. *Behavioral content.* Social and locomotor-rotational play include behavioral acts also present in two major forms of damaging animal conflict: intraspecific escalated fighting and predator-prey (including parasite-host) behavioral interaction. In such all-out conflict, the recipient of the behavior is not expected to cooperate in training the actor. Illness, injury, or death would be the most likely outcome of such interactions, especially when the actor is young, weak, and inexperienced. Instead,

structures and skills useful in potentially dangerous conflict situations should be exercised, rehearsed, and/or generalized in play.

8. *Effects of play.* It is important to distinguish between two different, but complementary approaches to the effects of play. One approach is mechanistic and serves to describe effects of play in anatomical, physiological, behavioral, or cognitive terms. The second approach is ecological and assigns a biological utility value to these effects by relating them to survivorship and fecundity.

On the mechanistic level, the twin roots of play behavior are the plasticity—somatic-neural, behavioral, cognitive—and resilience of developing organisms. Nearly all known or supposed mechanistic effects of play behavior involve one or both of these opposing characteristics. These effects span the entire spectrum of determinants of behavior from specific to general, with particular emphasis on general determinants.

a. Mechanistic effects. Play can have both beneficial and harmful effects on organisms. Time, energy, growth, and survivorship costs of play are generally recognized, if not rigorously demonstrated in all cases. Hypothesized benefits of play include simple, immediate behavioral thermoregulation, various forms of physical training, practice, and skill development, behavioral meshing and reciprocity, kin recognition, social cohesion, and even damage to competitors. These claimed effects vary according to time scale, mode of action, and impact on developmental plasticity or resilience. They often lack adequate conceptual and mechanistic bases. As yet, they enjoy little or no direct empirical support. Because the study of play as a possible biological adaptation assumes some knowledge of these effects, it is important to define mechanisms of action for play and to attempt to discriminate between adaptive and fortuitous consequences of play.

b. Physical training, motor and cognitive skill development. Exercise of body components increases the functional capacity of those components, an effect known as exercise adaptation. Exercise adaptation occurs in bone, muscle, connective tissue, the circulatory, respiratory and endocrine systems, and even in the central nervous system. Physically vigorous play could produce exercise adaptation. Is play equivalent to physical training for animals? If so, we must specify the type of training that occurs. Play is typically affected by the social environment and often consists of social "scrimmages" rather than solo "calisthenics." Indeed, it would be difficult in practice to separate the physical training effects of play from its mechanistically distinct effects on motor skill. Furthermore, an individual's play with objects and with parts of its own or another's body may influence the development of certain cognitive capacities. Recent research on exercise adaptation and on the development of motor and cognitive skills has shed much light on the biology of these effects, but certain questions remain. What is the importance of these effects in development and in evolution? Is play necessary for physical or skill development? To what extent does play provide benefits missing from calisthenics, rote practice, or exploratory manipulation? Which components of skill is play most likely to affect? In discussing developmental effects of play, might we separate plasticity (environmental tracking) from resilience (resistance against environmental influence) effects? And, from a more fundamental standpoint, under what circumstances should we expect plasticity or resilience in development to evolve?

c. Effects on flexibility. In play research, perhaps the single most pervasive and widely held belief about effects is that play makes the player behaviorally flexible: versatile, resourceful, creative, and able to cope productively with the novel and the unexpected. Play is said to develop generic learning skills that enable the player to adapt to new environments and to new situations. Whether the context is that of tool construction and tool use, problem-solving, dealing with socially stressful situations, or surviving and reproducing in a new habitat, it is felt that the animal with play experience is a flexible generalist, a "jack of all trades." The player can exploit new opportunities more quickly than competitors that lack play skill. In Konrad Lorenz's words, a playful animal is a "specialist in nonspecialization." According to this hypothesis, such general behavioral versatility is based on the individual's play-developed ability to manipulate inanimate and social objects having various degrees of responsiveness contingent on the individual's behavior. Versatility is said to result from self-control, inhibition of arousal, and the ability to alternate responses rather than persevering with an unsuccessful tactic. All these traits, skills, or characteristics supposedly develop as a result of play experience. Thus, for example, proponents of "flexibility" theories claim that an animal lacking social play experience is emotionally immature and thin-skinned, supersensitive, suspicious and rigid, that it takes things hard, is shy, and runs crying to Mama at the slightest provocation; and that it deserves the uncomplimentary epithets of crybaby, Mama's boy, bully, or big baby. On a slightly more sophisticated level, Geist (1978a) pursues this argument by viewing play as an environmentally controlled switch between two alternative grand strategies: narrow, efficient, but unstable specialization and broad, inefficient, but stable flexibility. Peter Reynolds (1976) terms this latter strategy the "flexibility complex."

This interesting hypothesis, or set of hypotheses, needs critical scrutiny. Experiments with human children have demonstrated indirect relationships of play to skill at object use. However, technically more rigorous experiments by psychologists on supposed behavioral flexibility of rats reared in enriched environments have so far failed to establish clear correlations among rearing environment, problem-solving skill, and global arousal or inhibition variables. The concept of the flexibility complex requires further analysis. Each term must be more precisely defined and appropriately qualified. Moreover, previous hypotheses have not adequately specified the likely evolutionary costs and benefits of the two hypothesized alternative grand strategies of efficiency and flexibility in their disparate contexts of social relationships, tool use, problem-solving, feeding, dispersal, and habitat utilization. After all, bullies, crybabies, shy people, and the like, however they may disgust, enrage, or be shunned by socially mature and experienced adults, may actually be displaying evolutionarily stable behavioral tactics adaptive in appropriate situations or environments.

d. Social skill development and related social effects. We can analyze likely effects of play on individual physical capacity or on certain skills without ever considering other animals or the social environment. However, many hypothesized effects of play are essentially social. In order to analyze play in the context of the player's social environment, we must carefully consider the level at which these effects are defined in order to avoid superorganism explanations. Furthermore, we must recognize that potential effects of play on qualities or outcomes of a social relationship involve the interests of both parties. In addition, we must always ask whether play is really capable of producing such effects and whether play is necessary for producing them. Some of the most intriguing effects hypothesized for play are higher-order components of social skill including assessment, timing, deception of others, recognition of attempts to deceive, and self-deception. Play has also been uncritically viewed as a mechanism for establishing dominance hierarchies, a mechanism for kin recognition, or even a mystical social glue bonding individuals together into a cohesive superorganism.

9. *Play and differential development.* A developing individual's own behavior and its social interactions affect not only somatic physiology, but also physical, cognitive, and social development. We are therefore compelled to view the play of many species in the context of a society of young organisms in which developmental rates (as well as learning and motivation, of course) are biological dependent variables. Early experience and social interaction can profoundly influence the timing of developmental events. These effects range from synchronization of development by behavioral meshing of parent and offspring or of littermates to feedback effects of social experience on individual physical and behavioral development. They affect interindividual differences within the social unit. Because ecological life-history theory predicts alternative developmental pathways and outcomes to be the rule rather than the exception (Fagen 1977, Schaffer & Rosenzweig 1977), we should expect the social interactions of developing organisms to affect the path taken by each organism through the developmental network as well as the rate at which any given path is traveled. We should expect both cooperation and conflict between organisms with regard to these potential effects, expressed in their social behavior within and outside play.

10. *Relationship of effects to components of fitness.* Effects of play are expressed in terms of such individual phenotypic character-

istics as strength, endurance, and skill or in terms of selected characteristics of a social relationship. Biological analyses of these effects seek to relate changes in these variables produced by play to changes in components of fitness. These fitness components include survivorship and fecundity of the individual and of its kin.

11. *Adaptiveness (function) of play.* It is difficult to avoid the conclusion that any currently fashionable, but poorly understood behavioral capacity having unknown environmental determinants and some metaphorical relationship to the structural characteristics of play behavior can be identified with the long-sought and mysterious "function of play." However, the actual situation is not so simple. Theoretical arguments and experimental evidence suggest the reality of at least some of these hypothesized effects on individual behavior. This evidence includes data from artificial intelligence research and from psychological experiments on rats and humans. Whether such mechanistic functional information is relevant to problems of adaptation and biological function is a question for evolutionary analysis.

The mechanisms discussed above can result either in positive or negative effects on fitness. For instance, physical exercise increases functional capacity, but too much exercise can drain energy reserves and, given sufficient duration and intensity, can produce pathological anatomical changes. Other negative effects include risks of injury, predation, or accidental death, as well as time and energy costs. In a competitive social environment, it is theoretically possible that under certain circumstances one play partner may behave so as to damage another by appropriate manipulation of the plasticity and resilience mechanisms discussed earlier. For example, if a strong animal could induce vigorous play in a weak animal, it could possibly drive its partner's energy reserves to dangerously low levels.

The evolutionary study of behavior as adaptation seeks evidence that a behavior pattern has been fashioned by natural selection to produce a particular effect. Indirect evidence may be provided by the structure of the behavior or by the circumstances in which it is performed. Comparative study across ages, sexes, populations, environments, or species furnishes strong indirect evidence of adaptiveness. Play may be an adaptation for somatic and skill development. Its function may vary depending on circumstances, and its current function may be different from that which play was selected to produce in the past. Historically, discussions of the function of play have been theoretically unsound, simplistic, and unsupported by acceptable indirect or direct evidence. There is need for tighter hypotheses and for more careful comparative study. It is not clear that better experiments alone will answer this need.

12. *Describing social play.* When demography, brain size and structure, and potential companions permit, play becomes a social phenomenon. Social play is the best-known form of play, but existing descriptions of social play may be criticized for lack of precision, integration, and theoretical legitimacy. The descriptive precision of accounts of play is limited by verbal descriptive techniques and by lack of quantification. Movement notation techniques are recommended. Reports on social play include separately reported data on rates of initiation, acceptance, refusal, and breakdown as well as separately reported data on results of these processes. One frequently reported result is a nonrandom composition of play dyads with respect to ages and sex. This piecemeal approach to the structure of social play results in a fragmentary, redundant, and often static picture of the phenomenon. Techniques of systems analysis, including flow diagrams, can be used to specify and to model the flow of a play bout, to specify relationships between processes and results, to present data concisely, and to suggest new experiments and observations. The parameters of such models are of developmental and evolutionary interest.

13. *Conventions of social play.* Social play is elicited and maintained or stabilized by remarkable social conventions ensuring reciprocity and fairness. These conventions include play-signals, role reversal, self-handicapping, mutual pauses, lack of dominance distinctions, and absence of damaging fighting tactics. In this sense, it resembles conventional fighting. However, in sharp contrast to conventional fighting, it is not always possible to identify the winner of a play interaction, and play, unlike agonistic fighting, does not appear to involve or to resolve conflict over a disputed resource.

14. *Control and breakdown.* Certain signals used in social play serve to interrupt the action or to bring a third party to the aid of the signaler. These signals are believed to indicate distress of some kind. They may occur when a partner is too rough physically or when the partner increases the intensity of the interaction too far. The partner may or may not respect this signal. When it does not, a conventional or escalated fight can occur. The significance of these breakdowns is not entirely clear. Typically very rare, they seem more frequent among older animals as the frequency of play decreases in ontogeny. They indicate conflicts of interest, probably over play itself. The direct relationship between play breakdowns and intervention by third parties must be viewed in the context of the dominance and coalition structure of the society at large. Sometimes social play seems to be little more than a medium for the expression and exploitation of this status and affiliation structure by infants and juveniles. Physical and skill effects become fortuitous as play is functionally usurped by the young animal's developing ability to manipulate its play interactions to probe, assess, and control the behavior of third parties.

15. *Sociobiology of play.* Sociobiological analysis of social play emphasizes previously unstudied aspects of this phenomenon. Conflicts of interest over play are predicted between potential playmates and between parents and offspring. Successful play initiation is viewed as the result of negotiation. Potential partners are expected to enter into a play bout only when such an interaction is mutually beneficial and to terminate the bout when mutual benefit ceases. When it is in only one individual's interest to begin, continue, or terminate play, conflict will occur. During an ongoing play bout, conflict is expected over the quality of the interaction, including intensity and behavioral content. Therefore, conflict over play is expected to occur whenever individual interests fail to coincide, especially when the quality of the interaction cannot be agreed upon through the use of play-signals before play actually begins.

Interruption and repetition frequencies, duration of play and of pauses within play, escalation, error correction, breakdown, and termination are results of the partners' coinciding or conflicting interests in play. These rates therefore depend on age, sex, relatedness, and past experience. To a greater or lesser degree depending on relatedness and demography, players' choice of companions will depend on the extent to which interests in play coincide, as measured by recent past experience. Use of play to harm potential competitors is conceivable.

Sociobiology also suggests the identity of some of the more complex cognitive and social skills that play might serve to develop. In the past, sociobiologists have concentrated on adult end points and on effects of behavior on survival and reproduction that could be convincingly argued, if not demonstrated and quantified. They have looked to developing organisms primarily for evidence of competitive behavior, especially escalated fighting and killing. Their emphasis mirrors that of the classical ethologists, who stressed sex and violence in animals, and is understandable in the light of Darwinian concepts of individual selection based on differential mortality and differential reproduction. However, this emphasis is mistaken, even within a narrow sociobiological context. The study of interspecific play can shed light on reciprocal altruism. Research on play in litters or avian clutches can shed light on kin altruism. Differing degrees of relatedness in juvenile societies of different species furnish an excellent means of exploring the implications of kinship theory. Additional aspects of play prove sociobiologically interesting: sex differences in play are widespread, play itself may sometimes be a competitive tactic, and breakdowns of play are one of the most easily studied natural instances of juvenile aggression. The cooperative social conventions maintaining play are themselves of considerable biological interest. Moreover, intervention in play by third parties is frequently a function of kinship. Assessment, deception, and self-deception may be practiced or employed in play. Finally and most importantly, play is profoundly implicated in human behavioral epigenesis. Play must therefore be viewed as a key element in the sociobiological study of culture.

16. *Play, creativity, aesthetics, and culture.* Rigorous analysis of animal play provides a basis for discussing possible biological roots of human artistic and intellectual creativity. It is necessary to recognize contrasting structural, causal, functional, developmental, and evolutionary schemes for analyzing relationships

between play and creativity. In order to define the issues precisely, we need better distinctions, or better behavioral categories, applicable to the borderland between play, exploration, and practice. We also need to distinguish between relatively specific and relatively general and between immediate and long-term effects of play, exploration, and practice.

The argument for possible causal relationships between play and creativity implicitly assumes that play is controlled by a motivational system generating behavioral variations in order to produce diverse, novel, or stimulating sensorimotor feedback effects. The analogous developmental argument implicitly assumes that play experience makes the player versatile, flexible, and creative in later life. The scientific status of these implicit assumptions is still uncertain. Both assumptions rest on hypotheses about play that remain poorly understood, loosely defined, and inadequately demonstrated. The pervasive idea that play and creative activity are causally and/or developmentally related cannot be rejected at this time. Clearer thinking and more data will be necessary, however, if this idea is to earn the status of a legitimate scientific hypothesis.

2

The
Problem

Play research is the ugly duckling of behavioral science. Marred by anecdotal evidence, by subjectivity, by excessive speculation, and by persistent disregard for the ground rules of hypothetico-deductive logic, the study of play remains controversial and esoteric (Beach 1945, Bruner, Jolly, & Sylva 1976, Müller-Schwarze 1978a, E. O. Smith 1978, E. O. Wilson 1975).[1] Critics of play research call for less speculation and more science (Beach 1945). They also argue that play behavior has been inadequately defined or even that the term "play" has no scientific meaning (Schlosberg 1947, Welker 1971, Lazar & Beckhorn 1974). I address each of these arguments below.

As stated, the problem of play behavior differs in kind from other problem statements in biology. To an evolutionary biologist, for example, the problem with iteroparous reproduction (Cole 1954, Charnov & Schaffer 1973), sexual reproduction (Maynard Smith 1978), conventional fighting tactics (Maynard Smith & Price 1973), altruism (E. O. Wilson 1975), and cooperative behavior (Hamilton 1971) is simply stated. Why should these behaviors exist at all, since they are all counterselected in simple mathematical models using natural selection theory? In order to ask fruitful evolutionary questions about play, we would need models like these. However, the problem with past play research is precisely that we have not had theory adequate to tell us what the problem is.

LESS SPECULATION
AND MORE SCIENCE

Because the same piece of research may be regarded as speculation by one investigator and as science by another, it is important to consider each of these terms more closely in the light of past criticisms of play, which sometimes amounted to little more than deliberately slanted, self-serving phraseology. What is speculation? What is science? The answers to these questions depend on the scientist.

Description

Some investigators (e.g., Beach 1945, Loizos 1966) saw a need for basic, objective ethological description of play. A decade of fine descriptive studies of play behavior marked by an increasing emphasis on field research might seem to have filled this need. As the information reviewed in Chapter 3 demonstrates, precise and detailed descriptions of play behavior in hundreds of species are available at the species, population, and individual levels. Entire books could be written describing the play of a single well-studied species, such as chimpanzee (*Pan troglodytes*), rhesus macaque (*Macaca mulatta*), savanna baboon (*Papio anubis*), squirrel monkey (*Saimiri* spp.), Norway rat (*Rattus norvegicus*), and polecats, ferrets, or their hybrids (*Mustela putorius, M. furo*). But description alone does not a science make. Empirical information actually needed in order to test new hypotheses about play is seldom found in existing literature. Extant (if implicit) theory, including the observer's own concerns and subjective biases, constrains any "objective" description of animal behavior. Theoretical constructs, whether explicit or unconscious, always determine the nature of the description made (Fagen 1978a, Gould 1977, Purton 1978). Believing is seeing.

Data

One scientist's descriptions can be another's data. Indeed, critics of play tend to see data in terms of dichotomies: soft or hard, qualitative or quantitative, bad or good, yours or mine. For example, Bekoff's (1976a) call for quantitative studies of play was quickly answered by studies of nonhuman primates (e.g., Symons 1978a), carnivores (Barrett & Bateson 1978, Hill & Bekoff 1977), and other species. And yet, had these excellent studies merely succeeded in reporting information in numerical rather than in verbal form, play research might have been no better off with numbers than it had previously been with words.

Experiments

From Beach (1945) to Welker (1971), critics of play research, especially those of a psychological or physiological bent, have called for scientific experimentation on play. These critics seem unaware that excellent science can be done without experiments. Witness current geoscience and astrophysics (black holes, continental drift), the entire tradition of the earth, planetary, and space sciences, and ecology's asymptotic approach to scientific maturity. Moreover, the quality of experiments, like that of data, can vary. Unfortunately, past experiments on play (e.g., Chepko 1971, Hansen 1974, Müller-Schwarze 1968) were unsatisfactory. These experiments were based on deprivation and isolation paradigms now known to be faulty (Bekoff 1976b, P. P. G. Bateson 1976b), and the interpretation of these experiments is questionable at best (Bekoff 1976b). Experiments as such have done no more to advance play research than numbers as such. Furthermore, advocates of manipulative experimentation tend to overlook the opportunities for rigorous empirical study furnished by "natural experiments," including comparisons of play in animals of different sexes (Symons 1978a), ages (Levy 1979), and species (Wilson & Kleiman 1974) and comparisons of play in populations of the same species in different environments (Berger 1979, 1980).

Theory

The problem with play research was not lack of descriptions, data, or experiments, but rather lack of adequate theory. Descriptions and data are inseparably wedded to underlying implicit theoretical assumptions. By themselves, the most careful descriptions and the most reliable and replicable data mean nothing. Experiments, however elegant their methodology, are only as good as the hypotheses they are designed to test. Beach (1945) correctly observed that play research has never been a hypothetico-deductive science. In my view, the problem with play research has been just this lack of adequate theory. Marler & Hamilton (1966) stated that it was not yet possible to ask the proper questions about play, let alone answer them. These authors mistakenly blamed lack of proper questions on inadequate observation and experiment. They correctly recognized that the problem is not plausible hypotheses and lack of data, but rather lack of "proper" (meaningful, important) scientific questions about play. In place of such questions were claims that play discharged surplus energy, regulated arousal, rehearsed adult behavior, or produced social cohesion. These statements could never have been adequately tested, even indirectly, because their underlying biological assumptions were incorrect, insufficiently specified, self-contradictory or vacuous. It becomes clear in retrospect that these and other "theories" of play were, as most critics pointed out (without truly understanding why), groundless speculation—groundless not only in fact, but also in theory.

What makes a scientific question "proper"? Propriety, of course, is in part culturally determined. To some scientists, physiological questions are proper whereas evolutionary questions are not. I submit that a proper scientific question about play would be one that not only generated a falsifiable hypothesis, but also linked the described rich phenomenology of play to "simple, continuous changes in underlying processes" (ecological, evolutionary, genetic) that biologists consider fundamental.[2] This, after all, is "the chief joy of our science" (Gould 1977).

Recent progress in behavioral ecology (Krebs & Davies 1978), sociobiology (Dawkins 1976b, Geist 1978a, E. O. Wilson 1975), developmental theory (P. P. G. Bateson 1978b, 1979, Gould 1977), and artificial intelligence (Winston 1977) can suggest some "proper" questions about play. Precise, testable, falsifiable predictions about play may be derived from fundamental theory. For example, we may now predict the algebraic signs, and perhaps even the relative magnitudes, of differences in certain quantitative characteristics of social play as a function of the age, sex, and genetic relatedness of the participants. Play research is finally ready to become respectable.

The phenomenology of play is almost too fascinating in itself, both as natural history and as one of the aesthetic pinnacles of field biology. The dedicated ethologists who spent decades observing play just for the great fun and sheer beauty of it all may have no desire to see the wild, glorious romance of animal play give way to social theory and mathematics—the spoiled child, selfish mate, and irritating vacuum cleaner of contemporary evolutionary ethology. In fact, appreciation of aesthetic aspects of play can only be enhanced by new theory about play. The theoretical achievements of Charles Darwin, G. Evelyn Hutchinson, Robert MacArthur, and E. O. Wilson are inseparable from the contributions of these men to the natural history of Galapagos finches, corixid waterbugs, New England forest warblers, and ponerine ants, respectively. Had it not been for evolutionary theory, our perception of these species would be severely limited in scope.

Of course, the real accomplishment of these and other naturalists of genius was precisely that they did not make the exotic banal. Rather, theory achieves an opposite effect, transforming the commonplace into the extraordinary. Play behavior, long regarded as trivial and inconsequential, offers ideal elements for transmutation.

I have argued that play research needs better theory in order to produce precise, falsifiable hypotheses worthy of the cost, time, and effort required to test them. Recent advances in evolutionary ethology and related fields have made such theory available. Why not view play in the light of these advances? The success of this approach will ultimately be measured not only by the scientific quality of the empirical and theoretical studies it fosters, but also by the extent to which it enhances aesthetic and intellectual appreciation of the natural history of animal play.

QUESTIONS OF DEFINITION
AND CATEGORIZATION

General Considerations

Past critics of play research (e.g., Schlosberg 1947, Lazar & Beckhorn 1974) argued that the word "play" fails to represent a meaningful category of behavior. I believe that these critics and others of their ilk are wrong and that play is no less well defined than such other molar (higher-level) behaviors as aggression, communication, social behavior, or sex.[3] However, important issues are involved, and it is therefore necessary to review the major criticisms previously directed at play as a behavioral category. Each of these criticisms is implicitly based on criteria for a meaningful behavioral category or for behavioral definition. Not all critics appear to assume the same criteria. As a result, we are immediately forced to consider the general problem of categorization and definition of behavior. It is impossible to address every aspect of this issue. Recent philosophical analyses of scientific definition, classification, and categorization include such accessible primary or secondary sources as Ayer (1970), Koestler & Smythies (1970), Kuhn (1962), Quine (1960), and White (1970). Philosophers have written entire books (e.g., Robinson 1950) on the single topic of definition. For current ethological views of the problems of definition and categorization, see Barlow (1977), Beer (1977), Fentress (1973), Hinde (1974), and especially Purton (1978).

1. *Real and nominal definition* (Robinson 1950). Nominal definition explains the meaning of a word. Real definition analyzes the object for which the word stands. The problems of defining words and things are different. The word "cat" is not a cat. Do critics of play research require a better definition of the word "play," or of the thing it denotes? The distinction between nominal and real definition is reflected in the following controversy: Should a topic of discussion be defined at the beginning or at the end of that discussion (Robinson 1950 pp. 3, 40)? Because it is unfair to leave the reader in doubt about the meaning of one's words, nominal definitions of a topic belong at the beginning (or wherever a given word is first used). However,

since real definition is the result of an effort to understand a topic better and to improve one's concept of that topic, it has been argued that the place for real definitions is at the end of a ` work. Literature on play furnishes examples of both approaches. Bekoff (1972), Fagen (1976a, 1977), Welker (1961), and E. O. Wilson (1975) nominally define their use of the word "play" as they begin their discussions. Aldis (1975) and Müller-Schwarze (1978a) state real definitions of the behavior called play at the end of their discussions, thus summarizing the improved understanding of the concept of play resulting from their analyses.

2. *Conceptual schemes.* Behavior may be categorized according to at least four connected, but different conceptual schemes (Hinde 1970, 1974, Purton 1978) having different meanings that should not be mixed inadvertently. Each of the four schemes can have value in the analysis of behavior. Definition by form (structure) is based on physical description of the movements involved and perhaps of their orientation or sequence. Functional definition classifies behavior by its adaptively useful consequences (Hinde 1974 p. 11, 1975) or "survival value" (Purton 1978 p. 656). Causal definition can refer either to immediate, short-term causation (external and internal eliciting factors, situational and motivational variables, internal states, or conditions that cause the behavior to terminate) or to long-term developmental causation. A fourth type of definitional scheme relies on common-language notions of human actions, including questions of motive or purpose. Failure to respect the differences between these conceptual schemes can lead to confusion. For this reason, Purton (1978) warns against mixing categories of different sorts. For example, investigators trained in psychology often find it natural to define behavior in terms of "definitive stimulus and response characteristics" (Schlosberg 1947, Welker 1971). Both Schlosberg's and Welker's critiques of play are confusing in large part because of a basic failure to recognize the distinction between structural (response characteristics) and causal (stimulus characteristics) conceptual schemes. Welker's (1971) critique is especially confusing because it mixes structural, external causal, and developmental conceptual schemes.

3. *Simplification and abstraction*. In addition to the general problems just mentioned, special difficulties plague biological and behavioral definition. Defining a behavioral category involves simplification and abstraction (including delineation of arbitrary boundaries). As a result, any behavioral category will exhibit some clear-cut instances along with a large number of ambiguous cases. Such fuzziness is unavoidable. It does not eliminate the heuristic need for behavioral categories. These categories have "a core about which there is little disagreement" and "very shady edges", but "this does not mean the categories are disreputable" (Hinde 1974 pp. 8, 283). However, we must not view our categories as real entities. "Stated briefly, and without much exaggeration, categories of behavior must be formed, but the investigator must not believe them" (Fentress 1973 p. 163).

The approach to definition taken by Hinde and by Fentress may surprise some behavioral scientists. However, it finds strong support in contemporary technical philosophy. Philosophers of science since Wittgenstein have accepted so-called conceptual aggregates, natural families of activities or phenomena, in lieu of formal definitions of a subject. For example, Kuhn (1962) states:

Though a discussion of *some* of the attributes shared by a *number* of games or chairs or leaves often helps us to learn how to employ the corresponding term, there is no set of characteristics that is simultaneously applicable to all members of the class and to them alone. Instead, confronted with a previously unobserved activity, we apply the term "game" because what we are seeing bears a close "family resemblance" to a number of the activities we have previously learned to call by that name. For Wittgenstein, in short, games, chairs, and leaves are natural families, each constituted by a network of overlapping and crisscross resemblances. The existence of such a network sufficiently accounts for our success in identifying the corresponding object or activity.

Minsky (1975) argues that in order to define concepts precisely to a computer it would be sufficient to program the computer to initiate and to improve conceptual aggregates. The structure of each aggregate would be hierarchical. (For one such hierarchical definition of play, see Müller-Schwarze 1978a). At each level of the hierarchy, the aggregates would have distinguished foci or capitols, stereotypes or typical elements that differ as little as possible from as many elements as possible of the aggregate they represent. One aggregate may even have several capitols, with no one attribute common to them all.

Ethologists and philosophers of science who prefer the classical Aristotelean method of nominal definition by genus and differentia might question the approach to behavioral classification just described. However, philosophers themselves currently disagree about definition and classification. I conclude that those critics of play research who blame its lack of progress to date solely on the absence of a formal Aristotelean definition have grossly oversimplified the issue.

In light of the foregoing discussion, it is not at all surprising that behavioral definitions are difficult. Consider, for example, the concept of aggressive behavior. Hinde (1974 p. 249) finds no definition of aggression wholly satisfactory. According to Andrew (1976), "a fully satisfactory definition of aggressive behaviour is perhaps never to be attained." H. Kaufman (1965) cites "considerable disagreement" over defining aggression, and Vowles (1970) finds aggression "undefinable." See also Izard (1975), Kummer (1967), McGrew (1972 pp. 4–5), and Suomi (1977). Similarly, Simpson (1978) questions the concept of sexual behavior. For Hinde (1974 p. 15), the term social behavior is not satisfactorily defined either in a causal or in a functional sense. Hinde agrees that this term can only be used as it is in common speech, that the category is admittedly a loose one, and that "further discussion of (its) definition is unprofitable." Of course, Robert Hinde's studies of social behavior span a quarter century and are internationally recognized. Yet many critics of the concept of play have argued that it is fruitless to conduct scientific research on a poorly defined behavioral category.

I doubt that the many comparative psychologists and sociobiologists who significantly advanced our understanding of aggression, sex, and social behavior during the past decade wasted much time worrying whether these topics had been adequately defined. If exact definitions were absolute prerequisites for scientific research progress, it would be hard to explain recent advances in the biology of social behavior, since we still cannot adequately define aggression, sex, or social behavior itself. This is not to say, of course, that we should not make a serious attempt to define whatever behavior we propose to study.

DEFINING PLAY

Is play even more difficult to define than other categories of behavior? Yes, at least according to Lorenz (1956), E. O. Smith (1978), and E. O. Wilson (1975). Hinde (1974 p. 226) finds play "impossible to define" (but inescapable in practice). How has play been defined, and on what grounds have these definitions been criticized? If defining play is truly an intractable problem, what appears to be the reason for this difficulty?

Previous definitions of play and their critiques represent a confusing mixture of nominal and real definition and of formal, functional, causal-motivational, developmental and purposive conceptual schemes. These typological and schematic distinctions will be used to clarify basic issues raised by previous definitions and critiques. Because the material to follow is lengthy and technical, I summarize its main points here.

Play behavior may be defined by its attributes, by operational means, by example, by enumeration of behaviors called play, or by contrast with similar behaviors. Evidence for the validity of the category so defined is that human observers without previous training agree on whether or not an animal is playing and that animals can recognize and respond in kind to play in other animals, even in those belonging to different species. It may be especially difficult to define play (compared to definition of other behaviors) because play is a molar category defined by other molar behaviors.

Structural, causal, functional, and purposive definitions of play have been freely and interchangeably proposed and criticized. Some of these discussions imply nominal definition, whereas others imply real definition. Many imply both simultaneously. Implicit and often questionable assumptions about behavior are almost always present in these discussions. As Purton (1978) commented in a related context, "The confusions here are multiple, and they interlock with one another in a way that almost defies analysis."

In Chapter 1, I nominally defined the word "play" to mean structural transformations and functional rehearsals or generalizations of behaviors or behavioral sequences also observed in other functional contexts where they increase the performer's inclusive fitness by resolving conflict over a disputed resource. In the light of the previous discussion, we see that this definition is logically inadequate because it invokes both structural and functional schemes. Perhaps, following Fagen (1972b) and Müller-Schwarze (1978a), we ought to give "play" behaviors different names depending on the number of schemes satisfied.

Given a nominal definition of "play," however unsatisfactory, we should seek a real definition of the conceptual aggregate "play," whose three capitols are (1) playfighting and play-chasing, (2) locomotor-rotational exercise, and (3) post-mastery manipulation. These three classes of behavior may appear to have little of biological importance in common, and a current trend in play research is to view them as categories of behavior, valid in their own right, whose structure, causation, and function are (at least for the time being) best analyzed separately. After all, lumpers are more vulnerable than splitters, and authentic unifying visions are rare in biology; grand syntheses are easier to proclaim than to achieve. However, I remain reluctant to dismiss the perceived unity of play. Evidence reviewed in this chapter and in those following suggests consistent relationships among the three major recognized classes of play behavior. As long as this question remains open, the analysis of play will need to continue accepting evidence of all kinds. To fragment the study of social, exercise, and item-oriented play into three disciplines might be as unwise as trisecting the study of predation into the behavioral ecology of feeding on small, medium-sized, and large food items.

Some Definitions of Play

Attempts to define play by its attributes span a wide range, from a three-level hierarchical classification (Müller-Schwarze 1978a) to pithy one-liners. For a representative sample of such attempts, see Appendix I, this chapter, and Gilmore (1966).

Lists of attributes often serve in lieu of a formal logical definition of play. These lists cite supposed structural, causal, and
functional characteristics of play that ultimately reflect the com-
piler's assumptions about the study of behavior. Subjectivity
lies closest to the surface in statements about causation, but
even characteristics of form assume basic if implicit and vaguely
suggested structural models of a behavioral act or sequence.

The nine lists given in Appendix II (Bekoff 1974b, Henry &
Herrero 1974, Loizos 1966, Marler & Hamilton 1966, Meyer-
Holzapfel 1956b, Rensch 1973, Symons 1978a, Wasser 1978,
XIII International Ethological Conference Play Round Table
1973) contain diverse material and exhibit certain areas of agreement. Judging solely from their inclusion on a majority of these
lists, only three characteristics of play seem generally accepted:

1. The behavioral acts occurring in play are similar, but not
   necessarily identical in form to acts that occur in well-
   defined adult and /or juvenile functional contexts, but a single
   play sequence may include acts from several of these functional contexts (Fig. 2-1).
2. The form of individual acts in play differs from their form in
   other functional contexts because play acts are exaggerated.
3. The form of play sequences differs from the form of corresponding behavioral sequences in other functional contexts
   because individual acts are repeated more often in play
   sequences.
   Some agreement on two other points is evident from their
   inclusion on three different lists:
4. Play sequences have a more variable (less predictable) order
   than do sequences from other functional contexts that share
   the acts of play.
5. Play sequences lack the consummatory acts and biological
   consequences of their nonplay counterparts.

Characteristics 2, 3, and 4 above cite the form or structure of
behavior, whereas characteristics 1 and 5 concern function or
consequences of behavior. The fundamental point seems to be
that similar acts (or bits of sequences?) can appear in two different contexts, one of which has a recognized biological func-

Fig. 2-1 Red fox cubs (*Vulpes fulva*) play-wrestling using their jaws. The cubs' facial expressions and their ear and tail positions indicate a lack of intent to inflict damage on the partner and clearly delineate playful from agonistic fighting. Alaska. (Stephen J. Krasemann/DRK Photo)

tion. The form of these acts and the structure of the sequences in which they appear differ in these two contexts. The value of these generalizations depends on the meaning of the term "functional context," on the type of comparison made, and on the ability or willingness of students of play behavior to distinguish play from other out-of-context behavior exhibiting the relative structural characteristics of exaggeration, repetition, and unpredictability.

Other general characteristics of the structure or causation of play cited in the nine lists of Appendix II include the following; the terms "relative" or "relatively" imply a comparison with the nonplay context.

STRUCTURE

1. Acts that involve several parts of the body may lack some components found in nonplay (instantaneous incompleteness).
2. Play acts having temporal structure may be relatively incomplete, e.g., a movement may be started, but not finished (incomplete execution, serial incompleteness).

3. Play sequences are relatively brief.
4. Play sequences may be relatively reordered or disrupted.
5. Play sequences show relatively rapid alternation of acts.
6. Play sequences are relatively incomplete.

CAUSATION

1. Play sequences may be interrupted by higher-priority behavior and fragmented by inclusion of motivationally irrelevant activities (Fig. 2-2).
2. Play occurs in a relaxed motivational field.
3. Play appears to be pleasurable to the performers.
4. Play patterns are relatively inhibited.
5. There is specific motivation to play.
6. Animals search for a play partner.
7. Non-conspecific items, including objects and living or dead organisms, may substitute for a conspecific partner in play.
8. Transitions from play to nonplay, or mixed forms of play, may occur.
9. Play occurs characteristically in immature animals.
10. Play sequences occur in different situations from nonplay sequences that include the same acts, or they occur as a result of different stimuli, including stimuli normally inadequate to elicit these acts.
11. In a given play sequence, the same behavior may be directed in turn at different stimuli.
12. The pattern of behavior in play is relatively less dependent on normal stimulus-response relationships.
13. In play, animals return repeatedly to the same stimulus source.

Some of these characteristics have been demonstrated empirically (Henry & Herrero 1974). In general, however, they seem vague and sometimes contradictory. Imprecise and theoretically suspect terminology, uncertainty about the exact philosophical meaning of the ever-present comparison of play with nonplay, and difficulty of application and interpretation render such lists unsatisfactory, both as formal definitions and as working outlines.

Fig. 2–2 Ease of interruption is a frequently cited characteristic of play. These three red fox cubs (*Vulpes fulva*) were all at play a moment before this photograph was taken. Now, one cub interrupts its play to sniff the anogenital region of a second playing cub. Could a play pheromone have elicited this behavior? Alaska. (Stephen J. Krasemann/DRK Photo)

Play is further characterized in a social context by a set of conventions that render the interaction reciprocal and "fair" (S. Altmann 1962b). Roles reverse, so that one animal is not always on top of or fleeing from another. Dangerous fighting tactics are never used, and bites are inhibited. Signals used in agonistic social communication, including stereotyped signals of threat and submission, are absent from play. A stronger or more skillful animal may self-handicap, running slowly or grasping its partner with only a fraction of its actual strength. Dominance distinctions fade in play: a subordinate animal may wrestle with or chase a dominant individual. Play-specific postures, facial expressions, gestures, gaits, vocalizations, and possibly smells accompany these behavioral elements, but only when performed in the "playful" manner described. These contextual considerations led Symons (1978a) to provisionally define social play of rhesus monkeys as behavior patterns that (1) are similar to agonistic patterns of fighting, chasing, and fleeing; (2) are inhibited compared with these agonistic behaviors; and (3) are not associated with stereotyped agonistic signals of threat and submission.

Operational Definitions

Play may be recognized by citing certain non-essential properties that are unique to play and universally present in play. Often, these definitions apply only to one or to a few related species, thus sacrificing generality for precision. Social play is defined operationally in a given species by the presence of play-signals and by the absence of agonistic communication signals. Specific independently defined signals or movements may likewise be diagnostic of play independent of the amount of social interaction occurring in play. Among these diagnostic signals and movements are the play-face, a variety of unusual locomotor movements, and particular forms of rotation of the body or of its parts.

Students of behavior, at least since Darwin (1898), recognized the human play-face and its nonhuman primate and carnivore counterparts. The primate play-face may occur in conjunction with particular body postures or movements (especially rotations) that form a phylogenetic link between primate play-signals and the play-signals of rodents and ungulates. These behavior patterns, termed locomotor-rotational movements, and identified in the play of many mammals and birds, are "exaggerated forms of normal locomotor and rotational body movements found in other functional contexts, such as anti-predator behavior" (Wilson & Kleiman 1974). The best-known locomotor-rotational movements are leaping, rolling, headshaking, body-twisting, neck flexion, rearing, and kicking. Wilson and Kleiman observed these behavioral acts in play of six species of mammals, including harbor seals (*Phoca vitulina concolor*), a pygmy hippopotamus (*Choeropsis liberiensis*), giant pandas (*Ailuropoda melanoleuca*), and three little-known species of caviomorph rodents. Common usage gives these play movements special status by employing unique terms: gambol, caper, romp, scamper, frolic, rollick, frisk, jink, cavort, ragrowster, *gambader* (French), and *balgen* and *tollen* (German).

To a professional ethologist, play behavior, just like other molar categories of behavior, can be defined operationally. Recent empirical studies of play behavior always include a precise, operational definition of play in the species studied. For example, Poole & Fish (1975, 1976) defined play in laboratory rats (*Rattus norvegicus*) as activity involving energetic, exaggerated movements (compared with those of adult fighting) and inhibited attacks, including the unique behavioral acts p-charge ("one animal rushes toward another with a vigorous, bouncing gait"), p-pounce ("all movements in which the body of one rat is moved over the back of the partner"), and p-jerk ("animal twitches violently and often repeatedly"). Sade (1973) defined rhesus macaque (*Macaca mulatta*) play behavior as sequences including a postural component that emphasized rotatory body movements in the transverse plane. Fagen & George (1977) defined play in horses (*Equus caballus*) as follows:

In solitary play, foals gallop rapidly back and forth, often between two physical objects such as tree trunks or clumps of tall grass; in doing so they may buck and kick, prop and rear, head-twist, head-toss, head-shake, or exhibit other locomotor-rotational movements. Occurrence of solitary play is defined by these behavioral acts and by the absence of vocalizations. In social locomotor play, foals chase and flee one another but do not vocalize, nor do these actions lead to injury or lasting dispersal of participants.

In theory, to ascertain whether or not an animal is playing we can always pay attention to the animal's messages or even ask the animal. When it is in an animal's interest to communicate that it regards its behavior to be play, it will signal accordingly. There is no reason to suppose that this approach cannot succeed, at least in primates and carnivores whose play-signals are similar and perhaps even homologous to our own (Andrew 1963, van Hooff 1962, 1967, Poole 1978). For example, the play-face (relaxed, open-mouth grin) is seen not only in social play, but also in primate and carnivore object and locomotor play (Aldis 1975, Angus 1971, Hutt 1966, Klinghammer pers. comm., Lethmate 1977, Loizos 1969, Rensch 1973, Tembrock 1957). These accounts suggest that the animals acted as if they regarded their behavior as play or as if they wished their behavior to be regarded as play. This observation casts serious doubt on the distinction between social, locomotor, and item-oriented play.

The meaning of "Ask the animal" (Griffin 1976) is actually even more radical than the interpretive approach outlined above. Griffin proposes use of two-way interspecific communication to study animal behavior. Because interspecific play is possible, we might seek to define play by playing with the animals concerned. Many descriptions of play in hand-reared animals are based essentially on this method. Like any experimental procedure, two-way communication has limitations. A potential problem with any study of animal behavior, to which this method appears especially vulnerable, is human elicitation of play behavior that would not normally occur within the subject's species.

Definition by Example

The concept of animal play is easily introduced to general audiences by citing the playful activities of kittens or puppies. We may thus define capitols of a conceptual aggregate. However, an element of vagueness remains, since common elements of these examples are never made explicit.

Definition by Enumeration

It might be possible to list examples of all the kinds of activities that one wishes to call play, thus furnishing a complete set of capitols along with sufficient variations and intermediate cases to indicate the shape and extent of the phenomenon. This approach and that previously cited exhibit the same deficiencies.

Definition by Contrast

In lieu of a formal definition, play has been demarcated by enumerating categories of behavior similar to play that are not considered playful and by drawing distinctions. This method of definition is complementary to the method of definition by enumeration. It can be informative to identify precise differences between play and similar nonplay behaviors. For example (Aldis 1975), agonistic fighting includes escalated fighting, in which dangerous tactics are used and injury or death seem likely, and tournaments in which two animals engage in repeated, vigorous, often fatiguing physical contact without tissue damage. Such aggressive competition resolves conflicts over access to or control of some limiting resource (Maynard Smith & Price 1973, Parker 1974, Popp & DeVore 1979, Treisman 1977). Exploration is behavior involving cautious, intent sensory inspection or examination of some newly encountered feature of the environment. Exploration yields information about that which is explored, and it is not immediately repeated unless the feature itself changes. Developmental precursors, vacuum activities, redirection, and additional nonplay categories are discussed by Aldis (1975), who defines play in part by contrast. Similarly, Müller-Schwarze (1978a) defines narrow and wide senses of play by contrast with all other forms of behavior not serving their usual immediate function.

Interobserver Reliability

As indicated by the specialized terms used (Appendix III), human observers acknowledge their awareness of animal play. Humans can correctly and confidently recognize animal play and can agree, even without specialized training, on whether or not an animal is playing (Fedigan 1972b, Loizos 1966, Mason 1965b, Miller 1973, Poole & Fish 1975, E. O. Smith & Fraser 1978, Weisler & McCall 1976). Zoo visitors even identify play in species that they have never seen before. Of course, sometimes members of the public make interesting mistakes, confusing play with courtship (which may, of course, exploit play), escalated fighting (but not for long), captivity artifacts, or various object-oriented exploratory behaviors; Aldis (1975) contrasts play with these other categories of behavior. Interobserver reliability scores for play recognition have been calculated by Anderson & Mason (1974), Bernstein (1975), Fagen & George (1977), Hansen (1966), Mitchell (1968), Rhine (1973), and Seay (1966). These scores reflect previous interobserver discussion and training; they tend to be satisfyingly high (90% agreement or more). Of course, these scores represent a positively biased sample, since observers who refuse to accept play as a behavioral category might score very low on reliability when asked to recognize play.

Identification of play in a species unfamiliar to the observer usually depends on the similarity of that play to play of related species. Could play be recognized in an organism whose anatomy differs radically from that of mammals—in an ant larva, a starfish, or a sponge, for example? As Müller-Schwarze (1978a) correctly observes, play may be identified by its structural and contextual characteristics after the general behavioral repertoire has been described. Thus, for example, we could recognize playful variants of aggressive behavior given information on the organism's repertoire of tactics used in aggressive competition.

Is Play Especially Difficult To Define?

Though no molar category of behavior is easily defined, play does appear to be especially difficult (Fig. 2-3). Although the reason for this difficulty is not clear, one possible source of problems is that play is a relative term. The sense of play is inextricably bound up with "comparison," and systematic differences from some other behavioral category sharing the same behavioral acts may constitute the sine qua non of play (Symons 1978a, see also G. Bateson 1955, 1956). With the possible exception of certain developmental precursors of adult behavior, no other molar behavioral category is only defined relative to other molar categories of behavior. Is this approach to categorization valid? (Of course, if we adopted a purely functional approach and looked at needs of developing animals for given types of experience, stimulation, information, or learning under certain environmental conditions, we might find that these needs could only be met by behavior that had the structural characteristics of play. The analysis might also reveal unanticipated relationships between some play behaviors and other behaviors not previously considered playful. Such an approach would be worth trying even though it might prove premature.)

The intriguing suggestion (e.g., Bavelas in Bateson 1956, p. 227) that play lies beyond the scope of ordinary human logical

Fig. 2–3 Play can be difficult to define because it sometimes grades into aggression. The young red fox (*Vulpes fulva*) in the on-top wrestling position has a playful, relaxed posture and play-face, but the on-bottom cub may have found its partner's play too rough. Its narrowed eyes and flattened ears signal defensive threat. Alaska. (Stephen J. Krasemann/DRK Photo)

analysis represents a philosophical challenge to play research. Alexander (1975) speculates that humans were selected to be unable to analyze certain phenomena. Could play be one such phenomenon? Radical brain lateralization theories even claim that control of certain nonverbal activities, including art and music (and play as well? What about verbal play?), reside in the right hemisphere, safe from the cold prying logic of the left hemisphere. I do not take any of these speculations very seriously, but as Dawkins (1976b) remarks about his own sociobiological work, some imaginative ideas are also serious science.

PLAY AS A BEHAVIORAL CATEGORY

Form

Employing structural schemes, we recognize play by the presence of certain unique behavioral acts and the absence of others, or by unique form, orientation, duration, or intensity of acts also occurring in other contexts. We also analyze and compare the structure of play sequences to that occurring in nonplay by using quantitative ethological methods (Fagen & Young 1978). Results of such analyses comparing playfighting with escalated fighting (Fagen 1978a, Hill & Bekoff 1977, Leresche 1976, Poole & Fish 1975, Schoen, Banks, & Curtis 1976) suggest that play sequences are predictable and have definite temporal structure, but that these sequences are less ordered (more variable) than their agonistic counterparts.

A very different structural approach to play behavior is that offered by Eshkol-Wachmann movement notation (Golani 1976), a precise descriptive system with great potential. Eshkol-Wachmann notation is not only a descriptive system, but also a philosophy of behavior. In this philosophy as in dance, movement is all. Other information (e.g., context) is irrelevant.

Causation

Proximate causation represents a second conceptual scheme used to define the behavioral category called play. Proximate causation includes contexts and situations; eliciting, inhibiting and terminating stimuli (both internal and external); motivation; and developmental causation on the long time scale of an entire life history or on the shorter time scale of development of a particular sequence of behavior.

One aspect of causation cited as negative evidence by critics of play (e.g., Welker 1971) is the lack to date of physiological or neurophysiological bases for play behavior. This cannot be a real problem, however. We may investigate learning and skill development using intact whole organisms or even computer models. Field ethology, behavioral ecology and sociobiology ask and answer interesting questions about behavior without recourse to lesions or intracranial electrodes. Even the control of behavior may be studied informatively from a "software" point of view, as Dawkins (1976a) argues.

Male-female differences in play behavior have a demonstrated physiological and neurophysiological basis. In immature humans (Symons 1979), rhesus macaques (Goy 1968, Goy 1970, Goy & Resko 1972, Phoenix 1974), and Norway rats (Olioff & Stewart 1978), sex differences in play "are not the result of direct hormonal action but of a central nervous system modified by prenatal testosterone so that normal males and pseudohermaphrodite females are predisposed to acquire male behavior patterns. . . . The evidence suggests that individual experience acts on a brain already biased in a male or female direction" (Symons 1978a p. 155).

Critics may also question the concept of play because the behaviors included in this category lack a demonstrated genetic basis. Popp & DeVore (1979) discuss this point in regard to sociobiology in general. Play alleles are easier to model than to demonstrate, although rigorous behavior genetic studies of play in dogs and other domestic animals may some day demonstrate genetic determinants of play for which anecdotal evidence has long been available. For example, the Nova Scotia duck-tolling retriever is a domestic dog breed said to attract the attention of waterfowl and even lure them towards land by playing with objects (Anon. 1976, Lunt 1968 p. 110). These retrievers are claimed to be especially playful dogs in all contexts (including social play) and at all ages. Furthermore, domestic cat responsiveness to the essential oil of the catnip plant *Nepeta cataria* appears to be controlled by a single autosomal dominant allele (Todd 1963). Because catnip elicits play behavior involving objects or humans as well as play alone in adult cats, the allele governing the catnip response is a genetic determinant of the tendency to play. Todd (1963) hypothesizes that catnip oil mimics a felid reproductive pheromone. This hypothesis is consistent with his observations that catnip also elicits courtship displays. No concrete evidence is available to support informal claims by horse fanciers that horses bred for dressage training (e.g., Trakehner, Lippizaner) develop slowly and are especially playful. The skilled movements of these horses include vertical leaps, rear-kicks, and sudden turns—the caprioles, caracoles and curvets of the dressage horse. These movements are precisely those used by all equid foals in play (Chapter 3 and Zeeb 1977). Artificial selection of dressage breeds could have favored tendencies to perform these movements and extension of these tendencies into later ontogeny as well as the ability to learn information concerning play movements.

In the past, naive homeostatic theories of behavior gave play the awkward status of an unmotivated act (Berlyne 1969, Welker 1971). Critiques of play continue to focus on supposed lack of evidence for play as a separate motivation. All such critiques face two problems. First, a separate motivation for play has been demonstrated using classical motivational paradigms. Play can act as a reinforcer (G. Bateson 1956 p. 237–239, Döhl & Podolczak 1973, Hick 1962, Mason 1967, Mason, Hollis & Sharpe 1962, Randolph & Brooks 1967, Yerkes & Petrunkevitch 1925), and short-term deprivation of play produces a post-deprivation rebound effect without changing the frequencies of other motivationally defined categories of behavior (Baldwin & Baldwin 1976a, Chepko 1971, Oakley & Reynolds 1976). That young animals are motivated to play is evident from experience with domestic animals as well as from observations of behavior in the wild, of which the following account (Blaffer Hrdy 1977 p. 295) is representative:

The "impulse to play," regardless of who the partners were, could be seen in the case of the lone Hillside infant, Miro. Because of the very high infant mortality in Hillside troop between 1971 and 1974, Bilgay's 10-month-old son was the only immature in the troop during the 1973–1974 study period. When I first observed Miro, I was struck by how extremely timid and dependent on his mother he appeared to be. When the troop moved, even a short distance, Miro would hang behind and whine until Bilgay returned, took him up, and carried him.
Because I had developed an image of Miro as a timid "only child," I was surprised to observe that Miro's behavior with strange langurs

was anything but timid. During encounters between the Hillside and Bazaar troops, Miro assertively sought out playmates of any age in the Bazaar troop. He would challenge parties of several alien adults and take on juvenile wrestling partners much larger and heavier than himself. Despite the trouncings Miro inevitably received, he remained undaunted. As far as I know, these contacts with Bazaar troop immatures were the only opportunities for play that Miro had.

A second, more serious (and ultimately fatal) flaw in all motivational critiques of play is that the classical concepts on which these critiques are based are now known to be incorrect (Hinde 1974, Popp & DeVore 1979, Purton 1978). Simpson (1973, 1978) questions the applicability of conventional motivational concepts to juveniles. McKay (1973) and Minsky (1977) question the applicability of these concepts to adults. Although there is no current agreement on proximate causes of behavior (see Dawkins 1976a, Fentress 1973, McFarland 1974, 1976, Simpson 1973, 1978), the authorities cited all argue that behavior is controlled in ways that resemble a sophisticated computer program, not Lorenz's flush toilet. If we can predict anything at all about the most sophisticated computer programs, it is that they will need to play in order to develop skills and to "debug" and maintain themselves (Sussman 1975, Winston 1977). Flush toilets do not play.

Sociobiologists (e.g., Dawkins 1976b, Popp & DeVore 1979, Trivers 1974) consider motivation to be nothing but a biological dependent variable totally at the mercy of the evolutionary strategies that control behavioral tactics as a function of whatever the environment and other animals happen to be doing at the moment and/or have done in the past. This point of view differs from the classical theory, but it marches well with contemporary research on causation of behavior. Hinde (1970) sought to place the study of behavioral causation on a quantitative, hypothetico-deductive basis. He argues that behaviors sharing common causal factors may be influenced in the same way by a particular stimulus and tend to occur consistently together in time. As Hinde (1970) also points out, this approach, however quantitative, has serious limitations. The stimulus of an item of food type A may elicit behaviors appropriate to gathering food type A, but not B; does this mean that be-

haviors appropriate to item B are causally not part of feeding
behavior? And suppose encounters with item A are temporally
statistically independent of encounters with item B? Our quantitative analyses of feeding behavior may well reveal different
clusters of behaviors and different temporal groupings for each
food item in the diet. Does this mean that dietary generalists
need separate causal systems for each food type? Quantitative
causal approaches to categorizing play, using sophisticated statistical techniques from temporal correlation and Markov chain
analysis to multivariate analysis, are flawed for this same reason. In fact, multivariate statistical analyses of play (Blurton
Jones 1967, 1972, Blurton Jones & Konner 1973, Fedigan 1976,
van Hooff 1970, 1972, E. O. Smith & Fraser 1978, P. K. Smith
1973, P. K. Smith & Connolly 1972, Strayer, Bovenkerk, &
Koopman 1975) all reassuringly identify one or two behavioral
entities that correspond to our intuitive concept of "play."
However, these results (1) depend crucially on the sampling
units, sampling universe, sampling techniques, and time units
employed and (2) all reflect two basic but questionable assumptions: that temporal correlation necessarily indicates shared
causal factors and that failure to demonstrate such correlation
means that "play" is causally heterogeneous.

The quantitative analyses above analyze behaviors occurring
either simultaneously or in time units large enough to include
several acts. Temporal contiguity may also be defined sequentially, although this approach is not foolproof. If behavior A
tends to follow behavior B with greater than chance probability, then we can infer that B directs or perhaps causes A (as suggested by serial correlation, time series analysis, Markov chain
methods, etc., Fagen & Young 1978). A and B would then
occur consistently together in time, in accordance with Hinde's
criterion for behavioral causation. However, when we speak of
moods, motivational states, and the like we imply that the
probabilities of behavior change over time. We therefore
require hierarchical models of behavioral sequences, such as
those recently formulated by Cane (1978), Dawkins & Dawkins
(1973), Dawkins (1976a), and Kuczura (1973). Results of Cane's
(1978) hierarchical analysis of a rhesus macaque repertoire that
included play show play to be a distinct behavioral category. (It

may prove important to distinguish between hierarchical structure of a behavioral sequence itself, hierarchical organization of the software generating these sequences, and hierarchical organization of brain hardware—e.g., cortical versus medullary structures. This distinction is not pursued here, but it is important to recognize that the principle of hierarchical organization applies to many fundamentally different kinds of conceptual schemes used in the analysis of behavior.) Stuart Altmann (pers. comm.) suggests that application and interpretation of Kuczura's model could aid in defining structural features unique to play sequences.

Weisler & McCall (1976) raise an additional aspect of the problem of inferring causation and categorization from temporal patterns of behavior. These authors question the distinction between play and exploration in human children because these activities tend to alternate too rapidly. The validity of this criticism is questionable. Would we include feeding and predator avoidance in the same causal category if a foraging baboon interrupted its search for edible roots every five seconds in order to scan for predators in the vicinity? On the other hand, an argument (see below) for the unity of play within the context of classical motivational theory cites observations that item-oriented, locomotor-rotational, and social play alternate (or substitute) very rapidly. Therefore, according to temporal causation analysis, they must belong to the same motivational control system. Although I prefer not to invoke classical motivational theory at any point, I cite this example because it demonstrates that the theory is sufficiently loose to be invoked in support of any point of view whatsoever.

Let us then play the game by the other side's rules temporarily. One type of play often appears to be an apparent motivational substitute for another (Baldwin & Baldwin 1972, Bekoff 1974c, 1976a, Bertrand 1969, Brazleton et al. 1975, Box 1975a, Dücker 1962, Gwinner 1964, Kaufman & Rosenblum 1965, Linsdale & Tomich 1953, Poole & Fish 1975, 1976, Pruitt 1976, Rose 1977b, Sinclair 1977, Ulrich, Ziswiler, & Bregulla 1972). According to Bertrand (1969), "shifts from solitary to social play, or vice versa, are abrupt and frequent" in the stumptail macaque (*Macaca arctoides*). Sinclair (1977) describes one buffalo calf (*Syncerus caffer*) bucking and frolicking during a break in a sparring match. A playfighting rhesus monkey (*Macaca mulatta*) may drop from a tree branch "after a minimal amount of play-fighting or even at the mere approach of a second monkey" (Symons 1978a p. 47), and during intervals between playfights rhesus may occasionally dangle from branches, adopt unusual postures or even, in one case observed, execute a standing back flip (Symons 1978a p. 151). A black baboon infant (*Papio anubis*) "lay on its back and mouthed a leafy twig which it held in one hand, sparred with another animal with the other hand, and at the same time made walking movements up a thin branch with its feet, ending up hanging bipedally and wrestling with the other infant with its forelimbs" (M. D. Rose 1977b). Although it may be argued that these observations prove nothing because they were not quantified, they suggest a pervasive interpenetration of social, object, and locomotor play that has not been more widely documented simply because there was no theoretical reason to look for it. Unfortunately the only quantitative data reported on this problem (Chalmers 1978) cannot be interpreted because of Chalmers's use of zero-one sampling, an invalid technique (J. Altmann 1974, Simpson & Simpson 1977). By appropriately choosing my sampling interval, I could readily demonstrate, using valid quantitative sampling techniques and data analysis methods, that play-wrestling and play-chasing were not only structurally, but also (gasp) causally distinct. Few ethologists would accept these conclusions in any but the narrow technical sense of the particular procedures employed. It would be ludicrous to argue, based on such evidence, that aggressive play represented an artificial, meaningless behavioral category that had produced paralyzing confusion and prevented all progress in the understanding of nonagonistic fighting behavior. The system controlling play, like that controlling feeding or mating, seizes the best available resource patch (in terms of benefits versus costs). In the case of play, this resource patch may be a conspecific, an inanimate object, a living or dead prey item, or even a suitably furnished space in the environment. Although this argument is independent of the assumptions of classical motivational theory, it is worth citing in the present context.

Some critics of the concept of play seek to explain it away by reinterpretation, citing even more vague physiological, motivational, or behavioristic concepts such as vacuum activity (Lorenz in Nice 1943, Lorenz 1956), displacement activity (Sladen in Bateson 1956, p. 188), lack of inhibitory mechanisms (Altman, Brunner, & Bayer 1973), or stimulus generalization (Schlosberg 1947). This approach sidesteps the basic question of definition and invokes seriously flawed explanatory concepts based on unsatisfactory intervening variables.

A final challenge to the concept of play from motivational theory states that play behaviors are inadequately motivated sequences produced by causal systems of feeding, sex, predator avoidance, or aggression. Low levels of motivation result in altered behavioral structure and in loss of function. We may expect to elicit such behavior experimentally, e.g., by overfeeding or by artificially altering hormonal levels. There is no reason to believe that the many assumptions about motivation implicit in this "explanation" are even approximately correct or that such experiments would actually achieve the result predicted. Neither Dawkins's (1976a), Fentress's (1973), nor McFarland's (1974, 1976) models of motivation would behave in this way. In order to obtain behavior resembling animal play from Lorenz's flush toilet metaphor, we would require many ad hoc patches and additional assumptions that even the most devoutly unreconstructed Lorenzian would find acutely embarrassing. If an inadequately motivated causal system is capable of producing even a small fraction of the elaborate, vigorous, intricate, and highly structured behavioral sequences described in Chapter 3, we would do well to reconsider our concept of an inadequately motivated causal system.

Developmental causation [considered by Tinbergen (1951) as a question distinct from that of immediate causation, but all possible intermediates exist; see P. P. G. Bateson (1976b)] represents another conceptual scheme within which play may or may not be well defined. To critics like Welker (1971), the only relevant schemes for considering the validity of play as a behavioral category are developmental, although, as we recognize (Purton 1978), this point of view is unnecessarily narrow.

Developmental critiques of the concept of play (Gentry 1974,

Lazar & Beckhorn 1974, Welker 1971) pose a different sort of challenge from that presented by causal–motivational critiques. Play behavior has its own developmental history. Play has multiple simple origins (Welker 1971), goes through a period of apparent motivational autonomy, and then, at least in the domestic cat (Barrett & Bateson 1978) splits into separate developmental tracks that merge with behavior immediately serving aggressive competition, feeding, and reproductive functions. Distinctness of motivational and developmental definitions is demonstrated by the fact that adult cats still play long after they can fight, hunt, and mate competently (Leyhausen 1973a). [The metaphorical merging and splitting of equally metaphorical developmental tracks may seem amusing to those readers still chuckling about flush toilets. Actually, the underlying theory, although still very new, is subtle and interesting: P. P. G. Bateson (1973, 1976a,b, 1978a,b, 1979), Goldberg (1978), Kagan (1978), Kagan, Kearsley, & Zelazo (1978), Simpson (1976, 1978). It remains to reconcile this theory with evolutionary concepts of development offered by Barash (1975), Daly (1976), Geist (1978a), Gould (1977), Konner (1975, 1977a), Trivers (1974), and others.]

Some developmental critiques of the concept of play are clearly without merit. Altman, Brunner, & Bayer (1973), for example, view play as behavior occurring before the development of inhibitory brain mechanisms. The idea of a primal behavioral chaos is amusing, but incorrect. Play only includes certain kinds of behaviors, many of which can also be performed in "inhibited" functional contexts at the given or at an earlier age; aggression is inhibited in play (S. Altmann 1962b); the concept of monolithic inhibitory systems is vacuous and unsupported by experimental evidence (Einon & Morgan 1978a, Morgan, Einon, & Morris 1977, Morgan, Einon, & Nicholas 1975); and the neurophysiological arguments upon which Altman, Brunner, and Bayer base their speculations about play are themselves incorrect (Nadel, O'Keefe, & Black 1975). Lazar & Beckhorn (1974) incorrectly claim that "the evolutionary approach to behavior" must emphasize innate behavior of mature adults [actually, as evolutionary biologists including Cole, Mayr, and Waddington have recognized, the entire life history is the target of selection; see Gould (1977)]. Possible trade offs and compromises in growth, survivorship, and fecundity between different life-history stages (and resulting immediate suboptimality, maladjustment, or inappropriateness of behavior of the young) are completely ignored. In addition, Lazar and Beckhorn make major errors in the basic biology of their own experimental animal and in ethological observation and interpretation (Poole 1978, Symons 1978a).

The more sophisticated developmental critiques of play (e.g., Barrett & Bateson 1978) are noteworthy, but limited in scope. Developmental causation is not the only conceptual scheme for approaching play. We do not yet know how much of current developmental theory will survive the scrutiny of rigorous evolutionary and cybernetic analysis. (Nor do we yet know how much sociobiology will survive current inquiries into developmental constraints on behavior.) Gentry (1974), whose point of view is thoroughly developmental and not evolutionary at all, reaches some of the same conclusions about play that an evolutionary perspective suggests. In Gentry's (1974 p. 403) words, play experience is "a vehicle by which the frequency, intensity and combination of behavioral patterns are changed over time." This developmental view is consistent with an interpretation of play as a behavioral result of evolutionary and behavioral control processes whose function is the generation of experience developmentally relevant to behavior in conflict situations, in order to benefit the performer, its kin, or unrelated animals who may be expected to reciprocate.

Function

Functional classification of play, understood in the sense of adaptive effects [note that E. O. Wilson (1975) uses the words "functional biology" to mean the study of physiological and developmental mechanisms], has been difficult because the adaptive significance of play has not been as obvious as the adaptive significance of feeding, fighting, or mating. (Of course, it may become obvious in the future as a result of developments in artificial intelligence or in other fields outside biology.) The study of adaptation (Lewontin 1978, Symons 1978a,b, Williams 1966) is a subtle area of evolutionary biology just now being applied to the problem of play (e.g., Symons 1978a). Because play has sometimes been categorized as nonfunctional behavior, or as behavior that appears useless in the eyes of a human observer, the concept of play has been criticized as a "wastebasket" category. Aldis (1975) and Müller-Schwarze (1978a) draw important distinctions between play and other apparently functionless behaviors. Although functional definitions of play have not been available, the concept of behavioral function has recently shifted, and initial reanalysis of play as a functional category in the light of these advances has been successful (Symons 1978a,b).

The criticism that functional analysis of play cannot proceed without exhaustive empirical information on function is incorrect. Assumptions that specify the function of play only incompletely can still generate testable hypotheses. In fact, we should always make as few assumptions as possible about the function of play. This naive approach to play has been highly successful. Fagen (1977) simply assumed that the costs of play were immediate in time relative to its benefits and was able to generate testable predictions about age schedules of play. Konner (1975) assumed only that play had some unspecified beneficial effect. From this simple assumption he successfully predicted many aspects of the temporal and social structure of interindividual play.

Purpose

Purton (1978) argues that a fourth conceptual scheme for behavior, distinct from schemes of structure, causation, or function, is one of purpose or motive, rooted in our everyday conceptual framework for describing and interpreting human activities. Many contemporary philosophers of mind accept the validity of this conceptual scheme and reject traditional accusations of subjectivity. Although this scheme will strike most behavioral scientists as alien, if not subversive, it appears to enjoy strong philosophical support at this time. I wonder about the relevance of the purposive approach to Lewontin's (1978) criticism that sociobiologists' behavioral traits are merely metaphors from human life and have no biological meaning. Is the concept of play just another of these metaphors? Loizos (1966) touches on this problem when she writes that the common-language notion of human play opposing work can have no rigorous application to animal behavior, since animals do not work in the (culturally defined) human sense.

How Should Behavior be Categorized?

Why not take advantage of as many different ways of categorizing behavior as possible? Evolutionary definitions of behavior represent a fifth major conceptual scheme not discussed by Purton. Altruistic behavior, for example, is rigorously defined by a particular pattern of individual fitness effects: altruistic behavior of an actor increases the fitness of another entity at the expense of the actor's own fitness. As defined, altruistic behavior cuts across functional, causal, and formal lines. This technical definition evokes, but is not identical with, the common-language concept of altruism. Students of altruism analyze hunting in social carnivores, nest defense in ants, cleaning symbioses in fish, and other behavioral phenomena using a single powerful theory. Similarly, cooperative behavior is defined by a particular class of patterns of fitness costs and benefits in a game-theoretic payoff matrix (Hamilton 1971, 1975). It is unlikely that the five conceptual schemes of behavioral classification discussed in this chapter are the only possible ones.

A Brief Definition of Play

I view play as behavior that functions to develop, practice, or maintain physical or cognitive abilities and social relationships, including both tactics and strategies, by varying, repeating, and/or recombining already functional subsequences of behavior outside their primary context. It is a matter of taste whether behaviors that do not simultaneously satisfy the structural, casual-contextual, functional, and developmental criteria of this definition are to be called play. In fact, we may define fifteen distinguishable categories of behavior based on the above concepts merely by enumerating the fifteen possible nonempty subsets of the four types of criteria listed in the definition above. The first five of these subsets are structural only; causal-contextual only; functional only; developmental only; structural and casual-contextual, but not functional or developmental. Much past confusion over the concept of play probably arose because different investigators implicitly chose different sets of criteria. Note in passing that if a fifth conceptual scheme, that of purpose, is also used to define criteria for play, then there are $2^5 - 1$ or 31 possible kinds of definitions of play behavior. I hope that no one will attempt the sterile exercise of classifying all recorded instances of "play" according to this ungainly scheme. Rather, I suggest that the study of play, like the study of animal communication (as Colin Beer 1977 observed), "needs openness in its concepts, both in the ostensive and connotative directions, so as to give sufficient room for the diversity of its subject matter and for the exercise of imagination in its investigation."

CONCLUSIONS

Because past critiques of the concept of play are generally invalid by current behavioral science standards, the continuing appeal of these criticisms is difficult to explain. Play might appear to be a frivolous research topic. Yet everyone is more or less familiar with the play of humans and of domestic animals. Apparent scientific frivolity of play and personal familiarity with play could tend to encourage scientists qualified in other behavioral specialties to disregard the inhibitions that normally prevent members of one specialty from criticizing specialized research in fields outside their competence. As Susan Wilson (pers. comm.) correctly notes, nearly every student of animal behavior, whatever his specialty or approach, considers himself qualified to speak with authority on the concept of play.[4] Yet, as argued above, past criticisms of this concept are based on incorrect assumptions about behavior or are empirically unsupported.

In the light of the previous discussion, therefore, I cannot expect that this chapter-length analysis of the problem of play will convince many scoffers. Philosophical positions in animal behavior research are deeply entrenched. No one who has based a life's work on the primacy of empirical physiological analysis of behavior will be led by this chapter to the gospel truths of evolutionary ethology, and no member of the older generation of epigenetic developmentalists will hesitate to defend a traditional territory now threatened both by the onslaught of sociobiological approaches to behavior and by evolutionary ecological approaches to behavioral development.

Ease of definition of play cannot be an all-or-none phenomenon. In some species, populations, or individuals, and at some ages, play will be easy to define; in other cases it will be almost impossible. What is the biological significance of this variation in the degree of difficulty of defining play? With what environmental (including social) factors does this degree of difficulty vary? Questions like these are likely to yield biological insight. To date no one really seems to have begun asking them.

The chief problem with past play research has been lack of adequate theory. Such theory is now becoming available. There is still no adequate real or Aristotelean nominal definition of the category of play, but in the light of current philosophical understanding of scientific definition, this lack is not crucial. It is most important to begin to test and to improve current biological theory on play. Such improved theory is expected to be based not only on ethology and sociobiology, but also on advances in such fields as artificial intelligence and behavioral ecology.

NOTES

1. It is only fair to point out that, not long ago, these same problems marked the study of other behavioral phenomena: aggression, communication, dominance, learning, motivation, sex, and territory. The series *Benchmark Papers in Animal Behavior* (M. W. Schein ed.; Dowden, Hutchinson and Ross, Stroudsburg, Pa.) offers ample documentation.

2. Mechanists may question my omission of physiology and development from this list of fundamental processes. I do not intend to slight physiology and development—after all, physiological and devel-

opmental processes mediate evolution. "Evolution is the control of development by ecology" (Van Valen 1973). For example, exercise adaptation is a fundamental physiological principle central to our understanding of play as physical training (Fagen 1976a). Differential development (Geist 1978a) is mediated by behavioral inputs to physiological and developmental processes. Entire developmental histories, rather than adult endpoints, are the target of natural selection (Gould 1977). At the risk of restating the obvious, we must simply recognize that physiological and developmental mechanisms constrain, but do not determine evolution. Functional (mechanistic) and evolutionary behavioral biologists tend to disagree about the importance of these constraints.

3. Whether these molar behavioral categories, interpreted as causal-motivational or as functional entities (Purton 1978), actually have objective meaning remains to be demonstrated. Like motivation, drive, consciousness, and sunrise and sunset, molar behaviors have common-language utility. If nonhuman animals themselves conceptualized behavior using molar categories despite the scientific vacuousness of such categories, what would we conclude about the significance of molar categories in the study of behavior?

4. Endemic opposition to play research is a worthy research topic in its own right. Two incidents from my own experience illustrate this point particularly well. It would be considered unethical for play researchers to enter a specialized round table discussion on orientation, acoustic communication, or the neural basis of feeding behavior in order to disrupt the learning process by challenging the very validity of these concepts and by sharply questioning the basic focus of the discussion. But a round table discussion on play at an international etho-logical conference (1977) suffered this very form of disruption. The abortive round table discussion gave some younger play researchers their first taste of combat. It was natural for them to sympathize with leading sociobiologists who were experiencing similar treatment at the time. If recent play research seems highly sociobiological, the reason for this emphasis may be found not only in the intellectual excitement of new evolutionary approaches to behavior, but also in a kind of historical coincidence not wholly consonant with older views of the dynamics of science.

The above incident is not the only one of its kind. At a biological society meeting in the 1960's, angry shouts from the audience punctuated the question period following a presentation on primate play. A sociobiologist reviewing these and similar incidents in the history of play research might well speculate that adult humans have been selected to direct moralistic aggression at those who dare to take play seriously. It would be very interesting to know just why this moralistic aggression should benefit the aggressor or his kin.

3
Natural History
of Play Behavior

Two snow leopards (*Uncia uncia*) bound playfully after one another along a mountain stream, then wrestle on the grass as if twisting up into knots (Ognev 1962 v. 3 p. 227). A young Hanuman langur (*Presbytis entellus*) scales a rocky outcrop, then does a handstand while a nearby infant watches (Blaffer Hrdy 1977, p. 140 and color figure 2). In the late-afternoon sunlight one Mongolian gazelle fawn (*Procapra gutturosa*) after another begins playing until the shadowed steppe seems alive with glowing, skipping forms (Andrews 1932 p. 445).

These accounts communicate some of the magic that makes the natural history of animal play endlessly fascinating.

DESCRIPTION AND INDUCTION

With increasing frequency, works on behavioral biology present the theory of a phenomenon first in sequence, then illustrate and test theoretical insights by citing appropriate evidence (e.g., R. R. Baker 1978, Popp & DeVore 1979, Schoener 1971, Trivers 1974). It is also highly informative to adopt a hypothetico-deductive approach to play, as I propose to do in succeeding chapters (Chapters 4 to 8).

In this chapter, however, I take a more traditional, inductive approach to play behavior. The chapter is organized phylogenetically and discusses representative species. Citations presented in tabular form offer extensive coverage of literature on animal play. But natural history and induction alone cannot yield fundamental insight. Existing descriptions cannot be considered in isolation from the implicit theory that formed them. Natural history can be complex, idiosyncratic, and even baffling. Play is multifaceted. Its characteristics appear to covary with many aspects of species and environment. Such correlations are individually seductive, and students of play have read entire theories into them post facto. Unfortunately, after the first five or six correlations, the mind reels. Observable characteristics of play include its frequency, duration, behavioral composition, sequential structure, and (variously defined) intensity, variability, diversity, stability, simplicity, complexity, interactiveness (with other individuals or with features of the environment), elaborateness, cooperativeness, reciprocity, and intricacy. These and other characteristics of play relate to demography, terrain, body form, body size and relative brain size, gender, age, social behavioral tactics, and social organization. If we consider known environmental and phylogenetic variation in feeding behavior and in social behavior, it is not at all surprising that play varies with all these factors and others. Predictive theory specifies what characteristics are important and suggests how they will vary with changing social and other environmental factors.

The inductive approach to play identifies phenomenological patterns that theory should at least consider. It is possible that some of these patterns lack biological significance and that others exist only in the mind of the observer. Four classical, if not always biologically correct hypotheses about play have molded the descriptive information available. These ideas are (1) play experience develops and maintains physical skill; (2) play is (metaphoric) glue producing interindividual bonds and social cohesion; (3) in a developmental analog of the cohesion hypothesis, play experience socializes the young; (4) play experience develops and maintains cognitive (including innovative) skill, and relative frequency, stability, and complexity of play therefore depend on relative brain size. According to (1), we interpret the structure of play in terms of practice for adult feeding, fighting, or predator avoidance. Ideas (2) and (3) suggest comparison of closely related species varying in adult group size, permanence, cooperativeness, or physical cohesion. Idea (4) suggests comparison of ecologically equivalent species differing in relative brain size.

None of these four hypotheses is acceptable as stated, even in theory (Symons 1978a,b). Thus (1) must be restated, qualified, and interpreted in terms of mechanisms of skill development and of natural selection; (2) and (3) are metaphoric, superorganismic, and probably group-selectionist; (4) involves many unstated assumptions about brains and about cognition, and it may only apply to humans. In addition, sharp distinctions between physical, social, and cognitive skill may be invalid (Mead 1976).

Existing information about play bears the mark of these and other equally unsatisfactory hypotheses. We might expect a very different natural history of play to result from a different set of biological views. In part, the question is one of emphasis. However, we should not underestimate the power of theory to ask new questions and to suggest new observations, to rescue phenomena or entire species from obscurity, and to place the familiar in an entirely new light.

SAFEGUARDS

Immediate recognition of play by human observers (Chapter 2) can be a problem if the observer misinterprets the animal's behavior. Unfamiliarity with an animal's behavior and environment can lead to misinterpretation. Ethologists, like all good naturalists, recognize this potential source of error and have built a variety of safeguards into their science. The professional stock phrase "getting to know your animal" fails to do justice to the subtlety and biological sophistication of these safeguards.

To illustrate the discipline of observing and interpreting animal behavior, consider some important factors a trained ethologist keeps in mind when studying an animal.

NONHUMAN PERSPECTIVES

A fundamental tenet of the ethological approach to behavior is the insight that each species inhabits its own universe defined by species-specific sensory and cognitive capabilities. This universe is called the *Umwelt* (von Uexküll 1909). For example, the richness and variety of a domestic dog's olfactory world are difficult for unaided intuition to comprehend, but a knowledge of dogs' remarkable abilities to detect and to discriminate between odors is essential to a full understanding of canid behavior. We may extend the concept of the *Umwelt* by adopting an ecological approach that recognizes the perceptive and cognitive primacy of evolutionarily relevant cues in the animal's physical and biotic environment and considers their implications for behavior. How are animals selected to perceive and learn about their environment? In what respects does this species-specific world differ from ours? Because even a close human relative like the chimpanzee must meet ecological, cognitive, and perhaps perceptual challenges differing fundamentally from those familiar to humans in Western technological society, it is not easy to imagine what it is like to be a champanzee, and it is impossible to think like one. Fortunately, as Darling (1937), Ewer (1975), Dillard (1973), and others have observed, study of an animal and familiarity with its environment can reveal flaws in our a priori hypotheses about the circumstances of that animal's life. Such study can often change our preconceptions so radically that it is tempting to conclude we have entered that animal's world and are perceiving with its senses. Whether or not such transmigration is possible in any but the most limited fashion, the differences between adult humans and young nonhumans must be recognized by students of play behavior in order to formulate hypotheses that make ecological sense and to design experiments whose results will be biologically meaningful. I will discuss some very obvious constraints the implications of which are not always obvious. Kummer (1971) presents a list of this type applicable to social primates.

Nonhumans Cannot Talk

The content of nonverbal communication is limited in temporal and spatial scope and in precision compared to that of human language. We may not adequately appreciate that animals lacking human speech may find it costly, difficult, or even impossible to obtain certain kinds of information. To imagine one's self unable to talk, write, or make signs is to imagine existence amid some of these constraints. Some glimpses of this world are possible if one travels in a foreign country without knowing the language, raises a human infant, or keeps pet animals. Even basic information on the health and well-being of another animal cannot be elicited by a simple query, such as "How are you feeling," but must be obtained by observing behavior or by initiating social interactions that produce behavior containing the required information.

Information May Be Imperfect and Limiting

We are not always willing to recognize that our own knowledge is imperfect and that the knowledge of different individuals may be imperfect in different ways. Students of nonhuman behavior sometimes recognize the existence of imperfect knowledge and uncertainty, for example by suggesting that young animals may learn preferred food types or hunting techniques by observing their parents' feeding behavior or by demonstrating that a female's ability to rear an infant improves from her first to her second offspring. But the consequences of imperfect information in early behavior have not always been pursued. It is possible that chance experiential factors (Cohen 1976) may influence development of juvenile knowledge, attitudes, and behavior toward parents and siblings, as do the basic genetic constraints cited by Trivers (1974), Alexander (1974, 1975), and Dawkins (1976b). If a young monkey notices a tree branch especially suited for acrobatic play, the juvenile may have no way of knowing that the tree harbors a hidden predator, but its mother may notice the predator or remember the site from previous years and snatch her young back. This incident might be misinterpreted by human observers as a case of restrictiveness or manipulation, but actually the "conflict" resulted simply from the juvenile's imperfect information about those trees in the home range most likely to harbor predators. Conflicts based on use of imperfect information by one or both participants in a social interaction are inevitable consequences of the fact that nonhuman animals, like humans, are neither omniscient nor infallible, even after a lifetime of experience. Selection of particular information-gathering and information-processing capabilities can produce considerable sophistication, including even adaptive self-deception. But over and above these obvious constraints, any single individual's experience is unique, limited, and occasionally misleading. The most successful animals will sometimes make mistakes even if their ancestors were selected to avoid mistakes.

Predation Occurs

Humans seldom appreciate the implications of predation in nature. Whereas adults of some species, for example, the African elephant and the giant panda, appear relatively immune to predation by nonhumans, young animals of all species (including those whose adults are not taken by predators) are prime targets for predation, for they are relatively small, weak, and defenseless. Predator avoidance requirements affect many aspects of behavior, from parental care and home range selection to activity schedules. The form, and occurrence in time and space, of behavior like play, which preempts participants' attention and which may result in enhanced conspicuousness to predators, can be influenced by predator avoidance considerations. Because almost all wild mammals harbor internal and external parasites (including microbial pathogens) at some time during their lives, parasite avoidance could be at least as important as predator avoidance in constraining behavior. Implications of pathogens and metazoan parasites for mammal and bird behavior are only now beginning to receive attention (e.g., Freeland 1976).

Other Constraints

Most nonhuman mammals are quadrupedal, use their mouths relatively frequently and their forelimbs relatively infrequently in feeding and in fighting, and rely as much on olfaction as on vision. Exceptions to this rule include many nonhuman primates, felids, and possibly ursids (Cartmill 1972, 1974; Henry & Herrero 1974). If we imagine ourselves blindfolded and required to walk on all fours through a buffet line, we can begin to appreciate some problems that other animals experience in selecting and in transporting food items. We might wish for a more sensitive nose, prehensile lips, or cheek pouches. If we fought with another individual over a particularly desirable piece of food we might wish for experience and training in the sport of wrestling, and we might find that certain holds or positions that would previously have seemed ludicrous actually made it especially easy to bite our opponent without being bitten. Again, as in the case of parasite avoidance, the commonsense, if nonintuitive implications of mouth use and quadrupedalism for mammal behavior have not, to my knowledge, been discussed in systematic fashion.

SURVEYING THE LITERATURE OF ANIMAL PLAY

The biological literature offers information on play in hundreds of mammalian and avian species. These accounts range in length from a few words to entire volumes. Their quality also varies. Many existing descriptions are verbal and anecdotal, some are brief or ambiguous, and only a few include quantitative data suitable for testing theory. However, even these anecdotes and short verbal descriptions serve to illustrate the range, variety, and coherence of animal play. Furthermore, as Blaffer Hrdy (1977) correctly emphasizes, such accounts can bring long-neglected, theoretically significant phenomena to light and can often reveal opportunities for further study.

To my knowledge, Welker (1961) is the only previous phylogenetic tabulation of mammalian play. Since 1961, the literature of mammalian natural history (especially field studies of behavior) has grown to such an extent that a new tabulation appears necessary. Ficken (1977) reviews avian play. I have made many additions to her excellent tabulation.

Although some short nineteenth-century accounts proved hopelessly ambiguous, I felt reasonably confident that, in the majority of cases, the behavior described would have been called "play" by most contemporary ethologists. Accounts that I consider ambiguous are identified as such. I attempt to use currently accepted nomenclature in all cases. Sources of taxonomic information included Corbet (1978), Hershkovitz (1977), Leyhausen (1979), Meester & Setzer (1971), Peters (1931), Repenning, Peterson, & Hubbs (1971), Ride (1976), and Walker (1975). Some species listed are subjects of taxonomic dispute. Also, ethologists cannot always identify material of uncertain provenance.

As this chapter demonstrates, I have considerably expanded Welker's and Ficken's brief lists. But no such summary, however extensive, can claim completeness. Some obvious taxonomic gaps result from inadequate behavioral study of the groups concerned. In some cases (e.g., Chiroptera, Insectivora) otherwise excellent studies of behavioral development and mother-offspring relations do not mention play or its counterparts, such as escalated fighting and erratic fleeing from a threatening disturbance in the environment. It is therefore impossible to make assertions about relative frequencies of play in different mammalian orders solely on the basis of existing accounts. White (1977b) correctly identifies a need for "more detailed and quantitative comparative work." Pending the results of such studies, we may only speculate that play behavior will eventually be found in all orders of mammals and birds under suitably favorable environmental conditions. Additional theory is needed to predict the form of play: nonagonistic companion-oriented or item-oriented chasing and wrestling, solo or parallel locomotor-rotational leaps, twists, rolls, and runs occurring in naturalistic environments in the absence of external threatening stimuli, or combinations of these two forms of activity.

To describe play of all species for which accounts exist would be needlessly repetitive. In many cases such descriptions are virtually identical. Even when a comparison of accounts by various authors suggests interspecific contrasts, the lack of adequate information on intraspecific variability and on methods and philosophies of observation makes it impossible to evaluate such differences. Tables in this chapter cite representative descriptions of nonhuman mammalian play currently known to me. In addition, I report the major known characteristics of play for at least one representative of each mammalian order. I also present accounts of play in representatives of major families in certain well-known orders. Sets of confamilial species studied by the same author under roughly similar conditions (making allowance for different species' requirements) were of particular interest. I also assumed the pleasant responsibility of reporting on play in exotic, but celebrated species, including Tasmanian devil (*Sarcophilus harrisii*), vampire bat (*Desmodus rotundus*), orangutan (*Pongo pygmaeus*), American beaver (*Castor canadensis*), royal antelope (*Neotragus pygmaeus*), and kea (*Nestor notabilis*).

The anecdotes we lack may be as important as those that we have. It was long fashionable in behavioral science to conceal play behind a screen of euphemisms and jargon or to neglect it altogether. The work of naturalists who ignored orthodoxy and published accounts of play has survived to be cited here. The behavioral science of their day no longer exists. I suppose that some of these naturalists, like the dying Marco Polo, called upon by friends and colleagues to recant, might well reply that they had not described even half of what they had actually observed, for fear they might not be believed.

PLAY IN MAMMALS

Although material reviewed below suggests several comparisons of major evolutionary importance, I must preface this synopsis of results by stating an obvious caveat. Such comparisons are probably premature because the relevant data are often not comparable or were gathered inappropriately. For example, suppose (for the sake of argument) that such relatively ungregarious species as the orangutan, gorilla, gibbon, woodchuck, and red fox tended to be shy, delicate, easily stressed, and behaviorally unresponsive under human observation in comparison to their uninhibited social counterparts, the chimpanzee, Olympic marmot, and domestic dog. Then, in the presence of the same field observer, or under the same controlled laboratory or captive conditions, play would be differentially affected, and the more gregarious species would play more during observation. Might a similar caveat apply to sex differences?

Although many uncertainties surround them, phylogenetic patterns of animal play represent the best available evidence that play is adaptive behavior. These patterns include marsupial-placental similarities and differences, apparent parallels between play and adult social organization in sets of closely related species, sex differences in play in some species but not in others, the shifting balance between social and individual play, occurrence of interspecific play, the complex ontogeny of play, relative amounts of wrestling and chasing in play, species differences in the apparent complexity or elaborateness of play, and species differences in play that reflect species differences in relative brain size. In addition, this material enables us to make biological comparisons between the play of mammals and that of birds. Finally, this material, taken as a whole, poses an important question: Why is play behavior near-universal in mammals and birds, but virtually absent from the known repertoires of all other organisms?

Monotremata

These fascinating, egg-laying mammals—the platypus and echidnas, or spiny anteaters—pose important biological questions (Griffiths 1978). What might we predict about play in species whose basic physiology and cranial anatomy diverge widely from that of all other mammals?

Verreaux (1848, cited by Owen 1848) observed mother-young interaction (called "play" by Owen) in platypus (*Ornithorhynchus anatinus*) inhabiting the New Norfolk River, Tasmania. This activity took place particularly when young were not feeding and occurred at a greater distance from the bank than did feeding. Bennett (1834, 1835, 1859, 1860a) kept young platypus in captivity. According to Bennett (1835), these animals played together, puppy-like:

One evening both the animals came out about dusk, went as usual and ate food from the saucer, and then commenced playing one with the other like two puppies, attacking with their mandibles and raising their fore paws against each other. In the struggle one would get thrust down, and at the moment when the spectator would expect it to rise again and renew the combat, it would commence scratching itself, its antagonist looking on and waiting for the sport to be renewed. When I placed them in a pan of deep water, they were eager to get out after being there only a short time; but when the water was shallow, with a turf of grass placed in one corner, they enjoyed it exceedingly. They would sport together, attacking one another with their mandibles, and roll over the water in the midst of their gambols; and would afterwards retire, when tired, to the turf, where they would lie combing themselves. It was most ludicrous to observe these uncouth-looking little beasts running about, overturning and seizing one another with their mandibles, and then in the midst of their fun and frolic, coolly inclining to one side and scratching themselves in the gentlest manner imaginable.

On 19 January 1836, near Hassan's Wells, N. S. W., Charles Darwin (1896 p. 441) observed platypus "diving and playing about the surface of the water." It would be wrong to read anything other than the colloquial sense of "play" as graceful, unencumbered motion into Darwin's remarks. However, we are free to imagine Darwin observing with amused delight the tussles and gambols of platypus at play in the wild.

In a recent treatise on monotreme biology, Griffiths (1978 p. 179) remarks, "One needs to read only George Bennett's (1835) charming description of two nestling platypuses at play to be absolutely convinced that the monotremes are mammals." However, despite Bennett's observation that his nestlings had eaten "as usual" before beginning to play, Burrell (1974 pp. 204–205) quotes Bennett's remarks (1860a) with skepticism, adding (p. 207): "From my experience with platypus in captivity, I think the so-called engaging antics of Bennett's captives were really the desperate struggles of slowly starving nestlings."

The platypus "has a large (if unconvoluted) brain for its size, proportionately much larger than those of many of the lower Didelphia and Monodelphia" (Burrell 1974 p. 15, see also Griffiths 1978). If play depended solely on relative brain size, platypus would be more likely to play than many (presumably playful) marsupials. There is obvious need for comparative study of this problem.

Marsupiala

Marsupial mammals are small-brained but elegant biological sophisticates, as adapted, specialized and "advanced" as their placental counterparts (Eisenberg 1975, Kirsch 1977, Low 1978, P. Parker 1977, Tyndale-Biscoe 1973). The marsupial alternative features an adaptive radiation of species that correspond to (but evolved independently from) placental rodents, lagomorphs, carnivores, and ungulates on other continents. This diversity offers excellent opportunities for comparative studies of play. If play is adaptive behavior, we would expect close marsupial-placental parallels in behavioral composition, degree of interactiveness, and social play stability in the two groups. How would marsupials' small brain size, relative to that of equivalent mammals, affect these differences? Tests of these predictions are of interest. Few data are now available. Existing evidence fully supports the hypothesis of gross convergence in the play of marsupials and placentals occupying similar ecological positions. Whether marsupial play is less elaborate, less frequent, or less stable than the play of corresponding mammals (as the brain-size hypothesis might predict) is not known.

DASYURIDAE

The family Dasyuridae has radiated widely. It includes species resembling mice or small rats, as well as carnivorous species. Among these carnivorous marsupials are the viverrid-like native "cats" (*Dasyurus* spp.) and the small, bearlike Tasmanian devil (*Sarcophilus harrisii*). Behavior of several captive-bred dasyurid species has been studied. Play of the young in small (head and body length of adults *ca* 100 mm), mouselike dasyurids

(*Sminthopsis murina* and perhaps *Sminthopsis crassicaudata*) is restricted to jerky running and brief dashes, and social play is absent (Ewer 1968b, Graham Settle pers. comm.). Young of the larger (120 to 250 mm), ratlike forms (*Antechinus stuartii, Dasyuroides byrnei*) exhibit "erratic leaping and bounding" (Aslin 1974) and sporadic, simple social play which principally involves chasing and which frequently escalates into agonistic fighting (Settle pers. comm.). Young of the larger (300 to 500 mm), viverrid-like dasyurids (*Dasyurus hallucatus, Dasyurus viverrinus*) play by themselves, chase and wrestle together, and play with objects (Fleay 1935, Nelson & Smith 1971, Settle pers. comm.). In *D. viverrinus,* the larger of the two species, mother and young continue playing with each other after weaning (Nelson & Smith 1971), and tame individuals may remain playful all their lives (Fleay 1935). *Dasyurus maculatus,* the tiger quoll, the largest (600 mm) and heaviest (6 kg) of the native "cats," is a mustelid-like stalking predator inhabiting heavily timbered areas and rain forests. Chasing, wrestling, and stalking social play (Fig. 3-1) is frequent, is structurally complex, and rarely escalates in these animals (Settle 1977, pers. comm.).

THE TASMANIAN DEVIL (SARCOPHILUS HARRISII)

General accounts of *Sarcophilus* behavior (Eisenberg, Collins & Wemmer 1975, Ewer 1968a, Golani 1976, Grzimek 1967, Hediger 1958, Turner 1970) indicate that play is complex and interactive in the Tasmanian devil. Chasing and wrestling are frequent, and play may incorporate features of the environment such as a water trough (Turner 1970) or a hollow log used in hiding, chasing, and ambushing (Hediger 1958). A hand-raised devil played vigorously with dusters and other toys, seizing them in its mouth and shaking them (Ewer 1968a pp. 40, 302).

PERAMELIDAE

Little is known about the behavior of these often rapidly developing marsupials. *Perameles gunnii,* the Tasmanian barred bandicoot, is a slender animal inhabiting open areas, whereas *Isoodon obesulus,* the Southern short-nosed bandicoot, is a more

Fig. 3–1 Tiger quoll (*Dasyurus maculatus*) juveniles play-wrestling and showing exaggerated, open-mouth play-face. This marsupial play-signal (an open-mouth gape) appears functionally equivalent to the eutherian play-face. A communicative display used for maintenance of play, it exhibits evidence of ritualization. The head is usually tossed backward, and especially when individuals are not in physical contact, the head is turned obliquely away, not directed toward the partner. Mystacial vibrissae are at most partly erected. By contrast, the agonistic gape in this species is directed at the opponent, and the mystacial vibrissae are always fully erected. Captive-bred males aged 92 days. Australia. (Graham Settle)

robust form. Both species are nocturnal and feed on small invertebrates. Heinsohn (1966) observed no play and few social interactions in four litters of *P. gunnii;* young active *I. obesulus,* however, exhibited nonagonistic chasing, leaping, burrowing, and climbing.

PHASCOLOMIDAE

Wombats, compact, thick-skinned, heavily built burrowing marsupials, enjoy a deserved reputation for playfulness.

Wünschmann (1966) described the locomotor play of a captive hairy-nosed wombat (*Lasiorhinus latifrons*). This animal would stand bipedally, leap vertically into the air, and even fall over backwards. Wombat social play frequently involves mutual butting; the animals may also jump, rear, and roll over (Troughton 1941 p. 140). A captive hairy-nosed wombat at Brookfield Zoo (J. Fagen pers. comm.) would play socially with its keepers, whom it butted, bit, chased, and leaned on. Keepers and the wombat had friendly shoving matches. The animal's bites seemed strong to the people bitten, but bites in play occurred at less than maximum intensity and would surely not have bothered another thick-skinned wombat.

MACROPODIDAE

Kangaroos and wallabies, undoubtedly the most familiar marsupials, occur in a variety of shapes and sizes. Many of these species are small and poorly known. Larger, grazing forms have been studied in the field, and their ecology has been compared with that of grazing placental herbivores.

Play of large kangaroos, genus *Macropus,* involves solo exercise and object manipulation, sparring, grappling, and "boxing." D. B. Croft (pers. comm.) has observed young-at-foot and subadults hopping "rapidly back and forth from their mother or in wide circles about her, sometimes making leaps into the air." Rapid hopping play is also known in the whiptail wallaby (*Macropus parryi*) (Kaufmann 1974). In the great grey kangaroo (*Macropus giganteus*) and in the euro (*Macropus robustus*), young-at-foot frequently spar, grasp, and grapple playfully with their mothers (Croft 1980, Herrmann 1971, Kaufmann 1975), but do not play at all with other young-at-foot and seldom interact with them. Interestingly, in these large kangaroos whose extreme sexual dimorphism in body size would be expected to correspond to a polygynous mating system, sparring with the mother occurs significantly more often in male than in female young (D. B. Croft pers. comm.).

Subadult and adolescent male *Macropus* spar harmlessly with one another in a manner precisely paralleling the sparring behavior of young male hoofed mammals and terrestrial primates. Kangaroo "boxing matches" are vigorous ritualized encounters involving repeated physical contact. These encounters usually terminate when one animal retreats, with the stronger (typically larger) animal usually winning by vigorous pushing. Mother-offspring play and both ritualized and damaging fights between older males contain many common behavioral elements (D. B. Croft pers. comm.). Mother–offspring play could develop the young's capacity for later fighting. Such play, as well as ritualized "boxing," could also develop the partners' social relationship in some way other than or in addition to that resulting from allogrooming and from suckling the mother. However, the latter hypothesis fails to explain the apparent sexual dimorphism in the frequency of mother-offspring play.

Playful object manipulation occurs commonly in young red kangaroos (*Macropus rufus*) and in other macropodids. Young-at-foot manipulate plant parts and occasionally even grapple with bushes (D. B. Croft pers. comm.). A captive young sub-adult great grey kangaroo repeatedly jumped up and down on an acetate filter and manipulated the filter with its forepaws, this game proceeding for about ten minutes (D. B. Croft pers. comm.). Previously, an adult and this subadult had both approached, sniffed, nibbled, and moved away from the object, but only the subadult showed any further interest in it.

Insectivora

This heterogeneous group includes hedgehogs, shrews, moles, solenodons, the aquatic desmans, the highly diverse Madagascar tenrecs, and various other small and poorly known mammals. Young hedgehogs (*Erinaceus europaeus*) playfight with inhibited bites and may lick, nose, or crawl under one another (Dimelow 1963, Poduschka 1969). They will chase and rub humans, apparently inviting play (Poduschka 1969). Their habits of running back and forth and of picking up and carrying objects also appeared playful to two authors (Dimelow 1963, Poduschka 1969). Simple locomotor play, including brief dashes and vertical leaps, has been reported in elephant shrews (family Macroscelididae) (Galen Rathbun pers. comm.). Studies of behavioral development in tenrecs (Eisenberg & Gould 1970) and shrews (Shillito 1963) fail to mention play. Erna Mohr, a respected European zoologist responsible for many excellent descriptions of animal play, observed *Solenodon paradoxus* in captivity. Mohr (1936a,b) looked specifically for play in captive solenodons, but failed to observe any behavior resembling play other than a short approach-hug sequence. Mohr finds it difficult to say whether these solenodons play when young.

Play may possibly occur in some species of shrew (Goodwin 1979, Lorenz 1952, Rood 1958). A few days after their capture, young masked shrews (*Sorex cinereus*) chased one another around their cage "in a nonaggressive manner and without audible vocalization" (Goodwin 1979). They spent half of their playtime on a ladder in their cage. This behavior may have served to develop adaptive relationships between the animals, because on the next day the shrews left their common nest and ceased forming head-to-tail caravans. They stayed together, foraging, drinking, and grooming together, but not playing, on this day outside the nest.

Young shrews' formation of head-to-tail caravans is a cohesive and highly organized social behavior present from early ontogeny. Perhaps these shrews need to achieve social cohesion or form bonds in advance in order to coordinate their common activity outside the nest. Such coordination could be adaptive for feeding, orientation, or predator avoidance. Suppose that previous social experience is necessary, for some reason, to achieve adequate bonding. Then why was the shrews' extensive previous experience in forming caravans not sufficient for this purpose? What might playfighting achieve for a relationship that caravanning cannot? This question calls for information on specific social developmental effects of play in comparison to general effects of social experience (or, if you prefer, specific effects of caravanning). Although we first encounter it in shrews, the problem is a recurrent one for cohesion or bonding theories of play: What is special about the hypothesized bonding effects of play compared to effects of grooming, sniffing, resting together, feeding together, fighting, or any other social activity of young animals?

Chiroptera

The bats form an enormous and highly diverse group of mammals having a variety of social systems, ecological niches, and parental and developmental styles. They figure prominently in several remarkable biological dramas: coevolutionary (Faegri & van der Pijl 1972), community ecological (Fleming, Hooper, & Wilson 1972), and sociobiological (Bradbury & Vehrencamp 1976a,b, 1977a,b). Whereas this diversity offers excellent opportunities for tests of evolutionary models of behavioral development (Bradbury & Vehrencamp 1977b), relatively little is known about play in bats. Existing accounts suggest surprisingly complex forms of social play. The group-living young of crevice-dwelling bats, especially freetailed bats, "all join together for the greater part of the day or night to play and tussle, stage sham battles and pursuits, and otherwise romp in a fashion which reminds one of a litter of puppies or kittens" (Leen & Novick 1969 p. 80; see Davis, Herreid & Short 1962 for information on the unusual "creche" method of parental care in the Mexican freetailed bat). Neuweiler (1969) describes mother-offspring play, including inhibited bites, in an Old World fruit bat (*Pteropus giganteus*). This behavior was clearly identifiable as play by virtue of the accompanying bite inhibition and silence. Later fights are serious and coincide with waning of the mother-offspring bond. At this time the young males begin to scuffle and romp together. Play apparently occurs in a well-known bat of the New World tropics, the vampire bat (*Desmodus rotundus*). Young vampires play together, chasing each other, slapping each other with their wings, and engaging in bouts of mutual sniffing (Schmidt & Manske 1973).

Dubkin (1952) reports young little brown bats (*Myotis lucifugus*) scrambling about and biting "playfully." At a later age, forty or more young bats would play in a group, squeaking and biting or crawling over and under each other. Welker (1961) includes this account in his list of play citations, but the behaviors described may or may not have been truly playful. (Differences from agonistic fighting were not specified.)

Bats' ecological and anatomical diversity (including ecologically interpretable variation in relative brain size, Wilson & Eisenberg 1978) ought to correspond to variety in play behavior. Varying degrees of sociality in bats of different species also offer an excellent opportunity for testing hypotheses that predict a correlation between frequency or stability of social play and permanence of adult groups.

Nonhuman Primates

Nonhuman primate play appears surprisingly uniform in broad outline. Locomotor-rotational exercises, reciprocal chases, non-agonistic wrestling and grappling, and relaxed open-mouthed grins or play-faces are reported in species after species, whereas solo object play is sporadic or entirely absent in most wild primates. (Chimpanzees are one possible exception.) Major features of play in squirrel monkeys (*Saimiri* sp.), rhesus macaques (*Macaca mulatta*), baboons (*Papio* sp.), and chimpanzees (*Pan troglodytes*) have become familiar to ethologists through chapter-length reviews (Dolhinow & Bishop 1970, Loizos 1967), symposium volumes (E. O. Smith 1978), and an entire book analyzing social play (Symons 1978a). Literature on rhesus play is extensive: by 1979, in addition to Symons's excellent book, at least three doctoral dissertations (Levy 1979, Lichstein 1973, White 1977a) and dozens of research papers or review articles dealt at length with play of this single primate species. Play of rhesus macaques has been examined more often and from a greater variety of points of view than that of any other nonhuman mammal.

Uniformity of descriptions of primate play is deceptive because play in the primates exhibits interspecific, interpopulation, interindividual, sexual, and developmental variation, especially in frequency and in behavioral composition. These sources of variety have become apparent as a result of in-depth studies on single populations, both free ranging and confined. In addition, results of field studies on multiple populations of a given species suggest that primate behavioral flexibility is as evident in play as it is in sexual, agonistic, and feeding activity.

TUPAIIDAE

Captive tree shrews of several species primarily play by chasing each other (*Tupaia longipes, T. gracilis*) or by solo running and rolling (*T. chinensis, T. tania*) (Hasler & Sorenson 1974, Sorenson & Conaway 1966). Young *T. longipes* race around and mount one another, sometimes many times within a few minutes. Both males and females mount. Arboreal and terrestrial chases involve up to eight individuals. Sometimes one animal will approach and surprise another from behind. The animal so approached leaps vertically one or two feet and then races away with its partner in close pursuit. These roles often alternate without any aggression occurring (Sorenson & Conaway 1966).

Wrestling play has not yet been reported in *Tupaia*, but *T. montana* box (Sorenson & Conaway 1968).

Tree shrew social development is unusual for primates. Twins are typically born in a nursery nest that the mother visits only infrequently (twice a day to once every 48 hours) (Martin 1968, Sorenson 1970). Existence of this absentee system of parental care led Wilson (1975 p. 347) to remark that "the tree shrews provide a minimum opportunity for the socialization of their young." Adult tree shrews are solitary, but *T. glis* nestmates form a well-defined society of their own. They remain together even after they leave the nest at about four weeks of age and only separate after attaining sexual maturity, nine weeks later (Martin 1968). Tree shrews therefore furnish an example of social play in an essentially solitary primate. The apparent absence of wrestling in all species studied is surprising, as this form of play is widespread in mammals. Further study will be needed in order to substantiate and to extend these findings.

Captive tree shrews apparently exhibit species differences in play style. These styles occupy the three lowest grades on a scale of interindividual distance culminating in actual body contact. Solo play, socially facilitated solo play, and chasing occur in tree shrews; boxing, mounting, and wrestling are apparently absent. This scale is frequently used, verbally or implicitly, to compare play in different sexes, at different ages or in different species. It may be uncritically interpreted as a scale of social distance in order to speculate about play and social cohesion. Actually, the major biological implication of this scale is that it measures the possible risk of direct physical injury to the individual if play turns into an escalated fight. Solo play by definition cannot have this outcome, but an agonistic chase may result in physical exhaustion or in displacement from a favored resource patch, whereas body-contact play may escalate into injurious agonistic fighting. (It is also possible to interpret a difference between chasing and wrestling by citing a species' body form or need to avoid predators. The latter "explanation" assumes skill development; the former implies cost-benefit considerations.)

LEMURIDAE, INDRIIDAE, DAUBENTONIIDAE, LORISIDAE

Most prosimians, or "lower" primates, belong to one of these four families, the first three of which occur exclusively on the island of Madagascar. Prosimian natural history is exemplified by the aye-aye (*Daubentonia madagascarensis*), which feeds woodpecker-like on insect larvae using a modified third digit, and male ringtailed lemurs (*Lemur catta*), which use their long, black-and-white striped tails as a means of spreading their personal scent during aggressive confrontations. Charles-Dominique (1977), Martin, Doyle, & Walker (1974), Petter, Albignac, & Rumpler (1977), and Tattersall & Sussman (1975) review prosimian ecology and behavior.

E. O. Wilson (1975) contrasts two extreme patterns of lemur sociobiology. Lesser mouse lemurs (*Microcebus murinus*) are "essentially solitary," omnivorous, nocturnal animals whose females nest communally, but forage alone. Ringtailed lemurs form troops usually consisting of several adult males, several adult females, and their offspring. Adult females are socially dominant over adult males. Lemurs of this species are diurnally active vegetarians. Despite this obvious difference in the adult social environment, there appear to be no striking differences between the play of these two lemur species. Lesser mouse lemurs (born in litters of one to four) chase each other from place to place, grab at each other and wrestle, and jump or climb on their mother, nipping at her hands and ears (Petter, Albignac, & Rumpler 1977 p. 64, Petter-Rousseaux 1964 p. 119, Pinto 1972). Ringtailed lemur play includes many variations on a theme of jumping on and wrestling. Animals may also sit facing each other and grapple or hang facing each other and kick or grab (Jolly 1966a p. 95, Klopfer & Klopfer 1970, Sussman 1977).

At least one major contrast exists in lemur play. Young ringtailed lemurs in large troops have a variety of play companions of different ages, sex, and degrees of relatedness. In *Lemur fulvus,* the brown lemur, multi-male groups occur, but may be so small that age-mates are absent and most infant play occurs with adults (Sussman 1977). Young *Lemur* in larger groups play among themselves (Petter, Albignac, & Rumpler 1977 p. 209). Long-term studies are required in order to compare demographic variability in these species.

Petter, Albignac, & Rumpler (1977 p. 469) hint at a second major contrast when they report relatively little play in two Madagascar prosimian species (*Varecia variegata* and *Indri indri*). It is difficult to accept this claim because these same authors discuss frequent, intense mother-young and young-young play in *V. variegata* (Petter, Albignac, & Rumpler 1977 pp. 272–273), whereas their information on *Indri indri* is based on another author's observations (Pollock 1975). Frequency data are not reported for any of these species.

TARSIIDAE

Tarsiers are small, delicate forest primates. An infant tarsier's exceedingly large eyes, slender limbs and filigree hands, enormous head, and tiny body create an indelible impression. A hand-reared individual played by mouthing, grasping, and wrestling with a human hand, and his facial expression during these antics resembled the characteristic primate play-face: mouth open, teeth not exposed, and features relaxed (Niemitz 1974). Captive adult tarsiers at Brookfield Zoo also made apparent play-faces at each other and at their keepers (J. Fagen pers. comm.).

CALLITRICHIDAE, CEBIDAE

These New World monkeys (Hershkovitz 1977) are structurally intermediate between prosimians and other Old World monkeys. Callitrichids (marmosets and tamarins) live in nuclear family units that may include older offspring as well as infants. They are among the least sexually dimorphic of all primates, and males take an active role in parental care. Play of these species includes wrestling, chasing, and hide-and-seek or peek-a-boo (Box 1975a,b, Frantz 1963, Stevenson 1976, Stevenson & Poole 1976). The latter game seems to have several forms (Box 1975a, Fitzgerald 1935, Frantz 1963). A young marmoset begins the game by hiding lightning-fast when in view of a potential playmate (parent or young), then slowly peeking out again. Alternatively, the initiator races up to its partner, touches it, and darts away ("feinting approaches and withdrawal," Hampton, Hampton, & Landwehr 1966). In captivity, adults play together and accept play invitations from their young (Stevenson 1976).

Cebid societies range from that of the well-studied, gregarious squirrel monkey (*Saimiri* sp.) to the marmosetlike parental family of the night monkey (*Aotus trivirgatus*). Supposedly, play-wrestling is "almost or completely lacking" in the latter species (Moynihan 1964 p. 12), despite the fact that *Aotus* is no less gregarious than marmosets or most prosimians. This report is often cited (e.g., Aldis 1975 p. 165), but scarcely credible, for it was not based on observations of undisturbed family groups. We would expect *Aotus*, like *Callicebus* (Mason 1971), whose social organization is similar to that of *Aotus* and the marmosets, to be very slow to establish social relationships with strangers outside the family. Play is especially vulnerable to disruption and would not be likely to occur in artificially formed groups of these monkeys even after other forms of social interaction had become established. It is hardly surprising that Moynihan (1964) failed to observe play "when several young Night Monkeys are kept together in the same cage."

I observed a juvenile *Aotus trivirgatus* in a family group at the Bronx Zoo, New York, apparently play-wrestling with an adult on December 26, 1979. The juvenile hugged, mouthed, pushed, and grasped the adult, who responded likewise. During most of their play, both animals lay facing each other on the cage floor. Because I had not previously observed these animals (although I had watched adult *Aotus* at Brookfield and Philadelphia Zoos), I consider these observations tentative, but suggestive.

CERCOPITHECIDAE

These Old World cercopithecine and colobine monkeys, ground dwelling and arboreal, include species well known to ethologists: robust baboons, gracile langurs, vervets, patas, and macaques. The rhesus macaque (*Macaca mulatta*) (Figs. 1-2, 3-2, 3-3, 3-4, 3-5) is a South Asian primate whose behavior is best known from the study of laboratory and provisioned, but free-ranging individuals.

Rhesus society typifies the highest social evolutionary grade among primates (Eisenberg, Muckenhirn, & Rudran 1972). Like the ring-tailed lemur, rhesus form multi-male troops containing several mature adult males and females as well as individuals of both sexes and all other ages.

Sexual dimorphism is a prominent feature of rhesus play-fighting. Males play more vigorously and use more body contact in play than do females, and males play until a later chronological age. The rhesus social system, in which females inherit their mothers' dominance rank and remain in their natal troop

Fig. 3–2 Rhesus macaque (*Macaca mulatta*) juveniles. Crouch and play-face invitations to play. According to Symons (1978a, pp. 25–26), rhesus macaques invite play in at least twelve different ways; one is illustrated here. La Parguera, Puerto Rico. (John M. Bishop)

Fig. 3–3 Playfighting between juvenile and infant rhesus macaques (*Macaca mulatta*). The juvenile rolls the infant onto its side and bites its neck. Same animals as in Fig. 1–2. La Parguera, Puerto Rico. (John M. Bishop)

for life, whereas males usually disperse to other troops in the company of peers or kin, appears well suited to maintain this dimorphism (Symons 1978a). As often described (Symons 1978a, White 1977b), rhesus play consists of nonagonistic chasing, wrestling, various forms of acrobatics, playful competition for objects, and almost no solo playful manipulation of inanimate objects. Some captive rhesus mothers play approaching-leaving games with their infants (Fig. 3-6). These games seem to have the qualities of human mother-infant eye contact games. The few rhesus mothers who play these games have especially sensitive and reciprocal relationships with their infants (Hinde & Simpson 1975).

Extensive study of rhesus play reveals two other interesting phenomena. The first is that juvenile rhesus occasionally play together in water, splashing and jumping in one after the other. Donald Symons (1978a pp. 53–54) describes these games:

Climbing in from mangrove roots soon gave way to jumping, first from the roots and then from progressively higher branches. The

Fig. 3–4 Rhesus macaque (*Macaca mulatta*) juveniles playing in mangrove trees. Both macaques hang upside-down, grasping the tree branches with their feet and grappling with their hands. La Parguera, Puerto Rico. (John M. Bishop)

only animals seen to jump were juveniles, and the only jumps of more than about ten feet were made by older juveniles—especially, but not exclusively, males—who sometimes jumped from as high as twenty feet. Jumping was strongly socially facilitated; as I approached a pool where juvenile monkeys were leaping and swimming frequently I heard staccato bursts of splashes interspersed with silence. Instead of leaping, monkeys sometimes hung by their hands or feet from a branch, and then dropped into the water. Monkeys usually leaped with limbs flailing but, even during rarely observed flips, twisted to land on all four feet.

Frequently a juvenile aimed its jumps at or near another juvenile who was swimming in the water, and occasionally scored a direct hit, pushing the swimmer under the water. Sometimes both monkeys disappeared beneath the surface and reappeared in a few seconds separated by several feet. In one case a three-year-old female (C0) sat on a

Fig. 3–5 Rhesus macaque (*Macaca mulatta*) juveniles grappling playfully in mangrove trees. One macaque is upright, the other hangs under the branch. La Parguera, Puerto Rico. (John M. Bishop)

branch watching a one-year-old female (L2) swimming across the pool in her direction. When L2 was directly below, C0 leaped, scoring a direct hit and pushing L2 under the water; when L2 surfaced C0 continued to cling to her back for a few seconds.

Monkeys swam both on and under the surface. Frequently a monkey jumped into a pool, disappeared beneath the surface, and reappeared perhaps twenty feet away on the other side, climbing out on a mangrove root. Swimming often was oriented to another swimming monkey. When one monkey swam toward a second, sometimes one or the other ducked under the surface just before contact would have been made and swam away under water. Occasionally monkey A would be seen swimming toward B; then, while still several feet from B, A would disappear under the surface and a few seconds later monkey B would suddenly disappear, suggesting that B may have been jerked under by A. Both monkeys would then reappear on the surface separated by several feet. Monkeys sometimes swam together, as well

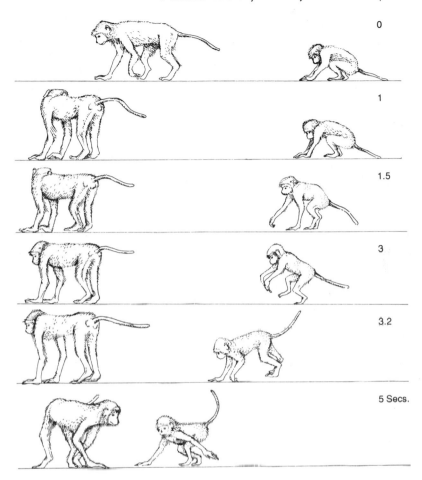

0

1

1.5

3

3.2

5 Secs.

Fig. 3–6 A 5-second sequence of a mother–infant leaving game of rhesus macaques (*Macaca mulatta*), based on film. Approximate times of the frames are shown (right). Confined animals. England. (Redrawn from Hinde & Simpson 1975)

as toward and away from each other. Occasionally one monkey would reach and grasp a second, but nothing like real playfighting was seen among monkeys in the water.

Monkeys usually sat and rested between periods of swimming, often watching the monkeys in the water, but playfights and playchases did occur in the mangroves surrounding the pool. Sometimes the water was merely one more kind of terrain for playchasing, the monkeys bounding, high-stepping through the shallows, or, rarely, swimming. A branch at the edge of the pool sometimes became the focus for a series of displacements, as one monkey climbing toward the branch triggered a leap into the water by a second monkey on the branch, the displacer then being itself displaced by a third monkey.

An excellent film on rhesus social play (Bishop & Symons 1978) includes footage of water play.

A second noteworthy aspect of play in rhesus is that group-living infant monkeys may play by themselves in confinement, repeating particular patterns of locomotor and manipulative activity again and again (Simpson 1976 p. 386):

One common early project is leaping vertically up to a low branch; later the monkey may leap horizontally from a post to the mesh side of the cage, and still later horizontally from the mesh side to some limited target like a pole. The "project" may also become complicated: climb up a sloping pole, jump across to the wire mesh, climb down, fiddle with some fixture like a bolt, climb up the sloping pole again. By 12 weeks, such sequences may involve other animals. The leaps may now be at peers, who must jump out of the way or be hit. The mother is usually incorporated in such sequences, in the rather passive sense that the infant regularly returns to her side.

Such projects may include leaping, lurching, hopping, or simply walking along the same pathway many times in sequence during a short period (Bishop & Symons 1978, Mears & Harlow 1975). Symons's observations of this behavior in free-ranging rhesus suggest that it should not be interpreted as a consequence of confinement.

As discussed in Chapter 2, classification of such repetitive behavior as "play" is a matter of taste. Simpson (1976) analyzes the apparent design of rhesus solo play for skill development. At this functional level, solo play exhibits the same design characteristics (Simpson 1978) described by Symons (1978a) for playfighting. In both cases the animal behaves so as to have informative experiences that tend to contribute to the development of its own skills, or skills of its close kin, at minimum risk.

A less well-known cercopithecine primate whose society is slightly less complex than that of rhesus is Lowe's guenon (*Cer-*

copithecus campbelli), a forest-dwelling, arboreal monkey from West Africa. These monkeys form so-called age-graded male troops containing males of different ages, but only a single mature adult male. The play of young Lowe's guenons combines the expected with the remarkable (Bourlière, Bertrand, & Hunkeler 1969, Bourlière, Hunkeler, & Bertrand 1970, Hunkeler, Bourlière, & Bertrand 1972). Solo play, including arboreal acrobatics and object manipulation, occurs frequently. A young guenon may carry some item (branch, heavy rock, or the like) up a tree, drop it, then descend, and repeat the game several times. Social play includes arboreal and terrestrial chasing and wrestling. A monkey may vault over another's back. One monkey may present itself as if to solicit mounting, then suddenly flee at the instant its partner attempts to mount. In an arboreal game two to five monkeys scramble up to a swaying tree branch and wrestle there. One monkey may abruptly fall off the swinging bough, which then springs upwards, knocking other monkeys off balance. Free falls are frequent in play. One monkey may drop down by stages through small trees or underlying bushes. Subadults appear to "seek thrills" (in the sense of Aldis 1975), breaking or tearing open the nests of stinging weaver ants (*Oecophylla longinodis*), jumping around frantically while the ants attack them, then returning immediately to the nest to repeat the process. Juvenile Lowe's guenons play interspecifically and reciprocally, chasing and being chased by wild squirrels (*Heliosciurus gambianus*) or fleeing from a charging (but tame) mongoose (*Crossarchus obscurus*), then returning to "tag" the mongoose and flee again. The monkeys also pursue other mammals and birds, apparently for the sole purpose of scaring them ("*apparemment dans le seul but de les effrayer*": Hunkeler, Bourlière, & Bertrand 1972). The habit of teasing hornbills (*Tockus fasciatus*) seems best developed. The hornbills roost in a large tree. Once they are settled, the juveniles leap toward them and shake the branches until the hornbills take flight. The monkeys then wait in the tree until the birds return. When the birds have again settled, the monkeys repeat their mock attack. This "hornbill game" brings to mind a domestic dog running, barking madly, into a crowd of pigeons with the apparent purpose of startling them into flight.

If Lowe's guenon represents an arboreal extreme among cercopithecine monkeys, the patas (*Erythrocebus patas*) represents a terrestrial extreme. Patas are cursorially adapted grounddwelling monkeys. They are highly sexually dimorphic; an adult patas male is almost twice the size of an adult female. Patas social demography is simpler than that of Lowe's guenon. Patas troops contain a single adult male, and sexually maturing males are not tolerated (Eisenberg et al. 1972).

Patas play, as described by K. R. L. Hall (1965), is unforgettable:

Social play interactions involve a most vigorous exercising or practising of speed and agility of ground locomotion which can be readily seen as an adaptation for survival from day-hunting predators such as cheetah or hunting dogs. The "arena" for this quite spectacular play of the young patas has almost always been a large open stretch of grassland with a few bushes or an erosion sheet or erosion gullies.

The components and sequence of play in the patas groups were:

(i) "Inviting" play—usually by bouncing, on the same spot, quickly from hands to feet to hands, and so on, with the limbs held straight, and glancing towards the other animal; the inviter, often the older of the two animals, then leaps away and races over the grass with the other chasing after it.

(ii) The play-chase—the usual sequel to (i), and carried out at full speed over the ground.

(iii) Play-wrestling—typically with each player on hindlegs, facing each other, a grappling with hands around shoulders and arms, then sometimes one, sometimes both, rolling over on the ground; two juveniles wrestled thus for two minutes, trying to "throw" each other; the play face almost always accompanies these wrestling bouts, the mouth being open, but with no threat of face or limb, and mock-biting often occurring.

(iv) Play-sparring—actual slapping movements with the hands may accompany or precede mock-biting and wrestling.

(iii) and (iv) are vigorous, but entirely non-threat replicas of actual fighting that has occurred between adults of the laboratory group.

(v) The play bounce is a very high-speed version of the on-the-spot "invitation" bounce—the animal, at full gallop, hurls itself sideways against a sapling or small bush so that the hands contact the bush first, being immediately followed by the feet, the effect being that the animal catapults itself sideways before continuing to gallop away. In the laboratory, exactly the same action is done against the walls of the room or the sides of cages, but, in the wild, the springy bushes were evidently preferred, because no side-bouncing was seen from tree trunks or cliff sides. The effect of this very fast movement is to bring about a very swift change of direction, but one which, in play-chase, is usually exactly followed by the chasing animal. Occasionally, the bounce is done from the top of a bush onto which the animal has leapt. What seems to be exactly this kind of action is carried out by the adult male patas in the totally different situation of diversionary or possibly threat display with reference to the observer. . . . In the laboratory group, it has been repeatedly and very noisily performed by a young adult female against the sides and top of a cage in which another adult female was confined. The young adult female had been continuously, and effectively, provocative and aggressive towards the older female when the two of them were free in the room.

The speed and grace of these play sequences in a wild group, often continuing amongst six or more of the young animals for periods of up to 30 minutes without respite, is such as to have no parallel in any other terrestrial monkey. Players may run at full gallop for 200 m or more away from the others before circling back to the grass and *Combretum* bush arena around which the play is centred.

Neither the bouncing "invitation" to play, nor the running side-bounce, have any known equivalent in baboons or vervets. Indeed, much of the character and tempo of the whole play sequence seems to be peculiar to the patas, though play-wrestling, biting and chasing occur in many monkey species.

Patas play (Hall 1965, 1968, Hall, Boelkins, & Goswell 1965, Struhsaker & Gartlan 1970) does exhibit some typical cercopithecine features including gamboling, grappling and sparring, climbing, and leaping about on bushes. Infant patas will approach one another quadrupedally and stretch out an arm toward the other animal. Whether or not actual contact is made, the toucher and usually the recipient as well then hop clumsily backward "away from the point of contact, raising themselves briefly on their hind legs at the peak of the hop." This pattern was repeated again and again (Struhsaker & Gartlan 1970).

In captivity, infant patas frequently playfight with their mothers (Baker & Preston 1973, Goswell & Gartlan 1965).

Baboons (genus *Papio*) are large, highly sexually dimorphic, ground-dwelling primates. Physically, they are far more robust than patas and have none of the ungulate-like locomotor adap-

Fig. 3–7 Young yellow baboon (*Papio cynocephalus*) and vervet monkey (*Cercopithecus aethiops*) in rough-and-tumble play. Amboseli National Park, Kenya. (From Altmann & Altmann 1970, Stuart Altmann)

tations of *Erythrocebus*. In general outline, the social structure of multi-male troops of baboons closely resembles that of rhesus troops.

Baboon play does not differ greatly from rhesus play (Altmann & Altmann 1970, Owens 1975a,b, 1976, Ransom & Rowell 1972, Rose 1977b). One remarkable feature of baboon play is its occasional incorporation of members of other species. Wild, free-living baboons are reported to play with individuals of at least five other species (vervet monkey, Altmann & Altmann 1970 and Fig. 3-7; bushbuck, Douglas-Hamilton & Douglas-Hamilton 1975; gelada, Dunbar & Dunbar 1974; bushbuck and impala, Grzimek & Grzimek 1960; chimpanzee, van Lawick-Goodall 1968a, Morris & Goodall 1977). Because baboons have probably been observed in the wild for more hours than any other nonhuman primate, and because they are most often studied in open, species-rich habitats, this relatively large number of reports is understandable, but still fascinating.

PONGIDAE

The great apes are among the longest lived and most highly encephalized primates, and by all rights they should, without exception, exhibit play. This is certainly the case for healthy, but confined animals in naturalistic environments, but paradoxically field workers have remarked on the low frequency of play in populations of at least three great apes: gorillas (*Gorilla gorilla*) (Schaller 1963), orangutans (*Pongo pygmaeus*) (Horr 1977), and, despite Carpenter's extensive description (1940) of their play in the wild, gibbons (*Hylobates lar*) (Ellefson 1968). The message of this literature is that the young of these three great apes are solemn, incurious, and unresponsive creatures. As anyone familiar with young orangutans, gorillas (Figs. 3-8, 3-9, 3-10, 3-11), or gibbons can testify, nothing could be farther from the truth. Consider, for example, the play of young orangutans in a family group (MacKinnon 1974a, Maple & Zucker 1978, Wallace 1962, Zucker, Mitchell, & Maple 1978, pers. obs.). These animals wrestle on the ground, chase, hang and wrestle, drape themselves with various inanimate objects, and exhibit play-faces. In confinement, an indulgent father may push his daughter over with his hand, roll her back and forth on the ground, and even tickle her. Siblings of different ages play together; the elder may take a passive role or may use less than its full strength in play.

Simplicity of social structure and of tool behavior in orangutans belie the complexity and sophistication of this species' adaptations to its tropical forest environment. Orangutans may be the world's premier field botanists (Galdikas 1978). Moreover, they boast a formidable array of sophisticated manipulative and locomotor skills used in arboreal life. Their intelligence is impressive even by great ape standards. A closer look at orangutan locomotor, manipulative, and rotational-vestibular play suggests that the complexity and elaborateness of these behaviors, if measured on an objective scale, would rival the complexity and elaborateness of chimpanzee social play (Rijksen 1978). The cohesion hypothesis of social play led commentators on great ape play to incorrectly downgrade orangutan play to rudimentary status and to ignore nonsocial play altogether. In actuality one might expect orangutans to play as elaborately and perhaps even as often as other great apes, in ways appropriate to their particular ecological adaptations. That orangutans given the opportunity to play socially do so readily further suggests that the tendency to play socially is not biologically distinct

Fig. 3–8 The gorillas sit in close proximity before beginning to play.

Fig. 3–9 The gorillas lie on their sides and grapple.

104

Fig. 3–10 Wrestling in a sitting position, one gorilla rises to its knees in order to obtain positional advantage.

Figs. 3–8 through 3–11 Lowland gorilla (*Gorilla gorilla*) play-fighting. Photographic sequence of a single playfight. San Diego Zoo, California. (John M. Bishop)

Fig. 3–11 Both gorillas are in an upright playfighting position.

from the tendency to play alone, but simply reflects the interaction of species-specific adaptive behavioral tendencies toward play and the resources for play provided by the environment.

The high frequency of chimpanzee play observed in the wild and the relatively low frequency of play observed in the field in gorillas and in Asian pongids may be attributed to differences in observers, to differences in environments, and very possibly to the reluctance of youngsters to play during human observation as well as the possible reluctance of their parents or older siblings to allow them to play at this time. This prediction has already been substantiated for gorillas. Marler & Tenaza (1977) review unpublished field data on gorilla behavior and conclude that gorillas may play much more often than has previously been suspected. In previous studies of gorillas, these authors suggest, "certain activities would have been inhibited during observation of nervous animals, notably play behavior" (Marler & Tenaza 1977 p. 997).

It is premature to conclude that any species of great ape is significantly more playful than any other. If social play is less frequent in a small family because of the absence of appropriate playmates, it would be appropriate to look for a compensatory increase in nonsocial play and to ask whether the infant or juvenile had previously solicited play from older family members, but had been rejected.

We humans may simply reveal a bias in favor of our closest nonhuman relative by citing the remarkable characteristics of chimpanzee play, but I speculate that the chimpanzee *Pan troglodytes* will always be regarded as the champion player among nonhuman primates. Play of juvenile chimpanzees with conspecifics, with baboons, and with manipulatable items, as well as chimpanzee solo and mother-infant play, have been described lovingly and at length by Jane Goodall (van Lawick-Goodall 1965, 1967, 1968a, 1970, 1971, 1973) and by other observers (Clark 1977, Loizos 1969, McGrew 1977, Menzel 1963, 1975, Savage & Malick 1977, Simpson 1978, Tutin & McGrew 1973). Play of wild chimpanzees is documented on film (Marler & van Lawick-Goodall 1971).

Chimpanzee play has challenged the scientific detachment of

students of animal behavior at least since 1835, when a young male chimpanzee at the London Zoo successfully initiated play with a human observer:

He soon showed a disposition to play with me, jumping on his lower extremities opposite to me like a child, and looking at me with an expression indicating a wish for a game of romps. I confess I complied with his wish, and a capital game of play we had (Broderip 1835).

Parallels between chimpanzee and human play also appeared obvious to R. L. Garner (1919), another early observer of chimpanzee play, who remarks "On several occasions I have seen the young ones romping and tumbling about on the grass, chasing and scuffling with each other, exactly as you see human children do."

The full complexity of chimpanzee play only became apparent as a result of Jane Goodall's close field observations of chimpanzee behavior at Gombe Stream, Tanzania. These observations have been continued by workers at the Gombe Stream Research Center. The fascinating picture of chimpanzee play now emerging is by no means complete. After all, chimpanzees are not expected to submit passively to scientific study. Bingham's (1929) young chimpanzees sometimes disrupted his anthropological research by playing when he was trying to measure them. A case of chimpanzees turning the tables on scientists with spectacular success involved young captive subjects who learned to use elongated objects, especially sticks and poles, as ladders after a period of play with the objects, then broke into an observation shed (apparently by using poles as ladders), did several hundred dollars worth of damage, carried off equipment and tools, and slept in the shed overnight (Menzel 1972).

Chimpanzee play shares certain characteristics with the play of other primates. These characteristics include a play-specific facial expression, nonagonistic modification of aggressive chasing and wrestling patterns, use of objects in social play, and vigorous repeated solo locomotor-rotational projects and exercises. Considerable tickling, frequent object play in the wild, and well-developed play between mother and offspring apparently distinguish chimpanzee play from the play of all other nonhuman primates. (However, it is not yet clear that these features are absent from orangutan or gorilla play.)

In the chimpanzees at Gombe (van Lawick-Goodall 1968a, 1971), as in other wild chimpanzee populations (Reynolds & Reynolds 1965), social play consists of

chasing, wrestling, sparring (when one individual hits towards another who fends him off, and vice versa), play biting, thumping and kicking (when the playmate is hit hard with the palm or knuckle of one or both hands or the heel or sole of the foot), butting with the head and a variety of "tickling" and poking movements. Tickling was done either with the mouth, when the animal made a series of nibbling nuzzling movements with the lips pulled inwards over the teeth or with the hands when the chimpanzee made prodding flexing movements of the fingers in the same way as does a human when tickling. A variant on this was "finger wrestling" when an individual (usually mature male) reached to the hand or foot of another and made gentle tickling, pulling and squeezing movements. During play sessions, chimpanzees made use of a wide variety of the objects of their environment: they climbed, jumped, swung and dangled from branches of trees, chased round tree trunks, broke off and waved or carried branches, leaves, or fruit clusters, grappled with each other for an assortment of small objects, dragged and hit each other with branches, and so on (van Lawick-Goodall 1968a).

Chimpanzees often chase each other round and round the base of a tree or a clump of vegetation (van Lawick-Goodall 1968a). Play-chases of this kind were first observed in the family of Flo, an especially playful female (van Lawick-Goodall 1971). "On two occasions Flo joined her three older offspring in a game which consisted of chasing each other round and round the base of a tree. The children sometimes played this game with Flo sitting in the middle" (van Lawick-Goodall 1968a). The film "Vocalizations of wild chimpanzees" (Marler & van Lawick-Goodall 1971) shows two infant males in a play-chase around a bush as they playfully compete for possession of a palm flower.

Chimpanzees invite play in many ways (van Lawick-Goodall 1968a, 1973). The chimpanzee play-face is typically accompanied by a rhythmic vocalization resembling (but not identical to) human laughter (Marler & Tenaza 1977). This play-specific vocalization is not known in other nonhuman primates. There are particular play-gaits as well (van Lawick-Goodall 1968a). Novel use of a facial expression as a play-signal is an example of a cultural element in chimpanzee behavior:

A two-year-old infant, during play sessions, consistently sucked in her cheeks instead of showing the normal play-face. After a few weeks other infants with whom she frequently played also began sucking in their cheeks during play sessions. The face itself was not novel, as it appears in most infants from time to time; the context in which it can be used, however, was new. Within the next few months the habit gradually disappeared (van Lawick-Goodall 1973).

Mother chimpanzees play with their infants, tickling them and nibbling at them. Flo often "lay on her back, held Flint above her with one of his ankles or wrists clasped in her foot, and tickled him. . . . Play behavior of mothers with their one-year-old infants included rolling the infants over, mock biting and tickling them, gently sparring with them, or pushing them to and fro as they dangled from a branch [Fig. 3-12]. The infants responded by grabbing and biting the mothers' hands, flinging themselves onto the mothers with flailing arms or tickling them" (van Lawick-Goodall 1968a). Repetitive mother-infant biting games, in which a mother would encourage her six-week-old infant to bite at her hand and fingers, occur in Gombe chimpanzees, accompanied one-third of the time by play-faces. These games continue in succeeding weeks, and "by twelve weeks the infant would initiate interaction by directing his bites especially towards the mother's hands" (Simpson 1978). In captivity, mothers playfully lift or walk their infants. These behaviors were performed repetitively and were accompanied by a play-face (Nicolson 1977).

Object manipulation by wild chimpanzee infants is much more common (McGrew 1977) than previously supposed, and most object manipulation by immature chimpanzees exhibits clear structural and functional hallmarks of play. Object manipulation may develop certain tool skills. Gombe chimpanzees use leaves as sponges to pick up or transfer liquids, they extract ter-

Fig. 3–12 Mother-infant play in wild chimpanzees (*Pan troglodytes*), showing play-face. Gombe National Park, Tanzania. (Baron Hugo van Lawick, © National Geographic Society)

mites from their mound nests by using specially prepared vegetation probes, and they insert plant tools into driver ant nests and collect the ants in one hand as the insects swarm up the tool. Leaf sponging behavior has playful antecedents that differ structurally from the final, skilled behavior. McGrew (1977) observed a 24-month-old male chimpanzee playing with components of leaf sponging skill:

12:02 FD finds tree hole full of water. (No need to sponge.) Dips hand and licks off water. Splashes water with hand. Has play face and slightly upturned upper lip expression.

12:03 Splashes and sucks fingers. Convenient potentially spongeable leaves within 30 cm of hole.

12:04 Stamps tree trunk. Sucks hand.

12:05 Leaves hole, then returns to stamp and splash. Licks water from hand.

12:06 Stamps and splashes. Has penile erection.

12:07 (Leaves hole.) Returns to hole, splashes.

12:08 Plays with water, with acrobatics and stamping. Inserts hand into hole.

12:09 Sucks and licks water from hand.

12:10 Moves back and forth between hole and branches. Play looks distracted. Splashes and sucks.

12:11 Returns and leaves hole. Mouths and carries piece of vine but does not use it in water.

12:12 Returns to splash, still with play face. Sucks.

12:13 Licks from hand. Attempts insertion into hole of two un-modified leaves (from vine). Removes these from hole after 2–3 seconds, then departs. Returns to hole carrying leaves in mouth.

12:14 Mouths and carries two green leaves.

12:15 Approaches FF. Still mouths and carries leaves. (Leaves hole.)

12:19 FF suckles FD.

12:22 Returns to hole and dips hand. Sucks from hand. Splashes.

12:23 Inserts hand and sucks finger. Play face.

12:24 Inserts hand again. Leaves hole. Returns with leaf.

12:25 Tries to insert leaf with mouth. Doesn't seem to know quite what to do. Drops leaf. Returns to dipping hand.

12:26 Sucks from hand. Mouths and carries twig to hole but drops it immediately.

12:27 Splashes and inserts hand.

12:28 Stamps. Inserts leaf into hole with lips. Has yet to make sponge by crushing leaves.

12:29 Tries to get lips into hole. Bites off leaf.

12:30 Leaves tree hole.

12:34 FF and FD leave area.

(Field notes, McGrew 1977 pp. 276–277)

Playful characteristics of this sequence (McGrew 1977) are ex-aggeration of dipping to splashing, and repetition of splashing;

interruptions to stamp or for acrobatics in the branches; incompleteness of the sponging sequence; use of an inappropriate object; abbreviation of some parts of the sequence (e.g., object handling) and prolongation of others (e.g., sucking); and insertion of the leaf with the lips rather than the hand.

Repetitive, variational performance of previously mastered termiting skill sometimes occurs in infant chimpanzees. The infant Flint, aged 32 months, was twice observed using grass tools out of context. "Once he pushed a grass carefully through the hair on his own leg, touched the end of the tool to his lips, repeated the movement and then cast the grass aside. On the other occasion he pushed a dry stem carefully into his elder sibling's groin three times in succession" (van Lawick-Goodall 1968a).

Of the three tool skills cited, ant dipping is mastered latest in ontogeny, apparently without previous practice of any kind. The differing ontogenetic histories of these three skills are fascinating. They suggest that practice, including object play, can facilitate some skills, but is not always safe. McGrew (1977) discusses this problem at length.

Chimpanzee play is equally interesting in its own right and as a neutral introduction, without political overtones, to the study of human play. From Köhler to Goodall, research on chimpanzee behavior has profoundly influenced the study of play just as it has influenced many other aspects of behavioral biology. To my knowledge, no monograph on chimpanzee play now exists, but a full-length study of chimpanzee play would be an immensely valuable document.

Humans (*Homo sapiens*)

To cite play behavior as *the* framework for hominization (P. Reynolds 1976) and for human culture (Huizinga 1950) makes us question what those other species have been doing with their playtime. Does human play differ more from the play of representative mammals like voles, dogs, rats, rhesus macaques, and deer than does the play of other, unusual mammals like elephants, bottle-nosed dolphins, and wombats? Stated another way, is the undeniable uniqueness of human play a uniqueness of a higher order than that afforded by any other species, however aberrant in morphology, ecology, and social behavior? And is this uniqueness one of process or one of form? Typological classification of juvenile play without regard to biology is unlikely to furnish a royal road toward understanding evolutionary constraints on human behavior. Play of human children may either resemble or differ from play in the young of other species. What evolutionary processes shaped the underlying patterns of development?

Like young of other mammals, juvenile humans chase and wrestle together, they perform varied locomotor and rotational exercises, and they investigate and playfully manipulate objects. Human play with constructed toys, mother-infant play, imaginative and sociodramatic play, older juvenile and adult object play, language play, and the intellectual "play" hypothesized to underlie artistic and scientific creativity are probably all unique. Each of these unique forms of play may be related to one or more human biological, behavioral characteristics: fetus-like helplessness and slow development of human infants; human use of tools and of language; and uniquely human components of cognitive ability based in our large, neotenous brains. For this reason studies of human play can buttress studies of other human life-history characteristics and adaptations, and may even be fundamental.

VIEWPOINTS

As Hutchinson (1976) and Blurton Jones (1967) suggest, one may either look "up" at human play from the perspective of nonhuman behavior or look "down" at nonhuman play using concepts derived from the study of human play. Consequences of the former perspective are suggested by Blurton Jones (1967 p. 350): investigative-manipulative object play, "the most 'human' side" of child behavior, is difficult for ethologists to investigate, but the ethologist "has more scope in studying the interactions of the children with each other and with adults." In both cases there appears to be something missing. It is difficult to decide which type of neglect is more pernicious: that of a human ethologist whose methodology excludes human object play (though ethology claims to describe phenomena objectively and to furnish essentially complete catalogs of behavior) or that of a naive cognitive-developmental psychologist whose dream of a biology of object play requires that precursors of this human play activity occur in the behavior of free-living baboons or rhesus monkeys. These points of view are useful once their limited scope is recognized, but it would be counterproductive to expect encyclopedic accounts of human play to emerge from single, narrow schools of thought. For this reason the following account of human play must draw on a number of sources, none comprehensive.

Human play behavior poses unique challenges to biologists (see also p. 125). As a pivotal epigenetic mechanism driving gene-culture coevolution, play furnishes unparalleled opportunities for hypothetico-deductive field work in cultural biology. Models like those formulated by Lumsden & Wilson (1981) offer novel insights into the rich empirical information on toys, games, and verbal play presented by Avedon & Sutton-Smith (1971), Garvey (1977), Opie & Opie (1967, 1969), Piaget (1962), Whiting & Whiting (1975), and others.

PLAYFIGHTS AND PLAY-CHASES

Nonagonistic chasing and wrestling (Fig. 3-13), accompanied by the relaxed open-mouth grin or play-face, are well documented in children of many cultures (Aldis 1975, Blurton Jones 1967, 1972, Garvey 1977 pp. 37–39, Konner 1975, P. K. Smith & Connolly 1972, P. K. Smith 1973). British children aged 3 to 5 years (Blurton Jones 1967, 1972, McGrew 1972b), American children and teenagers in California (Aldis 1975), and "Bushman" (Zhun/Twa) children in Botswana from about 1 year to late adolescence (Konner 1972) interact playfully by "running, chasing, and fleeing; wrestling; jumping up and down with both feet together; beating at each other with an open hand without actually hitting; and laughing." Besides the smile-like expression involved in laughing, a play-associated facial expression is exhibited by a child about to be chased by another. The child will stand "slightly crouched, side-on to the chaser and look at it with a 'mischievous' relaxed, open-mouthed play

Fig. 3–13 Human children play-wrestling. Sherpa boys. Melemchi, Nepal. (John M. Bishop)

face" (Blurton Jones 1967). Aldis (1975) offers impressions of possible differences between this human playfighting and play-fighting of other mammals: human play-chases tend to be brief (Aldis 1975 p. 207), and mouthing in humans is strikingly absent (Aldis 1975 p. 274). Whiting & Whiting (1975 p. 193) describe active sociable assaults or "horseplay" as nonagonistic "hitting, shoving, kicking, striking with a stick or other object, wrestling." They report such behavior in children of six cultures from India and Okinawa to New England (U.S.A.) and Kenya. Whiting & Edwards (1973) identify such playfighting as one of two best candidates for "innate characteristics" of human behavior. Playfighting (Fig 3-13) may involve friendly competition for positional advantage, in which one participant makes "strenuous efforts to throw another to the ground, to get on top of him, to hold him down, to flatten him, and sometimes to pin him to the ground" (Aldis 1975 p. 178). It may also consist of "attempts to push, pull or hold another child without further attempts to get into a superior position" (Aldis 1975 p. 189), which an ethologist might term "grappling" or "sparring" (described by Owens 1975b in baboons and Symons 1978a in rhesus). In this second type of body contact play, one partner pulls, pushes, or holds the other while standing, pulls the other's leg, or directs inhibited blows and kicks at the other (Aldis 1975). Like interactive social play in other animals, these behaviors are clearly nonagonistic: threat and submission signals are lacking; damaging tactics are avoided; the participants smile, laugh, and use the play-face; muscle tone is relaxed; roles reverse; and participants stay together at the end of a bout. Young humans likewise chase each other, with reversal of roles and self-handicapping by a stronger or more fleet player (Aldis 1975, Blurton Jones 1972, Smith & Connolly 1972).

Possible cultural and sex differences in human playfighting are discussed by Blurton Jones (1967, 1972), Blurton Jones & Konner (1973), Garvey (1977 p. 38), Konner (1972), P. K. Smith (1973), Smith & Connolly (1972), and Whiting & Whiting (1975). It was Aldis's (1975) impression that boys in California tended to play-wrestle for superior position more than did girls, and that girls tended to spar, grapple, and play-chase more than did boys; Blurton Jones & Konner (1973) also suggest this sex difference for London children. However, this hypothesis is yet to be tested rigorously. Age, the identity of the partner, the types of play apparatus available, and differences in developmental rates and in cultural standards may all interact with possible sex differences in human play (Aldis 1975).

Separate from the vigorous playfighting described above is a gentler form of playful social interaction in Zhun/Twa children (Konner 1972), who exhibit mutual touching, tangling of legs, clinging, and rolling on the ground. When so playing, children do not laugh; they move slowly and are unlikely to stand up. The closest nonhuman counterpart to such play is probably the dyadic rolling play of young common (*Phoca vitulina vitulina*) and grey (*Halichoerus grypus*) seals (S. Wilson 1974a). Both types of behavior include the rotational movements of head and trunk considered characteristic of play (S. Wilson & Kleiman 1974).

LOCOMOTOR-ROTATIONAL EXERCISE

Swinging, sliding, rolling, jumping, bouncing, and running in circles, as well as acrobatic maneuvers including somersaults, cartwheels, handstands, and spinning while standing are described in young humans by Aldis (1975). In some Western societies these activities take place with the aid of such playground apparatus as slides and swings, but a steep slope, tree branch, or hanging vine may suffice, as for many nonhuman animals. These play activities may either be performed alone (usually with a parent or older sibling nearby) or in a group.

Although laughter and smiling tend to occur during such human play, we have little detailed information about such other possible accompaniments as headshakes or body-twists that might be expected based on studies of other mammals (S. Wilson & Kleiman 1974). However, although these forms of play subjectively seem clearly differentiated from the tense, unadorned, tentative, and often clumsy behaviors involved in basic motor learning, there are no quantitative ethological data that might supply objective contrasts between the first, narrow exploratory or practice stages and the later, broader playful or refinement stages of human motor skill development. And whereas the child's social (including family) environment and physical environment, including object availability, are all likely to be important determinants of form, frequency, and possible sex differences in human locomotor-rotational exercise (Aldis 1975), next to nothing is known about precise effects of these factors, or possible interactions between them.

PLAYFUL MANIPULATION

Human object exploration and manipulative play exceed the expectations of ethologists familiar with the relative rarity and simplicity of object play in most wild primates. Human construction and manipulation of play objects (toys), use of found or constructed objects to explicitly symbolize or represent other objects, their incorporation into more elaborate constructions, and their differing roles in the play of different-aged children all attest to the complexity of object behavior in this aberrant primate.

Early human manipulative behavior is generally regarded as a continuum of patterns ranging from exploratory investigation to play. Investigation may be defined as perceptual examination of an object or a body part resulting in information about its physical and chemical characteristics, including size, weight, hardness, texture, taste, and elasticity. Investigation is object-oriented receptor activity designed to answer the question "what does this object or body part do?" The child's approach to the item inspected is slow, cautious, and vacillating; muscle tone is tense; receptors are fixated on the item; the item is observed and manipulated frequently at first, but responses decrease monotonically with time; and both play-signals and loco-motor-rotational movements are absent from the exploring child's behavior (Aldis 1975 pp. 67–71, Hutt 1966, 1970, Weisler & McCall 1976).

Behavior of a child at play with an object or a body part is not investigation (Aldis 1975, Hutt 1966, 1970). The child plays with items that have already been investigated and are familiar. The item is approached rapidly and in a relaxed manner, the child's muscles are not tense, receptors may diverge from the item, the behavior is often physically vigorous, responses increase and then decrease over time, and the child decorates its behavior with locomotor-rotational movements and even with play-signals. These signals are not random, functionless events. A human child will not be left alone for a long time to play, nor will it play if left alone for a long time. These signals could serve to inform a caregiver of the child's state and attitude. They tell a watching parent or other kin that the child is comfortable and well-fed, is neither uncertain nor fearful, and is currently engaged in play. Manipulative play is self-oriented effector activity designed to answer the question "What can I do with this item?"

Dunn & Wooding (1977), Fenson, Kagan, Kearsley, & Zelazo (1976), Fenson, Sapper, & Minner (1974), McCall (1974), Rosenblatt (1977), and a host of other observers, including Charles Darwin and Jean Piaget, have described development of manipulative behavior in human children. Very young children scrutinize and manipulate body parts repeatedly as in the familiar cases of hand regard or grabbing and handling the toes or genitals. They bob and push known objects up and down in water. They bang, wave, shake, and throw around familiar toys. These actions can be performed tensely and intently, with all receptors focused and with cautious body movements, or they can be repeated with slight variation in a relaxed, diffuse manner, without sensory synchrony and perhaps even accompanied by a play-face, by laughter, or by locomotor-rotational movements.

A feature of human play whose uniqueness cannot be questioned is construction and use of toys, objects fashioned with varying degrees of sophistication for the specific purpose of use in play. Although individuals of many species, including wasps, blue jays, and chimpanzees, construct and use tools (Beck 1975, 1978), no nonhuman species is known to modify or fashion objects for the use of its young in play.

Like young of other species, human children perform acrobatics on or in appropriate objects, and they use objects as lures in chasing games (Aldis 1975). They manipulate objects, parts of their own body, and parts of other animals in other ways as well. With these manipulative behaviors the distinction between human and nonhuman manipulative play begins to sharpen as the distinction between investigation and play blurs. But the concept of object play may not apply to infant or child behavior. Does the infant distinguish between an inanimate object, a part of another animal's body, and a part of the infant's own body? The question must be rephrased. What manipulative experiences has natural selection favored at different points in ontogeny? What manipulanda are optimal for obtaining these experiences? How does the young organism's physical and social environment affect the range and variety of available experience? Do human brain hypertrophy and delayed development require departures from a general primate or mammalian pattern of early manipulative experience?

MOTHER–INFANT PLAY

Comparative data on gestation length and on infant behavior suggest that human infants are born at a young age post-con-

Fig. 3–14 Approach.

Fig. 3–15 Withdrawal.

Figs. 3–14 and 3–15 Human child-infant "looming" (approach-withdrawal) game. Two phases. California. (Owen Aldis)

120

Fig. 3–16 Mutual gazing during human mother-infant play. California. (Owen Aldis)

ception and at an early developmental stage compared to the young of all other primates (Gould 1977). Body size alone cannot account for this difference. Humans also develop more slowly than other primates. The dependence of the human infant on its caregiver is therefore extreme, and clues to the role of play in early development may be found by analyzing play of mothers with their very young infants (Stern 1974a,b).

Human infants begin game-like interactions (Figs. 3-14, 3-15) with a caregiver (usually the mother) as early as the second month of life. In these interactions, the mother adjusts the tempo, sequence, and coordination of her activities to the changing behavior patterns of her infant, "timing her own interventions . . . in such a way that she supports and extends" the exchange (Dunn 1976). Mothers exaggerate and prolong their facial expressions, movements, and vocalizations during these episodes (Stern 1974b). Mutual gazing (Fig. 3-16) is marked at this early age.

American mothers of 3- to 4-month-old infants direct "unusual variations of intra-adult interpersonal behavior" toward their offspring. They may employ a special form of speech ("baby talk") marked by an expanded range of pitch, changes in rhythm and stress, and altered range of loudness and speed of changes in loudness; they gaze repeatedly and for long periods of time at their infant; and they make extraordinary (exaggerated, slowly forming and prolonged) facial expressions, including a "mock surprise" expression with raised eyebrows, wide-open eyes, and open and pursed mouth. The infant also performs "an array of facial, vocal and gaze behaviors," notably gaze alternation during which the infant turns toward and away from the mother's almost constantly gazing face (Stern 1974b). To my knowledge such behavior has never been described in mother-infant dyads of any nonhuman species; Hinde (1974 p. 184) views it as "an almost qualitative advance over non-human forms."

In older infants the pattern of these interactions extends beyond its initial context to object play (Dunn & Wooding 1977) and to the game of peekaboo (Bruner & Sherwood 1976). The near-uniqueness (Hinde 1974), complexity, and delicacy of "the game" (Watson 1972) and its effectiveness as a framework for cognitive development (Dunn 1976) suggested to several authors (Brazelton, Koslowski, & Main 1974, Brazleton et al. 1975) that these interactions were necessary for the development of cognitive ability. Further investigations (reviewed by Dunn 1976) revealed, however, that the developmental process is more flexible than this simplistic view would imply and that children can compensate for variations in their early rearing environment. Like so much else in mammalian behavioral development, early interactive play appears to be a social mechanism regulating developmental paths or rates.

GAMES WITH EXPLICIT RULES,
IMAGINATIVE PLAY, AND FANTASY

In older children and in adults, play can take imaginative or symbolic forms: children agree to play at being something, e.g., Indians, hunters, or astronauts. For example, Doughty (1923) described desert Bedouin children who played at being horses. We can imagine chimpanzees pretending to be leopards, baboons, or field primatologists. However, such games have never been observed in years of intensive field and laboratory research on chimpanzee behavior, suggesting that this level of symbolic complexity most likely lies beyond the reach of any species on this planet but our own. There is, however, an anecdote of apparent imaginative play in a home-reared chimpanzee (Hayes 1952, reprinted in Bruner, Jolly, & Sylva 1976).

An important level of abstraction separates playfights, playchases, simple object manipulation, and the mother-infant "game" from games with explicit rules (hopscotch, marbles, ring-a-rosy and the like, Avedon & Sutton-Smith 1971, Caillois 1961, Opie & Opie 1969, Piaget 1962), as well as from representational play (make-believe, sociodramatic play, symbolic play, imaginative play, thematic play, fantasy play) (Cobb 1977, Dunn & Wooding 1977, Garvey 1977, Lieberman 1977, Piaget 1962, Singer 1973, Singer & Singer 1976, Smilansky 1968, Smith 1977), from language play (Garvey 1977, Opie & Opie 1967, Weir 1976), and from quiet fantasy and daydreaming (Klinger 1971, Singer 1973, Winnicott 1971). In these higher-order human play activities the literal rules of the game or the imaginative structures established by the players allow individuals of different sex, sizes, and ages to verbalize and to discuss the rules of the interaction, a distinctively human trait. It would be interesting to compare the origin, control, and outcome of breakdowns and misunderstandings in games with explicit rules and in imaginative play with the bases of breakdowns in chasing and wrestling. We are familiar with the conventional and escalated fights that can erupt from children's games or imaginative play, often over differing interpretations of the rules, over actual cheating or accusations of cheating, or, in the case of imaginative play, over role assignments or over failure to make desirable responses. How do these breakdowns vary with the relative ages, sex, sizes, and genetic relatedness of the players, and do these patterns of variation parallel those established for nonhumans?

Some games with explicit rules are little more than formal versions of chasing and fleeing patterns, or of other recognizable nonhuman social play configurations. Similarly, children may chase, climb, and jump with abandon while imaginatively playing at cowboys or at monsters or while imaginatively transforming objects into cars and space ships (Appendix to P. K. Smith 1977). It is not known whether the human wrestling and chasing episodes described by Aldis (1975) had symbolic content, because Aldis observed children at a distance and tried to remain undiscovered by his subjects. On the other hand, play that has imaginative content can be quiet and involve little or no physical activity (Appendix to P. K. Smith 1977).

PLAY, CREATIVITY, AND INNOVATION

Numerous experiments on, and longitudinal studies of, human children support the hypothesis that play facilitates object, concept, and language use (Cropley & Feuring 1971, Dansky & Silverman 1973, 1975, Feitelson & Ross 1973, Golomb & Cornelius 1977, Hutt & Bhavnani 1972, Lieberman 1977, Sutton-Smith 1975, 1979, Sylva 1977). However, this hypothesis may not hold for all species (Symons 1978a). It is tempting to seek to relate play experience to later innovation in the arts, in literature, in the sciences, or to general behavioral flexibility and the ability to cope with the unexpected in everyday life. Play of a sort has been viewed as the root of literary creativity (C. Bartholomew 1975) and as fundamental to scientific originality (Einstein 1945, Hadamard 1945). Playfulness is termed the essential element in the style of numerous authors, musicians, and painters (see pp. 467–471).

A social science of creativity, i.e., of play at its most distinctively human, is still far off, at least according to three psychologists (Bruner, Jolly, & Sylva 1976 p. 531), who state that human artistic activity, and related behavioral tendencies generally called "creativity," have baffled psychology and anthropology. These disciplines can as yet offer few useful insights into dance, music, art, or literature. Schools of literary or aesthetic criticism based on Freudian and similar psychologies, like biologists' attempts to study the origins of art by giving captive chimpanzees drawing material (Morris 1962b), promise the general reader little but embarrassed amusement.

To my knowledge, not even the most audacious social

theorists have yet attempted sociobiological artistic criticism. However, biological ideas help explain the fact of human creativity, though they hardly begin to suggest why it should take the particular forms that it does (E. O. Wilson 1978).

It is difficult to find animal play behavior that departs from the mammalian mean in as many dimensions as does human play. Human mother-infant play, construction of toys by humans for their young, and imaginative or symbolic play are all, as far as is known, unique to *Homo sapiens*. Human play is not one exception among many, but seems to possess characteristics that single it out for scientific attention. To make biological sense of each of these characteristics is perhaps the single greatest challenge facing contemporary students of play behavior.

Edentata

New World armadillos, sloths, and anteaters have received relatively little attention from developmental ethologists despite the fact that certain armadillos produce genetically identical littermates (Taber 1945), a possible opportunity for tests of social theory (Alexander 1974, Dawkins 1976b). A hand-reared two-toed sloth (*Choleopus didactylus*) play-fought vigorously with humans, who were impressed by its quickness and agility (McCrane 1966). The giant anteater (*Myrmecophaga tridactyla*), a large, slowly developing, long-lived, solitary edentate and dietary specialist, forms enduring relationships with its single young. Only fragmentary accounts of its play behavior are available (Honigmann 1935, Schmid 1939). Honigmann's (1935) zoo-born giant anteater reared on its hindlegs and hit at objects, and it was observed manipulating various objects, including a coal shovel, chunks of coal, shoes, and a branch for hours at a time. An adult female giant anteater at Brookfield Zoo bounded around her cage and rolled on the floor after bathing, when new straw was placed in the cage, or when given grapefruit or oranges to eat (J. Fagen pers. comm.). These behaviors, all elicited by changes in the olfactory environment, may have had scent-marking functions.

Giant anteaters have certain characteristics (large size, slow development, long life, highly developed parental care) that supposedly correlate with playfulness, along with other characteristics that supposedly do not (solitary habits, dietary specialization on insects). One consequence of this fact is that whether or not giant anteaters are ultimately found to be playful, a post facto hypothesis based on exactly one of these two sets of characteristics will inevitably be "substantiated" by the data. A preferable comparative analysis would contrast play in anteater-like mammals from several phylogenies. At least three such mammals are known (aardvark, pangolin, and *Myrmecophaga*), and three more are likely (E. O. Wilson 1975 p. 183). This analysis would measure the effect of phylogenetic constraints (including, perhaps, relative brain size) on play of ecologically similar mammals.

Pholidota

Behavior of pangolins is relatively little known. One African species, *Manis tricuspis,* has been studied in confinement and in the field (Pagès 1972, 1975). This species exhibits simple mother-young play consisting of mock attacks. Juveniles becoming independent of their mothers often encounter other juveniles, and playfights may ensue. Pagès's impression, confirmed by field observations, is that juvenile pangolins form temporary bonds with other juveniles and that these mainly playful social relationships are remarkably rich and varied. Older, subadult pangolins continue to play, but the nature of their social relationships changes gradually and comes to resemble that of adult pangolins, who seem generally intolerant of one another.

Lagomorpha

Despite Hugh Hefner's successful efforts (symbolized by the familiar rabbit's head) to make public property of private play, we can cite relatively few accounts of play in real, four-legged bunnies. An excellent if brief description of play in the Mountain hare (*Lepus timidus*) indicates that in northeast Scotland these animals leave their resting places about dusk, jumping,

chasing, rolling, and sandbathing (Flux 1970). This play occurred most often during good weather "and did not occur if
the hares had seen the observer entering the hide even three to
four hours previously." European rabbits (*Oryctolagus cuniculus*)
are said to play both as kittens (juveniles) and as adults, frisk-
ing, taking little jumps into the air, twisting sideways in mid-
air, or running in wide circles, pausing to mark the ground
with the chin glands (Lockley 1974). There seems to be surprisingly little information on the locomotor-rotational gymnastics
and nonagonistic fighting of these evidently playful creatures.
The apparently regular occurrence of adult play in free-living
animals is particularly striking.

Rodentia

Most known inter-order contrasts in mammalian play also exist
between or even within taxonomic families of rodents. Some
young rodents chase and wrestle like puppies or kittens. Others
run, rear, prance, buck, and flee and chase one another like
deer fawns or horse foals. Some young aquatic rodents roll,
wrestle, and splash in the water like seal pups. There are even
rodents that chase and sandbathe like hares. Rodent interactive
play can be nonagonistic and reciprocal, or it can be virtually
impossible to distinguish from escalated and tournament fighting. Finally, young of some rodent species always play non-in-
teractively by performing solo body-twitching, by leaping vertically, and by darting about in various directions. Littermates
in these species may play in parallel or simultaneously, but they
do not wrestle with, chase, or flee from one another once a play
session begins.

Because rodents exhibit nearly all major forms of mammalian
play, it might be most correct to reverse the comparison and to
say that puppies play like rats or that hares play like salt desert
cavies. Rodent play has been neglected compared to that of
primates or carnivores. Until very recently, rodent play was
considered to be rare, unstable, and virtually absent in all but a
few large-bodied forms. Existing accounts of rodent play were
viewed with skepticism. It was felt that these accounts were
based on inexact observations of agonistic fighting, captivity ar-

tifacts, or miscellaneous agonistic behaviors related to olfactory or ultrasonic communication.

It may be difficult to describe and to correctly interpret every aspect of the behavior of small, nocturnal, fossorial animals whose chief modes of communication are chemical and ultrasonic, but not all rodents present this frustrating combination of characteristics. The behavioral variety present in rodent play presents an attractive opportunity for comparative analysis. Many rodents are easily bred in quasi-naturalistic conditions in confinement. Physiological data on environmental enrichment effects in rodent brain development suggest one possible effect of play. Scientifically rigorous experimental studies of the biology of play might well be pursued in rodents.

SCIURIDAE

Tree, ground, and flying squirrels are relatively well-studied rodents whose adult social organization may vary across closely related species and along environmental gradients. Social marmots of the genus *Marmota* roll, push, rear and box, pounce on one another, hug one another around the waist, and chase or wrestle with reversal of roles and inhibited bites (Armitage 1974, Barash 1973a,b, 1974, 1976, Müller-Using 1956, Münch 1958). Mouth-sparring, "characterized by a jabbing of the head at the head, shoulders, or chest of the other animal while the other animal does likewise" (Armitage 1974), is a tactic used in attempts to deliver an inhibited bite without being bitten or to sniff the opponent in body areas where scent glands may occur. Marmots turn somersaults (Münch 1958) in the best locomotor-rotational manner and have been known to play in captivity with humans and with dogs (Koenig 1957). Free-living young of the year, yearlings, and even adults (in the highly social Olympic marmot, *M. olympus*) play with each other. The woodchuck or groundhog, *M. monax,* perhaps the least social of all marmots, represents an unsolved mystery. Barash (1974) observed woodchuck families in the laboratory, and T. K. George (pers. comm.) made casual observations on a mother woodchuck and her young in the field over a period of several weeks. In neither instance did these trained observers record a single bout of play. On the other hand, Chuckles, a tame female woodchuck at Brookfield Zoo, often played with keepers, chasing them or rearing up to play-fight with them. Social play can occur frequently in young of many sciurid (e.g., *Sciurus* spp., *Tamias striatus, Tamiasciurus hudsonicus*), nonsciurid rodent, and nonrodent species that are not group-living as adults. It is surprising, therefore, that woodchuck play has so far proved difficult to observe.

The ground squirrels (*Citellus, Spermophilus, Cynomys*) of central and western North America play both in their first year of life and as yearlings, although yearling play may "become more aggressive" (Betts 1976) and grade into agonistic fighting as its "violence and disruptiveness" increase (Steiner 1971). Younger yearling Columbian ground squirrels play cooperatively, pouncing, rushing, pursuing and chasing each other, leaping in the air, turning somersaults, and playing leapfrog (Steiner 1971). Black-tailed prairie dogs (*Cynomys ludovicianus*) are permanently social ground squirrels whose complex, quasi-cultural land tenure system is of particular sociobiological interest (King 1955, Smith, Smith, Oppenheimer, DeVilla & Ulmer 1973, E. O. Wilson 1975). When prairie dog pups emerge from their burrows in the spring they solicit play from each other and from adults, who respond infrequently to these invitations (Smith et al. 1973).

The arboreal and terrestrial play-chases of young temperate zone tree squirrels (*Sciurus, Tamiasciurus*) are familiar to residents of Europe and North America. Sideways springs or high, bouncing leaps initiate spiral chases around tree trunks or pursuits from tree to tree (Eibl-Eibesfeldt 1951a, Frank 1952, Horwich 1972). Solitary locomotor play may also occur (Ferron 1975, Horwich 1972).

Temperate zone squirrels are well known, but form a relatively small assemblage compared with the far more diverse sciurids of tropical regions. Many of these species are easily bred and easily studied in confinement. They represent an excellent opportunity for research on animal play.

Unfortunately, nothing appears to be known about play in tropical tree squirrels and almost nothing about play in tropical ground squirrels, which also exhibit considerable faunal diversity. Ewer (1966) described the dashing and jumping play of two wild-caught infants of the African ground squirrel (*Xerus erythropus*). During play, the young squirrels cheek-rubbed on objects and uttered a vocalization that Ewer terms a play squeak.

CASTORIDAE

The genus *Castor* includes two beaver species, *Castor canadensis*, of North America, and the European *Castor fiber*. Although they commonly symbolize diligence and industry, beavers are also especially playful. Young beavers ("kits") remain with their parents as yearlings and disperse when they are about two years old. Interactive play of beaver kits includes wrestling, mutual upright pushing in which each partner places its forepaws on the other's shoulders, and aquatic rolling bouts. The highly buoyant young roll, tumble, and somersault with one another in the water and seem especially fond of these rotational movements, which they may perform repeatedly or in combination (Kalas 1976, Schramm 1968, Wilsson 1968, 1971).

CRICETIDAE

The cricetid rodents, including hamsters, gerbils, jirds, microtines (voles, lemmings, and muskrats), and allies, are known for their roles in the household pet trade and in scientific research on hormones and on population cycles.

Short-tailed voles (*Microtus agrestis*) may play when young, exhibiting typical mammalian locomotor-rotational movements (S. Wilson 1973). These voles are multivoltine (they have more than one generation per year). Juvenile voles born in the spring and early summer frequently nose one another's fur, especially in the rump area, but also on the nose and mouth, and on the back of the head. Autumn-born young do not play. When nosed, young short-tailed voles often play, "jumping, and running very fast, with uneven movements, often not in a straight line" (Fig. 3-17) (Wilson 1973).

A vole rarely attempted to involve another in play. Sometimes a vole ran or jumped toward another, barged into it, vigorously thrusted its

Fig. 3–17 Short-tailed voles (*Microtus agrestis*) playing, a juvenile nosing-contact sequence resulting in a play-jump. England. (Redrawn from Wilson 1973, © Linnaean Society of London)

nose into the other's fur, and dashed away again. Occasionally a "play interaction" developed, when two voles made play movements while nosing each other several times. (Wilson 1973)

Wilson found that a particular scent, secreted by spring young, at the back of the head, tended to stimulate play. Autumn young did not secrete this scent, although application of a chromatographically isolated fraction of spring juvenile back-of-head secretion to the back of the head of autumn juveniles produced significant amounts of play.

The golden hamster (*Mesocricetus auratus*), a somewhat phlegmatic cricetid, produces playful young at frequent intervals, and the behavior of these animals represents a socially interactive extreme of cricetid play. Hamster pups aged two to six weeks wrestle with, nibble, and bite each other gently. One pup may run several feet to initiate play with another in a special "fighting corner" where the pups may later build a secondary nest. Compared with agonistic fighting, play is slightly slower, with gentler movements, may or may not be accompanied by vocalizations, and typically terminates with a grooming bout. Thelma E. S. Rowell (1961), whose early studies of golden hamsters formed the basis for one of the first biological approaches to primatology (e.g., Rowell 1972), and Dieterlen (1959) present essentially identical descriptions of golden hamster play. However, Rowell states that play bouts are unaccompanied by chirping and squealing, whereas Dieterlen describes hamster play as "noisy." Neither author reports solo or parallel locomotor-rotational play in golden hamsters, but this behavior occurs frequently in other cricetids. "Jerky running, darting, shivery hops, and whole-body shivers," as well as brief, nonagonistic playfights occur in sandrats (*Psammomys obesus*) (Daly & Daly 1975), in short-tailed voles (*Microtus agrestis*) (S. Wilson 1973), in the tamarisk gerbil (*Meriones tamariscinus*) (Rauch 1957), and in the Mongolian gerbil (*Meriones unguiculatus*) (Ehrat, Wissdorf, & Isenbügel 1974). Ehrat et al. also describe social play of the "bump, jump, and run away" type in 3-week-old *M. unguiculatus*. One animal approaches another and noses the base of its tail. The recipient immediately flees with exaggerated jerky movements to a third animal, and the previous events are repeated. After sniffing a partner the initiator remains standing in place as its partner flees. "King of the castle" games also occur.

MURIDAE

The familiar laboratory rat (*Rattus norvegicus*) and house mouse (*Mus musculus*) are but two of many murid species exhibiting play behavior. No mammalian family could be more accessible for behavioral study, but comparative play research on the group has been uneven. Detailed information is available only for the two well-known domestic species named above and for an interesting suite of Australian murids, the conilurines.

Play in laboratory rats and in house mice offers interesting contrasts (Poole & Fish 1975). Both species' solo locomotor-rotational play includes vigorous and often-repeated body twitches, exaggerated vertical leaps, and vigorous, erratic running. Young rats box, wrestle, pounce, charge or stand, push, and crawl over a partner. They may lie on their backs, kicking and pawing at a partner with whom they had been boxing. Norway rats exhibit individual differences in playfulness, and the most playful individuals in a group of rats often play together, but it is not known whether this difference is simply due to differing individual tendencies or to nonrandom formation of particular partnerships at frequencies over and above those predicted from individual differences (Poole & Fish 1976). The playful, but noninteractive mice did not respond to rats' play invitations and did not play socially even when raised with rats (mothers and pups) as their only companions (Poole & Fish 1975). Laboratory rat play exhibits sex differences highly reminiscent of those found in cercopithecine primates (Olioff & Stewart 1978).

The conilurine rodents of Australia (subfamily Conilurinae) are believed to have had a monophyletic origin on that continent, where they underwent a spectacular adaptive radiation comprising at least four different types of social organization (Happold 1976a,b). According to field and laboratory studies of representative conilurines, *Pseudomys desertor,* the brown desert mouse, is found in sandy deserts, where it inhabits oasis-like patches having moderate microclimates. This animal does not live in permanent groups or form long-lasting pair bonds. *Pseudomys shortridgei,* the blunt-faced rat, is found in cool temperate dry sclerophyll forests. Males and females pair for long periods, but adults of the same sex do not tolerate each other in close proximity. *Notomys alexis,* the northern hopping mouse, from deserts, and *Pseudomys albocinereus,* the ashy-grey mouse, from semi-desert heathlands, form large, mixed-sex groups (Happold 1976a).

From an early age, young of all four species "twitch spasmodically and have bouts of head-jerking and body-rocking which last from a few seconds to more than a minute in duration" (Happold 1976b). Young of *N. alexis* and *P. albocinereus,* the two group-living species, playfully crawl under, walk over or leap over, ride, follow, and chase littermates and adults. These actions are characterized by role reversal, incomplete versions of adult sequences, and behavioral contexts differing from those eliciting roughly similar responses in adulthood (Happold 1976b). For these conilurines, occurrence of juvenile social play in the laboratory appears to correlate with the tendency of adults to form permanent groups.

GLIRIDAE

Little is known about play in this family, a lack of information apparently due to inadequate study. Despite the impression conveyed by the Dormouse in *Alice in Wonderland,* glirids do not sleep all the time and are capable of mustering considerable behavioral variety. Koenig (1960) observed relatively little play in the greater dormouse (*Glis glis*). Young were seen chasing and mounting each other several weeks before weaning, and a hand-reared female playfully chased the human who raised it (Koenig 1960). One-month-old hazel mice (*Muscardinus avellanarius*) playfought, pursued, and bit each other (Zippelius & Goethe 1951).

HYSTRICIDAE AND ERETHIZONTIDAE

Old and New World porcupines both appear to be relatively playful rodents judging from the limited information available. *Hystrix cristata* in captivity run around their enclosure, push off raised objects, hop and skip together, and spring high (Mohr 1965; also seen in two *H. cristata* at Brookfield Zoo). North American porcupines (*Erethizon dorsatum*) play in a manner similar to that of their Old World counterparts, rearing, whirling in circles, or wrapping themselves around a roughly porcupine-sized object (human leg, furniture) and biting it. Either form of play may lead to the other, and both are easily elicited by a familiar human or conspecific (Shadle 1944, R. A. Powell pers. comm.).

CAVIOMORPH RODENTS (CAVIIDAE, HYDROCHOERIDAE, DINOMYIDAE, DASYPROCTIDAE, CHINCHILLIDAE, CAPROMYIDAE, OCTODONTIDAE)

The domesticated guinea pig (*Cavia porcellus*) is the most familiar representative of this group of South American rodents,

Fig. 3–18 Choz-choz (*Octodontomys gliroides*) playing, play-jump and open-mouth play-face. Confined animals. U.S.A. (Susan Wilson)

which includes rat- and mouse-like forms, in addition to some unusual species that appear to represent morphological and ecological counterparts of lagomorphs or of artiodactyls. Play in these animals covers nearly the entire spectrum of mammalian play forms, from chasing and wrestling to lamblike gambols and gallops, and from continuous interactive play with frequent body contact to mainly solo locomotor-rotational exercise. Diversity of social systems in this group is at least as great as diversity of play patterns (Kleiman 1974, S. Wilson & Kleiman 1974).

As a general rule, long-legged caviomorphs, whose locomotor patterns are ungulate- or lagomorph-like (*Dasyprocta aguti, Cuniculus paca, Dolichotis patagonum, D. salinicola*), perform occasional social play-chases, whereas short-legged rat- or mouse-like forms (*Octodon degus, Octodontomys gliroides*) (Fig. 3-18) play interactively with greater likelihood and with frequent and sustained body contact. Solo locomotor-rotational play is widespread in caviomorphs and includes headshakes and body-twists, often accompanied by prancing, erratic or skittish flight following arc-shaped paths, kicking up the heels, and rearing up on the hindlegs. Paca (*Cuniculus paca*), spotted forest rodents that resemble very small deer, hop in circles, run, and stop suddenly (Pilleri 1960). Green acouchis (*Myoprocta pratti*) prance in play, leaping "repeatedly into the air," taking off as soon as they land, and often twisting their bodies while in mid-air so that they land facing in a new direction (Morris 1962a). Bell's (1830) observations that acouchi "leap occasionally in play to a considerable height, and frequently on springing from the ground to an elevation of two feet, descend on the spot where they rose" probably represent one of the first scientific accounts of rodent locomotor-rotational movements. Maras (*Dolichotis patagonum*) gallop, perform high arching leaps, kick up their heels, headshake, and run in straight lines or in circles (Dubost & Genest 1974). Among the species that engage in social play, but exhibit relatively little body contact, are the desert cavy (*Microcavia australis*), in which young frisk together, jump in the air, climb over each other, climb on each other's backs, and chase each other back and forth (Rood 1970, 1972), and the guinea pigs (*Cavia aperea* and *C. porcellus*), which run and caper or join in jumping games in which butting is a play invitation (Coulon 1971, Kunkel & Kunkel 1964). Rat-like or marmot-like riding, mutual upright postures with pushing and boxing, supine quadrupedal (stand up/belly up) positions, and mutual circling are major features of reciprocal playfighting in degus (*Octodon degus*) and in choz-choz (*Octodontomys gliroides*) (S. Wilson & Kleiman 1974). Young of the caviid *Galea musteloides* perform reciprocal chases and lunge at one another (Rood 1972).

Interactiveness (solo, companion-oriented, or social play), body contact, and content (behaviors used in intraspecific fighting and behaviors used in predator avoidance) are the chief descriptive dimensions of caviomorph play, and perhaps of rodent play in general. Little or no information is available on a fourth potentially important parameter, that of stability (defined as the probability that play will escalate into agonistic fighting or into earnest, decisive escape). Two nonexclusive hypotheses seem to be indicated. One possibility is that interactiveness, contact, content, and stability of play are explained by an animal's body form, locomotor adaptations, environment (open or closed), and risk of suffering predation. Alternatively, adult or juvenile social systems may directly determine these characteristics of play. As stated, both these hypotheses lack an adequate evolutionary mechanism. They are excellent examples of the strengths and weaknesses of correlational analysis.

Cetacea

The cetaceans, or baleen and toothed whales, are almost all social, and many species exhibit elaborate parent-offspring communication and/or cooperative hunting behavior. Their behavior at sea is difficult to observe in all but the most favorable conditions, and with one exception, the bottle-nosed dolphin (*Tursiops truncatus*), accounts of play are brief. Nothing at all is known of play in baleen whales. Sperm whales (*Physeter catodon*) tumble, surf-ride, and were seen to push and dive back and forth under a plank (Caldwell, Caldwell, & Rice 1966). Orcas (*Orcinus orca*) are social and group-hunting predators and should be interesting cetaceans in which to study play. One report (Hewlett & Newman 1968) of play in a captive orca states that the whale picked up a circular floor brush from the bottom of its pool and swam away with the brush balanced on its head. In captivity, the bottle-nosed dolphin (*Tursiops truncatus*), a small odontocete known for its unusual ability to learn and to communicate, plays with conspecifics, with humans, and with objects. Social play-chases are frequent, and there is some indication of partner preferences (Bondarchuk, Matisheva, & Skibnevskii 1976). Object play interested the first observers of the behavior of captive dolphins: animals would push their snouts through the ring of an inner tube and hurl the tubes into the air (McBride & Kritzler 1951); balance items on their snouts and drop the objects and catch them again; and repeatedly suspend small objects on underwater water jets (Tavolga 1966). Cetacean play is important because these animals' specialized body forms and aquatic habitats require general concepts of play free from implicit assumptions about quadrupedalism or terrestrialism. The study of avian play and the study of human play pose analogous challenges. Observations on cetaceans repeat the familiar history of play research in other groups. Vigorous motor behaviors representing specific adaptations to locomotion, orientation, or communication in an aquatic environment have appeared playful to observers lacking adequate background and contextual information. Breaching, lobtailing, spinning, surf-riding, riding the bow wave of ships or of larger whales, "playing around the bows" of ships, and various contact and approach-withdrawal behaviors occurring before or during mating can also be performed in a playful context. Difficulty of making detailed contextual observations at sea and resulting unfamiliarity with contrasting forms of social and locomotor activity leave interpretation of many reported behaviors in doubt.

Carnivora

Two well-known carnivore species, domestic dogs and domestic cats, share prime responsibility for our awareness of the play of nonhuman mammals. Abundant descriptive material covers all seven taxonomic families of carnivores. Excellent field studies, augmented by thorough observations of captive animals, serve to document carnivore play in considerable detail.

Play in young carnivores persists into adulthood in healthy animals and in confinement. At all ages, play behavior assumes various shapes depending on the opportunities available. Social play involves littermates, siblings, parents, or group-mates; solo, object, or item-oriented play occurs in their absence. Prey animals serve both as playthings and as food. Behavioral composition of carnivore play is similarly varied. Play includes behavioral acts used in intraspecific escalated fighting, in predator avoidance, and/or in predatory behavior. Despite this considerable diversity, carnivore play is more coherent than rodent or ungulate play because contrasts in body form and ecology, so marked in these latter orders, are present to a somewhat lesser extent in the carnivores. For this reason carnivore play, like primate play, exhibits considerable interspecific similarity.

Nearly all carnivores give birth to more than one offspring. Littermates form complex if temporary societies in which play is often the most frequent form of interactive behavior. Most adult carnivores do not form permanent groups, but there are many exceptions to this rule, including timber wolf (*Canis lupus*), blackbacked jackal (*Canis mesomelas*), African wild dog (*Lycaon pictus*), coati (*Nasua narica*), European badger (*Meles meles*), dwarf mongoose (*Helogale parvula*), meerkat (*Suricata suricatta*), spotted hyena (*Crocuta crocuta*), and lion (*Panthera leo*).

Young of those group-living carnivores in which several females breed at the same time, and young of carnivores whose offspring remain with the mother during the development of the next litter, potentially have access to a wide variety of play companions. Infant mortality can, however, reduce the number of offspring surviving to the age at which play occurs. Populations or individuals of polytokous species may in effect be monotokous if only one member of a litter is likely to survive to that age. Furthermore, even like-aged siblings or group-mates need not simultaneously have the same size or shape or occupy the same developmental stage. Asynchronous breeding or divergent parental care in group-living species, genetic variation, sexual dimorphism, or unequal access to milk, warmth, or grooming can all result in phenotypic heterogeneity at a given age both within and between litters. Parents, older siblings, or littermates at a different stage of growth or development may represent important social companions for a young carnivore that lacks playmates similar to itself. In monotokous, nongroup living carnivores, the mother (and father, if present) appear to play with their offspring to the greatest extent known outside the primates. Carnivore parents may even take the initiative in stimulating their offspring to play.

CANIDAE

Canids exhibit various degrees of sociality (Fox 1972a, Kleiman & Brady 1978, E. O. Wilson 1975). African hunting dogs (*Lycaon pictus*) are virtually obligate group-dwellers. Timber wolves (*Canis lupus*) live in packs that form, break up, and reform as prey availability and weather conditions dictate. At the other extreme, the largest social unit of some South American canids [for example, the maned wolf (*Chrysocyon brachyurus*)] is the temporary society formed by a female and her most recent litter (Kleiman & Brady 1978). If we take into account species-specific degrees of shyness and adaptability to observation in confinement, we find that canids of all species are extremely playful animals, both as puppies and (often) as adults.

The timber wolf (*Canis lupus*) can run tirelessly over long distances in pursuit of ungulate prey, which it brings down and kills using its powerful teeth, jaws, and neck (Mech 1966, Mech 1970). Both of these sets of predatory tactics are reflected in wolf play (Zimen 1972). Running play, including chasing and ambushing, is frequent both in captive and in free-living wolves. Another common form of play in wolves is biting play or muzzle-wrestling. Wolf play also includes zigzag jumps, head tossing, and side-to-side shoulder swaying as well as the relaxed, open-mouth play-face; all of these movements may initiate play (Bekoff 1974c, Krämer 1961, Zimen 1972). Wolves compete playfully for objects, play with objects by themselves, and play together for extended periods. Erik Zimen (1972 p. 224) once watched five young wild wolves at a lake in British Columbia, Canada, play almost without interruption for five hours. The wolves chased back and forth across the beach, jaw-wrestled, played "King of the castle" on a large stone in the water, and took occasional breaks to play-bite each other or play with objects.

The blackbacked jackal (*Canis mesomelas*) inhabits brush country in eastern and southern Africa. These canids live in family groups composed of a father and mother with offspring of different ages (Moehlman 1979). Pups play much like domestic dogs, chasing, tugging at, jumping onto, and wrestling with one another (van Lawick & van Lawick-Goodall 1971), but their play has some interesting additional features. For example, as observed by P. D. Moehlman in the Serengeti, an older pup may play a kind of "blind man's buff" with its younger siblings. The young pups dart in at their older sib, who will then turn and chase them. If a pup is "caught" it flings itself on the ground, belly up, pawing at its older sib (Moehlman pers. comm.). Participation in play is one of several ways in which older pups contribute to the welfare of their younger siblings (Moehlman 1979).

URSIDAE

The bear family, among whose numbers we may perhaps include the giant panda (*Ailuropoda melanoleuca*), cannot match the canids in social cooperation, but mother-cub and cub-cub play are prominent features of bear behavior both in the wild and in confinement (Figs. 3-19 through 3-26). Indeed, ursid and canid play are remarkably alike in many respects (Henry & Herrero

1974). American black bears (*Ursus americanus*) play with various items and with parts of their own bodies (Leyhausen 1949), and they wrestle with and chase each other. Bears' two major weapons are the teeth and the claws, and playfighting involves maneuvers that would permit offensive use of both of these weapons in a serious fight, as well as defensive tactics that serve to neutralize these weapons by preventing the opponent from using them or by blocking an attack. Bears seize each other's necks or muzzles, paw and lunge at each other's head, arm, or shoulder, butt heads, roll, and rear (Henry & Herrero 1974):

The typical play-fighting sequence usually exhibits four phases: initiation, pushing-pulling, biting, clawing, and finally, termination. One bear approaches another bear and initiates play by rearing, pawing, biting, head butting, or infrequently, lunging at that animal. Frequently, both animals rear and an extended pawing match follows. The purpose of pawing appears to be to knock the social partner off balance so that a biting or clawing action can be delivered. Biting intention movements and face-pawing actions are repeatedly observed during these extended rearing-pawing matches. Once the social partner is pushed or pulled into a vulnerable position, a play-bite or neck bite-hold is frequently delivered. The bitten animal then rolls over and may hind-leg claw in order to break the neck-bite hold. Once freed, both bears frequently return to the rearing-pushing-pulling phase until one of the bears once again becomes vulnerable to a bite or clawing action. The alternation between the pushing-pulling and biting-clawing phases continues on the average for about 20 sec until one of the bears closes its mouth, frequently gives a cut-off display, pivots, and flees. Flight behavior by one of the bears usually terminates the play-fighting sequence. The initiation and termination phases of the play-fighting sequences are comparatively regular. The pushing-pulling phase alternating with the biting-clawing phase shows a great amount of variability in the temporal sequencing of motor patterns.

Interesting and unusual features of polar bear (*Thalarctos maritimus*) play include maternal participation and object manipulation. Polar bears give birth in midwinter to one, two, or occasionally three or four tiny cubs in ice or rock dens. The cubs grow rapidly from rat size to the size of a small dog and eventu-

Fig. 3–19 Initial face-off.

Fig. 3–20 Mutual open–mouthed approach.

Fig. 3–21 Mutual upright position, with body contact—one bear achieves an advantageous (superior) position.

Fig. 3–22 A play-bite from a superior position.

Figs. 3–19 through 3–22 Kodiak bears (*Ursus arctos*) playfighting in water. Photographic sequence of a single playfight. Bronx Zoo, New York. (Philip Teuscher)

Fig. 3–23 Polar bear (*Thalarctos maritimus*) cubs play-wrestling. Brookfield Zoo, Illinois. (A. Blueman)

Fig. 3–24 The cub bites its mother's neck, the mother shows an exaggerated play-face.

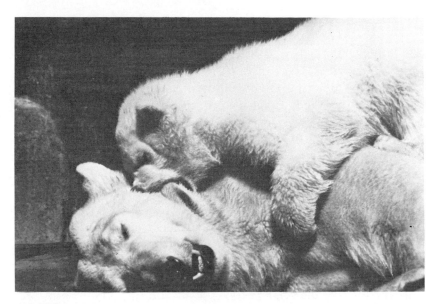

Fig. 3–25 The cub seizes a fold of skin on its mother's neck.

Fig. 3–26 The cub hugs its mother's neck, the mother shows a play-face.

Figs. 3–24 through 3–26 Polar bear (*Thalarctos maritimus*) mother-offspring playfighting. Photographic sequence. Brookfield Zoo, Illinois. (Robert Fagen)

145

ally emerge with their mother from the winter den. Often only one cub survives.

A mother polar bear will roll over on her back with a playful expression (relaxed, open-mouth grin) on her face. Her cub then lunges at her, seizes a fold of neck skin, and pulls it back and forth (Figs. 3-24, 3-25, 3-26). The two may spar with their muzzles for a while. If there is water nearby and the cub has learned to swim, the two may play water games, rolling and diving together, or the cub may bob vertically up out of the water and paw at or play-bite his mother on the bank. Twin cubs play together much as do black bear cubs, and after they can swim, a considerable proportion of their play takes place in the water. Large objects, including pieces of wood and beer kegs, are hugged, rolled over, climbed on, picked up, and thrown by captive polar bears. A bear cub may try to pull such an object underwater and to bite or even to throw it (author's observations at Brookfield Zoo, Pluta & Beck 1979).

Bear play reflects offensive and defensive tactics used in agonistic fighting. Because polar bears prey on large mammals (particularly seals) and are highly aquatic, the play of these animals includes a number of elements not found in other ursids. As Henry & Herrero (1974) point out, social play in bears seems to have diverged to a lesser extent from its canid counterpart than ursid morphological or ecological characteristics, but the polar bear may represent an exception to this rule.

Play of captive giant pandas (*Ailuropoda melanoleuca*) includes rolling movements (especially somersaults), headshakes, head tosses, body-twists, and running (S. Wilson & Kleiman 1974). The pandas that Wilson and Kleiman observed were housed in adjoining enclosures and could see, smell, and hear, but not touch each other. Judging from the play of the pandas with items including basketballs, plastic hoops, and metal kegs, pandas probably butt and swat each other in play as do American black bears (S. Wilson & Kleiman 1974).

PROCYONIDAE

The raccoon family includes polytokous, monotokous, group-living, and solitary species. Behavior of few procyonids other than the American raccoon *Procyon lotor* has been studied in any detail. Young crab-eating raccoons (*Procyon cancrivorus*) invite play by mock flight, with body-twists, and by hopping. One raccoon will throw itself on its adversary and the two will wrestle in a ring, mouthing or biting at each other's backs (Löhmer 1976). The ringtail or cacomistle (*Bassariscus astutus*), is known to play quite frequently in captivity, and the rotatable, grasping hindfoot is used in playfighting (Poglayen-Neuwall 1973, Trapp 1972) [just as in the margay *Leopardus wiedi* (Leyhausen 1963)]. According to Poglayen-Neuwall (1973), play is less evident in two other (monotokous) procyonids, the kinkajou (*Potos flavus*) and the olingo (*Bassaricyon* sp.).

Solo play is rare in olingos, who may jump up or back, spring vertically in the air "like young mice," stalk and wrestle each other, but seldom chase or flee (Poglayen-Neuwall & Poglayen-Neuwall 1965). The kinkajous at Brookfield Zoo were among the most playful mammals I observed there. In a family group of males, females, and offspring, playful wrestling, chasing, and grappling while hanging head-down from a swinging rope were almost constant features of daily social behavior. Two of Poglayen-Neuwall's (1962) kinkajous, seemingly less playful, were hand-reared, whereas a third was left with the mother.

Play in coati bands (*Nasua narica*) includes wrestling, chasing, climbing while chasing, and sparring with the forepaws. Coati play has some interesting sociobiological features (Kaufmann 1962). Siblings wrestle and spar frequently with each other and seldom have serious disputes, although they "are quite belligerent" toward other coatis. They have determined and very serious fights with any juveniles who are not family members. Older juveniles engage in playfighting, as do subadult males and adults of both sexes during the breeding season. These fights usually begin and end in silence, but vocalization occurs if they should escalate, and when play gets too rough a coati will cover its eyes with its paws. In addition, older juveniles will play with the new young of the year (Kaufmann 1962), but it is not known whether these older juveniles are playing selectively with their younger siblings.

The safest statement that can be made about procyonid play

is that more study is required, especially on the tropical species. In coatis, relationships between play and juvenile sociobiology, and possible existence of various intermediate modes of inhibited fighting, are important topics for study. The play of single, mother-reared and twin, mother-reared olingo and kinkajou kits would also be interesting to examine and to compare.

MUSTELIDAE

Mustelids, like humans, chimpanzees, ravens, large parrots, bottle-nosed dolphins, goats, red deer, and felids, seem to have adopted play as a way of life. The view that martens, badgers, and especially otters are among the champion players of the animal kingdom is widespread (e.g., Hodl-Rohn 1974, Murie 1954 p. 74, Rosevear 1974 p. 137). Although this impression cannot be far wrong, quantitative data on relative frequency, stability, and elaborateness of play in different carnivore families do not yet exist.

Mustelid diversity defies simple generalizations. Weasels and ferrets (genus *Mustela*) are long, thin, morphologically specialized predators. Included in this group are a domestic weasel (the ferret *Mustela furo*), an aquatic weasel (the mink *Mustela vison*), and a fossorial weasel (the black-footed ferret *Mustela nigripes*) that lives almost exclusively in prairie dog burrows, feeding on the young of these rodents and on other small prey.

Play in the genus *Mustela* (weasels, stoats, ermines, and mink) includes catlike stalking, rushing or springing from ambush, and body-contact interactions (Müller 1970, Poole 1966, 1978, Svihla 1931). Ermine (*Mustela erminea*) and ferrets (*Mustela furo*) solicit play from a partner by hopping stiff-legged around it, back arched and tail high (Müller 1970, pers. obs.).

Hand-raised animals without conspecific play partners and older individuals are known to exhibit considerable amounts of exercise play. Ermine run, collide with objects or walls, leap vertically in place, or roll on their backs; these behaviors can be stimulated by "interesting" substrates like leaf litter or snow, by sounds, or by the presence of the human who reared them (Müller 1970). A hand-raised mink chased her tail (Herter 1958). Object play is not often described for young captive

Mustela, who seem to prefer human hands, fingers, feet, and legs as playthings (Aldous 1940, Herter 1958, Müller 1970). It is curious that some weasels seem to reverse the ungulate and rodent ontogenetic pattern for play. In ermine, solo locomotor-rotational play is said to appear at (or persist to?) a later age than social play (Müller 1970). This observation is interesting, but requires confirmation.

The conventional wisdom that playfighting is silent may require some modification in the case of *Mustela.* Ermines "mucker" (grumble softly, purr, mutter) in play (Müller 1970), the play of two young mink "was always accompanied by a great deal of hissing, squealing and growling from both participants" (Svihla 1931), and ferrets keep up a continual chattering as they play-chase, wrestle, and explore, but hisses and squeals do not occur unless fights escalate. Ferrets also utter a vocalization ("Klaffen," Goethe 1940) when play becomes "half in earnest." Because Goethe worked at a time when sound spectrographs were not available, it is impossible to identify this vocalization with certainty, but the onomatopoeic name suggests chattering and clicking sounds.

Transitions and intermediate stages between play and fighting are easily produced in polecats (*Mustela putorius*), in ferrets, and in their hybrids. To provoke an escalated fight between two playful young ferrets, one need only toss a food item halfway between them and an escalated contest will ensue. After ownership is decided and the food consumed, play often begins again (N. Mankovich pers. comm.). "Transitional aggression" in polecats "resembles play in the patterns which it displays, [but] the whole tempo of play and transitional aggression differ and a definite sustained attack is carried out in the latter case. Threat on the part of the victim in the form of hissing, yelping, squealing, and crying are ignored by the aggressor which persists in its attack; whereas in play these vocalizations lead to an immediate cessation of attack. In addition, jumping on the opponent and rapid reversal of roles of attacker and defender are absent from transitional aggression" (Poole 1967). Furthermore, polecats can engage in ritual aggression, "a ritualized and abbreviated form of aggressive play"; males in the breeding season have severe, escalated fights; strangers fight savagely; and males experimentally caged together for protracted periods of time exhibit "companion fighting," which resembles play in that biting is inhibited and neither participant is intimidated, but differs from play in being sustained and in lacking the typical extravagant (locomotor-rotational?) movements associated with playfighting (Poole 1966, 1972, 1973, 1974). Companion fighting represents a kind of "dear enemy" phenomenon, previously described in birds (Fisher 1954, E. O. Wilson 1975 p. 273).

To what extent is the sharp contrast between play and agonistic fighting in other species due to a discontinuous distribution of behavioral types and to what extent does it result from a discontinuous distribution of social situations and contexts or of physical features of the environment (e.g., cover or places to hide)? Poole was able to produce intermediate cases artificially, using ingeniously designed encounter, rearing, and housing paradigms. This experimental approach to play appears very promising, and polecats and ferrets seem ideal subjects for such studies.

Many larger-sized members of the weasel subfamily Mustelinae are exuberantly playful, and it is unfortunate that we lack detailed descriptions of the play of martens, fishers, and their Neotropical relatives, the tayra and grison. In captivity, martens (genus *Martes*) spring on one another, tumble and wrestle, and play like cats with apples, tennis balls, and dead prey, flinging and jumping on these items (Herter and Ohm-Kettner 1954, Remington 1952, Schmidt 1934, 1943). Adult martens play at all times of the year, chasing and stalking one another; mothers play with their litters, as do fathers in captivity (provided there is enough food to go around). The mother marten's play with her offspring may continue through their first winter of life, when they are already full grown. Adults may respond to a younger conspecific's play solicitations, and males and females may play before mating (Schmidt 1943).

Play of yellow-throated martens (*Martes flavigula*) is worthy of special note. These martens, inhabitants of East and South Asia from Siberia to Kashmir, southern India, and the Indo-Malayan archipelago, are robust-bodied, supple mustelids, strikingly patterned in rich chestnut brown above, with deep golden fur on the cheeks, neck, throat, and chest. A pair of yellow-throated martens at Brookfield Zoo frequently play-chased each other around their enclosure, jumping down on one another from tree branches or rolling, kicking, and pawing. Chases often occurred aloft, where the martens dodged or jumped over one another as they ran up and down branches and tree trunks in their enclosure. If one marten succeeded in out-maneuvering the other and ran or jumped above and over it, the second marten would paw or bat at the first as it passed overhead. Two other martens housed without cagemates played running and biting games with hoses, balls, and their human caretakers. In the wild and in the home, yellow-throated martens are extremely playful (Jerdon 1874 p. 82); a litter of three young chewed each other's ears and tumbled about in fights (Roberts 1977 p. 115).

The fisher (*Martes pennanti*) inhabits mountainous and forested areas in western and northern North America. A generalized feeder, it is best known for its habit of preying on porcupines. Two fisher kits hand-raised by Roger Powell (pers. comm.) played with each other and with their caretaker, attacking his arms and wrists. The female would attack, then wrap her body around his arm, and hold firmly with all four feet while she changed her mouth grip. She liked to be thrown through the air, and she once played with a small branch, bouncing it, hitting it, and biting it. Sometimes she would jump on the branch in ways that resulted in her being thrown off, and at other times she would bat at it, making it bounce up and down, and then try to seize it in her mouth (R. Powell pers. comm.).

Tayra (*Eira barbara*) and grison (*Galictis cuja*) resemble other Mustelinae in their propensity to play even as adults, in their relative lack of interest in inanimate objects as playthings, and in the tendency of mothers to play with their young (Brosset 1968, Dücker 1968, Vaughn 1974). Young tayra may adopt a mutual upright position and stand facing each other in play, and they seem to enjoy chasing their mother's tail (Vaughn 1974). The motivational distinction between play and hunger is well illustrated by a tame grison that interrupted its play with a live mouse to eat a piece of meat, then recommenced dribbling, picking up and dropping and carrying, and flinging away and retrieving the mouse, which was apparently a substitute play companion (Dücker 1968).

The Mellivorinae (ratel or honey badger) and Melinae (true badgers of the Old and New World) are reputed to be quite playful animals. Ratels (*Mellivora capensis*) somersault and slide together (Roberts 1977 p. 125). The European badger (*Meles meles*), one of the few social mustelids, lives in multi-generation family groups in burrow systems that may be far older than any living family member, and these badgers may play somersaulting or leapfrog games together above ground in the evening (Neal 1948, Walker 1975 p. 1206). Tame badgers discover and repeatedly perform movement patterns (e.g., somersaulting, Eibl-Eibesfeldt 1950b), lie in wait to attack their human caregiver (Naundorff 1929), invite people to play by bringing an object and shaking it, and jump into the air in the midst of the ensuing tug-of-war game (Schmid 1932). Important new information about badger play can be expected to result from field studies of this species, which potentially offers excellent material for a study of play in a free-living social carnivore.

Young skunks (subfamily Mephitinae) play to some degree, but it is not known whether their play reflects their adaptations as omnivorous searchers and their divergence from other mustelids. Six-week-old striped skunks (*Mephitis mephitis*) are said to "hop about" in play (Stegeman 1937), and the play of young spotted skunks (*Spilogale putorius*) at a campsite in the Nantahala Mountains of North Carolina interested a bobcat who sat for ten minutes on a fallen log and watched them and then sauntered off (Gates 1937).

Otters (subfamily Lutrinae) are a group of semiaquatic mustelids famous for their play. One species, the sea otter (*Enhydra lutris*), is marine and almost wholly aquatic. Popular and semitechnical books on otters (e.g., Davis 1979, Maxwell 1961) contain scientifically useful accounts of otter play.

Otter play includes sliding, wrestling, "tussling" (Liers 1951), chasing, and fleeing. Smooth-coated otters (*Lutra perspicillata*) invite each other to play by springing back and forth, much like ferrets. They jump on one another, wrestle, and chase (sometimes in the water), hide from and rush at a partner, and defend particular spots from which other otters try to pull or push them. Vigorous play occurs under, on top of, or next to novel items (Hodl-Rohn 1974). River otters (*Lutra canadensis*) slide in snow or mud, and several otters may take turns climbing up and playfully sliding down a snowy stream bank (Murie 1954). The sliding habits of otters are commonly associated with play, but the otter's body plan makes land locomotion difficult and costly, and sliding is also an easy way to travel on level or on gently sloping ground. It may conceivably serve a scent-marking function as well.

Sea otters use stones as anvils to break hard-shelled molluscs. Possible developmental precursors of this behavior have been described (though not in sufficient detail to detect possible playful variants) (Ewer 1973 p. 349, Hall & Schaller 1964), but sea otter social and solo play is little known. The single cub has the opportunity to play with other cubs in the herd as well as with its mother and with other adults. Does group play occur, or is play mainly dyadic? Do mothers take an active part in the play of their offspring, and do they intervene in playfights between young?

Mustelid play poses some unsolved ethological questions. These questions include the degree to which skunk play diverges from the play of other mustelids; the possibility of highly complex forms of social play in free-living European badgers; the comparison of European badger play with that of the apparently solitary American badger; the extremely agile play of yellow-throated martens (which should serve as the ultimate test of movement notation techniques); possible species differences in object and in solo play; the experimental study of ferrets and polecats; and the sociobiology of sea otter play.

VIVERRIDAE

The Viverridae (mongooses, civets, genets, and allies) represent an ancient and diverse carnivore lineage. The group includes permanently, facultatively and seasonally group-living species (Ewer 1973, E. O. Wilson 1975). An East African array of no less than five sympatric mongoose poses a number of interesting problems in behavioral and community ecology (Dorst

1970). The dwarf mongoose (*Helogale parvula*) is a possible mammalian paedomorph (Rasa pers. comm.). On the island of Madagascar, adaptive radiation of viverrids into many different carnivore niches furnishes rich material for comparative socio-biological and ecomorphological analysis (Albignac 1973).

Genets and civets are small viverrids and are often compared with cats, whom they resemble both behaviorally and morpho-logically (Ewer 1973). However, these viverrids are generally less specialized for hunting prey by stealth than are their felid counterparts. The black face masks, omnivorous diets, and general body plans of certain civets (e.g., *Civettictis civetta, Paradoxurus hermaphroditus*) suggest that procyonids, especially the raccoons and ringtails, might be the best New World analogues of these Old World species. The relatively long muzzles and cat-like forepaws of civets and genets support the hypothesis that these carnivores have evolved past the canid level of paw-eye coordination, but that the role of the muzzle in seizing and in handling prey is greater in civets and genets than it is in felids. Play in the African civet (*Civettictis civetta*) includes doglike gripping and shaking components, and large-spotted genets (*Genetta tigrina*) head-spar and grapple (Wemmer 1977) to a greater extent than one might expect to find in a felid. On the other hand, the play of civet and genet kittens is catlike in many respects. Captive two-spotted palm civets (*Nandinia binotata*) played with small, movable objects including paper balls and corks, and they frequently lay on their sides or backs, hugging, chewing, or kicking at objects (Dücker 1971, Vosseler 1928). Adult play was relatively rare in two captive-reared small-spotted genets (*Genetta genetta*), but on one occasion when the genets were given a tree trunk they spent more than an hour playing on and around it, jumping, climbing, and running with a great variety of movements. In fact, this occasion was the only instance in which social play was ever observed between these two animals (Gangloff & Ropartz 1972). A dead rat appeared to stimulate play in the African civet; the play continually switched over into mounting attempts, and although mounting during play had not been seen previously, the civets played again the same evening without any mounting (Ewer & Wemmer 1974). Three different authors suggest that self-restraint in play appears weak in some genet and civet species: African civet play easily escalates into serious fights (Ewer & Wemmer 1974); a two-spotted palm civet was initially careful when first catching a human hand in play, but then tended to play roughly; and in a pair of the same species (adult female, immature male ten months old) the adult played so roughly with the juvenile that the pair had to be separated (Dücker 1971). The proposed intermediate position of civet and genet play on a canid-felid continuum extending from muzzle tactics to paw-eye tactics must be weighed against the view that viverrids are archaic, unspecialized carnivores whose play represents a general, primitive or ancestral pattern from which canid and felid play is derived. Although data are not yet available to support either of these hypotheses, we could consider instability of viverrid play a primitive character, while viewing its positional and tactical content as functions of both the performer's head and limb morphology and of its feeding and fighting tactics. These ideas can be tested by comparing play of young civets and genets with play in like-sized felids, canids, and perhaps procyonids.

If civets and genets may be compared to felids and procyonids, mongoose (Hinton & Dunn 1967) (subfamily Herpestinae) can be likened to mustelids. Mongoose have elongated bodies, and their legs appear quite short in proportion to their overall size. In the same sense that genets fall somewhat short of the felid extreme, mongoose never approach the extreme level of morphological or dietary specialization represented by the weasel. Moreover, many mongoose live in permanent family groups or in seasonal bands having no musteline counterpart.

The broad outlines of mongoose play are similar in several species studied to date (Dücker 1962, Ewer 1963, Rasa 1977, Rensch & Dücker 1959, Wemmer & Fleming 1974, Zannier 1965). Behavior patterns used in predation and in escalated fighting occur in play as well. Mongoose seem especially fond of hiding and chasing games played in burrows, crevices, or under pieces of paper and cloth; such "tortoise play" is known in a number of species (Rasa 1977, Rensch & Dücker 1959).

African dwarf mongoose (*Helogale parvula*) chase and reverse chase, hug, wrestle, and roll over and over together, rear and maintain a mutual upright position, and paw and lunge at each other (Rasa 1977). During solo play a dwarf mongoose springs in the air, somersaults, or chases its tail in circles. Small objects are favorite play items and elicit springing, biting, tugging, and shaking (Rasa 1977, Zannier 1965). Dwarf mongoose, like banded mongoose (*Mungos mungo*) (Simpson 1964) and meerkats (*Suricata suricatta*), play as adults; all three species live in permanent groups. Play of fathers with their children is a striking feature of social behavior in dwarf mongoose and in meerkats. However, in these two species the mother mongoose remains aloof and takes virtually no part in the family's play (Rasa 1977, Wemmer & Fleming 1974).

Among the viverrids of Madagascar are monogoose-, fox-, and cat-like forms (Albignac 1973). The cat-like fossa (*Cryptoprocta ferox,* subfamily Cryptoproctinae) is large, slowly developing, and highly playful. The mother initiates play with her single or twin cubs by biting them. They usually respond by playfighting with her, while she continues to bite and swat them playfully (Albignac 1969a, 1975). *Galidia elegans,* the ring-tailed "mongoose" (subfamily Galidiinae), is a smaller and slightly more rapidly developing animal. The single young chases its mother, chews on her tail, jumps on her, and bites her from the time it is one month old; she remains tolerant, but passive.

Like the ring-tailed "mongoose," the narrow-striped "mongoose" (*Mungotictis lineatus*) is actually a mongoose-like member of the subfamily Galidiinae, a taxonomic group whose distribution is restricted to the island of Madagascar. *Mungotictis lineatus* live in dry, seasonally severe environments; temporary social groups of five to seven individuals form during the wet summer, when the young are born, and break up during the winter dry season. Young *M. lineatus* develop more rapidly than young *G. elegans*. From two weeks of age they gallop and leap about on the ground, play with objects, and wrestle and chase nonagonistically with their mothers and with other adult group members including, occasionally, their fathers (Albignac 1971, 1976). The variety and extent of adult play behavior are greatly reduced in the falanouc (*Eupleres goudotii*) and in the fanaloka (*Fossa fossana*), two Madagascar species belonging to a third viverrid subfamily, the Hemigalinae. The single young is relatively playful, galloping, jumping, somersaulting, biting its mother's mouth, head, neck, and tail, climbing on her back, or chasing her, but she remains indifferent if tolerant. Play with food items occurs in both species, and a young *Eupleres goudotii* repeatedly pursued and captured a dead leaf, but object play is rare in the falanouc (Albignac 1970a, 1974).

Albignac's studies of Madagascar viverrids reveal levels and kinds of play that seem to vary in a regular pattern across subfamilies. *Cryptoprocta ferox* mothers actively solicit, stimulate, and cooperate in play with their young. In the mongoose-like Galidiinae, mothers are passive and tolerant when their offspring chase or attack them, and play continues into adulthood. Some of the characteristics of play in this group suggest the play of true mongoose who live in permanent groups, and it would be interesting to pursue this comparison further. Play in Madagascar Hemigalinae seems restricted to the young, but these young are still relatively playful in contrast to their elders, and more information about possible differences between play in young Hemigalinae and play in young Galidiinae or *Cryptoprocta* might aid in documenting to what if any extent juvenile play is reduced in the two species (*E. goudotii, F. fossana*) whose adults are not known to play frequently. Although it is not forbidden to view these results with the same skepticism that we earlier brought to bear on published reports of the rarity, simplicity, or even absence of play in other species, it should be pointed out that all the studies of Madagascar viverrids cited above, although based on small samples, were conducted by the same investigator, both under naturalistic conditions and in the field. For this reason the results of these studies are more nearly comparable than results obtained on many other mammalian species. They suggest some directions for future investigations of Madagascar carnivore species, some of which are rare and endangered. Variation in the degree to which the mother assumes an active role in play is particularly interesting, for few parallels can be cited in other mammals. The Madagascar carnivores represent especially good material for a study of this problem, because all five species cited have small litters (one to two offspring per birth). The principal quantities associated with the mother's role in eliciting and maintaining play appear to be developmental rate, body form and feeding tactics, taxonomic affiliation, and the degree to which the species forms groups larger than the mother-offspring dyad. Viverrid play is as underrated as it is fascinating. The role of play in the complex social organizations of group-living mongoose, the active play of dwarf mongoose and meerkat fathers and fossa mothers with their offspring, the apparent instability of play in some civets and genets, and the fondness of mongoose for "tortoise play" are all of considerable theoretical interest, suggesting major needs for future studies of play in this family of carnivores.

HYAENIDAE

The hyaenids—striped, brown, and spotted hyenas and the aberrant, termite-eating aardwolf—are all known to play to a certain degree, and field studies of the spotted hyena (*Crocuta crocuta*) (Kruuk 1972, 1975; see also van Lawick & van Lawick-Goodall 1971) reveal unusually complex social and object play in this group-living, cooperatively-hunting species. Play of spotted hyenas, wrestling and chasing on a rocky den ledge, is illustrated in an excellent short film (Apfelbach 1969a). Aardwolf play (*Proteles cristatus*) appears to consist mainly of vigorous running with rapid turns and zigzags (Von Ketelhodt 1966), as one might expect in an insectivorous animal with reduced dentition and little need to practice the skills of feeding on large, difficult, and well-armed prey. Apparent absence of playfighting in this species is most likely the result of very limited observations on an isolated animal in an enclosure. Additional study of aardwolf play is required.

Spotted hyena cubs play together, chasing and biting each other. Littermates as well as different-sized siblings form play-dyads. Adults, especially males, play often with the cubs, and adults also play among themselves (Kruuk 1972 p. 249). Four adult hyenas observed by Kruuk in the Serengeti walked from their den about 1 km to a deep river pool, where they "swam about, pushing each other underwater with their forepaws, biting, and splashing hard. Every so often they chased and ran after each other out of the pool for several hundred meters in a big circle before jumping back into the water again" (Kruuk 1972 p. 249). Adults also play together around kills, where the first hyenas to eat their fill "sometimes seem to invite others to chase them" (Kruuk 1972 p. 250).

Spotted hyenas frequently use items in social play. They play a kind of "keep-away" with sticks, bones, or fragments of prey (e.g., wildebeest tail), and a tug-of-war may result if the pursuer gets hold of the item (Kruuk 1972 p. 125, van Lawick & van Lawick-Goodall 1971 p. 174). Playful competition for objects is especially well documented in this species.

Spotted hyena play, like other aspects of spotted hyena social behavior, is elaborate compared to that of most mammals. Aspects of this complexity include gamelike play patterns, incorporation of objects, active involvement of older cubs and adults, play between adults, and apparent stability of social play. There is need for detailed study of spotted hyena play to measure possible selectivity of play companions by size and kinship, roles in initiating and maintaining play, prolongation and termination of play, and play initiation and refusal rates. A study of sex differences in spotted hyena play would be especially fascinating. Adult female spotted hyenas are larger than adult males, and the genital regions of male and female spotted hyenas are morphologically similar (Kruuk 1972). Sex differences known to characterize rhesus macaque and baboon play (Owens 1975a,b, Symons 1978a) may be found to disappear, or even to occur in reverse, in spotted hyenas.

FELIDAE

In a phylogenetic sense, cats are creatures of the middle distance. Not so far from man as to afford science easy objectivity, yet not so near as to provide fertile ground for intuition, felids were studied little by ethologists even when small rodents and higher primates had been subjected to thorough investigation. The cat family proves well suited to bridging existing gaps in man's understanding of mammalian behavior (Lorenz & Leyhausen 1973 p. xiii). An excellent recent book presenting results of contemporary feline research for a general audience is Beadle (1977).

The three dozen species of wild cats (Ewer 1973, Guggisberg 1975, Leyhausen 1973c, 1979) are specialized and dangerously armed predators that rely on stealth, use environmental cover whenever possible, and exhibit an extreme degree of pedal dexterity along with highly sophisticated paw-limb-eye coordination. In the Felidae, tactics that depend on the paw-eye system tend to replace more "primitive" or "generalized" seizing, biting, and shaking techniques in which the muzzle is the chief anatomical feature used to handle prey (Cartmill 1972, 1974). Compared with the adults of most other carnivores, adult felids have large eyes, short muzzles, and smooth, round, wide skulls that rest in a relatively erect position on the vertebral column. In all these respects felids resemble juvenile carnivores of other families and are therefore said to be paedomorphic, or juvenile-like in morphology (Fagen & Wiley 1978). These cranial-facial features are best explained as adaptations related to felid predatory tactics (Cartmill 1972, 1974, Fagen & Wiley 1978).

The highly specialized felid body plan admits little gross variation in shape, but cats have radiated along the body size dimension, forming classical size-graded arrays or "hunting sets" (Rosenzweig 1968) wherever the environment is sufficiently diverse to include sympatric felids. Because of their ecological position as top carnivores, felid evolution must contend with severe trophic-dynamic constraints in addition to the morphological constraints expressed earlier. Assuming that at most 10 to 15% of the energy reaching one trophic level reaches the next trophic level, cats preying on herbivores (e.g., domestic cats on voles, Canada lynx on snowshoe hare, or lions on wildebeest) can count on less than 0.5% of primary production. For this reason, sympatric felids are potentially in a competitive trap. Constrained to radiate along the single, not always effective (Anderson 1977, D. S. Wilson 1975) dimension of body size, subject (particularly outside the tropics) to fluctuating prey abundance (e.g., snowshoe hare or vole population cycles), felids are highly vulnerable both to intraspecific and to interspecific competition. An intermediate-sized felid sandwiched between two sympatric competitors is in an especially difficult position, for it faces a serious form of competition (MacArthur 1972) whose impact can be profound even if the intensity of competitive interactions with any single species is low.

What has the felid response been to this complex of selective pressures? In some cases, partitioning of the habitat occurs, and apparently sympatric felid species are found to hunt in different microhabitats and even to be interspecifically territorial (e.g., leopard and tiger, Seidensticker 1976). In many cases this kind of resource partitioning is suggested by coat color and pattern differences (stripes or deep, richly ocellated patterns for species inhabiting mature forest; pale rosettes, solid red or grey coats, or even red-grey polymorphism in second-growth, forest edge and scrub species). Many felid adaptations serve to broaden what would otherwise be a narrow ecological niche. Felids show considerable intraspecific size variation allometrically linked to developmental variation (Hemmer 1976). Single populations may exhibit year-to-year morphometric variation (apparently in response to varying food abundance) that often encompasses most recognized "subspecies." As Paul Leyhausen (1973b, 1979) has demonstrated, cats exhibit profound behavioral and developmental plasticity as well. For example, differential development within a litter can produce wide differences in growth and in developmental rates, so that some kittens are nearly independent and feed mainly on solid food, whereas others still suckle frequently. The effect of this differential development may be equivalent to that of staggered clutches in birds. Finally, felid morphological conservatism notwithstanding, the cat family includes a rapidly accelerating sprinter, the cheetah (*Acinonyx jubatus*); at least two species with unusual arboreal adaptations, the margay (*Leopardus wiedi*), and to a lesser extent, the marbled cat (*Pardofelis marmorata*); two species adapted to exploit aquatic prey, the web-footed fishing cat (*Prionailurus viverrinus*) and especially the minklike flat-headed cat (*Ictailurus planiceps*); a cat with morphological specializations for life on sand dunes, the sand cat (*Felis margarita*); a social, plains-dwelling, cooperatively hunting species, the lion (*Panthera leo*); the lynx (*Lynx lynx*) and the Pallas cat (*Felis manul*), both specialized for life in harsh northern or montane environments; and the facultatively social, behaviorally plastic and opportunistic domestic cat. The felid fossil record offers much additional diversity. As Gould (1977 p. 409) notes, "There may be nothing new under the sun, but permutation of the old within complex systems can do wonders."

The remarkable grace and elaborateness of feline play inspired discussions of the topic by ethologists from Bastian Schmid (1919 pp. 9–19) to Konrad Lorenz (1955 pp. 154–160) and E. O. Wilson (1975 p. 166). Cats of all ages play with other cats, with living and dead prey, with inanimate objects, and by themselves. In nature, cats are most likely to play together when young (Schaller 1972, West 1974). Mother cats of many species play with their offspring (Schaller 1967, pers. obs.). It would be interesting to describe the role of mother cats in stimulating and prolonging play with their offspring in species like margays (*Leopardus wiedi*) having 1 to 2 kittens per litter, or in those domestic cat mothers who give birth to a single kitten.

Mates may play during consortship, as I observed in one pair of jaguarundi (*Herpailurus yagouaroundi*) at Brookfield Zoo. Play between an adult male and adult female tiger (*Panthera tigris*) has been reported from the wild (account cited in Schaller 1967 p. 245). In captivity and in the wild, felids play regularly with objects (Biben 1979, Leyhausen 1979, Schaller 1972, West 1977) and with prey (Biben 1979, Leyhausen 1973b, Schaller 1968). Some "play" with living prey may serve to test (Schoener 1971) an unfamiliar and possibly dangerous item. A second explanation of play with prey may apply in certain situations: an adult felid may catch animals simply in order to play with them. This hypothesis seems plausible for cats whose need for play exceeds the opportunities provided them by conspecifics or inanimate objects and is consistent with my impressions of well-fed domestic and captive wild felids playing with living and dead prey. These cats can play with other adults if given the opportunity. However, a relatively defenseless bird or rodent is a less dangerous partner, especially if it can be induced to escape from the cat. A play item of this sort offers training and response-contingent stimulation relevant to predatory ability. It offers a reduced risk of injury as well as freedom from the social constraints associated with adult cat-cat play. In fact, Barrett & Bateson (1978) and Wasser (1978) suggest that older juvenile cats perform two structurally and causally distinct types of play, one to train predatory skill and one to train skill in intraspecific aggression. If this hypothesis is true, adult cats should gain specific benefits from playing with prey, and such behavior need not be interpreted merely as a motivational mistake. However, the issue of play with prey may be somewhat more complex than this simple hypothesis suggests. Predatory specializations are reflected in play in a number of ways, but these variations are not necessarily evidence that play specifically trains *predatory* skill, because even phylogenetically new predatory behavior may also be used in intraspecific escalated fighting (Leyhausen 1963). The cheetah paw-slap (Prater 1935) and margay hindfoot push and grasp (Leyhausen 1963, pers. obs.), as well as a number of specialized techniques that lions use to bring down big game (Leyhausen 1979, Schaller 1972), all appear in social play of these species. Furthermore, play with prey has one additional complexity that cat aficionados may accept more easily than most scientists. When an oncilla (*Leopardus tigrinus*) killed a brown rat after a difficult fight, a veritable orgy of play followed. The cat vigorously flung, chased, sprang on, pawed, and kicked at the dead rat for half an hour (Leyhausen 1953, Leyhausen 1979). This behavior is reminiscent of humans laughing and slapping each other in relief after escaping from a very tense or dangerous situation. Leyhausen (1979) finds this behavior, which he has termed "play of relief," widespread in felids.

By far the best-studied felid species is the domestic cat *Felis catus*. Recent studies of young domestic cats at play reveal previously undescribed complexity and variety (Barrett & Bateson 1978, Leyhausen 1979, West 1974). From kittenhood to adolescence and even into adulthood, domestic cats play together (Fig. 3-27). A typical play encounter begins when one kitten crouches, holds its head near or against the ground, brings its hindlegs underneath its body, extends its tail straight behind or curls it around its body, and treads with its hind feet or sways its hindquarters from side to side. The kitten then springs forward, pushing off with its hindlegs. A second kitten, at whom the spring may have been more or less accurately aimed, falls or rolls belly-up with all four limbs held in a semi-vertical position, and with its tail held straight back and possibly moving back and forth. The hindlegs reach and the front legs reach at or paw at the first kitten, who stands up near or over its playmate, its head oriented toward the playmate's head and neck region.

Fig. 3–27 Domestic cats (*Felis catus*) playfighting. Cat (left) the full-grown son of other cat, hugs her and kicks with his hindfeet. (Robert Fagen)

While in this stand–over, belly-up position, the two kittens hold their mouths open, exposing their teeth. They may paw at and attempt to mouth each other. The open-mouth display is a typical example of the mammalian play-face; kittens hold their mouths visibly open in play, and this pattern can be performed independently from the mouthing attempts that occur. Two kittens at play may exchange roles, or both may lie on their sides, pawing and kicking at each other.

Play may also begin when one kitten arches its back, raises its tail into a recurved position over the back, dorsiflexes its neck, and walks sideways toward or around another kitten. The conspecific replies in kind, sometimes leaping vertically with back arched and tail curled.

In kittens several months old, a belly-up position may elicit the following "face-off" response: the second kitten sits and hunches its body toward its partner while moving its tail back and forth; it lifts a front paw and moves it in the direction of the other kitten. Two kittens may face off simultaneously. Or they may rear back on their hindquarters and stand erect bipedally while striking up and out with the front paws.

Play encounters usually result in a chase in which one kitten runs after another. At times the roles of pursuer and pursued reverse. A leap into the air may also terminate a play bout. (The foregoing account of domestic kitten play is based on Barrett & Bateson 1978, West 1974, and pers. obs.)

The behavior just described is "typical" only in a probabilistic sense. Although play may begin with a spring or with a side-step, it may also begin with a belly-up, stand-up, vertical stance, or chase. In fact, fewer than one-half of the play bouts observed by West (1974) began with the "typical initiation" spring. And although a springing kitten usually elicits a belly-up response, its playmate may sidestep or adopt a vertical stance almost as frequently as it will turn belly-up.

These social play behaviors of domestic kittens are shared by their wild relatives (Lindemann 1955) and seem not to have been materially altered by domestication. On the other hand, play with objects, frequently seen in domestic kittens and in grown cats, is difficult to observe in small, wild felids, even in captivity, although it may occur when the observer is absent: a student of wild felid behavior may present toys that are completely ignored during the observation session, only to return the next day and find the objects strewn around the animals' cage!

Under certain circumstances, domestic cats (and their wild relatives) will play with a prey item as they might with an inanimate object. Occurrence of this behavior is influenced by the cat's level of hunger and by the type of prey item (Biben 1979).

Although mother-young groups form complex if temporary societies in all felid species, the domestic cat and the lion (*Panthera leo*) appear to be the only truly social cats. Social play of lion cubs (Figs. 1-1, 3-28, 3-29) was observed in the field by George Schaller (1972 pp. 159–161):

Social play usually involves two animals although three or more may join briefly in such activities as chasing each other. Several gestures are used to convey a readiness to play; these include an approach with exaggerated bounds, lowering the front part of the body, rolling on the back, or direct contact in the form of a nip or a push with the head. The play face is often assumed in such situations. Because play pat-

Fig. 3–28 Mouthing.

Fig. 3–29 Use of paws in defense.

Figs. 3–28 and 3–29 African lion (*Panthera leo*) cubs playfighting. Brookfield Zoo, Illinois. (A. Blueman)

terns are so variable, I found it difficult to classify them into discrete categories, but four main types of social play were recognized—chasing, wrestling, pawing, and stalking and rushing. These were used both by cubs and adults except that the latter also exhibited some of these patterns in other contexts, such as hunting.

1. Chasing. One animal bounds after another, swats at its legs, and tries to grasp its rump with the forepaws. Occasionally it runs beside the other, puts a forepaw on its shoulder, and pulls it down. In such a situation the fleeing animal sometimes collapses in an exaggerated manner by falling on its side with a thump. Wrestling and pawing often follow a chase, but some end without further contact when the fleeing animal turns and faces its pursuer. Chasing is especially prevalent when the group travels.

2. Wrestling. While one cub lies on its back, another is on top of it, mauling the throat, pulling a leg. Contact is often brief, but occasionally two animals wrestle for several minutes, rolling over and over as they paw and bite. This type of play is common when groups rest.

3. Pawing. Animals sit or stand facing each other and swat with a forepaw; occasionally they rear up on their hind legs and slap; or they lie on their sides and paw each other gently.

4. Stalking and rushing. On seeing another animal, a lion may crouch and wait and then advance in the stalking or crouching walk. On about half of the occasions no further actions occur, but on the other half the lion rushes as far as 10 m. Usually contact is brief, a simple token touch on the rump with a paw, but sometimes it ends in a pawing or wrestling bout.

The domestic cat and the lion are both highly social species. Cooperative hunting and rearing of offspring among lions (Bertram 1976, Schaller 1972) are no more impressive than the evolved behavioral adaptations of domestic cats to long-term cooperation and elaborate communication with other domestic cats (Fagen 1978b, Laundré 1977) and with humans (Leyhausen 1979) in environments favoring such cooperation. In contrast to the two species just cited, most wild cats are believed to form relatively impermanent social groups. The only enduring social unit in most Felidae consists of a mother and her offspring from one or more litters (Kleiman & Eisenberg 1973). (Of course, so little is known about most felid species that this impression may prove erroneous.)

Play in the Felidae reflects adult foraging techniques rather than adult social specializations and bears no apparent relation to grades of adult social structure. This generalization, which appears to hold whether we consider absolute frequency (Barrett & Bateson 1978, Fagen pers. obs., Schaller 1972, Wasser 1978), sexual dimorphism (Barrett & Bateson 1978), interactiveness, complexity, or stability of play, directly contradicts the widely believed prediction, based on the cohesion hypothesis, that the amount of social play performed by young should increase with the sociality of adults (Bekoff 1974c, S. Wilson 1973; see also p. 78 and Chapter 5).

The degree of adult sociality in tigers (*Panthera tigris*) differs from that of lions at least as much as that of woodchucks from Olympic marmots, red foxes from timber wolves, or orangutans from chimpanzees. However, play of tiger cubs (Schaller 1967, Wasser 1978) differs little in broad outline from that of lions. Tiger cubs stalk, rush toward, ambush, and pounce playfully on each other; they spar and wrestle, rolling over one another on the ground and mock-biting each other's head, shoulders, or other body parts; or they may face each other, rear on their hindlegs, and swat with their paws or grapple.

Margay (*Leopardus wiedi*) play illustrates independence of felid play from adult patterns of social organization. Young of the margay, an arboreal specialist with single-kitten litters, long gestation, and delayed postnatal development, play in ways that accurately reflect their species' adaptations to arboreal life. Casual observations of the development of play in a single, hand-reared female margay kitten at Brookfield Zoo indicate possible evidence of the mother's role in stimulating play of her single kitten. These indications must be confirmed by study of mother-reared margay kittens. I believe, however, that a margay mother has an especially important role in the development and maintenance of her kitten's play. This view is based on the observation by the keepers who raised the kitten that, although healthy, she appeared rather serious during her first weeks of life compared to domestic cat and Pallas cat (*Felis manul*) kittens (Fig. 3-30) at comparable developmental stages. At this age the kitten rarely solicited play from humans or played on its own. Reasoning that in the wild a single margay

Fig. 3–30 Pallas cat (*Felis manul*) kitten inviting human to play. Brookfield Zoo, Illinois. (Johanna Fagen)

kitten might well be stimulated to play by its mother, the keepers began a regular program of play with the kitten in which they tickled her, grasped her, touched her, and made play-faces at her in addition to their customary frequent feeding and grooming activities with her. The kitten's response to this treatment was a sudden increase in playfighting (Figs. 3-31, 3-32, 3-33). Whereas either the quantity or the quality of the additional stimulation (or even maturation) could have been responsible for this apparent change in the kitten's behavior, her response was striking.

On several occasions when I observed the kitten, she played in three distinct ways. She playfought with my hand, she slid down my arm head first, and she jumped from my shoulder to a burlap bag hanging from the wall of her cage. She would also hang upside-down from the bag and swing back and forth.

At three months of age the kitten was introduced to her older sister (aged seven months and also hand reared) in Brookfield Zoo's large, naturalistic exhibit for margays. Within a few

Fig. 3–31 Margay (*Leopardus wiedi*) kitten inviting human to play. Brookfield Zoo, Illinois. (Johanna Fagen)

weeks of their introduction the two margays were playing together (Fig. 3-34). They wrestled in a ball on the ground, lying on their sides and pushing or grasping each other with their mobile hindfeet exactly as described and photographed by Paul Leyhausen (1963).

The familiar stand-up/belly-up play position of domestic cats is also found in many wild felids including lynx (*Lynx lynx*) (Lindemann 1955), tigers (Schaller 1967), and European wildcats (*Felis silvestris*) (pers. obs.). In my experience it is uncommon in margay play. Leyhausen's (1963) photograph of two playfighting margays lying on their sides in ventro-ventral orientation, grasping each other with the hindfeet, illustrates the typical margay playfighting position. Apparently a unilateral stand-up position is disadvantageous because it prevents the

Fig. 3–32 Margay (*Leopardus wiedi*) kitten playfighting with human. Brookfield Zoo, Illinois. (Rodger Phillips)

Fig. 3–33 Margay (*Leopardus wiedi*) kitten play-biting human. Brookfield Zoo, Illinois. (Johanna Fagen)

171

Fig. 3–34 Margay (*Leopardus wiedi*) orienting to partner's tail in play. Brookfield Zoo, Illinois. (Robert Fagen)

margay from fighting with its hindfeet, so that when one cat falls on its back soliciting play in typical cat fashion, the other immediately rolls over on its side and begins kicking, pawing, and grasping at its partner using both front and hind paws. However, margays, like many other felids, will adopt mutual stand-up positions in play, grappling, batting, and biting at each other (Petersen 1979, pers. obs.).

The margay sisters at Brookfield Zoo play-chased in addition to their play-wrestling. Rapid arboreal chases interspersed with wild leaps were frequent. One margay would hang upside down 50 feet above the ground and play-bite or hit out at the other as it leapt by. These aerial leaping games evolved into elaborate exercises of margay locomotor skill. For example, the younger kitten would stand on a branch and watch as her sister prepared to leap through the air onto her. When the older kitten jumped, the younger would wait until the last possible moment to jump away, leaving her "surprised" sister clinging precariously to the branch just vacated by the intended "victim." These chain-reaction jumps were one of the most dramatic features of margay social play.

Petersen (1979) describes social and solo play of his male and female margay. The margays would drop a ball from a chair and chase the ball as it bounced away, toss objects in the air and catch them, or dribble objects across the floor using the front paws. Flinging objects held in the mouth high in the air and then catching or chasing them was also a favorite game of the Brookfield Zoo margays.

When not fighting or in estrus, felids are generally quiet animals. Their play is usually quiet as well, but there are exceptions to this rule. Vocalizations have the effect of controlling play, and it is always possible for an observer to detect that a margay play-bout is becoming rough when the animals begin making a low growling or rough purring sound whose intensity may rise if roughness continues (Petersen 1979, pers. obs.). When play actually escalates, domestic cats, Pallas cats, and European wildcats may hiss at a partner (pers. obs.). In play, snow leopards (*Uncia uncia*) (Freeman 1975), margays (Petersen 1979), and Pallas cats (pers. obs.) may jump heavily and noisily down to the floor from a shelf or nest box, but leaping is always quiet and light in these species when they are stealthily moving about their environment.

Cats' affinity for water varies, both among individuals and interspecifically. The most aquatically specialized felids, the fishing and flatheaded cats, as well as leopard cats (*Prionailurus bengalensis*) and tigers (*Panthera tigris*), play-wrestle and splash together in water (Birkenmeier & Birkenmeier 1971, Muul & Lim 1970, Ulmer 1968, pers. obs.).

Both from a scientific and from an aesthetic point of view, the study of felid play is one of the most satisfying areas of play research. There is great need for additional field studies of felid play and for comparative observations of mother-reared kittens in confinement.

Pinnipedia

Sea lions and fur seals, walruses, and "true" or earless seals form the order Pinnipedia. Pinniped social organization varies both intraspecifically and interspecifically in regular fashion as functions of diet, of temperature and of the type and availability of breeding substrates (Marlow 1975, Peterson 1968, E. A. Smith 1968, S. Wilson 1974a). For example, Australian sea lions (*Neophoca cinerea*) breed on rocky beaches, often fight with conspecifics, attack strange young, sleep at some distance from one another, and develop slowly as pups, whereas New Zealand sea lions (*Neophoca hookeri*) breed on sandy beaches, tolerate strange young, sleep in contact with conspecifics, and develop rapidly as pups (Marlow 1975). *N. hookeri* pups, but not *N. cinerea* pups, suckle from several different lactating females ("milk-stealing," Marlow 1975 p. 220).

OTARIIDAE

The eared seals (sea lions and fur seals) are social pinnipeds, and studies of their behavior in localities from California to the Antarctic reveal a general social developmental pattern of early mother-pup social relations followed later by integration into a mixed-age or peer group "pod." This pattern is subject to great variation. Parent-offspring bonds may form only briefly (*Callorhinus ursinus*, Bartholomew 1959) or may endure for months (*Neophoca cinerea*, Marlow 1975). Mother-offspring play occurs in several species, including the New Zealand fur seal (*Arctocephalus forsteri*) (McNab & Crawley 1975) and the Steller sea lion (*Eumetopias jubatus*) (Farentinos 1971). Playfights between otariid pups are pushing, lunging, and shaking contests in which the goal appears to be to bite the opponent without being bitten (Gentry 1974). Chases may also occur. In the Antarctic waters around Amsterdam Island, *Arctocephalus gazella* pups 3 to 4 months old play "keep-away" games using seaweed lures. One pup will seize a piece of kelp in his teeth, then rapidly dive or swim away, pursued by several playmates who attempt to wrest the item from him. If one challenger succeeds in doing so, he flees with the item and the game continues (Paulian 1964). Object play is also known to occur in the wild in other otariids (Farentinos 1971).

ODOBENIDAE

Walrus behavior is less well known than that of other pinnipeds. E. H. Miller carried out a field study of walrus ethology (1975a,b) and was able to observe play only infrequently. Play often occurred underwater and consisted of mock fighting bouts (Miller 1975a).

PHOCIDAE

Of the three pinniped families, true seals exhibit the greatest social variety. Some relatively little-known species, such as the leopard seal (*Hydrurga leptonyx*), appear "nearly solitary" (E. O. Wilson 1975 p. 464). Others, including the grey seal (*Halichoerus grypus*) and elephant seal (*Mirounga leonina*), form social groups in which males fight each other and defend groups of females during the breeding season. *Phoca vitulina*, a phocid of northern hemisphere cool temperate and subarctic waters, inhabits the coasts of Britain, Ireland, and Iceland (common seals), New England and eastern Canada (harbor seals) (Figs. 3-35, 3-36), western North America (Pacific harbor seals), and Japan (Kurile Island harbor seals). Susan Wilson, who has analyzed social behavior in several *P. vitulina* populations in Europe and North America (S. Wilson 1974a,b, S. Wilson & Kleiman 1974), describes the play of the subspecies *P. v. vitulina* along rocky shores as follows (pers. comm.):

The programme offered by common (harbor) seals along rocky shores will usually include a divertissement by pairs of small seals.
At first the scene is still and empty. Then, as the tide begins to ebb and fronds of seaweed attached to the tidal rocks just become visible,

Fig. 3–35 Harbor seal (*Phoca vitulina concolor*) mother-pup play, showing mother delivering a play-bite to the nape of pup's neck. Sable Island, Canada. (Susan Wilson)

Fig. 3–36 Harbor seal (*Phoca vitulina concolor*) juveniles in typical play-fighting position. Sable Island, Canada. (Susan Wilson)

two round heads appear, as if from nowhere. Silently they approach, meet, and submerge into privacy for a few moments. When they reappear, each seal may have its nose to its companion's tail as together they form a continuously splashing wheel, leaping backwards in and out of the water. This is their theme. By way of variation they may twist around each other's face and neck, splashing not quite so enthusiastically. From time to time they will break apart and gaze around as if each has quite forgotten the other. Then one will follow the other at a discreet distance, all of a sudden catch up, and leap on top of its partner again. The climax of their display comes when the two lie quietly at the surface with their mouths or cheeks in contact, eyes closed, flippers round each other.

Several pairs may play in this way until finally they crawl out on to the rock exposed by the tide, and fall asleep.

Sometimes there is an unpredictable encore, when all the herd join together for a spirited conclusion. This might be initiated by one seal leaping on to a rock, looking expectantly back into the water, and diving in again as soon as a companion swims beneath it. Suddenly seals all around are leaping and streaking round the rock, blowing bubbles, splashing, tossing seaweed into the air, and continually leaping out on to the rock only to tumble back into the water when another seal passes by and nudges it. The herd of youngsters plays this game in relays, newcomers joining in as the starters tire one by one and clamber higher on to the rock where they remain. Gradually the water empties of seals and the tidal rocks become shapeless and seemingly dead to the world.

On Sable Island, an isolated sandy locality in the Atlantic Ocean off Nova Scotia, play patterns of young harbor seals (*P. v. concolor*) often differ strikingly from those employed by young conspecifics inhabiting rocky shores. Juvenile Sable Island harbor seals are more individualistic (or perhaps inventive?) in their play than are common seals (S. Wilson & Kleiman 1974), with each play bout assuming a distinct and unique character. Different play patterns occur in different relative proportions, regardless of whether play took place on the sandy beach or in the water. These very diverse patterns included repeated porpoising and splashing, lunging and rolling, bouncing, flopping, and head tossing. One adolescent even leapt out of the water and rolled on its own axis while airborne (S. Wilson & Kleiman 1974 Appendix 4).

Play behavior and breeding systems of grey and common seals both appear to differ, although the exact relationship between the two is not yet clear. Male grey seals compete for space on shore, the male with the best shore space will attract and keep the most females, and one male may therefore inseminate many females. Common seals mate in the water without overt male-male competition. Grey seals are highly sexually dimorphic, common seals less so. In principle, then, the members of grey seal mixed-age play groups could be descendants of a single male, whereas common seal young play with individuals less closely related to themselves.

Common seals, unlike grey seals, play in peer groups; grey seals appear to play only in dyads, which may be of mixed age. Grey seals and Sable Island harbor seals, but not common seals, play on the beach using behaviors that resemble adult agonistic patterns. Both species' dyadic play in the water is similar, combining muzzle-to-body and body contact with exuberant somersaulting movements (S. Wilson 1974a).

Proboscidea

The African (*Loxodonta africana*) and Asian (*Elephas maximus*) elephants are sole survivors of a diverse and spectacular order whose nearly 400 species included the mastodons, mammoths, and other giants as well as pygmy elephants standing less than 1 m high (Accordi & Colacicchi 1962, Hooijer 1967). African and Asian elephants form matriarchal family units, several of which may be associated in larger kinship groups or clans (Douglas-Hamilton 1972, Douglas-Hamilton & Douglas-Hamilton 1975, Laws, Parker, & Johnstone 1975, McKay 1973).

Play in *Loxodonta africana* begins at an early age and may, at least in males, persist into adulthood. Calves of all ages form constantly changing play groups that include members of different families. They push, roll, gambol, and use their trunks to slap and wrestle (Bolwig 1965, Douglas-Hamilton & Douglas-Hamilton 1975, Kühme 1961, 1964, Sikes 1971). According to Douglas-Hamilton & Douglas-Hamilton (1975), when partners of uneven sizes play, the stronger matches its strength to that of the weaker. Mothers or adolescent females may intervene in playfights in support of young calves if the calf is knocked down by a larger playmate or if the play becomes too rough and threatens to escalate. A young calf's earliest play consists of solo charges and rushes. These locomotor routines occur within the safe circle of the calf's family, but they may also extend the calf's range to nearby areas, exposing the calf to accidental injury or to predation. In such cases older female members of the calf's family seem to respond quickly to the danger.

Play of Asian elephants is less well known. Its major features appear to correspond to those previously described for African elephant play and include chasing, wrestling with heads, mouths, and trunks, and other forms of mock fighting (McKay 1973). Precopulatory behavior resembling juvenile head-wrestling is known (Eisenberg, McKay, & Jainudeen 1971). Adults may intolerantly kick playing infants (McKay 1973).

Sirenia

Dugongs and manatees are seldom-studied aquatic mammals. Little is known of their play. A captive female dugong (*Dugong dugon*) played in its tank with a human diver who would hold onto its tail as it towed him through the water. The dugong sometimes rolled over during these performances in apparent attempts to shake the diver off. If the diver lost his hold the dugong "would wait for him to come up and take hold of its tail, whereupon it would repeat the whole performance again" (Oke 1967). Field observations of manatees (*Trichechus manatus*) in cold-induced aggregations (Moore 1956) indicate that immatures may nuzzle, make rolling surface dives, and splash while apparently wrestling together. Calves did not play during these periods of severe environmental stress. Unfortunately all observations were made during a cold wave when manatees sought warmth in a heated power plant outflow, and conditions of stress are not favorable for observing play in any species.

Hartman (1979) observed congregations of manatees in warm-water refugia on the west coast of Florida (U.S.A.). Despite harassment by powerboats and by divers, and despite occasional cold stress, manatee populations in this area actually seem to be increasing. Behavior of calves and juveniles, easily studied at close range, includes several forms of play. Playing calves would twist, tumble, and barrel-roll through the water, and "one calf completed several minutes of play by 'rocketing' to the surface so that his chest and flippers broke water" (Hartman 1979 p. 110). Social play involved nibbling as well as "mutual kissing, mouthing, embracing, bumping, nudging, and chasing." Juvenile males engage in caressing play with juvenile and adult females; during these play sessions the young males mouth and embrace the females.

Some social behavior that Hartman (1979) called "play" may actually have been social grooming, a behavioral category not discussed by Hartman. Manatee hide undergoes "constant epidermal sloughing" and "supports an assortment of plant and animal associates" (Hartman 1979 p. 62). Manatees "make use of their flippers to scratch but are restricted by the limited reach of the appendages and must confine scratching to the chest, ventral neck, and head regions. Irritations on the remainder of the body are relieved by rubbing on objects" (Hartman 1979 pp. 87–89). Some well-habituated animals "were attracted to divers and actively solicited caresses from them." One young cow "gave the impression that she returned . . . as much to be scratched by divers as to linger in the warm water" (Hartman 1979 p. 130). To me, these observations suggest that the manatee social behavior described as "play" by Hartman may include instances of simple social grooming, as well as gentle contact play consisting of playful variations of grooming behavior that are not primarily behavioral adaptations for cleaning a conspecific's skin.

Perissodactyla

Equids (horses, zebras, and wild asses), tapirs, and rhinoceroses are hoofed mammals having an odd number of toes. The equids are open-country, group-living grazing forms and feed on abundant, but coarse and low-quality plant matter. Some equids, like the feral asses of Death Valley (U.S.A.) and Grevy's zebra (*Equus grevyi*), are desert-dwellers. The tapirs are solitary inhabitants of tropical forests, whereas rhinoceroses form shifting social aggregations that may display complex, if temporary structure.

Play in horses (*Equus caballus*) (Fagen & George 1977, C. C. Hamilton 1967, Schoen, Banks, & Curtis 1976, Tyler 1972) involves solo running decorated by locomotor-rotational movements (Figs. 1-6, 1-7, 1-8, 1-9), as well as nonagonistic chasing and various forms of body contact similar to those occurring in olfactory investigation and serious fights. Foals in their first year of life frequently approach and sniff each other. Mutual sniffing may turn to play when one partner takes action to evade the other's sniffing attempts, bobbing and weaving its head and neck. The foals may neck-wrestle as each attempts to grip the other's face and mane, and they may rear or mount from the side in an apparent attempt to gain superior position. Often when rearing occurs the second foal may rear as well, neutralizing the momentary positional advantage gained by its partner. While rearing, they can push each other or even lock legs. Or their interaction may break off into a chase, in which the roles of pursuer and pursued frequently reverse as each partner tries to seize the other's mane in his teeth. Especially in male yearlings, playfights may escalate, becoming one-sided. Previously quiet, the interaction becomes noisy as both horses squeal and snort. If one horse gains an advantageous position and pursues this advantage, the other kicks viciously in defense.

Young tapirs' play resembles young horses' solo running near their mothers. The tapir gambols and gallops, tossing its head. The mother rarely takes part in this activity, although her offspring may push her lightly with its head in an apparent attempt to elicit play (Frädrich & Thenius 1972, Richter 1966).

Rhinoceros calves at play are pleasingly agile (Leuthold 1977). Black rhinoceros calves (*Diceros bicornis*) gallop back and forth near their mother, butt her (Schenkel & Schenkel-Hulliger 1969), and captive-bred calves have been observed pushing and chasing adult males (Dittrich 1967). Square-lipped rhinoceros (*Ceratotherium simum*) calves run around in the vicinity of the mother, galloping back and forth or in circles, and frequently coming to sudden halts. Social play in this rhinoceros species consists of horn-wrestling bouts that are often preceded by

Fig. 3–37 Plains zebra (*Equus burchelli*) foals playfighting. Confined animals, Smithsonian Institution, National Zoological Park, Conservation and Research Center, Front Royal, Virginia. (Joel Berger)

head-tossing or prancing gestures. Compared to agonistic fights, play was easily interrupted, involved relatively gentle horn-body contact, and was quiet (Owen-Smith 1973, 1975).

Play of Asian rhinoceros (*Rhinoceros unicornis*) has been observed in two zoos. A calf and its mother played, pushing head to head (Tong 1962); a mated female tossed her head and ran from the male, inviting him to chase her (Buechner, Mackler, Stroman, & Xanten 1975).

Play in perissodactyls (Figs. 3-37, 3-38) follows a uniform pattern: solo locomotor-rotational exercise in the very young, with the mother a more or less passive participant, followed at later ages by interactive play when appropriate social companions are available. Play among subadults may be a regular occurrence. In horses and in square-lipped rhinoceros, play of older animals may escalate into agonistic fighting.

Fig. 3–38 Feral domestic horse (*Equus caballus*) group play involving an entire band. Nevada. (Joel Berger)

Artiodactyla

Artiodactyls are an order of ungulate or hoofed mammals having an even number of toes, a character that distinguishes them from perissodactyls. Swine, hippopotamus, giraffes, deer, antelopes, sheep, and goats are only some of the better-known members of an order in which morphological, ecological, and social adaptive radiation has stimulated analysis of social evolution in its ecological context (Bradbury & Vehrencamp 1977a, Estes 1974, Geist 1974, 1977, Jarman 1974, Leuthold 1977).

Artiodactyl play includes solo locomotor-rotational running (often in arcs, circles, or figure–eights or with rapid turns at high speed), leaping, kicking, and jumping with headshakes, head-tosses, or body-twists. These and other characteristic and well-known artiodactyl patterns of movement are commonly termed gambols and capers. Indeed, the etymological root of the word "caper" is the Latin *capra,* goat. Solo gambols and capers are not the only form of artiodactyl play, however. Interactive play may take the form of chasing and reverse chasing and may include body contact in the form of pushing, butting, neck-wrestling, mutual rearing, or riding. This form of play is especially frequent in group-living species and may grade imperceptibly into low-level agonistic fighting. Social interactions involving physical contact between young male artiodactyls may present serious difficulties to typological concepts of play, as they sometimes combine structural and functional characteristics of play with clear indications of low-intensity agonistic fighting and have been hypothesized to determine position or rank in animals just entering subadult society.

Manipulation of or contact with items other than conspecifics constitutes another interesting set of artiodactyl play or playlike behaviors. Inanimate objects, plants, or parts of plants may be tossed, shaken, pawed, pushed, jumped over, or butted repeatedly. As in other mammals, these item-oriented play behaviors may substitute for conspecifics, may elicit play from a conspecific, may be designed for physical training, and may be confused with or even grade into autogrooming, scentmarking, or investigation.

SUBORDER SUINA (FAMILIES SUIDAE, TAYASSUIDAE, AND HIPPOPOTAMIDAE: PIGS, WARTHOGS, PECCARIES, HIPPOPOTAMUS)

Several days after they have established their order of preferential access to teats by engaging in savage, unrestrained, and damaging fighting (McBride 1963), domestic piglets (*Sus scrofa*) begin to play. One piglet may play solo in the locomotor-rotational, running and leaping manner typical of ungulates and of ungulate-like rodents. Piglets also play together, approaching each other and meeting head-on or moving shoulder to shoulder for a pushing match. They ride, chase, and flee each other (McBride & James 1964, Ian Taylor pers. comm.). Wild swine exhibit forms of play very much akin to those described in the highly domesticated breeds of *Sus scrofa*. For example, young of the warthog (*Phacochoerus aethiopicus*), an African suid with a broad, grotesque face and great curved tusks, head-toss, spring, hop, jump, and run by themselves and wrestle and chase with and flee from one another. They may also manipulate stones and other objects (Frädrich 1965).

Both social behavior and life-history parameters vary interspecifically in Suina. Most swine exhibit some degree of sociality (E. O. Wilson 1975), and all appear to be polytokous, an unusual condition for an artiodactyl. Hippopotamus have only one calf per birth as a rule. The common hippopotamus (*Hippopotamus amphibius*) (Figs. 3-39, 3-40) forms large herds. In contrast, the largest social unit in the pygmy hippopotamus (*Choeropsis liberiensis*), may be a mother and her calf. Pygmy hippopotamus mothers and calves play together (S. Wilson & Kleiman 1974). A mother stimulates her calf to play by splashing, lying on her side, or gaping (opening wide and shutting the mouth). The calf plays using a variety of body movements that include headshaking, head-tossing, pirouetting on the hindlegs, and leaping out of the water. Tactile and olfactory contact occur in play as consequences of muzzle-to-muzzle sniffing, cheek and chin rubbing, and jaw-nibbling (S. Wilson & Kleiman 1974).

SUBORDER RUMINANTIA (CAMELIDAE, TRAGULIDAE, CERVIDAE, GIRAFFIDAE, ANTILOCAPRIDAE, BOVIDAE)

These artiodactyls, ruminant herbivores, pose a variety of questions to students of play. Deer fawns (Cervidae) and bovid calves (especially goat kids and sheep lambs) are considered to be especially playful, but young giraffes are reported to play relatively little by several field observers of giraffe behavior (Coe 1967, Dagg & Foster 1976, Innis 1958, Langman 1977). Pronghorn (Antilocapridae) play is well documented, though the behavior of this animal is not as familiar as that of the giraffe. A clear and unambiguous description of play even exists for one of the primitive and little-studied tragulids (mouse deer and chevrotains). What little is known about play in Old and New World camels suggests that young Camelidae could represent a condition midway between the giraffe and the deer. It seems likely that some of this variation is only apparent, resulting from differences among observers, among conditions of observation, and among research projects.

Fig. 3–39 Calf play-nibbles conspecific's lips.

Fig. 3–40 After rearing, vigorously waving its front legs, and leaping over its playmate, the hippopotamus calf does a back flip with all four feet waving.

Figs. 3–39 and 3–40 Hippopotamus (*Hippopotamus amphibius*) calf loco-motor play. Photographic sequence of a single play bout. Kazinga Channel, Uganda. (Maitland A. Edey)

The water chevrotain (*Hyemoschus aquaticus*) is a tragulid whose solitary and archaic status (Dubost 1975) among the artiodactyls gives its play special interest. All the play that Dubost observed was solo activity, occurring in the early morning hours just after midnight. An animal would run and bound in play. Running occurred over a brief distance, in a straight line, in a circle, or in a figure-eight. While running in play, chevrotains kicked up their heels, turned in mid-air, and dashed back and forth making sudden pivoting half-turns (Dubost 1975).

New and Old World camels spring, roll, leap, and chase in play using behaviors similar to those seen in intraspecific fighting and in predator avoidance (Gauthier-Pilters 1959, Pilters 1954). In young guanacos (*Lama guanacoe*), playfighting occurs, but is unstable, according to Pilters (1954), who stresses that play of young guanacos is apt to break out into serious fighting.

The most complete ethological information available to date on any camelid concerns the vicuna (*Vicugna vicugna*), a strictly territorial ungulate inhabiting the high Andes of west central South America. Juveniles of both sexes and immature males chase and playfight, wrestling with the necks or biting gently (Koford 1957). Play-chasing is frequently a group activity:

Often a group of young run away from the adults and back, several times in succession, one then another in the lead. At two weeks of age the young rarely race more than 50 yds. from the family group, but at one month of age they run away as far as 100 yds., and at several months of age, twice as far. On occasions the galloping youngsters make high leaps over rocks, ditches, or other obstacles. Sometimes a juvenile acts as if chasing an invisible partner, running fast and recklessly, jumping high, and kicking its hindfeet back. After jumping down over a bank, one fell and turned a somersault. Another fell on a road and skidded along on its side.

Baby giraffes (*Giraffa camelopardalis*) are active creatures, "often running for no apparent reason except excess of energy" (Innis 1958 p. 263). A giraffe calf born in Budapest Zoo galloped playfully around its mother from age 2 days (Szederjei & Fábián 1975). Solo and contagious locomotor play occur regularly in giraffe calves in the Serengeti (Pratt & Anderson 1979). Giraffes playfight using behavior patterns that prefigure the unique fighting techniques of adults (Coe 1967, Innis 1958). In an escalated fight between adult male giraffes, the head and long neck act as a club or bludgeon. The head is whipped back and swung with tremendous force, delivering blows to an opponent's chest, neck, belly, or legs. Wounds may result, especially in the neck area. An effective defensive tactic is to wrap one's neck together with that of one's opponent, and in so doing deprive him of the use of his weapon. In this position, fighting giraffes push against each other with their bodies (Innis 1958). As in other mammals (Geist 1972), pushing tactics may represent an attempt by one giraffe to free itself in order to attack, whereas its opponent strives to keep the first giraffe "pinned" (Innis 1958). Gentle neck-sparring occurs as a kind of non-injurious agonistic fighting (tournament) between adult males and is most frequent when opponents are unevenly matched, whereas fights between equally matched males tend to escalate (Coe 1967). Baby giraffes may spar with adult males, with young males, or with each other. A baby male sparred between two adult bulls, who "hit him gently from both sides, and the baby between them waved his head about in all directions, sometimes completely missing both of them" (Innis 1958 p. 261). Mounting occurs frequently in play between male giraffes, and the smaller or younger often mounts or rides his partner.

Another giraffid, the okapi (*Okapia johnstoni*), is also known to play. Okapi play is less well-known than that of the giraffe. An adult male okapi was seen to self-handicap in play (Walther 1962). Walther observed a young bull and the bull's father together in captivity. They solicited play by using exaggerated forms of adult threat postures. A bout of play included chasing and neck-wrestling. The young bull chased his father, who would adopt a less advantageous wrestling position. A play bout ended with mutual grooming.

Three different types of "simple" play, none quite so simple as the name would indicate, exist in ruminants. The solo loco-motor-rotational play of the water chevrotain, the unstable playfighting of the guanaco, and the cooperative neck-wrestling play (which appears to represent an extreme version of sparring occurring between comically mismatched or harmless opponents) of the giraffe are all precursors, in a sense, of more complex forms of play that begin to appear in cervids, in antilocaprids, and in bovids. What is added may include such complex "games" as tag and "King of the castle"; increasing emphasis on reciprocal chasing (approach-withdrawal); participation of females in social play; and use of locomotor-rotational movements as play signals in a social context.

ANTILOCAPRIDAE

The sole living representative of this family (perhaps only a subfamily of the Bovidae) is the pronghorn (*Antilocapra americana*), of western North America. Pronghorn form large herds of allied female-offspring units, and pronghorn play is highly complex and interactive. Groups of fawns race together in wide arcs for great distances (a hand-reared fawn aged 1.5 months covered 1000 to 1200 m in its play, Müller-Schwarze & Müller-Schwarze 1973). They leap and bound, jump on their mothers' backs, and play tag and "King of the castle" (Autenrieth & Fichter 1975, Bruns 1977, Einarssen 1948, Kitchen 1974); sparring also occurs, but seems to be a relatively minor component of pronghorn play (Bruns 1977).

CERVIDAE

Deer span virtually the entire spectrum of artiodactyl ecology, social organization, developmental patterns, and play. They offer several practical advantages to the field biologist, including economic importance. Field studies (e.g., Darling 1937) of cervid behavior in relation to its environment and to human management demonstrated interactions among behavior, survivorship and reproduction, and quality of deer populations.

Muntjacs (*Muntiacus muntjac* and *M. reevesi*), or barking deer, are small, primitive cervids of tropical Asia. Their play does not differ greatly from that of other ungulates that live in small family groups. Solo play includes running and jumping, standing vertical leaps, bounds, and kicking up the heels; a muntjac fawn may butt his mother's side. Muntjac play contains other social elements because young may run around adults or stimulate other young to play (Dubost 1971). Male-male sparring, usually involving an adult and a juvenile, is playful (Barrette 1977). Bouts of playful sparring are initiated by antler pointing, a serious threat in other contexts. Sparring deer "establish contact slowly before pushing at and turning around each other while attempting to twist each other's neck" (Barrette 1977).

Another deer of generally solitary habits is the moose (*Alces alces*). Although moose may congregate in casual groups to feed, the only enduring social unit is the cow moose and her calf or twin calves. Therefore, depending on local food distribution and maternal age and nutritional condition, a moose calf may have no peer playmates at all; it may have a full sibling as playmate and constant companion; or it may have brief contacts with a number of peers whose genetic relationship to it is more or less distant. Moose calves play alone, rushing and kicking around their mothers, running back and forth making sharp turns and circles, splashing through water, hopping, and lifting their legs high; peers play with each other, fighting, charging, and shaking their heads (M. Altmann 1958, De Vos 1958, De Vos, Brokx, & Geist 1967, Geist 1963, Stringham 1974). Moose calves may also playfight with their mothers (Fig. 3-41) or with objects (Fig. 3-42).

Reindeer and caribou (*Rangifer tarandus*) are typically herd-living deer whose young interact socially to a limited extent in play, without obvious "games," mainly chasing and fleeing from each other. Calves may also jump up onto high objects and jump down again, butt, and mount or ride each other (Espmark 1971, Lent 1966). Forest-dwelling, solitary individuals also occur; in these populations the mother and calf are alone for most of the year, and the calf plays by bounding in place (Pelosse 1977) and presumably by running. Gamboling and bucking are characteristic of very young calves in all known populations (Espmark 1971, Lent 1966) and may be predicted in forest reindeer. That caribou play may sometimes exceed the degree of complexity assigned to it by these authorities is suggested by the following anecdote (Stefansson 1944 p. 624), if in fact the behavior described was mutually playful:

Three large bulls were feeding quietly and near them were eight yearlings. When I was half a mile away and my head was showing above a ridge, several of the yearlings started on a sudden run. At first I

Fig. 3–41 Young of the moose (*Alces alces*), a giant northern deer of generally solitary habits, find ample opportunity for play even in the absence of peers. Here, a moose calf play-butts with its mother. Alaska. (Stephen J. Krasemann/DRK Photo)

thought they had seen me, though that was really not possible, and when I looked at them through the binoculars I saw they were chasing a fox. Three or four of the yearlings chased it for two or three hundred yards and then returned to where the rest were feeding. The fox now waited a little, as if to see if the caribou had given up the game, and then ran in among them again and was chased by others. Sometimes the fox dodged in and out among them and when it had secured the interest of the yearlings and induced them to the chase, it would run around the three old bulls using them as a sort of protection from the yearlings. The old bulls paid not the slightest attention to either fox or yearlings and went on feeding quietly. I watched the game about half an hour, after which I mounted the ridge and approached till they saw me. The first to see me were the old bulls which ran off at full speed. They had perhaps a hundred-yard start over the young bulls which were, however, so much fleeter that they

Fig. 3–42 An environmental feature such as this tree branch readily elicits play-butting from a moose calf. Alaska. (Stephen J. Krasemann/DRK Photo)

caught up and passed the old bulls within half a mile. I have seldom seen caribou run so far without stopping, but these must have run nearly a mile. After two or three stops the yearlings came back and approached within fifty yards of me several times. All ran with their mouths open, though chiefly at a trot after the first mile sprint. The big bulls kept their course for two or three miles and then commenced feeding.

Roe deer (*Capreolus capreolus*) are Old World cervids whose social unit is a family consisting of one buck, one doe, and one or two fawns of the year. Single fawns' solo play and the play of twin fawns are similar, involving rushing, vertical springing, and butting (Bubenik 1965, Espmark 1969, Prior 1968). Fawns may chase and flee from their mothers, and whole families or even several families at once may play together at sunrise on a cold morning. Rare group play consists of chasing and springing with reversal of the roles of pursuer and pursued, and a kind of tag may develop from which two or three deer "break off to circle a tree at dizzy speed" before rejoining the chase (Prior 1968 p. 49).

In permanently group-living deer, such as the wapiti or elk (*Cervus canadensis*) and the red deer (*Cervus elaphus*), intricate social play resembling the rare group play of roe deer is the rule rather than the exception. Red deer fawns begin playing by themselves on their first day of life, running in straight lines or arcs, springing in the air, capering, and springing backward (Bubenik 1965, Burckhardt 1958). Very young elk play in similar fashion, kicking, jumping, and circling (M. Altmann 1952). Older fawns, yearlings, and even adults play together, racing, contacting, and fleeing each other in a kind of tag, rearing up and striking at each other, running and doubling around hillocks, or climbing and defending these hillocks (red deer, Bubenik 1965, Burckhardt 1958, Darling 1937, Gossow & Schürholz 1974, Olivier 1958; wapiti, M. Altmann 1952, 1963, Struhsaker 1967a). A red deer fawn accidentally discovered how to somersault while butting leafy branches. After its first accidental somersault, the fawn somersaulted again and again for a period of about two weeks (Bubenik 1965). This reported discovery of a movement, which then became a prominent feature of the animal's play for a period of time, is strikingly similar to

Eibl-Eibesfeldt's (1950b) account of his badger's discovery and use in play of somersaulting movements.

Deer seem to exhibit the same dimensions of ontogenetic, intrapopulation, and interspecific variation in their play. A fawn of a given species seems capable of responding "correctly" to a wider variety of social contexts than it would be likely to encounter in its lifetime. Socially interactive play is not equally likely in fawns of all species, but any individual seems capable of playing at a level of complexity several grades of social organization above its own. For example, roe deer were observed playing in large groups like red deer, and young caribou may have played with a fox. The use of play as a kind of minor league preparation for adult male competition, so prominent in other artiodactyl species, seems to have faded into the background in deer, although careful reading of Darling (1937), of Gossow & Schürholz (1974), and of Struhsaker (1967a) suggests that this element is still present. These lines of increasing complexity in play parallel Geist's (1966, 1972, 1977) analysis of relationships between the evolutionary elaboration of horn-like organs and social complexity. These parallels suggest interesting questions about play, social organization, and mechanisms of competition, which are further illustrated by comparative analyses of cervid play. Some of these questions are the following: To what extent are the relative stability and complexity, elaborateness, and degree of interactiveness of wapiti or red deer social play merely a demographic phenomenon? Do these parameters depend on the number and on the age and sex distribution of potential partners? In what ways does the genetic relationship of these partners affect these considerations about play?

Opportunities exist for comparing play in single calves and in three kinds of twin pairs (male-male, male-female, female-female), in small and in large herds, or between like-aged partners and between partners of increasingly different age or size. Hypotheses based on one family may be tested on several others. For example, a hypothesis suggested by observations of deer could be tested in bovids, perissodactyls, caviomorph rodents, ground squirrels, phocids, primates, bats, or carnivores. Although existing data and interpretations suggest any number of hypotheses, these data are seldom adequate for testing even the simplest theoretical ideas. Furthermore, although intraspecific comparisons between different populations may be as informative as interspecific comparisons, interpopulation studies have only recently begun.

BOVIDAE

The bovids are a spectacularly diverse family of 44 genera and 111 species. Moreover, several taxonomic groups within the bovids, particularly the antelopes (subfamily Antelopinae) and the sheep and true goats (subfamily Caprinae, tribe Caprini), offer comparable variety. Patterns of play vary along lines roughly parallel to those recognized in cervids.

Bovinae The play of domestic cattle (*Bos taurus*) and that of their wild relatives is a combination of body contact and chasing interactions. Very young calves may frolic on their own in the vicinity of their mothers. American buffalo (*Bison bison*) push, butt, mount or ride, and race as calves (Lumia 1972); wisent (*Bison bonasus*) fathers play seemingly good-natured butting games with their calves (Mohr 1968); and kudu (*Tragelaphus* spp.) run and jump, playfight, and participate in interactive games (Leuthold & Leuthold 1973, Walther 1964). Immature males and perhaps young adults of the African buffalo (*Syncerus caffer*) spar playfully together, bucking and frolicking like small calves (Sinclair 1977 pp. 110–111). Young gayal (*Bos frontalis*) play-chase and horn-wrestle. Gayal cows may join their calves' chasing games briefly, and gayal bulls may horn-wrestle with calves (Scheurmann 1975). This form of sexual dimorphism, both in adult play with immatures and in the play of the young, should be investigated in other ungulates, for it parallels the familiar sexual dimorphism of play (males wrestle, females chase) seen in such non-ungulates as rhesus (e.g., Symons 1978b).

Play in the wild cattle (*Bos taurus*) of France's Camargue region includes solo frolics by young calves, who run around, leap vertically and twist their head or body while in midair, and kick up their heels (Schloeth 1961). They may butt or try to mount their mothers (Schloeth 1958). Older calves and year-lings in the Camargue herd played together around a small hill-ock. They would butt, rear, and "box" with their front legs and chase and flee. Playfights were decorated by springs and capers, and boxing and butting alternated with chasing and fleeing. Calves also horned bushes and hummocks (Schloeth 1961).

Hippotraginae This bovid subfamily of African and South-west Asian ungulates includes highly gregarious large-bodied species well-known to sociobiologists. These bovids include the stout-bodied, long-horned oryxes (*Oryx* spp.), whose effective defensive tactics and aggressive communication techniques pre-vent opponents from skewering each other in a fight, and the blue wildebeest or brindled gnu (*Connochaetes taurinus*), with its unique, almost playful male territorial challenge ritual and its social system that varies adaptively in a changing environment.

Play patterns for all species studied include solo running and vertical leaping decorated with assorted locomotor-rotational movements, interactive chasing and some wrestling (pushing and shoving) among calves, and more or less playful sparring matches or brief chasing bouts among young males in bachelor herds. Ecology, developmental rates, and demography interact to shape these patterns differentially in different species. Defassa waterbuck (*Kobus defassa*) calves join their peers at two to three weeks of age, and group chasing occurs interspersed with bounds, rear-kicks, and capers (*"ces courses folles et cabrioles de toute espèce,"* Verheyen 1955) (Spinage 1969). Bontebok (*Damaliscus dorcas*) calves grow up in small nursery herds of sizes that average three females. Often, only one calf occurs in these small herds (average 1.5), and although play does sometimes occur between calves, it is neither frequent nor prolonged (David 1973, 1975):

Sometimes young calves chase each other round in big circles for a few minutes, giving "frolic" jumps on the way, which consist of leap-ing into the air and kicking out with the hindlegs and one may walk up to another calf and waggle its head at the other in an invitation to play. Even before horns have appeared they face each other and push lightly head to head—a forerunner of later playfights as yearlings (David 1975 p. 219).

At about one year of age, calves of both sexes leave their mothers and join bachelor herds. Calf solo play and yearling social play occur in this species as in other hippotragines (David 1973). It would be interesting to determine if either or both of these forms occur at a higher frequency in young bontebok than in the young of related species, compensating for a bontebok calf's lack of like-aged play companions.

Another interesting aspect of bontebok behavior is that male bontebok, like male blue wildebeest (Estes 1969), may defend territories using a physically vigorous, ritualized display of lo-comotor-rotational movements. "Aggressive cavorting" in bontebok includes running, vertical leaping, kicking up the heels, and vigorous headshakes. Use of these ubiquitous mammalian play movements in the agonistic displays of these two species appears unparalleled, although movements used in play behavior frequently recur in non-injurious agonistic fighting and in escalated fighting.

The interspecific contrast between Defassa waterbuck and bontebok recurs intraspecifically in topi (*Damaliscus lunatus*). In Rwanda plains populations, calves aged eight to nine months and independent of their mothers form large juvenile herds containing distinct, recognizable subgroups. Group play is frequent and includes chases, vertical twisting leaps, and kicking up the heels. Dyadic playfighting is also very frequent. By contrast, hill populations of Rwanda topi contain small harems in which calves have few social companions, and the young stay with their mothers until a later age. Group play appears to be relatively rare in these societies (Monfort-Braham 1975).

The contrasts between solo, dyadic, and group play in hippotragines recall similar contrasts found in grey and common seals and in caviomorph rodents by S. Wilson (1974a) and by S. Wilson & Kleiman (1974). It is not known how demography and individual tendencies to play interact in any of these species. Quantitative comparisons would be valuable. Another curious aspect of hippotragine play is its apparent lack, judging by the few available reports, of intermale climbing, mounting, and riding behavior.

Antelopinae Antelopes, some medium-sized and some very small, inhabit Africa and parts of Asia. The small forms feed on a number of high-quality food items selected from various plant species and are found in small family groups of one, two, or three individuals, whereas the larger antelopes occur in herds and include grass shoots and seeds as well as other plants in their diet (Jarman 1974). Antelope play follows general ungulate patterns, possibly resembling bovine play more closely than hippotragine play.

The world's smallest living ungulate, the royal antelope (*Neotragus pygmaeus*), stands only 25 cm high at the shoulder and weighs less than 3 kg when full grown. A dainty creature the size of a rabbit, this antelope takes over a year to attain full size and is seldom observed in captivity or in the wild. A young royal antelope raised by Owen (1973) ran and capered in play, leaping with a lateral body-twist in mid-air and a backward kick of the hindlegs. From the age of five months, "it became increasingly active and venturesome in the garden, describing figure-eights on the lawn at a fast run with frequent leaps" (Owen 1973).

Larger antelopes play solo, dyadically, and in groups. Decorated running and chasing are frequent. In blackbuck (*Antilope cervicapra*), mounting, riding, and climbing on a partner's back all occur, and young males challenge adult males to play-fights (Schaller 1967 p. 159, Schmied 1973). Blackbuck calves also tease and chase birds including pheasants and ducks (Schmied 1973). Young impala (*Aepyceros melampus*) gallop together away from the herd, circle back, and run away again; adult females may join these racing games (Schenkel 1966b). Young male impala playfight (Leuthold 1970, Schenkel 1966b). Gerenuk (*Litocranius walleri*) are desert antelopes known for their habit of standing on their hindlegs to browse on tall shrubs. In play they chase, jump, and spring back and forth; these play bouts usually occur between a young male and an adult male (Walther 1961), but may also take place between adult males and females in courtship (Walther 1958). Grant's and Thomson's gazelle fawns (*Gazella granti* and *Gazella thomsoni*) play around their mothers, running, galloping in circles, and making high, arching jumps; running and jumping games also occur in older fawns (Walther 1965b, 1969, 1972, 1973).

Caprinae The very word "caper" is testimony to the playful habits of goats and their allies. Sheep, chamois, musk oxen, and mountain goats are Caprinae. So are the takin, serow, and goral, three curious goat-antelopes inhabiting mountain bamboo and rhododendron forests, cliffs, screes, and talus slopes of eastern Asia. Caprinae are northern hemisphere ruminants exhibiting great diversity of form and habitat along with considerable diversity of social organization.

Play in Caprinae begins with solo locomotor-rotational exercise (Fig. 3-43). Chamois fawns (*Rupicapra rupicapra*) hop, run back and forth at top speed, or run away from their mothers and then return at a gallop. Distinct from their running play is a set of caprioles, Walther's (1961) "chamois dance," in which fawns spring vertically into the air, run at a rocking gallop, make sudden stops, toss their heads high and headshake, and whirl on their own axis, often falling down (Krämer 1969). Musk ox calves (*Ovibos moschatus*) romp running and whirling in circles, stiff-legged and with heads flung up (Tener 1965), whereas young Corsica mouflon (*Ovis ammon*) gallop around their mothers, leap from rock to rock, spring vertically, or chase dead leaves. One lamb played for almost ten minutes with a broken branch, tossing it into the air, retrieving it, and tossing it again (Pfeffer 1967). Feral goat kids (*Capra hircus*) break into a "jerky skipping dance" when they see moving objects (including butterflies) (Rudge 1970).

Social play in the Caprinae is as elaborate as solo play. Mountain sheep lambs (*Ovis canadensis*) chase, butt, and ride each other, or gallop back and forth along a slope, jumping and sliding (Geist 1971). Ibex (*Capra ibex*) play "King of the castle" on rock ledges; they may jump on and off their mother's back, and race or bound. Both males and females playfight (Meyer-Holzapfel 1958). Mounting, riding, and sparring occur in many Caprinae, including feral goats (Shank 1972), bharal (*Pseudois nayaur*) (Schaller 1972), chamois (Krämer 1969), mountain

Fig. 3–43 Siberian ibex (*Capra ibex*) kid play-rearing. Brookfield Zoo, Illinois. (Brookfield Zoo)

sheep (Geist 1971), and mouflon (Pfeffer 1967). Aoudad (*Capra lervia*) (Haas 1959) and chamois (Krämer 1969) exhibit group play, in which many young or even an entire herd run and caper together. Whether group play occurs in all Caprinae or only in these species and perhaps a few others is not known.

It is tempting to suggest that the Caprinae, believed to be the most "advanced" group of bovids, also exhibit the most elaborate and antic play, and literature cited certainly tends to support this impression, but even the most rudimentary evidence is lacking. Comparative measurements of diversity of locomotor-rotational movements and definition and measurement of the elaborateness and complexity of games or of the interactiveness and cooperativeness of social play have not been made. And although these data would be interesting purely from a descriptive point of view, no existing theoretical hypothesis and no measurable parameter (with the possible exception of relative brain size) shows much promise for generating testable predictions. But evidence that play of Caprinae surpasses that of all other bovids in diversity, complexity, or stability would definitely stimulate interest in formulating some testable hypotheses.

PLAY IN BIRDS

If ecological and ethological research were ever to be arrayed in a taxonomic Social Register, bird studies might be found to constitute the high aristocracy. This classic research includes Julian Huxley's observations on great crested grebe, Darwin's studies of finches and pigeons, Lorenz's accounts of greylag geese, Tinbergen's gull experiments, and MacArthur's studies of sympatric warblers.

In most temperate habitats, birds are the most conspicuous animals. Many sing, often quite melodiously, and some have bright, colorful plumage.

Unfortunately, the ease with which certain aspects of avian social and feeding behavior can be rigorously analyzed has led in some cases to premature generalization and to theory applicable only to the organisms upon which it was based. Ralls (1977), for example, points out that empirically substantiated theories of sexual dimorphism in birds led to major confusion about the biological meaning of mammalian sex differences.

Play research is not one of the many feathers in avian biology's cap. Although the play of corvids, parrots, or raptors is often as elaborate and as interesting to watch as that of primates or carnivores, and although play behavior may prove to be as widespread in birds as it is in mammals, avian play has attracted relatively little attention in the past. Many available descriptions are unsatisfactory by contemporary standards, so that it is sometimes difficult to ascertain just what is being described. In addition, published information on play in birds is extraordinarily scattered, to such an extent that even Millicent Ficken's (1977) excellent review of avian play fails to cite a significant fraction of the available literature on play in birds.

No bird, living or otherwise, is ancestral to mammals. Therefore, play evolved independently in the two groups (barring possible play in some common reptilian ancestor). For this reason the comparative study of avian play is extraordinarily important. Because of this importance we must identify major biological aspects of avian-mammalian differences.

Birds and mammals evolved from separate reptilian stocks and have been taxonomically distinct entities for at least 180 million years (Colbert 1955, Romer 1966). The members of both taxa are endothermic (except newborns in altricial species and certain species who exhibit winter torpidity in seasonal environments). Birds have a higher weight-specific metabolic rate (Schmidt-Nielsen 1975) and smaller brain (Jerison 1973) than like-sized mammals. Flying and gliding are common, though not universal modes of locomotion in birds; mammals, however, excepting bats and a few gliding species, are generally earthbound. Vision and audition are as important in birds as in primates and carnivores, but olfaction is less important in most birds than in most mammals. Mating is far more likely to be monogamous and the father is more likely to invest heavily in his offspring in birds than in mammals, but there are some exceptions.

Avian societies and mammalian societies tend to converge in similar environments (Geist 1977). Patterns of avian postnatal development also resemble those of mammals. Altricial and precocial species are both common in each taxonomic class. Altricial birds are generally semi-independent of their parents by the time they leave the nest and are able to fly, and they become fully independent soon afterward. A young mammal at an equivalent stage of locomotor and thermoregulatory development may still suckle or be fed for several months. Again, there are many exceptions to this pattern, especially in migratory birds and in certain species with small clutches and prolonged parental care (Ashmole & Tovar 1968). Developmental rate is a biological dependent variable in birds just as it is in mammals, and food fights among nestlings are widespread, with initially small size differences affecting the outcome of competition between sibs. Some birds (e.g., owls, Burton 1973) stagger their clutches over a period of weeks. The first birds to hatch can easily outcompete later-born nestmates for food. This distribution of hatching times appears to be to the parent's advantage, as it allows a larger clutch to be raised at the maximum feeding rate and minimizes losses due to variable food abundance (Brockelman 1975, O'Connor 1978). No mammal adapts to low or varying resource levels in exactly this way, although widespread occurrence of mammalian runts and even an instance of intrauterine sib competition by anatomical means (O'Gara 1969) suggest that different competitive mechanisms may produce analogous differential allocation of resources to members of a mammalian brood.

Avian play described by early naturalists and others often bears little resemblance to animal play as we currently understand it. These accounts make no mention of such pertinent details as locomotor-rotational movements, existence of two forms or modes of performance of the same behaviors, accompanying signal patterns, nonrandom selection of objects or environments, and likely biological effects of nonplayful prototypes. Failure of human intuition to appreciate the constraints of self-powered and gliding flight and to understand how birds court, feed, fight, and mate while airborne made some observers prone to call all spectacular aerial behavior "play" (cf. the description of feeding swallows in Book 3, Chapter 12 of Chateaubriand's *Mémoires*). Moreover, the term "play" has been applied to threat, courtship, and pair-bonding displays (Chisholm 1958, Groos 1898), including the decorating of the bower by male bowerbirds, to comfort movements, and to other behavior, including bathing (Lebret 1948) and, as Simmons (1966) points out, anting. This last-named phenomenon involves placement of ants by the bird on its own body.

Ficken (1977) correctly observes that, for these reasons, assessing descriptions of avian play from the literature "presents numerous problems." She further clarifies the issues involved by noting that in birds (as in mammals), various captivity artifacts and maturational behavioral intermediates lacking sequentially distinct counterparts have been incorrectly identified as play.

Play in birds differs little from play in mammals once we factor out the constraints imposed by aerial locomotion and by a relatively small brain. Medium-sized passerines and larines exhibit rodent-like exercise play, complete with locomotor-rotational movements, and they occasionally perform ambiguous playfights and play-chases. Raptors exhibit varied carnivore-like play with objects and prey items as well as simple social play. Corvids and certain parrots, the champion players of the bird world, play as do primates or carnivores, and their play includes elaborate acrobatics, intricate object manipulation, and complex, variable social play including game forms and play-signals. These diverse forms of avian play will now be considered.

Miscellaneous, Primarily Aquatic Orders (Sphenisciformes, Pelecaniformes, Anseriformes, Charadriiformes)

Play in these aquatic birds focuses on feeding and on predator avoidance techniques. It is found relatively often in gulls and terns and sporadically in pelicans and allies (but commonly in those species that develop slowly and exhibit long periods of parental care), and it appears to be rare (at least relatively) in waterfowl. Only a few accounts of play exist for any of these orders.

Frigate-birds (Fregatidae) are large, long-winged, scissor-tailed, soaring birds found in and around tropical seas. Juvenile Ascension Island frigate-birds (*Fregata aquila*) catch feathers and strands of seaweed from each other in midair while flying together (Stonehouse & Stonehouse 1963). Small mixed-age juvenile groups of great frigate-birds (*Fregata minor*) play follow the leader over rock pools, "each in turn swooping down over the water and pausing just above its surface." They also manipulate objects, "pulling up pieces of dead creeper and trying to break prominent branches off low trees" (Gibson-Hill 1947).

Various duck species have been briefly described rising from and immediately splashing down into or across bodies of water and making sudden turns (Lebret 1948, Lorenz in Nice 1948). Common eider (*Somateria mollissima*) in Iceland ran rapids (Roberts 1934) by riding a tidal current along a narrow sandspit, repeating the performance over and over again. It is impossible to place these accounts in context as they stand. These activities, which Lorenz (in Nice 1943) abruptly dismisses as "emotion-dissociated fleeing movements," may be equivalent to the exercise play, decorated by locomotor-rotational movements, of young rodents and ungulates. The problem calls for further study and more detailed descriptions.

Oyster-catchers (Haematopodidae) are large shorebirds whose massive, bright red bills aid them in opening hard-shelled molluscs. Norton-Griffiths (1966, 1967, 1969) and Tinbergen & Norton-Griffiths (1964) describe the oyster-catcher *Haematopus ostralegus* in an estuarine environment. The shell-opening behavior of young oyster-catchers develops slowly and in the course of considerable interaction with objects in this environment. When about ten days old, the young peck at many different kinds of objects (both edible and inedible), including empty mussel shells. At a later age they chisel at a variety of targets including plants and (in one case) an observer's toes (Tinbergen & Norton-Griffiths 1964). Skill at opening bivalves has a long developmental history, and six-week-old chicks, who feed almost as efficiently as adults when probing for worms, are unable to open bivalves and may have to be fed for months if molluscs and crabs are the chief available food items (Norton-Griffiths 1969). Whether the development of this skill includes playful components is not known, but I consider it likely. If so, three predictions follow. Young oyster-catchers should nonrandomly select objects to peck, they should repeat and vary behavior at given stages of skill after mastery, and they should continue to peck at nonfood objects after having mastered intermediate stages of feeding skill, perhaps selecting certain objects or environments separate from those in which the probability of gaining food is high. These predictions require further empirical study.

In young gulls and terns (Laridae: Larinae and Sterninae), play occurs in social groups, but centers on the manipulation of objects. Juvenile Inca terns (*Larosterna inca*) rest together and then take flight, vigorously plunging and dipping at the surface of the water, picking up inedible objects, flying with them, and dropping and retrieving them in bursts of activity (Ashmole & Tovar 1968). Herring gulls (*Larus argentatus*) and lesser black-backed gulls (*Larus fuscus*) manipulate and fling away objects, or they may carry them to the water, shaking the objects in their bills (Delius 1973, Goethe 1955). They may pull at objects held by others, whereupon a tug-of-war results (Goethe 1955). Neither the tern behavior described by Ashmole and Tovar nor the gull behavior described by Delius and by Goethe could be confidently and without question said to represent play, but these accounts are suggestive.

Playful object-dropping from the air by gulls, first described by Wheeler (1943), was later investigated by Beck (1976) in herring gulls and in great black-backed gulls (*Larus marinus*). Beck found that even though the behavioral topography of playful (inedible object) and predatory (edible object) dropping was very similar, no less than six criteria clearly differentiated the dual forms of shell-dropping: the soft substrates used in play; the minimal amount of object inspection and manipulation following a play drop; midair play catching and redropping of objects; simultaneous, close and noncompetitive dropping by several playing animals; play dropping in strong winds; and random distribution of play dropping, but not of predatory dropping, with regard to tide time. The gull described by Wheeler (1943) repeatedly dropped and caught a mussel in midair, against a strong wind. Beck found that immature herring gulls who have mastered predatory dropping did not play drop, whereas great black-backed gulls did not drop in predation at any age but play dropped both as yearlings and when older. Play dropping by both species involves rapid vertical ascent and descent, as does herring gull predatory dropping, but this type of flight is rare in all other contexts, suggesting a possible role for play dropping in physical training.

Miscellaneous, Primarily Terrestrial Orders (Galliformes, Columbiformes, Cuculiformes, Apodes, Coraciiformes)

The status of play in these orders is uncertain and the few published reports fragmentary or enigmatic. Jungle and domestic fowl (*Gallus gallus*) chicks run briefly, jump up and down, and engage in brief, incomplete, and uncoordinated social interactions (Guhl 1958, Kruijt 1964). Red-legged partridges (*Alectoris rufa*) in confinement run at full speed for a short distance, making brief flights and sudden turns (Goodwin 1953). After the nest-building season has ended, chimney swifts (*Chaetura pelagica*) circle and strike at willow twigs, then bound backward and upward (Ball 1943). Immature collared doves (*Streptopelia decaocto*) exercise their wings while grasping the perch and have been observed manipulating sticks (Miller & Miller 1958). Hummingbirds of many species perform flights that resemble courtship or agonistic flights, but the number, sexes, ages, and species of participants may vary. These incidents are observed relatively frequently and aggressive signal patterns are absent. Participants neither mate nor drive one another away, and occasionally birds of different species will so interact (Wagner 1954). An Anna hummingbird (*Calypte anna*) repeatedly floated down a small stream of water flowing from a hose (Stoner 1947); compare the rapid-riding eider of Roberts (1934). A young kingfisher (*Alcedo atthis*) was seen repeatedly diving for and fetching a twig floating in water (Swanberg 1952). Young hornbills of various species are briefly described "barging each other, wrestling on a branch," "hustling each other, worrying each other with their bills" (Moreau & Moreau 1944, Ranger 1950, Swynnerton 1908). The most intriguing account of play in these orders concerns *Turaco fischeri,* the turaco (plantain-eater). Young birds who could walk, but not fly apparently play-fought. They would jump up, seize a conspecific's tail, and try to pull the bird off a perch, chase each other wildly up and down the perches, "play tricks on each other monkey fashion," and even tease neighboring fowl (Moreau 1938). Although contextual information necessary to distinguish this play from early agonistic fighting was not reported by the author, the subjective impression of play imparted by the account is strong. This species and its close relatives merit additional study.

Birds of Prey (Orders Falconiformes, Strigiformes)

Hawks, owls, and eagles, especially when young, play with living and dead prey and with inedible objects in a way that suggests a mammalian carnivore at play. A harrier (*Circus cyaneus*) caught a horned lark (*Eremophila alpestris*) and clumsily, but repeatedly dropped it and pounced on it while it was still alive (Sumner 1931). Several apparently distinct play routines may be repeated and alternated: jumping on and striking, soaring and swooping, dropping and catching in midair, tossing and pouncing, flinging up or away and attempting to catch, grasping and biting, and tearing and shaking. These behaviors have been observed in free-living (Chisholm 1958, Huxley & Nicholson 1963, Munro 1954, Parker 1975, Sumner 1931) and in confined adults and juveniles (Table 3-29). Captive birds seem especially fond of paper wads for this game.

Many of these accounts, although fragmentary, contain indications that the behavior seen was play: performance by well-fed animals at certain times of day (Brosset 1973), accompaniment by possible play-signals (turning the head nearly upside down, Bond 1942; jumping and turning in the air, Todd 1974), higher probability of occurrence in young birds than in adults (Meyer-Holzapfel & Räber 1976), and occurrence in captive adults, but only in young birds in the wild (Bond 1942).

The behavior of a young, wild-caught California condor, *Gymnogyps californianus* (Todd 1974, Todd & Gale 1970), is interesting because of this species' especially slow development, long period of parental care, and endangered status. These reports describe the bird as soliciting play from humans, jumping about for several minutes "with numerous runs, jumps, gyrations and aerial revolutions." Some of these acts suggest a courtship display. They may be used in both contexts by free-living condors or the behavior may possibly have been misinterpreted. Additional information on condor behavior would be valuable.

Social play, not always satisfactorily distinguished from fighting in these accounts due to lack of necessary contextual information, is suggested in free-living peregrines (*Falco pere-*

grinus), which typically perform "aerial mock-battles" when five to eight days out of the nest (Parker 1975), and in captive barn owls (*Tyto alba*), which when fully fledged engage in vigorous running, pushing, and wrestling activity marked by role reversal and rapid interchange of participants (Trollope 1971). Contextual information provides strong evidence for social play in buzzards (*Buteo buteo*). These large, social raptors perform spectacular aerial maneuvers in several contexts, including territorial defense, driving away fledged young, and encounters with other groups. Immatures become involved in these communal displays of up to 14 birds in autumn, but they have their own playful forms of this activity as well. Soaring over slopes in groups of three to six birds, they drop and catch objects, dive, and swoop (Weir & Picozzi 1975).

These accounts indicate that raptors may exhibit social play-fighting and play-chasing distinct from low-intensity or incomplete maturational forms of fighting behavior. Ficken (1977) was apparently unaware of these reports and additional references on parrots when she stated that social play "is rare in birds except for some corvids and their relatives."

Smaller Passerines and Parrots; Woodpeckers [Orders Passeriformes (in part), Psittaciformes (in part), and Piciformes]

Object manipulation exhibiting at least some playful characteristics is known in warblers, finches, and parakeets. Garden warblers (*Sylvia borin*) repeatedly pick up, drop, and catch small stones, an activity that may be socially facilitated (Sauer 1956). Cactus finches (*Camarhynchus pallidus*), a tool-using bird, will push mealworms into crevices and then pry them out (Eibl-Eibesfeldt & Sielmann 1962, Millikan & Bowman 1967). Young budgerigars (*Melopsittacus undulatus*) pick up and drop objects, including feathers and pieces of newspaper. Several birds may perform these activities simultaneously (Engesser 1977).

At least one well-documented experimental developmental, field and laboratory study of passerine object manipulation indicates that the maturation of avian skilled behavior may depend on environmental feedback and that this interaction of developing behavior with its environment in the service of skill need not be playful. Adult loggerhead shrikes (*Lanius ludovicianus*) impale captured prey on thorns. Most components of this behavior mature without environmental feedback, but proper orientation requires experience (S. M. Smith 1972). Development of impaling behavior, observed in the field, includes incomplete and preliminary impaling and manipulation of inanimate objects. In these birds, skill appears to develop monotonically and unimodally, with no play evident. Necessary details are lacking in almost all cases. However, because information on object-oriented behaviors in small birds is of general biological interest, these details may eventually emerge from future study.

Information on vigorous locomotor activity and social interaction in young woodpeckers (Kilham 1974), small parrots and passerines (Quaker parakeet, *Myiopsitta monachus*, Shepherd 1968; babblers, *Turdoides striatus* and *T. malcolmi*, Gaston 1977; sedge warbler, *Acrocephalus schoenobaenus*, Howard 1907; chaffinch, *Fringilla coelebs*, Marler 1956; a blackbird, *Turdus merula*, Messmer & Messmer 1956; song sparrow, *Zonotrichia melodia*, Nice 1943; common whitethroat, *Sylvia communis*, Sauer 1954) indicates that these young birds play like rodent and ungulate young. They chase and dodge, make sudden, rapid dashes with sharp turns, dart at one another, pull each other's tails, and jump or spring into the air. Flights may be wild and erratic. This impression of play not greatly different from that of many mammals is strengthened by a relatively complete verbal description of social play in young budgerigars (*Melopsittacus undulatus*). Carrying an object in its beak, one bird flees from another, stimulating a chase; or one bird can approach another and try to bite it, whereupon the other parries with its beak and a beak-wrestling bout develops. Compared with the agonistic fighting also seen in these birds, such playfighting is much slower, one blow is not immediately answered with another, but rather with an apparent invitation to "hit me again"; and although the entire interaction is physically vigorous, actual blows delivered are gentle (Engesser 1977).

Except for Engesser's interesting account of budgerigar play, these anecdotes of social interaction in young passerines, parrots, and woodpeckers fail to mention particular locomotor-rotational movements or play-signals. Context and background information are absent. Whether fledgling birds at this age have escalated fights or flee in earnest from environmental disturbances is not stated. For these reasons, it would be premature to pass judgment on such observations. If budgerigar play is like that of other small birds, these accounts indicate play comparable to that of many rodents and ungulates.

Corvids and Large Parrots (Orders Passeriforms [in part]; Psittaciformes [in part])

The spectacular acrobatic, object, and social play of corvids is well known (white-winged choughs, *Corcorax melanorhampos,* Chisholm 1958; white-necked raven, *Corvus albicollis,* Moreau & Moreau 1944; hooded crow, *Corvus corone,* Persson 1942; yellow-billed magpie, *Pica nuttalli,* Verbeek 1972; and especially common ravens, Budich 1971, Dawson & Bowles 1909, Gwinner 1964, 1966, Højgaard 1954, Hutson 1945, Mech 1970, Thorpe 1966) and is frequently cited (Ficken 1977, E. O. Wilson 1975). These activities often combine locomotor activity with object manipulation, and playful contests for positions or objects are frequent.

The common raven, a large, social, long-lived member of the crow family, is a bird that has stimulated the imaginations of artists from the unknown Icelandic author of *Njal's Saga* to Edgar Allan Poe. Observations of ravens in captivity and in the wild indicate that ravens of all ages play with each other and with objects. Their play behavior suggests that of primates or carnivores, but it occupies a unique position in the natural history of play behavior because the birds frequently use their flying ability in play.

Gwinner (1964, 1966) observed a captive raven colony. When young ravens have learned to fly, they begin chasing and contact play. In a frequent form of play, one raven flew carrying an inedible object, and another raven would chase the first; if the first raven dropped this object, the second animal often caught it before it reached the ground. Play between ravens on a perch or on the ground included playfights in which one animal would jump on the back of another; play partners would also lie on their sides and kick at each other.

I observed behavior that appeared to be play for about 30 minutes in a band of five ravens flying above the Franconia Ridge in the White Mountains, New Hampshire (U.S.A.). The ravens dived on one another, chased each other in a straight line, or soared higher only to repeat these activities. Because individual ravens were easily followed from my vantage point on the ridge crest, I could see that the roles of attacker and defender, and of pursuer and pursued, alternated frequently and that the band continued to play while flying the length of the ridge. The animals were silent while playing and did not disperse or separate for more than a few seconds. For these reasons, and because the aerial acrobatics and "dogfights" of the birds matched Gwinner's description, I judged them to be playing.

Mech (1966 p. 159) describes play between ravens and wolves (*Canis lupus*):

As the pack traveled across a harbor, a few wolves lingered to rest, and four or five accompanying ravens began to pester them. The birds would dive at a wolf's head or tail, and the wolf would duck and then leap at them. Sometimes the ravens chased the wolves, flying just above their heads, and once, a raven waddled to a resting wolf, pecked its tail, and jumped aside as the wolf snapped at it. When the wolf retaliated by stalking the raven, the bird allowed it within a foot before arising. Then it landed a few feet beyond the wolf and repeated the prank.

Crisler (1958 p. 283) reports similar play between a raven and her free-ranging wolf pups; Mech (1970 p. 288) views this relationship as a cooperative bond formed between two "extremely social" species. Existence of this tradition of interspecific play on Isle Royale and in Minnesota (U.S.A.) is interesting, because wolves in this geographical area do not appear to feed on ravens. Wolves occasionally catch ravens in Ontario (Pimlott, Shannon, & Kolenosky 1969). Whether wolf-raven play occurs in areas where wolves feed on ravens is unknown.

Gwinner's (1964) observations of raven social play indicate that wind stimulates play. His observations also serve to compare raven play with agonistic fighting. In play, agonistic displays are absent, behavior sequences are loosely linked, and object and acrobatic play are intermixed with chasing, jumping, and side-by-side kicking.

If ravens were the sole birds to exhibit elaborate play like that of primates or carnivores, their behavior might be dismissed as an interesting, but isolated case of convergence. However, large parrots may rival, if not surpass the common raven in playfulness.

Keas (*Nestor notabilis*) are New Zealand parrots. They inhabit forested mountains in the spectacular glacier and fjord country of South Island. These large, green birds use their powerful hooked bills and supple grasping feet in perching, feeding, object manipulation, and social interaction. Impressed by this and other observations of their manual dexterity, G. A. Smith (1971) argues that keas occupy an ecological niche and practice a life-style like that of large corvids in other localities. (Ravens and crows, as well as all mammalian carnivores, were not present in New Zealand before Western colonization.)

The first published account of play in this exotic bird (Derscheid 1947) appeared posthumously. It was written in Brussels while the author was a prisoner of war, with materials smuggled in by a fellow prisoner, who "was able to pass paper and pencils under a bowl of soup and by other devious means." Derschied's keas played with snow and with chunks of ice, fishing the ice out of a stream; the birds used a cup and a metal box to empty water out of their bath, and they manipulated and attempted to dismantle complex objects. In a companion article to Derscheid's, Porter (1947), who succeeded in breeding these "extra-ordinarily intelligent and ingenious" animals, described their play with objects and with each other. Keas roll an empty metal can back and forth to one another by pushing it along the ground with their heads. They float small objects on moving water in their cage gutter and trot alongside until the objects disappear down the drainpipe. Doubt about the relevance of these observations to free-living birds was dispelled by Jackson (1963a), who conducted field observations at a kea's nest. Although interpretation of most of his observations seems unexceptionable, Jackson's account of the behavior of a fledged chick in flight with its "stepmother" raises some questions: "The chick dives at her like a falcon. She rolls over and parries a blow. Then the tables are turned. They play in the gusts of a storm, swing around a spur, plunge down into the shelter of a gully and back into the wind on the turbulent air." If the male had remated with a female unrelated to his former mate we might naively expect his new mate to drive the chick away, not to play with it. It is unclear from this account whether this interaction was isolated or repeated, how "playful" it really was, whether the chick was in fact eventually driven away, the role of the mate, and the degree of relatedness of the new mate to the chick. Like the early accounts of langur aggression cited by Blaffer Hrdy (1977), this anecdote of a newly mated female playing reciprocally and cooperatively with her mate's chick by a previous hen suggests need for detailed study of the phenomenon described.

Over a period of nearly four years Keller (1975, 1976) observed two kea families and three additional adults in the kea breeding colonies at Zurich Zoo and at Jersey Zoological Park. Keller's study placed particular emphasis on observations of a group of four young keas. Young keas are true clowns. They stand on their heads and turn somersaults on branches, on the ground, and very often in deep water, where the birds may then swim around on their backs or perform pirouettes. Play also includes behaviors that exercise muscles and skills used in flight—skimming the surface of newfallen snow, or landing upside down. The birds hang by their bills from branches, or hang upside-down, holding on with their feet. A favorite form of play makes use of a swing (a branch suspended from two parallel chains). The keas stand on the seat of the swing and set it in motion by wing flapping or by moving their center of gravity back and forth. Often, two or three birds swing together, exhibiting an astonishing degree of coordination in their joint efforts to put the swing in motion. Birds also swing from a hanging chain, performing body gymnastics.

Keller describes rich and diverse forms of play with objects and food items. Keas play with snow, making and pushing snowballs. A bird often lies on its back, balancing an object with its feet. Keas throw objects and chase after them. Keller's observations confirm earlier reports of keas floating objects on water, but of all individuals he observed, only one was seen playing this game.

Social play in young captive keas is as frequent as it is complex. Numerous postures and body movements serve to elicit play. These play-signals include stiff-legged walking, "teasing" the partner by manipulating objects, rolling on one's back, and adopting a defensive posture with head down and foot raised. If the partner responds to these solicitations, playfighting and play-chasing follow. Two animals may wrestle with their beaks, or they may adopt a stand-up/belly-up position. Social play is silent in keas, but when one bird occupies an elevated position and the others try to knock or pull it down ("King of the castle"), all birds vocalize. Hide-and-seek is a second kea game having mammalian counterparts.

So spectacular and so well-documented is kea play that we may confidently predict similarly complex behavior in other large-brained, large-bodied, long-lived parrots having known manipulative skill including macaws, Amazons, and the African grey parrot. Brief reports of play in the kea's close relative the kaka (*Nestor meridionalis*) (Jackson 1963b), in golah (*Kakatoe roseicapilla*) (Brereton 1971), and in captive hyacinthine macaws (*Anodorhynchus hyacinthinus*) (Hick 1962), African grey parrots and blue-fronted Amazons (Braun 1952) confirm this expectation.

The magnificent hyacinthine macaw, of central Brazil, is perhaps the largest and finest of all parrots. Its deep cobalt blue plumage is set off by a bright yellow eye-ring and accented by a narrow band of bare yellow skin bordering the base of the powerful bill. Play of this fashion-plate bird, reported from the Köln Zoo by Uta Hick (1962), is likewise remarkable. A macaw turns on its back and bites the hand of its human playmate very gently. The partner pulls its feathers, pulls its foot, then pushes it away, and the macaw hops or flies back for another round of play. A macaw invites play by hopping toward its partner, head obliquely inclined, and then springing. If its partner pulls the macaw's feathers too roughly or does not let the bird up, or if noises from a neighboring cage disturb the macaw, it threatens its playmate—an instance of play breaking down in a bird.

Hyacinthine macaw play has also been observed by Mr. Ralph Small (pers. comm.). Mr. Small's aviary is the first in the United States to successfully breed this species. One female macaw would sit on a branch next to the male bird, then jump back and forth over him in an apparent attempt to solicit play from him. Hand-reared macaw chicks would playfight with Mr. Small:

The hand-raised babies would put their upper bill over my wrist and the lower under and then gradually squeeze . . . . He[one chick] would keep one eye on my face and watch my other hand which I would raise. When he thought I was going to hit him he would take off like a puppy, with a big smile on his face. (R. Small, pers. comm.)

Subsong

Subsong is a developmental stage of bird song. Young birds produce quiet, prolonged vocalizations similar to adult song, but differing from it in structure and in frequency range. Analysis of subsong in its ontogenetic environment is an important aspect of current developmental research (p. 264). Applying classical structuralist criteria to subsong, and observing that subsong appears to serve no function other than that of practice for adult activity, Ficken (1977), following Thorpe's (1966) suggestion, chose to consider subsong "as at least sharing some important characteristics with play."

Is subsong play? Young birds that produce subsong cannot produce primary song, and subsong is very rare or absent in adult males when they produce primary song during the breeding season. Yet both its elaborateness and lack of simple formal relationships with primary song indicate that subsong is more of an independent behavioral entity than the pre-impaling behavior (incomplete and preliminary impaling behavior and inanimate object manipulation) of young loggerhead shrikes or the seemingly random pecking of young oyster-catchers. There is need for an objective scale on which we may define, measure, and compare structural and contextual properties of different developmental intermediates.

Avian Play: Conclusions

Because birds, unlike mammals, are primarily visual and aerial, object play is common and vigorous, whereas aerial maneuvers of various sorts confuse the uninitiated. In addition, subsong is an interesting, but typologically annoying phenomenon that seems to be on or close to the arbitrary border between play and other kinds of behavior occurring in development. Given the evolutionary opportunity, avian phylogeny can produce play like that of the kea, hyacinthine macaw, and common raven that appears to surpass most mammalian play in elaborateness and complexity. On the other hand, most avian play seems somewhat ambiguous by mammalian standards, either because it has not fully emerged from maturational intermediate status or because existing accounts relied on inadequate criteria and failed to provide necessary information. Birds' high weight-specific metabolism and low relative brain size may fail to favor evolution of clear distinctions between play and other adult or developing behavior. Higher metabolic costs may make play economically less likely. Low relative brain size may also affect the need to play, the ability to play, the benefits and costs of play, or the evolutionary stability of social play. Such postulated effects require identification and analysis of mechanisms that phylogenetic correlations cannot possibly reveal. Avian play provides a comparative base line essential for understanding the evolution of animal play behavior. Birds could become as important to the study of play as they have been to developmental studies of such phenomena as imprinting and species-specific vocalization.

PLAY IN TAXONOMIC GROUPS
OTHER THAN MAMMALS AND BIRDS

Play is a minority phenomenon in nature. The evolution of play precisely mirrors the evolution of the brain. Play and a highly

developed cerebral cortex go together. Despite the highly sophisticated forms of social (including parental) and feeding behavior found in many species of fishes and social insects, play is
virtually nonexistent in these small-brained and relatively de-
corticate species (Table 3-30). Even in reptiles, where play
might confidently be expected purely on phylogenetic grounds,
especially in the large, long-lived varanid lizards or in family-
living crocodilians, accounts of play are rare and restricted to
interactions with objects (Table 3-30).

Accounts, chiefly anecdotal, of object and social play in lizards, crocodilians, mormyrid and cichlid fish, and even ants and
cockroaches (though these reports of insect play are unconvinc-
ing), suggest that the question of play in "lower" animals is still
open. Brain changes occur in response to social experience in
jewel fish (Coss & Globus 1979). Could environmental feedback obtained in play stimulate such responses, as has been
hypothesized (Fagen 1976a, Ferchmin & Eterovic 1979) for
mammals? Far too frequently, conclusions based on a few observations that individuals of some species failed to play have
been prematurely elevated to the status of dogma. Sclater's
(1863) observation that slow loris (Nycticebus coucang) never
played did not stop later observers from describing unmistak-
able play in this species (e.g., Ehrlich & Musicant 1977). Similarly, play in taxonomic groups other than mammals and birds
cannot be ruled out at this time despite the current lack of evidence for its existence.

ABOUT THE TABLES

These tables cite published naturalistic descriptions and analyses
of animal play. Major and definitive works are in italic type. I
report all such references known to me, with the exception of
information on three species: rhesus macaque (Macaca mulatta),
chimpanzee (Pan troglodytes), and human (Homo sapiens). Literature on play in these three species is voluminous. I have tabulated studies that I consider most important for understanding
the natural history of play in these three species. Films are denoted by an asterisk (e.g., *Bishop & Symons 1978).

As previously discussed, literature on the natural history of any behavioral topic (including play) includes verbal or anecdotal accounts, observations of confined animals, and some ambiguous reports [identified here by a (?) following the citation]. These tables cannot, of course, be considered complete.

Table 3-1 PLAY IN NONHUMAN MAMMALS

Order Monotremata (Echidnas, platypus)	See text
Order Marsupiala (Pouched mammals: opossums, koala, kangaroos, and allies)	See text and Table 3-2
Order Insectivora (Shrews, hedgehogs, moles, and allies)	See text
Order Dermoptera (Flying "lemurs")	No information
Order Chiroptera (Bats)	See text
Order Primates (Lemurs, monkeys, apes, and allies)	See text and Tables 3-3, 3-4, 3-5, 3-6, 3-7, 3-8, 3-9
Order Edentata (Sloths, anteaters, and armadillos)	See text
Order Pholidota (Pangolins)	See text
Order Lagomorpha (Hares, rabbits, and pikas)	See text
Order Rodentia (Gnawing mammals: squirrels, mice, rats, and allies)	See text and Tables 3-10, 3-11, 3-12
Order Cetacea (Baleen and toothed whales, the latter including dolphins and porpoises)	See text

Order Carnivora (Dogs, bears, cats, and allies)	See text and Tables 3-13, 3-14, 3-15, 3-16, 3-17, 3-18, 3-19
Order Pinnipedia (Seals, walruses, and sea lions)	See text and Table 3-20
Order Tubulidentata (Aardvark)	No information
Order Proboscidea (Elephants)	See text
Order Hyracoidea (Hyraxes)	No information
Order Sirenia (Dugongs and manatees)	See text
Order Perissodactyla (Odd-toed hoofed mammals: horses, zebras, tapirs, and rhinoceroses)	See text and Table 3-21
Order Artiodactyla (Even-toed hoofed mammals: swine, deer, sheep, antelopes, cattle, and allies)	See text and Tables 3-22, 3-23, 3-24, 3-25

Table 3-2 MARSUPIAL PLAY

Family Dasyuridae

Dasyuroides byrnei, kowari	Aslin 1974
Dasyurus maculatus, tiger quoll	Settle 1977
Dasyurus viverrinus, quoll	Fleay 1935, Nelson & Smith 1971
Sarcophilus harrisii, Tasmanian devil	Ewer 1968a, Hediger 1958, Roberts 1915, Turner 1970
Sminthopsis crassicaudata, fat-tailed dunnart	Ewer 1968b(?)

Family Peramelidae

Isoodon obesulus, Southern short-nosed bandicoot	Heinsohn 1966

Table 3-2 MARSUPIAL PLAY (*Continued*)

Family Phascolomidae	
Lasiorhinus latifrons, hairy-nosed wombat	Wünschmann 1966
Other accounts	Grzimek 1967, Hediger 1958, Troughton 1941
Family Macropodidae	
Macropus giganteus, great gray kangaroo	Herrmann 1971, Kaufmann 1975
Macropus parryi, whiptail wallaby	Kaufmann 1974
Macropus rufus, red kangaroo	Cicala, Albert, & Ulmer 1970, Lamond 1953

Table 3-3 PRIMATE PLAY: FAMILY LEMURIDAE*

Cheirogaleus major, greater dwarf lemur	*Petter & Petter 1963*
Daubentonia madagascarensis, aye-aye	Petter et al. 1977, Petter & Peyrieras 1970a
Galago alleni, Allen's galago	Charles-Dominique 1971
Galago crassicaudatus, greater bush-baby	Newell 1971, Roberts 1971, Rosenson 1972, Welker 1977
Galago senegalensis, lesser bush-baby	Andersson 1969, Bearder 1969, Doyle 1974a, Gucwinska & Gucwinski 1968, Lowther 1940, Pinto 1972, Sauer & Sauer 1963
Hapalemur griseus, bokombouli	Petter et al. 1977, Petter & Peyrieras 1970b
Indri indri, endrina	Petter et al. 1977, Pollock 1975
Lemur catta, ringtailed lemur	Jolly 1966a, Klopfer & Klopfer 1970, Martin 1972, Petter et al. 1977, Sussman 1977
Lemur fulvus, brown lemur	Chandler 1975, Conley 1975, Harrington 1975, Petter et al. 1977, Sussman 1977, Tattersall 1977, Vick & Conley 1976

Lemur macaco, black lemur	Jolly 1966a, Klopfer 1972, Petter et al. 1977, Petter & Peyrieras 1970b
Microcebus coquereli, Coquerel's mouse lemur	Petter et al. 1977
Microcebus murinus, mouse lemur	Petter 1965, Petter et al. 1977, Petter-Rousseaux 1964, Pinto 1972
Nycticebus coucang, slow loris	Ehrlich & Musicant 1977
Perodicticus potto, potto	Anderson 1971, Charles-Dominique 1971, Walker 1968
Propithecus verreauxi, sifaka	Jolly 1966a, 1972, Petter et al. 1977, Richard 1974, 1976, Richard & Heimbuch 1975
Varecia variegata, ruffed lemur	Petter-Rousseaux 1964, Petter et al. 1977

* General: Bishop 1962, Bishop 1963, Doyle 1974b, Jolly 1966a, Petter 1965, Petter et al. 1977

Table 3-4 PRIMATE PLAY: FAMILY CEBIDAE

Alouatta seniculus, red howling monkey	Carpenter 1934, 1969, Neville 1972a,b
Alouatta villosa, mantled howling monkey	S. Altmann 1959, *Baldwin & Baldwin 1973a,* 1976b, Bernstein 1964, Carpenter 1965, Eisenberg 1976, Richard 1970
Alouatta sp., howling monkey	Coelho et al. 1976, Schmid 1934
Aotus trivirgatus, night monkey	English 1934 (?), Moynihan 1964, 1970
Ateles belzebuth, Humbolt's spider monkey	Bopp 1968, Klein 1971
Ateles geoffroyi, red spider monkey	Carpenter 1935, 1969, Eisenberg & Kuehn 1966, Klein 1971, Richard 1970, Rondinelli & Klein 1976
Ateles sp., spider monkey	Coelho et al. 1976
Callicebus moloch, dusky titi	Moynihan 1966

Table 3-4 PRIMATE PLAY: FAMILY CEBIDAE (*Continued*)

Callicebus torquatus, yellow-handed titi	Kinzey et al. 1977
Callicebus sp., titi	Mason 1974
Cebus albifrons, white-fronted capuchin	Bernstein 1965
Cebus apella, brown capuchin	Gehring 1976, Nolte 1958, Weigel 1979
Cebus capucinus, white-faced capuchin	Oppenheimer 1968, 1969, *1974
Cebus nigrivittatus, weeper capuchin	Oppenheimer & Oppenheimer 1973
Chiropotes albinasus, white-nosed saki	Hick 1968 (?)
Lagothrix lagothricha, woolly monkey	Kavanagh & Dresdale 1975, Schifter 1968
Saimiri spp., squirrel monkey	Baldwin 1968, 1969, 1971, Baldwin & Baldwin 1971, 1972, *1973b, 1974, 1976a,* DuMond 1968, Hopf 1971, 1974, Kaplan 1972, *Latta et al. 1967,* Mason 1974, Ploog & Maurus 1973, Ploog et al. 1967, 1975, Rosenblum 1968, Rosenblum & Lowe 1971, Strayer et al. 1975, Thorington 1968, Winter 1968, 1969

Table 3-5 PRIMATE PLAY: FAMILY CALLITRICHIDAE

Callithrix argentata, silvery marmoset	Stevenson 1976
Callithrix jacchus, white tufted-ear marmoset	Box 1975a, Chartin & Petter 1960, Fitzgerald 1935, Lucas et al. 1927, Stevenson 1976, Stevenson & Poole 1976
Cebuella pygmaea, pygmy marmoset	Christen 1974
Leontopithecus rosalia, lion tamarin	Frantz 1963, Kleiman 1977, Snyder 1972

Saguinus fuscicollis, saddle-back tamarin	Moody & Menzel 1976, Stevenson 1976
Saguinus midas, black-handed tamarin	Christen 1974
Saguinus oedipus, cotton-top marmoset	Hampton et al. 1966, Willig & Wendt 1970(?)

Table 3-6 PRIMATE PLAY: FAMILY CERCOPITHECIDAE

Allenopithecus nigroviridis, Allen's swamp monkey	Pournelle 1962 (?)
Cercocebus albigena, grey-cheeked mangabey	Gautier-Hion & Gautier 1971
Cercocebus atys, sooty mangabey	Bernstein 1971, 1976, Glickman & Sroges 1966
Cercocebus sp., mangabey	Hinde 1971
Cercopithecus aethiops, vervet	Altmann & Altmann 1970, Bramblett 1978, *Fedigan 1972b,* Gartlan 1968, 1969, *Rose 1977a,* Struhsaker 1967b, 1971
Cercopithecus ascanius, redtail monkey	Galat-Luong 1975, Struhsaker 1975
Cercopithecus campbelli, Lowe's guenon	*Bertrand et al. 1969, Bourlière et al. 1969, 1970, Hunkeler et al. 1972
Cercopithecus diana, diana monkey	Mörike 1973
Cercopithecus erythrotis, russet-eared guenon	Struhsaker 1969
Cercopithecus mitis, Sykes's monkey	Dolan 1976, Struhsaker 1975
Cercopithecus neglectus, De Brazza monkey	Mörike 1976, Stevenson 1973
Cercopithecus nictitans, spot-nosed guenon	Gautier-Hion & Gautier 1974, Struhsaker 1969
Cercopithecus pogonias, crowned guenon	Gautier-Hion & Gautier 1974, Struhsaker 1969
Cercopithecus sabaeus, green monkey	Bernstein 1971, McGuire et al. 1974

Table 3-6 PRIMATE PLAY: FAMILY CERCOPITHECIDAE (*Continued*)

Colobus angolensis, Angola colobus	Groves 1973
Colobus badius, red colobus	Clutton-Brock 1972, 1974, Struhsaker 1975
Colobus guereza, black and white colobus	Horwich & La France 1972, Oates 1974, *Rose 1977a,* Struhsaker 1975
Colobus polykomos, king colobus	*Kirchshofer 1960*
Colobus satanas, black colobus	Struhsaker 1969
Erythrocebus patas, patas	Baker & Preston 1973, Bolwig 1963, *Gartlan 1975, Goswell & Gartlan 1965, Hall 1965, 1968, Hall & Mayer 1967, Hall et al. 1965, Struhsaker & Gartlan 1970
Macaca arctoides, stumptail macaque	Bertrand 1969, Chevalier-Skolnikoff 1973a, 1974b, Parker 1977, Rhine 1972, 1973, Rhine & Kronenwetter 1972, Trollope & Blurton Jones 1975
Macaca fascicularis, Kra macaque	Angst & Thommen 1977, Baker & Preston 1973, Bernstein 1971, *Fady 1969,* Flower 1900, Van Hooff 1972, Kurland 1973, *Oakley & Reynolds* 1976, Pitcairn 1976, Poirier & Smith 1974a, Schlottman & Seay 1972, Schmid 1934, Soczka 1974
Macaca fuscata, Japanese macaque	Alexander 1970, Candland et al. 1978, *Carpenter 1971,* Casey & Clark 1976, Fedigan 1976, Hanby 1974, Hanby & Brown 1974, Itani 1959, Itoigawa 1975, Kawai 1965, Kurland 1977, *Mori 1974,* Norikoshi 1974, Sugiyama 1976b, Takeda 1965, Yamada 1963
Macaca mulatta, rhesus macaque	S. *Altmann 1962a,b, 1965, *Bishop & Symons 1978, Breuggeman 1978,* Hansen 1966, Hinde & Davies 1972, Hinde, Rowell, & Spencer-Booth 1964, *Hinde & Simpson 1975,* Hines 1942, Koford 1963, *Levy 1979,* Lichstein 1973, *Mears & Harlow 1975,* Mukherjee 1969, Oakley & Reynolds 1976, Redican & Mitchell 1974, Simpson 1976, Southwick 1967, Southwick et al. 1965, *Symons 1978a, White 1977a,b*

Macaca nemestrina, pigtail macaque	Bernstein 1967, 1969, 1970, 1971, 1972a, Kaufman & Rosenblum 1965, 1966, 1967a,b, Rosenblum 1971a,b, Rosenblum et al. 1969, Stynes et al. 1968
Macaca nigra, Celebes black ape	Bernstein 1971, Dixson 1977, Sparks 1967
Macaca radiata, bonnet macaque	Kaufman & Rosenblum 1965, 1966, 1967b, Nolte 1955a,b, Rahaman & Parthasarathy 1969, Rosenblum 1971a,b, Simonds 1965, 1977, Stynes et al. 1968, Sugiyama 1971
Macaca sylvanus, Barbary ape	Burton 1972, Deag 1973, Lahiri & Southwick 1966, MacRoberts 1970
Mandrillus leucophaeus, drill	Gartlan 1970, Struhsaker 1969
Mandrillus sphinx, mandrill	Emory 1975, 1976, Jouventin 1975, Sabater Pi 1972
Miopithecus talapoin, talapoin	Dixson et al. 1975, Gautier-Hion 1970, 1971, Gautier-Hion & Gautier 1971, Hill 1966, Rowell 1975, Wolfheim 1977, Wolfheim & Rowell 1972
Nasalis larvatus, proboscis monkey	Pournelle 1967
Papio anubis, olive baboon and *Papio cynocephalus,* yellow baboon	Altmann & Altmann 1970, Berger 1972, Babault 1949, Buirski et al. 1973, Chalmers 1978, Douglas-Hamilton & Douglas-Hamilton 1975, Dunbar & Dunbar 1974, Grzimek & Grzimek 1960, van Lawick-Goodall 1968a, Morris & Goodall 1977, Nash 1978, *Owens 1974, Owens 1975a,b,* 1976, Ransom 1971, Ransom & Ransom 1971, Ransom & Rowell 1972, Rose 1977b, Rowell 1967, Rowell, Din, & Omar 1968
Papio hamadryas, hamadryas baboon	Kummer 1967, 1968a,b, Kummer & Kurt 1965, *Leresche 1976*
Papio ursinus, chacma baboon	Bolwig 1959, Cheney 1978, De Vore 1963, Hall 1962, 1963, *Hall & Carpenter 1967,* W. J. Hamilton et al. 1975, Saayman 1970
Presbytis cristatus, lutong	Bernstein 1968

Table 3-6 PRIMATE PLAY: FAMILY CERCOPITHECIDAE *(Continued)*

Presbytus entellus, hanuman langur	Beck & Tuttle 1972, Blaffer Hrdy 1977, Dolhinow & Bishop 1970, Jay 1963, 1965, McCann 1933, Rahaman 1973, Ripley 1967, Sugiyama 1965a,b, 1976a, Vogel 1976
Presbytis geei, golden langur	Mukherjee & Saha 1974
Presbytis johnii, Nilgiri langur	Poirier 1968, 1969a,b, 1970a,b
Presbytis obscurus, spectacled langur	Horwich 1974, McClure 1964
Presbytis sp., langur	Flower 1900, Milne 1924
Pygathrix nemaeus, Douc langur	Hick 1972, Kavanagh 1978
Theropithecus gelada, gelada	Bernstein 1971, 1972b, 1975, *Crook 1967, Dunbar & Dunbar 1974, 1975, Emory 1975, 1976, Fedigan 1972a, van Hooff 1972

Table 3-7 PRIMATE PLAY: FAMILY PONGIDAE, AFRICAN SPECIES

Gorilla gorilla, gorilla	Carpenter 1937, R. C. Elliott 1976, Fossey 1972, *1979,* Freeman & Alcock 1973, Golding 1972, Hess 1973, Kirchshofer et al. 1967, 1968, Maple & Zucker 1978, Marler & Tenaza 1977, Redshaw & Locke 1976, Roth 1967, *Schaller 1963*
Pan troglodytes, chimpanzee	Angus 1971, *Bierens de Haan 1952, Bingham 1927,* 1929, Birch 1945a, Broderip 1835, Budd, Smith, & Shelley 1943, C. B. Clark 1977, Fouts, Mellgren, & Lemmon 1973, Fouts & Rigby 1977, Gardner & Gardner 1969, Garner 1919, Goodall 1965, Hayes 1952, Hayes & Hayes 1952, Hladik 1973, van Hooff 1962, 1967, 1970, 1971, 1972, Köhler 1926, van Lawick-Goodall 1967, *1968a,b,* 1970, *1971,* 1973, *Loizos 1969,* McGrew 1977, Maple & Zucker 1978, *Marler & van Lawick-Goodall 1971,* Marler & Tenaza 1977, Mason 1965a,b, 1967, Mason, Hollis, & Sharpe 1962, Menzel 1963, 1969, 1971, 1972, 1973, 1974, 1975, Menzel, Davenport, & Rogers 1961, 1970, 1972, *Merrick 1977,* Milhaud, Klein, & Chapouthier 1973, Morris & Goodall 1977, Nicolson 1977, Nissen

1931, Randolph & Brooks 1967, Reynolds &
Luscombe 1976, Reynolds & Reynolds 1965,
Riss & Goodall 1976, Rumbaugh 1974, *Savage
& Malick 1977,* Savage, Temerlin, & Lemmon
1973, Schiller 1952, 1957, Sheak 1924, Simpson
1978, Sugiyama 1969, Tutin & McGrew 1973,
M. L. Wilson & Elicker 1976, Yerkes 1943

Table 3-8 PRIMATE PLAY: FAMILY PONGIDAE, ASIAN SPECIES

Hylobates lar, lar gibbon	Baldwin & Teleki 1976, Bernstein 1971, Bernstein & Schusterman 1964, Brody & Brody 1974, *Carpenter 1940,* *1974, Delacour 1933, Du Mond 1970, Ellefson 1966, Ibscher 1967, Paluck et al. 1970
Pongo pygmaeus, orangutan	Davenport 1967, Du Mond 1970, *Freeman & Alcock 1973, Galdikas 1978,* Harrisson 1963, Horr 1977, Jantschke 1972, Lethmate 1976, 1977, *MacKinnon 1974a, Maple & Zucker 1978, Rijksen 1978,* Tempel 1971
Symphalangus syndactylus, siamang	Aldrich-Blake & Chivers 1973, Bennett 1976, Chivers 1973, 1974, 1976, G. Fox 1972, McClure 1964

Table 3-9 PRIMATE PLAY: FAMILY HOMINIDAE

Homo sapiens, humans	Aldis 1975, Blurton Jones 1967, 1972, 1975, Blurton Jones & Konner 1973, Draper 1976, DeVore & Konner 1974, Eibl-Eibesfeldt 1973, Fenson et al. 1974, 1976, Goldberg & Lewis 1969, van Hooff 1972, Hutt 1966, Hutt & Bhavnani 1972, Konner 1972, 1976b, McCall 1974, Opie & Opie 1967, 1969, P. K. Smith 1973, P. K. Smith & Connolly 1972, Stern 1974a,b, Whiting & Whiting 1975

Table 3-10 RODENT PLAY: FAMILIES SCIURIDAE, CASTORIDAE

Family Sciuridae

Cynomys gunnisoni, Gunnison's prairie dog	Fitzgerald & Lechleitner 1974, Tileston & Lechleitner 1966
Cynomys ludovicianus, black-tailed prairie dog	*King 1955,* W. J. Smith et al. 1973, Tileston & Lechleitner 1966

Table 3-10 RODENT PLAY (*Continued*)

Cynomys mexicanus, Mexican prairie dog	Pizzimenti & McClenaghan 1974
Eutamias sp., chipmunk	Jaeger 1929
Glaucomys volans, flying squirrel	Svihla 1930 (?)
Marmota bobak, Bobak marmot	Andrews 1932
Marmota caligata, hoary marmot	Barash 1974c
Marmota flaviventris, yellow-bellied marmot	Armitage 1962, 1974
Marmota marmota, Alpine marmot	Barash 1976, Koenig 1957, Müller-Using 1956, Münch 1958
Marmota olympus, Olympic marmot	Barash 1973a
Otospermophilus beecheyi, California ground squirrel	*Linsdale 1946, McDonald 1977*
Otospermophilus lateralis, mantled ground squirrel	Gordon 1943
Petaurista petaurista, giant flying squirrel	Krishnan 1972
Petaurista sp., flying squirrel	Minett 1947
Sciurus carolinensis, grey squirrel	Horwich 1972
Sciurus niger, fox squirrel	Svihla 1931 (?)
Sciurus vulgaris, European red squirrel	David 1940, Eibl-Eibesfeldt 1951a, Frank 1952, Horwich 1972
Spermophilus armatus, Uinta ground squirrel	Balph & Stokes 1963
Spermophilus columbianus, Columbian ground squirrel	Betts 1976, *Steiner 1971*
Spermophilus richardsoni, Richardson's ground squirrel	Yeaton 1972
Tamias striatus, Eastern chipmunk	Henisch & Henisch 1970 p. 86, Horwich 1972, Thoreau 1949 p. 508
Tamiasciurus hudsonicus, American red squirrel	*Ferron 1975*, Horwich 1972, Klugh 1927
Xerus erythropus, African ground squirrel	Ewer 1966

Family Castoridae

Castor canadensis, American beaver	Cahalane 1947, Kalas 1976, Leighton 1933, Schramm 1968, Warren 1927
Castor fiber, European beaver	Wilsson 1968, 1971

Table 3-11 RODENT PLAY: FAMILIES CRICETIDAE, MURIDAE, GLIRIDAE

Family Cricetidae

Cricetus cricetus, common hamster	Eibl-Eibesfeldt 1953
Dicrostonyx groenlandicus, collared lemming	Kay 1978
Meriones persicus, Persian jird	Eibl-Eibesfeldt 1951b
Meriones tamariscinus, tamarisk gerbil	Rauch 1957
Meriones unguiculatus, Mongolian gerbil	DeGhett 1970, Ehrat et al. 1974
Mesocricetus auratus, golden hamster	Dieterlen 1959, Goldman & Swanson 1975, Kalacheva 1965, Ponugaeva 1961, Rowell 1961
Microtus agrestis, short-tailed vole	S. Wilson 1973
Neotoma albigula, white-throat wood rat	Richardson 1943
Neotoma floridiana, Eastern wood rat	Kinsey 1977
Ondatra zibethicus, muskrat	Steiniger 1976
Psammomys obesus, sand rat	Daly & Daly 1975

Family Muridae

Apodemus flavicollis, yellow-necked field mouse	Zippelius 1971
Apodemus mystacinus, broad-toothed field mouse	Dieterlen 1965
Apodemus sylvaticus, wood mouse	Dieterlen 1965
Cricetomys gambianus, African giant rat	Ewer 1967
Micromys minutus, harvest mouse	Frank 1957
Mus musculus, house mouse	van Abeleen & Schoones 1977, Poole & Fish 1975, Wahlsten 1974, Williams & Scott 1954
Notomys alexis, northern hopping mouse	Happold 1976a, Stanley 1971

Table 3-11 RODENT PLAY (*Continued*)

Pseudomys albocinereus, ashy-grey mouse	Happold 1976a
Pseudomys desertor, brown desert mouse	Happold 1976a
Pseudomys shortridgei, blunt-faced rat	Happold 1976a
Rattus norvegicus, Norway rat	Barnett 1958, 1969, 1975, Bolles & Woods 1964, Draper 1967, *Müller-Schwarze 1966*, 1971, *Olioff & Stewart 1978, Poole & Fish 1975, 1976*
Rattus rattus, black rat	Ewer 1971, Mohr 1928
Family Gliridae *Glis glis*, greater dormouse	Koenig 1960
Muscardinus avellanarius, hazel mouse	Zippelius & Goethe 1951

Table 3-12 RODENT PLAY: FAMILIES HYSTRICIDAE, ERETHI-ZONTIDAE, CAVIIDAE, DINOMYIDAE, DASYPROCTIDAE, CAPROMYIDAE, OCTODONTIDAE

Family Hystricidae *Hystrix cristata*, African porcupine	Mohr 1965
Family Erethizontidae *Erethizon dorsatum*, North American porcupine	Shadle & Ploss 1943, Spencer 1930
Family Caviidae *Cavia aperea*, guinea pig	Rood 1972
Cavia porcellus, domestic guinea pig	Coulon 1971, Gerall 1963, Kunkel & Kunkel 1964
Dolichotis patagonum, mara	Dubost & Genest 1974
Dolichotis salinicola, salt desert cavy	*Wilson & Kleiman 1974*
Galea musteloides, yellow-toothed cavy	Rood 1972
Microcavia australis, desert cavy	Rood 1972
Myoprocta pratti, green acouchi	Bell 1830, Morris 1962a
Family Dinomyidae *Dinomys branickii*, pacarana	Collins & Eisenberg 1972

Family Dasyproctidae
Cuniculus paca, paca — Pilleri 1960

Dasyprocta aguti, agouti — Roth-Kolar 1957

Family Capromyidae
Geocapromys brownii, Jamaican hutia — Oliver 1975

Myocastor coypus, coypu — Eibl-Eibesfeldt 1952

Family Octodontidae
Octodon degus, degu — Fulk 1976, *Wilson & Kleiman 1974*

Octodontomys gliroides, choz-choz — *Wilson & Kleiman 1974*

Table 3-13   CARNIVORE PLAY: FAMILY CANIDAE

General (*Canis*)	*Bekoff 1974c,* Fox 1969a,b, 1970, 1972a, Fox & Cohen 1977, *Fox et al. 1976*
Alopex lagopus, arctic fox	Stefansson 1944, *Tembrock 1960,* Donald 1948, Heimburger 1961
Canis aureus, golden jackal	van Lawick & van Lawick-Goodall 1971, Neelakantan 1969, Roberts 1977, Rodon 1898, Seitz 1959, Wandrey 1975
Canis dingo, dingo	Meshkova 1970
Canis familiaris, domestic dog	Anon. 1976, Brownlee 1974, Ebhardt 1954, *Ludwig 1965,* Lunt 1968, Seitz 1955, Zimen 1972
Canis latrans, coyote	Bekoff 1978a, Borell & Ellis 1934, Fichter 1950, Fox & Clark 1971, *Hill & Bekoff 1977,* Lehner 1978, Snow 1967, *Vincent & Bekoff 1978*
Canis lupus, timber wolf	Fentress 1967, Fox 1973, Heimburger 1961, Krämer 1961, Lockwood 1976, Mech 1966, 1970, Mowat 1963, Schönberner 1965, Zimen 1972
Canis mesomelas, blackbacked jackal	van Lawick & van Lawick-Goodall 1971, Moehlman 1979
Canis sp., jackal	Saayman 1970
Chrysocyon brachyurus, maned wolf	D. Altmann 1972, Encke et al. 1970, Hämmerling & Lippert 1975, Kleiman 1972
Cuon alpinus, dhole	Sosnovskii 1967

Table 3-13 CARNIVORE PLAY (*Continued*)

Dusicyon sechurae, Peruvian desert fox	Birdseye 1956
Lycaon pictus, African hunting dog	*Apfelbach 1969b*, Estes & Goddard 1967, Kühme 1965, van Lawick & van Lawick-Goodall 1971, Pfeffer 1972
Nyctereutes procyonoides, racoon-dog	Meshkova 1970, Seitz 1955
Otocyon megalotis, bat-eared fox	von Ketelhodt 1966, van Lawick & van Lawick-Goodall 1971, Loveridge 1923(?)
Speothos venaticus, bush dog	Bates 1944, *Drüwa 1977*, Jantschke 1973, Kleiman 1972
Urocyon cinereoargenteus, grey fox	Fox 1969a,b, Lunt 1968
Vulpes corsac, corsac fox	D. Altmann 1971
Vulpes fulva and *V. vulpes*, red fox	Anon. 1976, Dodsworth 1913, von Frisch & von Frisch 1971, Gianini 1923, Hurrell 1962, Meshkova 1970, Seitz 1950a,b, Sheldon 1921, *Tembrock 1957, 1958*, Vesey-Fitzgerald 1965
Vulpes rueppelli, sand fox	Petter 1952
Vulpes zerda, fennec	Gauthier-Pilters 1962, *1966*, Koenig 1970, Melchior 1976, Vogel 1962

Table 3-14 CARNIVORE PLAY: FAMILY URSIDAE

Ailuropoda melanoleuca, giant panda	Haas 1963, Kleiman & Collins 1972, Morris & Morris 1966, *Wilson & Kleiman 1974*
Thalarctos maritimus, polar bear	Ehlers 1964, Faust & Faust 1959, Gorgas 1972, *Koenig 1972, Perry 1966, Steinemann 1966
Tremarctos ornatus, spectacled bear	Eck 1969
Ursus americanus, black bear	Aldous 1937, Burghardt & Burghardt 1972, *Henry & Herrero 1974*, Howard 1935, Leyhausen 1949, Pruitt 1976
Ursus arctos, brown and grizzly bears	Egbert & Stokes 1976, Herrero & Hamer 1977, Krott 1961, Krott & Krott 1963, Meyer-Holzapfel 1968, *Stoinitzer 1959
Other ursids	MacKinnon 1974b, Volmar 1940

Table 3-15 CARNIVORE PLAY: FAMILY PROCYONIDAE

Ailurus fulgens, red panda	Keller 1977, Roberts 1975
Bassaricyon sp., olingo	Poglayen-Neuwall & Poglayen-Neuwall 1965
Bassariscus astutus, ringtail	Toweill & Toweill 1978, Trapp 1972
Bassariscus sumichrasti, cacomistle	Poglayen-Neuwall 1973
Nasua narica, coati	Bennett 1860b, Kaufmann 1962, Wallmo & Gallizioli 1954
Potos flavus, kinkajou	Poglayen-Neuwall 1962
Procyon cancrivorus, crab-eating raccoon	Löhmer 1976
Procyon lotor, raccoon	Cole 1912, Roth 1970, Welker 1959, Whitney & Underwood 1952

Table 3-16 CARNIVORE PLAY: FAMILY MUSTELIDAE

General (otters)	Davis 1979, Harris 1968, Maxwell 1961
Arctonyx collaris, hog badger	Jackson 1918
Eira barbara, tayra	Brosset 1968, Vaughn 1974
Enhydra lutris, sea otter	Ewer 1973, Fisher 1940, Hall & Schaller 1964, Limbaugh 1961
Galictis cuja, grison	Dalquest & Roberts 1951, Dücker 1968
Gulo gulo, wolverine	Behm 1953 (?), Krott 1953, 1960
Lutra canadensis, river otter	Audubon & Bachman 1851, Liers 1951, Lyon 1936, Murie 1954
Lutra lutra, European otter	Schreitmüller 1952, Ziems 1973
Lutra maculicollis, spotted-necked otter	Procter 1963
Lutra perspicillata, smooth-coated otter	Ansell 1947, Hodl-Rohn 1974, Yadav 1967
Martes foina, beech marten	Schmidt 1943
Martes flavigula, yellow-throated marten	Roberts 1970 (?), 1977
Martes martes, pine marten	Herter & Ohm-Kettner 1954, Remington 1952, Schmidt 1934, 1943
Martes zibellina, sable	Schmidt 1934

Table 3-16 CARNIVORE PLAY: FAMILY MUSTELIDAE *(Continued)*

Meles meles, European badger	Eibl-Eibesfeldt 1950b, Hardy 1975, Meshkova 1970, Naundorff 1929, Neal 1962, Schmid 1932, Walker 1975
Mellivora capensis, ratel	Hoesch 1964 (?), Roberts 1977
Mephitis mephitis, striped skunk	Stegeman 1937
Mustela erminea, ermine	East & Lockie 1965, Müller 1970
Mustela frenata, longtail weasel	W. J. Hamilton 1933, Pearce 1937
Mustela furo and *Mustela putorius,* ferrets, polecats and ferret-polecat hybrids	Goethe 1940, Herter & Herter 1955, Lazar & Beckhorn 1974, Poole 1966, 1967, 1974, 1978, Weiss-Bürger 1975, Wüstehube 1960
Mustela nigripes, black-footed ferret	Aldous 1940
Mustela nivalis, least weasel	East & Lockie 1964, Hartman 1964, Heidt et al. 1968, *Sedlag 1973*
Mustela vison, mink	Herter 1958, Poole 1978, Svihla 1931
Pteronura brasiliensis, Brazilian giant otter	Autuori & Deutsch 1977, Skeldon 1963
Spilogale putorius, spotted skunk	Gates 1937
Vormela peregusna, marbled polecat	Yate 1898
Other mustelids	MacKinnon 1974b

Table 3-17 CARNIVORE PLAY: FAMILY VIVERRIDAE

General (mongoose)	Hinton & Dunn 1967
Arctictis binturong, binturong	Kuschinski 1974, Landowski 1972
Civettictis civetta, African civet	Ewer & Wemmer 1974
Crossarchus obscurus, cusimanse	Bourlière, Bertrand, & Hunkeler 1969
Cryptoprocta ferox, fossa	Albignac 1969a, 1973, 1975
Eupleres goudotii, falanouc	Albignac 1973, 1974
Fossa fossana, fanaloka	Albignac 1970a, 1973
Galidia elegans, ringtailed mongoose	Albignac 1969b, 1973
Genetta genetta, common genet	Faugier & Condé 1973, Gangloff & Ropartz 1972

Genetta rubiginosa, rusty-spotted genet	Rowe-Rowe 1971
Genetta tigrina, large-spotted genet	Wemmer 1977
Helogale parvula, African dwarf mongoose	Babault 1949, *Rasa 1973, 1977,* Zannier 1965
Herpestes edwardsi, Indian grey mongoose	Fischer 1921, Frere 1929, *Rensch & Dücker 1959*
Herpestes ichneumon, ichneumon	Dücker 1960, *Rensch & Dücker 1959*
Ichneumia albicauda, white-tailed mongoose	Dalton 1961, Loveridge 1923
Mungos mungo, banded mongoose	Neal 1970, Rood 1975 (?), Simpson 1964
Mungotictis lineatus, narrow-striped mongoose	Albignac 1971, 1973
Nandinia binotata, two-spotted palm civet	Ball 1955, Dücker 1971, Vosseler 1928
Prionodon linsang, linsang	Gangloff 1975
Suricata suricatta, meerkat	Dücker 1962, Ewer 1963, *Wemmer & Fleming 1974*

Table 3-18 CARNIVORE PLAY: FAMILY HYAENIDAE

Crocuta crocuta, spotted hyena	*Apfelbach 1969a, *Kruuk 1972,* 1975, van Lawick & van Lawick-Goodall 1971
Hyaena brunnea, brown hyena	Inhelder 1955
Hyaena hyaena, striped hyena	M. W. Fox 1971
Proteles cristatus, aardwolf	von Ketelhodt 1966

Table 3-19 CARNIVORE PLAY: FAMILY FELIDAE

General	Leyhausen 1965, 1973a,b, 1979, Rosevear 1974
Acinonyx jubatus, cheetah	Benzon & Smith 1975, Corkill 1929, Dominis & Edey 1968, Eaton 1974, Editors 1935, Florio & Spinelli 1967, 1968, Prater 1935, Rodon 1898, Schaller 1968, Veselovský 1975, Wrogemann 1975

Table 3-19 CARNIVORE PLAY: FAMILY FELIDAE (*Continued*)

Felis catus, domestic cat	*Barrett & Bateson 1978*, Bateson & Young 1979, Biben 1979, Collard 1967, Egan 1976, Herter & Herter 1955, Kuo 1930, Langfeldt 1974, Rosenblatt 1972, Rosenblatt et al. 1962, *West 1974*, 1977, Wilson & Weston 1946
Felis libyca, African wildcat	Ognev 1962
Felis nigripes, black-footed cat	Armstrong 1975
Felis silvestris, European wildcat	Lindemann & Rieck 1953
Herpailurus yagouaroundi, jaguarundi	Cutter 1957
Ictailurus planiceps, flat-headed cat	Muul & Lim 1970
Leopardus tigrinus, oncilla	Leyhausen 1953
Leopardus wiedi, margay	Leyhausen 1963, Petersen 1979
Leopardus sp.	Wilson 1860
Leptailurus serval, serval	Bolwig 1963
Lynx lynx, lynx	Lindemann 1955, Ognev 1962
Neofelis nebulosa, clouded leopard	Elliot 1871, Fellner 1965, 1968, Geidel & Gensch 1976, Hemmer 1968
Panthera leo, lion	Adamson 1960, Babault 1949, Bolwig 1959, Cooper 1942, Dominis & Edey 1968, Forbes 1963, Haas 1967, Hubbard 1963, Rudnai 1973, *Schaller 1972*, Schenkel 1966a
Panthera onca, jaguar	Brukoff 1972
Panthera pardus, leopard	Brukoff 1972, Editors 1935, Hingston 1913, Walther 1969, Wilson & Child 1966
Panthera tigris, tiger	Ognev 1962, *Schaller 1967*, *Wasser 1978*
Prionailurus bengalensis, leopard cat	Birkenmeier & Birkenmeier 1971, Dathe 1968, Pohle 1973
Prionailurus rubiginosus, rusty-spotted cat	Jerdon 1874
Prionailurus viverrinus, fishing cat	Ulmer 1968

Puma concolor, puma	Greer 1965
Profelis aurata, African golden cat	Tonkin & Kohler 1978
Profelis temmincki, Asian golden cat	Gee 1961
Uncia uncia, snow leopard	Freeman 1975, Ognev 1962

Table 3-20 PINNIPED PLAY

Family Otariidae

Arctocephalus australis, South American fur seal	Murie 1869
Arctocephalus forsteri, New Zealand fur seal	McNab & Crawley 1975, E. H. Miller 1975a, 1975b, Stirling 1970
Arctocephalus gazella, Kerguelen fur seal	Paulian 1964
Arctocephalus pusillus, Cape fur seal	Müller-Using 1972, Rand 1967
Arctocephalus townsendi, Guadalupe fur seal	Peterson, Hubbs, Gentry, & DeLong 1968
Callorhinus ursinus, Alaska fur seal	Bartholomew 1959, Rowley 1929
Eumetopias jubatus, Steller sea lion	*Farentinos 1971, Gentry 1974*
Neophoca cinerea, Australian sea lion	Marlow 1975
Neophoca hookeri, New Zealand sea lion	Marlow 1975
Otaria flavescens, southern sea lion	Hamilton 1934, Vaz-Ferriera 1975
Zalophus californianus, California sea lion	Lorenz 1969, *Peterson & Bartholomew 1967*
Zalophus wollebaeki, Galapagos sea lion	Eibl-Eibesfeldt 1955, Orr 1967

Family Odobenidae

Odobenus rosmarus, Pacific walrus	E. H. Miller 1975b

Family Phocidae

Halichoerus grypus, grey seal	Cameron 1967, *S. Wilson 1974a*
Leptonychotes weddelli, Weddell seal	Cline, Sinniff, & Erickson 1971, Kaufman, Siniff, & Reichle 1975
Mirounga angustirostris, northern elephant seal	*Rasa 1971*
Mirounga leonina, southern elephant seal	Angot 1954, Ring 1923
Phoca vitulina, common seal and harbor seal	*S. Wilson 1974a,b, S. Wilson & Kleiman 1974*

Table 3-21 PERISSODACTYL PLAY

Family Equidae	
Equus caballus, domestic horse	Antonius 1939, Ebhardt 1954, *Fagen & George 1977*, Feist & McCullough 1976, Grzimek 1949, Hafez 1962, *C. Hamilton 1967*, Jaworowska 1976, *Schoen, Banks, & Curtis 1976, Tyler 1972*
Equus burchelli, plains zebra	Klingel 1967, 1974a, Wüst 1976
Equus grevyi, Grevy's zebra	Klingel 1974b, Schuller 1961
Equus zebra, mountain zebra	Joubert 1972a, 1972b, Klingel 1968
Family Tapiridae	
General	Frädrich & Thenius 1972, von Richter 1966
Family Rhinocerotidae	
Ceratotherium simum, square-lipped rhinoceros	Owen-Smith 1973, 1975
Diceros bicornis, black rhinoceros	Dittrich 1967, Ritchie 1963, Schenkel & Schenkel-Hulliger 1969
Rhinoceros unicornis, Indian rhinoceros	Buechner et al. 1975, Tong 1962

Table 3-22 ARTIODACTYL PLAY: FAMILIES SUIDAE, TAYASSUIDAE, HIPPOPOTAMIDAE, CAMELIDAE, TRAGULIDAE, GIRAFFIDAE, ANTILOCAPRIDAE

Family Suidae	
Phacochoerus aethiopicus, warthog	Frädrich 1965, *Koenig 1959
Sus scrofa, domestic swine	*Gundlach 1965*, McBride 1964, McBride & James 1964
Family Tayassuidae	
Dictyles tajacu, collared peccary	Dobroruka & Horbowyjova 1972
Family Hippopotamidae	
Choeropsis liberiensis, pygmy hippopotamus	*Wilson & Kleiman 1974*
Hippopotamus amphibius, hippopotamus	Dittrich 1962, Noble 1945
Family Camelidae	
Camelus dromedarius, dromedary	Gauthier-Pilters 1959, Pilters 1954
Lama guanacoe, guanaco	Pilters 1954

Lama peruana, llama	Pilters 1954
Vicugna vicugna, vicuna	*Koford 1957*
Family Tragulidae	
Hyemoschus aquaticus, water chevrotain	Dubost 1975
Family Giraffidae	
Giraffa camelopardalis, giraffe	Dagg & Foster 1976, Innis 1958, Pratt & Anderson 1979, Szederjei & Fábián 1975
Okapia johnstoni, okapi	Walther 1962
Family Antilocapridae	
Antilocapra americana, pronghorn	*Autenrieth & Fichter 1975*, Bruns 1977, Buechner 1950, Canfield 1866, Einarsen 1948, Kitchen 1974, Müller-Schwarze & Müller-Schwarze 1973, Murray 1932, Voss 1969

Table 3-23 ARTIODACTYL PLAY: FAMILY CERVIDAE

Alces alces, moose	M. Altmann 1958, 1963, De Vos 1958, De Vos et al. 1967, Geist 1963, Oberholtzer 1911, Stringham 1974
Axis axis, chital	Graf 1966, Kennion 1921, Schaller 1967
Axis porcinus, hog deer	Kennion 1921
Capreolus capreolus, roe deer	Bubenik 1965, Espmark 1969, Page 1962, Prior 1968, Thornburn 1921, Toepfer 1971
Cervus canadensis, elk	M. Altmann 1952, 1963, *McCullough 1969*, Struhsaker 1967a
Cervus dama, fallow deer	Gradl-Grams 1977, Hassenberg 1977, Meier 1973 (?)
Cervus duvauceli, barasingha	Hassenberg & Klös 1975
Cervus elaphus, red deer	Bubenik 1965, Burckhardt 1958, *Darling 1937*, Gossow & Schürholz 1974, Meshkova 1970, Meyer 1972, Olivier 1958
Cervus nippon, sika deer	Meshkova 1970
Hippocamelus antisensis, taruca	Roe & Rees 1976 (?)

Table 3-23 ARTIODACTYL PLAY: FAMILY CERVIDAE (*Continued*)

Mazama gouazoubira, grey mazama	Frädrich 1974
Muntiacus muntjac, muntjac	Barrette 1977, Dubost 1971
Odocoileus hemionus, black-tailed deer	Caton 1877, Dasmann & Taber 1956, *Linsdale & Tomich 1953, F. L. Miller 1975, Müller-Schwarze 1968,* Müller-Schwarze & Müller-Schwarze 1969
Odocoileus virginianus, white-tailed deer	Caton 1877, Hirth 1977, *Michael 1968,* Severinghaus & Cheatum 1956, Skinner 1929, Townsend & Smith 1933
Pudu pudu, pudu	Hick 1969, Junge 1966, Vanoli 1967
Rangifer tarandus, reindeer and caribou	De Vos 1960, Espmark 1971, Lent 1966, Pelosse 1977, Stefansson 1944, Voss 1963

Table 3-24 ARTIODACTYL PLAY: FAMILY BOVIDAE (EXCEPT CAPRINAE)

Aepyceros melampus, impala	Grzimek & Grzimek 1960, Jarman 1972, Jarman & Jarman 1973, Leuthold 1970, Schenkel 1966b, 1966c
Alcelaphus buselaphus, red hartebeest	Backhaus 1959
Antidorcas marsupialis, springbok	Fitzsimons 1920
Antilope cervicapra, blackbuck	Schaller 1967, Schmied 1973
Bison bison, bison	Fuller 1960, *Lumia 1972,* McHugh 1958
Bison bonasus, wisent	Mohr 1968
Bos frontalis, gayal	Scheurmann 1975
Bos gaurus, gaur	Frädrich & Klos 1976, Rodon 1894
Bos taurus, domestic cattle	*Brownlee 1954,* Linsdale & Tomich 1953, *Schloeth 1961,* Schuller 1961, Stephens 1974, Weiland 1965
Connochaetes taurinus, wildebeest	Blaauw 1889, Estes 1969, van Lawick & van Lawick-Goodall 1971, Walther 1965a
Damaliscus dorcas, blesbok and bontebok	David 1973, 1975, Walther 1968

Damaliscus lunatus, topi	Monfort-Braham 1975
Gazella dorcas, dorcas gazelle	Ghobrial & Cloudsley-Thompson 1976
Gazella gazella, mountain gazelle	Grau & Walther 1976
Gazella granti, Grant's gazelle	Estes 1967, Walther 1965b, 1972
Gazella thomsoni, Thomson's gazelle	Brooks 1961, Estes 1967, van Lawick & van Lawick-Goodall 1971, Walther 1964, 1969, 1973
Hippotragus equinus, roan antelope	Backhaus 1959
Kobus defassa, Defassa waterbuck	Hanks et al. 1969, Spinage 1969, Verheyen 1955
Kobus leche, lechwe	De Vos & Dowsett 1966, Lent 1969
Litocranius walleri, gerenuk	Leuthold & Leuthold 1973, Schomber 1963 (?), Walther 1958, 1961
Madoqua kirki, Kirk's dik-dik	Dittrich 1967, Ziegler-Simon 1957
Neotragus pygmaeus, royal antelope	Owen 1973
Oryx gazella, oryx	Walther 1965a
Procapra gutturosa, Mongolian gazelle	Andrews 1932
Syncerus caffer, African buffalo	Sinclair 1977
Tragelaphus imberbis, lesser kudu	Leuthold & Leuthold 1973, Mitchell 1977
Tragelaphus scriptus, bushbuck	Douglas-Hamilton & Douglas-Hamilton 1975, Grzimek & Grzimek 1960
Tragelaphus spekei, sitatunga	Bartikova 1973
Tragelaphus sp., kudu	Walther 1964

Table 3-25 ARTIODACTYL PLAY: FAMILY BOVIDAE (SUBFAMILY CAPRINAE)

Budorcas taxicolor, takin	Editors 1959, Wallace 1913
Capra cylindricornis, East Caucasian tou	Meshkova 1970
Capra falconeri, Kashmir markhor	Schaller 1977

Table 3-25 ARTIODACTYL PLAY: FAMILY BOVIDAE (SUBFAMILY CAPRINAE) (*Continued*)

Capra hircus, domestic goat	Blauvelt 1956, Chepko 1971, McDougall 1975, Rudge 1970, Shank 1972, Schaller 1977, Yocom 1967
Capra ibex, ibex	*Byers 1977,* Meyer-Holzapfel 1958, Steinborn 1973
Capra lervia, aoudad	Haas 1959
Capra sp., goat	Ali 1927
Hemitragus jemlahicus, tahr	Heath 1908, Kinloch 1926 (?), Pillai 1963
Ovibos moschatus, muskox	Freuchen 1915, Jensen 1904, Tener 1965
Ovis ammon, urial and mouflon	D. Altmann 1970, Pfeffer 1967, Schaller 1977
Ovis aries, domestic sheep	Banks 1964, Morgan & Arnold 1974, *Sachs & Harris 1978*
Ovis canadensis, mountain sheep	Berger 1979, *1980,* Forrester & Hoffmann 1963, *Geist 1971,* Sheldon 1921, Spencer 1943, Welles & Welles 1961
Ovis dalli, Dall sheep	Murie 1944
Ovis polii, Marco Polo wild sheep	dePoncins 1895
Pseudois nayaur, bharal	Schaller 1977
Rupicapra rupicapra, chamois	Briedermann 1967, Krämer 1969

Table 3-26 AVIAN PLAY

Sphenisciformes	Müller-Schwarze 1978b
Pelecaniformes	Gibson-Hill 1947, Meischner 1959, Stonehouse & Stonehouse 1963
Anseriiformes	Baker 1899, Lebret 1948, Roberts 1934
Falconiformes	See Table 3-27
Galliformes	Goodwin 1953, Guhl 1958, Kruijt 1964, McBride, Parer, & Foenander 1969, McCabe & Hawkins 1946
Charadriiformes	Ashmole & Tovar 1968, Delius 1973, Goethe 1940, Knappen 1930, Norton-Griffiths 1969, Tinbergen & Norton-Griffiths 1964, Wheeler 1943

Columbiformes	Miller & Miller 1958
Psittaciformes	Braun 1952, Brereton 1971, Derscheid 1947, Durrell 1966, Engesser 1977, Hick 1962, Jackson 1963a,b, Keller 1975, 1976, Porter 1947, Shepherd 1968, Ulrich, Ziswiler, & Bregulla 1972
Cuculiformes	Gurney 1909, Moreau 1938, Young 1929
Strigiformes	Heinroth & Heinroth 1928, Hubl 1952, Meyer-Holzapfel & Räber 1976, Trollope 1971
Apodes	Ball 1943, Stoner 1947, Wagner 1954
Coraciiformes	Moreau & Moreau 1941, 1944, Ranger 1950, Swanberg 1952
Piciformes	Kilham 1974
Passeriformes	See Table 3-28

Table 3-27 AVIAN PLAY: ORDER FALCONIFORMES

Accipitridae	Bond 1942, Brosset 1973, Gunston 1971, Herrick 1924a,b,c, 1934, Huxley & Nicholson 1963, Johnson & Gayden 1975, van Lawick-Goodall 1970, Mohr 1960, Neelakantan 1952, Pakenham 1936, Rowe 1947, Shelley 1935, Sumner 1931, Todd 1974, Todd & Gale 1970, Weir & Picozzi 1975
Falconidae	Balgooyen 1976, Battersby 1944, Brosset 1973, Cade 1953, Koehler 1966, Munro 1954, Mueller 1974, Parker 1975, Thorpe 1963

Table 3-28 AVIAN PLAY: ORDER PASSERIFORMES

Corvidae

Corcorax melanorhampos, white-winged chough	Chisholm 1958
Corvus albicollis, white-necked raven	Moreau & Moreau 1944, Sclater & Moreau 1933
Corvus corax, common raven	Budich 1971, Dawson & Bowles 1909, Gwinner 1966, Højgaard 1954, Hutson 1945, Manns 1978, Mech 1966, 1970, Thorpe 1966
Corvus corone, hooded crow	Persson 1942
Corvus splendens, house crow	Biddulph 1954, Panday 1952

Table 3-28   AVIAN PLAY: ORDER PASSERIFORMES (*Continued*)

Corvus sp., jungle crow	Neelakantan 1952
Garrulus glandarius, jay	Goodwin 1951
Pica nuttalli, yellow-billed magpie	Verbeek 1972
Pica pica, white-rumped magpie	Radcliffe 1909
Pyrrhocorax hybrid	Thaler 1977
Other families	Barker 1924, Eibl-Eibesfeldt & Sielmann 1962, Gaston 1977, Howard 1907, Marler 1956, Messmer & Messmer 1956, Millikan & Bowman 1967, Morris 1977, Nice 1943, Sauer 1954, 1956, Skutch 1951, Sprunt 1944

Table 3-29   OBJECT PLAY IN EAGLES, HAWKS, AND OWLS

Eagles	
Aquila audax, wedge-tailed eagle	Chisholm 1958
Aquila chrysaetos, golden eagle	Gunston 1971
Gypaetus barbatus, lammergeier	Huxley & Nicholson 1963
Haliaeetus leucocephalus, bald eagle	Herrick 1934, Johnson & Gayden 1975
Hawks	
Accipiter spp., accipiters	Bond 1942, Brosset 1973, Mohr 1960
Buteo spp., buzzards	Battersby 1944
Circus cyaneus, harrier	Sumner 1931
Falco spp., falcons	Battersby 1944, Brosset 1973, Cade 1953, Koehler 1966, Mueller 1974, Munro 1954, Parker 1975, Thorpe 1963
Milvus migrans, milan, kite	Brosset 1973
Owls	
Asio otus, long-eared owl	Meyer-Holzapfel & Räber 1976
Athene noctua, little owl	Hubl 1952
Bubo bubo, eagle owl	Heinroth & Heinroth 1928
Tyto alba, barn owl	Hubl 1952, Trollope 1971

Table 3-30 "PLAY" (?) IN OTHER SPECIES,
REAL AND OTHERWISE

Class Reptilia

Alligator mississippiensis, American alligator	Lazell & Spitzer 1977
Varanus komodoensis, Komodo monitor	C. Hill 1946

Comment: These accounts cite object contact and manipulation in single individuals. However, large crocodilians (e.g., *Alligator*) have postnatal parental care and slow development (Burghardt 1977, 1978), whereas varanid exercise physiology resembles that of mammals rather than that of other reptiles (Fagen 1976a). These factors are not sufficient for play to evolve, but they may be necessary.

Class Pisces

General (review)	Gunter 1953, Meyer-Holzapfel 1960
Gnathonemus petersii, tapir snout fish	Meder 1958
Mormyrus kanname, tapir snout fish	Meyer-Holzapfel 1960

Comment: The two species cited are both mormyrids, fish with exceptionally large brains relative to those of other fish. These accounts are based on observations of a few captive animals. Meyer-Holzapfel (1960) considers multiple hypotheses about the behaviors observed. She concludes that the mormyrids studied were playfighting, were behaving nonagonistically, and were using objects as substitute play-partners.

Class Insecta

Eciton drepanophora, army ant	Bates 1969
Formica sp., ant	Huber 1810
Formicoxenus nitidulus, ant	Stumper 1921
Periplaneta americana, American cockroach	Olomon et al. 1976

Comment: The ant behavior originally called "play" was later found to be aggressive competition (E. O. Wilson 1971b p. 218). Olomon et al. (1976) discuss a possible maturational precursor to adult agonistic behavior, but do not give a complete ethological analysis of the behaviors cited.

Literary or mythical descriptions

The Kenneth Grahame story *The Reluctant Dragon* alludes briefly to play of dragons. Vladimir Nabokov's *Pale Fire* describes a playful butterfly *Vanessa atalanta* (note to lines 993–995). James Stephens's *The Centaur* describes a centaur at play. And Munchausen's fictional Travels (xxiv. 104) include an account of "The noble sphinx gamboling like a huge leviathan."

4

Biological Bases
of Animal Play

The Elizabeth Baldwin translation of Groos's *Die Spiele der Thiere* (Groos 1898) includes a brief preface by James Mark Baldwin, a prominent evolutionist (and Elizabeth's husband). In a few pages Baldwin comments on animal play, on the role of phenotypic flexibility in evolution, on play and artistic behavior, and on self-deception—no small achievement, especially when we consider that Mendel's genetical research had not yet been rediscovered.

Behavioral biologists who consider developmental and evolutionary analyses incommensurate, if not antithetical, might be surprised to learn that Baldwin only followed biological tradition in considering evolutionary and developmental processes together. In fact, juxtaposition of ontogenetic and phylogenetic patterns "may be the most durable analogy in the history of biology" (Gould 1977 p. 13). Natural selection theory now gives ecological and genetic analyses of development increased precision and insight. To approach animal play without making use of this knowledge would be to deny the relevance of evolution to the study of behavior.

This chapter will emphasize theoretical bases for the study of play. I cannot review the entire evolutionary biology of behavior development here. The subject deserves an entire book (and a catchier name. The most accurate designation, developmental behavioral ecology, is far too long to emblazon on a banner.)

FUNDAMENTAL EVOLUTIONARY PRINCIPLES

The basis of all evolutionary thinking about development is the concept of natural selection acting on heritable variation in populations. The unit of the variation on which such selection acts is the gene. These units persist through the lifetime of a given individual. Different individuals may have identical copies of the same unit. These two properties of genes (persistence in individuals and replication between individuals) give rise to two evolutionary principles important for understanding development.

Allocation (Fisher 1958)

Selection operates at each stage of the life cycle, weighing benefits at one age against costs incurred at another age. Because an individual continues to have a given gene at all ages, evolution may favor genetically based characteristics of the young individual that cause it immediate harm while yielding delayed benefits. Such acts benefit an older individual who shares not only the same gene but also the same identity.

Inclusive Fitness and Kin Selection (Hamilton 1964)

Because different individuals may share a given gene by descent from a common ancestor, evolution may favor a genetic determinant of helpful acts that one individual performs at some harm to itself and that benefit another individual who shares the gene in question. In contrast to Fisher's allocation principle, the immediate cost to a young individual of having a given gene is repaid not by later benefits to the same individual but by benefits to a different individual with the same gene.

Frequency-dependent natural selection is an important aspect of the general principle (which also underlies the idea of kin selection) that the adaptive value of behavior depends on the social environment of the performer. Any behavior affecting more than one individual must be considered in terms of what other individuals do to the individual actor. This principle is most evident for overt social behaviors, such as fighting and sex (or play), but other interactions are equally important. For example, age-dependent schedules of thermoregulation, of energy and nutrient requirements, and of motor development have social impact in mammals and in birds. They affect both the parent-offspring relationship and relationships among siblings.[1]

The evolutionary approach to behavior development amounts to two very simple but powerful principles that follow directly from the central dogma of frequency-dependent natural selection of genetically based variation. First, ontogenies result from biological adaptations keyed to the demands of particular environments. Natural selection acts on genetically based variation in ontogenetic patterns to change the frequency of genes affecting developmental timing and developmental mechanisms. Second, genotypic and phenotypic composition of a given population determine the rate and direction of social behavioral evolution, including evolution of the ontogeny of social behavior. Therefore, selection of any social behavioral trait (defined broadly in the sense of the above discussion) is frequency dependent and is affected by the degrees of genetic relatedness among all individuals affected by the trait. The first principle is central to ecological life-history theory (Cole 1954, Fisher 1958, Schaffer & Gadgil 1975, Stearns 1976, 1977), to the evolutionary theory of plasticity and resilience (G. Bateson 1963, Bradshaw 1965, Emlen 1975, Levins 1968, 1969, Schmalhausen 1949, Waddington 1957, 1965), and to the evolutionary study of changes in developmental timing (Gould 1977). The second principle is the source of two important concepts in social theory. The first concept, inclusive fitness (Hamilton 1964), is the basis of sociobiological altruism theory and theory of the family (Dawkins 1976b, E. O. Wilson 1975). The second (frequency-dependent natural selection of evolutionarily stable strategies) is the basis of evolutionary theories of animal conflict (Maynard Smith & Price 1973, Maynard Smith & Parker 1976) and of parental care (Maynard Smith 1977). Only Konner (1977a) has attempted both to summarize some of these ideas and to apply them to the development of behavior.

Ecological life-history theory offers a systematic exploration of the evolution of age-specific patterns of resource allocation and therefore of behavioral maturation. Sociobiological theory predicts patterns of parental care, of learning and social development, and of group composition. The central insight of both approaches to development is the same: behavior that benefits my genes, whether resident in my future self or in another individual, may be selected even if it decreases my immediate personal fecundity or probability of survival.

ONTOGENY AND ADAPTATION

Why develop? Organisms like bacteria, who reproduce by binary fission, have no well-defined infant or juvenile stages, but they were the earliest and must be ranked among the successful life forms. In principle, even a multicellular organism could grow in size without exhibiting behavioral changes or increases in behavioral complexity other than those simple (allometric) changes in frequency, duration, or intensity of movement that were direct consequences of changing body size. The evolutionary significance of the development of behavior is that for a number of different and highly idiosyncratic reasons, young organisms exhibiting behavioral development survived and reproduced better than other young organisms whose behavior changed less fundamentally with age and with size. Developmental patterns can therefore be viewed as reflections of complex adaptations. These adaptations include one or more components (called "strands" by Mason 1965a and "abilities and skills" by Simpson 1978) that are identifiable as distinct entities and that may interact. For example, an adult individual's sexual behavior "may be understood in terms of the operation of a number of component skills, including attentional, tactile, and social ones, the development of each of which could have been followed through his whole life, in many situations" (Simpson 1978). Each component is characterized by its own rate of development and by its particular susceptibility to environmental influence during the lifetime of an individual. Ecological life-history theory offers evolutionary predictions about developmental rates. In the past, life-history theorists have not predicted the extent to which an individual's life history should be modified by its environment as a function of age, although life-history theorists have recently become concerned with this question. There is, however, a separate ecological theory of plasticity and resilience in which this problem is considered in more general terms.

Ontogenies reflect adaptations. Developmental rates are biological dependent variables keyed to environmental demands in a way that results in the greatest possible number of surviving offspring. Developmental rates evolve. The patterns of response of developing organisms to environmental variation are themselves adaptations molded by natural selection. These fundamental principles are the basis of the evolutionary approach to development. In particular, they serve to analyze development of behavior.

The Allocation Principle

Developing organisms have evolved to budget limiting resources differentially to intraindividual biological processes or activities that compete for these resources, in a way that maximizes gene survival. Diverse factors can act as limiting resources, and the complex causal networks affecting juvenile survivorship or age at sexual maturity may involve energy (including food and warmth), nutrients, time, information, or attention. Social interactions with parents, kin, and unrelated animals affect resource availability to the young. Alexander (1974) suggests that mates might be a limiting factor in expanding populations that are not limited by factors affecting viability. Survival and age at sexual maturity of immature individuals in such populations could similarly be regulated by the demographic environment. For example, if social play were essential for future reproduction (which it probably is not, Symons 1978a), an animal growing up without playmates would never produce offspring.

By definition, juveniles devote resources to growth, survival, and information gathering, not to reproduction. Sexual maturity is delayed only because animals who waited to reproduce (or their kin) left more surviving offspring than those who reproduced at once. Resource budgets are also age dependent. An individual who devoted its time, energy, and attention to feeding and play when young may stop eating entirely and risk physical injury in order to reproduce as an adult.

It follows directly from the allocation principle that traits for which benefits occur at one age and costs at another can evolve by natural selection (Emlen 1970, Hamilton 1966, Konner 1977a, Williams 1957). Selection may favor certain forms of behavior having immediate neutral or negative effects and delayed positive effects. Migratory birds travel long distances even though their own immediate survival is thereby endangered. A tree squirrel may expose itself to ground-dwelling predators while it stores its winter food. A young primate may play to physical exhaustion, decreasing its ability to keep up with the movements of its social group, but obtaining experience that might benefit it as an older juvenile or as a reproducing adult.

The converse effect may also occur (Konner 1977a). Selection may produce optimal viability at one age "even though the adaptation necessary may, in the life of the individual," result in less than maximal viability or fecundity later on. To understand this, consider a simple three-age life-history model in which all reproduction occurs at age three, after which the animal dies. At age one, suppose an animal has resource E, which can either be used to improve immediate viability (survival from age one to age two, denoted by p_1) or can be stored as fat and metabolized at age two to improve later viability (survival from age two to age three, denoted by p_2). The model serves to answer the following question: If genetic variants exist that devote different fractions x of the resource E to improve immediate viability, which variants will be selected? We can always scale the units of E so that later viability is linear in x. Immediate viability will be some function $p_1 = Pf(x)$, where P is the maximum possible immediate viability if all E units are used up. Thus P will depend on E and on the level of environmental factors causing mortality. Function $f(x)$, with values between 0 and 1, measures the effectiveness of resources devoted to increased survivorship. Later viability $p_2 = C(1-x)$, where C is the maximum possible viability at age two. Selection will favor animals whose x maximizes p_1p_2 because these animals are most likely to survive to breed. Evolution may have the additional option of allowing the animal to reproduce at earlier ages, a possibility that Konner (1977a) does not discuss; see Bell (1976) and Charnov & Schaffer (1973) for the argument.

Evolution will maximize $p_1 p_2 = Pf(x)C(1-x)$. By inspection, we can see that the optimal x, x^*, depends only on f(x) and not on P or C. Thus Konner's statement that selection operates most intensely at "a stage of the life cycle in which mortality is very high" must not be interpreted simply as a stage with low P, since x^* is independent of P. In fact, x^* satisfies the equation $df/dx \,|\, x^* = f(x^*)/(1-x^*)$. So, for example, for f(x) $= x^2$, $x^* = \frac{2}{3}$, and for f(x) $= \sqrt{x}$, $x^* = \frac{1}{3}$. Thus we can actually find *high* x^* associated with *low* p_1 if we choose P sufficiently small. The only component of the model on which x^* depends is the shape of the function f(x). For given values of P and C, x^* will be high (and p_2 correspondingly low) if f(x) is highly convex, that is, if the environment is so severe that near-total commitment of the limiting resource is necessary to survive at all. In this case survival to age two is possible only at the cost of a sharp reduction in later viability. This reduced viability is *optimal* (contra Konner), but not *maximal*.[2]

The general problem of the evolution of survivorship curves is analyzed by Hamilton (1966), by Emlen (1970), and by Caswell (1978). They model the response of selection to deleterious effects caused by environmental change and to mutations having positive effects at one age, but negative effects at another. E. O. Wilson (1975 p. 98) states the chief result of survivorship theory: "In general, age-specific mortality in optimal life cycles should be high at or near conception, fall to a minimum during later prereproductive life, and then, after the age of first reproduction, rise steadily with age." This prediction holds for mammals (Caughley 1966) and probably for birds (Lack 1966, Ricklefs 1977). The implications of this result for play are twofold. First, ages at which play occurs tend to coincide with the minimum in age-specific mortality. Second, juvenile survivorship is one determinant of the number of play companions potentially available to an individual. (Other factors include sex ratios, litter or clutch size at birth, birth synchrony and spacing, developmental synchrony of age-mates, age-specific dispersal rates, and the number of breeding females in the animal's social group.)

Life-History Patterns

The diversity of life-history patterns in nature is striking. Age-specific patterns of mortality and fecundity vary greatly among mammals (Walker 1975). To some extent, this variation is a simple consequence of variation in body size (Gould 1966, Millar 1977). Mammalian body weights range from 2 g (the shrew *Suncus etruscus*) to 112,500 kg (the blue whale *Balaenoptera musculus*). Life-spans may be a few weeks in some rodents, several decades in humans and elephants. Litter sizes range from one (most primates, hoofed mammals, marine mammals, and many other species) to over thirty in the tenrecid insectivore *Tenrec ecaudatus* and over twenty in certain rodents. Gestation periods may be as short as ten days in certain marsupials, sixteen days in the hamster genus *Mesocricetus,* or as long as two years in the large whales. Age at sexual maturity, reproductive life-span, degree of development at birth, brain size, parental investment in offspring, and virtually every other component of mortality and natality vary both within and between taxa. If life histories truly are adaptive, certain general life-history patterns ought to occur in certain types of environments. Much previous work in life-history theory amounted to an attempt to classify environments, to classify life-history patterns, and to obtain simple predictive rules linking environmental type with life-history type. This quest proved difficult (Wilbur, Tinkle, & Collins 1974, Stearns 1976, 1977). The consensus is that regular relationships do exist, but fundamentally different environmental regimes can produce the same life-history pattern, and a particular life-history characteristic, such as parental care, small litter size, long life-span, slow development, or delayed sexual maturity, can evolve for a number of fundamentally different, highly idiosyncratic reasons (Bradbury & Vehrencamp 1977b, Grime 1977, Schaffer & Gadgil 1975, Stearns 1976, 1977, Whittaker & Goodman 1979, Wilson 1975). For example, consider the history of the influential r-selection, K-selection concept (reviewed by Stearns 1976). It was first suggested that "organisms exposed to high levels of density-independent mortality, wide fluctuations in population density, or repeated episodes of colonization will evolve towards a combination of earlier maturity, larger broods, higher reproductive effort, and shorter life-spans than will organisms exposed to density-dependent mortality or constant population density" (Stearns 1977 p. 145). The former type of environment is said to exert r-selection, defined as selection for traits that maximize the rate of population growth; the latter type of environment is said to exert K-selection, defined as selection for traits that maximize carrying capacity. It was then found that a second evolutionary process would produce the same evolutionary result: in a stochastically fluctuating environment that causes highly variable juvenile mortality, a syndrome of delayed maturity, smaller reproductive effort, and greater longevity should also evolve. This multiplicity of reasonable explanations for the same observable phenomenon is a common difficulty in population biology, as discussed by E. O. Wilson (1975 p. 29).

Attempts to integrate the natural history of play with life-history theory at the level of gross life-history patterns are suggestive at best (Fagen 1977, Gould 1977), especially since life-history theory as such has little to say about the degree to which environmental variation within the life of the individual should modify that individual's development. Indeed, two opposite speculations exist. Lorenz (1956), Morris (1964), and Geist (1978a) claim that play is most likely in morphologically unspecialized, ecologically opportunistic organisms, "specialists for non-specialization." In life-history terms, these organisms would be r-selected. Fagen (1974a, 1977) and Gould (1977) see play as a behavior of K-selected species, species expected to have large brains, slow development, and intense parental care. The latter point of view is also implicit in Happold's (1976a,b) comparison of play in conilurine rodents whose life-history patterns differ.

Life Histories, Paedomorphosis, and Neoteny

The above gross comparisons may be confusing because of failure to distinguish between different processes that select for the same juvenile-like phenotype. In simplified form, the argument is as follows.

The size and shape of an animal can influence its fecundity and its survivorship in many ways. In sexually dimorphic species having large males, a male's chance of winning a fight can depend in part on his size or on the size or shape of some part of his body, such as horns or teeth (Barrette 1977, Darling 1937, Geist 1966, Struhsaker 1967a). There are many other biologically important relationships between size, shape, and fitness both in males and in females. A seeming paradox in the study of form is that certain species lose highly developed adult characteristics of their evolutionary ancestors and exhibit a unique, relatively undifferentiated morphology that actually resembles that of an immature animal. In these so-called paedomorphic species, the head and brain are large relative to the body, the eyes are large, and the muzzle is short, giving the organism a child-like phenotype (Lorenz's *Kinderschema*). The physical appearance of paedomorphic species is striking, and fully adult members of these species are often mistaken for juveniles at first sight.

The phenomenon of paedomorphosis challenged a traditional view known as the Biogenetic Law: ontogeny recapitulates phylogeny. Gould (1977) reviews the history of this controversy and offers a new biological interpretation of this phenomenon. According to Gould, paedomorphosis becomes biologically orthodox within the context of ecological life-history theory if we distinguish between different kinds of developmental mechanisms that produce the same morphological result, namely a juvenilized phenotype. By the principle of allocation, an animal or a plant whose rate of sexual maturation is accelerated by evolution puts its resources into reproduction and accordingly diverts these resources from growth and differentiation of other body components, whose development then slows down or ceases completely. This process of acceleration of sexual maturation relative to the development of the rest of the body, known as progenesis, yields a paedomorphic adult adapted to rapid reproduction. It is an opportunist, designed to exploit localized and transient resource patches. Its fecundity is high, and its ability to withstand environmentally induced stresses, such as extreme temperatures or predation, is correspondingly low. A prediction of life-history theory well supported by available data (Gould 1977) is that such organisms will evolve in severely fluctuating environments in which even adults cannot withstand these perturbations. Here age and size at first reproduction are expected to be relatively "lower and smaller, reproductive effort higher, size of young smaller, and number of young per brood higher, than in constant environments, where the opposite trends should hold" (Stearns 1976 p. 42).

But can paedomorphic species also evolve by selection for other developmental patterns in other kinds of environments? Gould argues that this is indeed the case and that this second evolutionary complex is particularly important for human evolution. In order to understand this argument, we must consider a number of factors influencing the life history.

In environments that are relatively constant and crowded with conspecifics or with predators, immediate reproduction becomes less important than the ability to survive and to compete, because a young adult that has just become independent will only rarely escape predation or find resources that have not already been preempted by another older adult. Therefore, selection in such environments should favor late maturity, small litter, (clutch, or brood) sizes, large young, parental care, and low reproductive effort.

What patterns of development will be selected under these conditions? Gould (1977) argues that in saturated environments of the type described above, somatic development will under certain conditions be retarded relative to sexual maturation (which is itself delayed). The resulting adult is juvenile-like in morphology. In this case, unlike that preceding, paedomorphosis is a consequence of the process of neoteny (retardation of the rate of somatic development with respect to the time of sexual maturation). Among mammals, this process seems to have been important in primates (especially humans), in felids, and possibly in mountain sheep.

Cartmill (1972, 1974, 1975) presents a second reason for the paedomorphic morphology of both primates and other mammals (e.g., felids) that rely on both vision and use of the forelimbs for feeding. In these animals muzzle length is reduced because the hands or forepaws take over many functions served by the mouth in other species. Skilled forelimb-eye coordination is crucial, and special perceptual and motor skills are required. The result, again, is paedomorphosis, certainly as a consequence of muzzle (rostral) reduction and possibly also through selection for increased brain size relative to the size of the body.

K-selection may act either to produce a neotenous phenotype or to produce a non-neotenous phenotype, given the proper circumstances. K-selective regimes usually favor changes in developmental timing that result not simply in delayed maturation, but also in complex morphological specializations accompanied by restricted flexibility (Gould 1977 pp. 343–345). We must consider factors in addition to an animal's position on the r-K continuum if we wish to understand the adaptive significance of its pattern of development.

Clearly, just as there is more than one way to evolve a given pattern of life-history traits, there is more than one way to be an opportunist. The topic has been a confusing one because modifiability arguments have not been kept distinct from arguments about patterns of maturation, because different processes can yield the same result, and because the concept of an r-K continuum is really too crude to explain very much about life histories.

Bridges between Life-History Theory and a Theory of Optimal Modifiability

An important recent result of theoretical life-history analysis is that not one, but several fundamentally different, evolutionarily stable patterns of development may be selected in the same species and in the same environment (Fagen 1977, Schaffer & Rosenzweig 1977). Does this result mean that only historical factors can explain life-history evolution? If so, life-history theory is in big trouble as a predictive science. Schaffer & Rosenzweig (1977) snatch victory from the jaws of defeat by ingeniously linking life-history theory to a theory of optimal modifiability that had grown up independently of it. They argue that if organisms could switch life-history patterns in response to environmental cues, it would always be possible to pick the life-history peak in the adaptive landscape that had the highest fitness, in effect performing global rather than local optimization.

Gould (1977) tackles the related problem of indeterminate K-selection. If morphological adaptations are selected, then body size often increases and the absolute time of maturation is delayed. Juvenile features will now be exposed to selection at ancestral adult sizes. Under these circumstances, what selective pressures favor retarded somatic development? When is it advantageous to preserve unspecialized juvenile structures in this way? One answer may be a requirement for continued modifiability of behavior. Such a requirement would impact the rate of maturation of the brain and favor "prolongation of embryonic growth rates and stages to later ages and sizes" (Gould 1977 p. 350 uses this phrase when discussing Portmann's theory of ontogenetic type; my argument, however, is distinct from his).[3]

A third complication in life-history theory also points toward modifiability theory. Life-history characteristics tend to be norms of response. Particularly when adults survive better under harsh conditions than do juveniles, an individual may defer reproduction until conditions improve (Charnov & Schaffer 1973, Nichols et al. 1976). When conditions are favorable, animals mature rapidly, reproduce heavily, and die young. When conditions are unfavorable, for example if population density is high and stress, crowding, or resource depletion ensue, animals genetically identical to those just discussed mature slowly, reproduce at a low rate, and live longer (Geist 1978a, Laws et al. 1975, Nichols et al. 1976). Social behavior and therefore social structure change in step with these changes in life-history pattern. The phenomenon of behavioral scaling (E. O. Wilson 1975 pp. 19–21), adaptive changes in an individual's social behavior in response to a change in environment [which Konner (1977a) terms "facultative adaptation"], is well known in nature. Wildebeest, yellow-bellied marmots (Armitage 1977), wild canids (Kleiman & Brady 1978), raccoons (Schneider, Mech, & Tester 1971), equids (Klingel 1969, Moehlman pers. comm.), mountain sheep (Geist 1978a), short-tailed voles (S. Wilson 1973), timber wolves (Mech 1966, Mech

1970), common and harbor seals (S. Wilson pers. comm.), domestic dogs (A. Beck 1973), and domestic cats (Laundré 1977) all exhibit social flexibility in response to environmental change. Hamadryas baboon (Kummer 1968a) and especially chimpanzee (E. O. Wilson 1975) show adaptive social flexibility on a day-to-day basis. Indeed, Wilson considers such flexible, open, variably cohesive societies to occupy the highest grade of vertebrate social organization.

Theoretical models in ecology predict behavioral scaling in response to changes in the competitive environment (Mac-Arthur & Pianka 1966, MacArthur & Wilson 1967, Schoener 1974a,b), and ecological data demonstrate the validity of these predictions (Diamond 1975, Schoener 1974c). The relevant prediction, known as the compression hypothesis, states that increased competition will cause individuals to use a decreased variety of habitats, but not to reduce the number of types of food eaten. Upon release from competition, animals should use a greater range of habitats, but not a greater variety of food types. This prediction applies only to short-term changes, not to genetic change. Therefore, for most species, it is by and large a prediction about changes in behavior. In the long run, of course, a change in the competitive environment may cause gene-frequency change, and behavior may change as a result of changing gene frequencies, but such changes lie beyond the scope of the compression hypothesis.

How are individual animals in a complex, open society such as those mentioned able to modify (and to learn to modify) their life-history patterns and social behavior in accordance with their environment? The foregoing incomplete list of species known to exhibit behavioral scaling is noteworthy for its inclusion of many of the animal kingdom's most remarkably playful members. However, lest we jump to conclusions, as has so often been the case in play research, it is worth noting that behavioral scaling of various kinds is widespread in the animal kingdom (in fact, it was first described in canaries), that grey seals and the playful spotted hyena fail to scale their social behavior to a changed environment (E. O. Wilson 1975 pp. 33–34), and that the long-lived, large-brained African elephant has similarly failed to change its social behavior adaptively in response to a changed environment (Laws, Parker, & Johnstone 1975).

Modifiability Theory

The evolutionary theory of modifiability offers explanations of the ultimate causation of developmental and phenotypic plasticity and resilience. To what extent shall the environment affect development on a time scale less than that of genetic change? G. Bateson (1963), Emlen (1975), Levins (1968, 1969), Waddington (1957, 1965), E. O. Wilson (1975 Ch. 7), and others point out that the degree of modifiability of the phenotype is a genetically based trait subject to natural selection. Some aspects of the phenotype (e.g., eye color) are highly resilient; no amount of environmental perturbations will change them. Such resilient characters are buffered against environmental effects by the developmental system. On the other hand, some characters (e.g., body weight) are notably plastic and respond rapidly and profoundly to a change in the environment. Waddington (1965) formulated this contrast as follows. Selection may favor individuals that are good at producing one particular phenotype under almost any circumstances, relying on the environment always offering a possibility for this phenotype to get by. This leads to the evolution of systems of developmental resilience and canalization of the phenotype. A second alternative is to produce switch mechanisms between canalized phenotypes. A third is plasticity: "to allow the environment to have a strong influence on individual ontogeny, provided it is ensured that the environmental modifications are toward the selection optimum for that particular environment. This leads to the evolution of developmental systems which are highly adaptable." A fourth option, in a genetically variable species, is a developmental pattern relatively unaffected by normal environmental variations, but in which most genetic changes come to phenotypic expression. Waddington invoked group selection to explain the origin and maintenance of these four different patterns of development, but we are not required to do so.

As P. P. G. Bateson (1976a) points out, developmental flexibility and redundancy in development are two separate concepts. Developmental flexibility means plasticity—modifiability of behavior in response to environmental change, leading to behavioral divergence between individuals and therefore to individual differences. Bateson contrasts this conceptualization with canalization, in which "different developmental control mechanisms generate the same behavioral end-product." Developmental pathways may indeed be self-stabilizing, enabling animals to become functioning adults and to perform the evolutionarily essential tasks of feeding, predator avoidance, and offspring production despite exposure to the worst conditions their rearing environment had to offer. Dunn (1976) presents examples of such equifinality in human development. Bateson describes possible self-stabilizing developmental mechanisms based on rules and on procedures for changing rules. A rule may be a negative feedback control policy, as in the case of Bateson's imprinting model (P. P. G. Bateson 1976a pp. 411–412) in which behavior ensures that an animal will visually scan and move around until it encounters a visually conspicuous object having specified characteristics. Animal playfighting suggests another example of a hypothetical behavioral rule: behave so as to mouth your partner without being mouthed by it and without incurring physical pain. The imprinting model, like the template hypothesis (Marler 1975), incorporates rules for changing rules. For instance, experience with particular visually conspicuous objects may change the characteristics that specify whether the animal will approach or avoid other objects encountered in the future.

When the object encountered is another animal whose behavior is also developing, whether the hypothetical dyad consists of a parent and its offspring or two immature individuals, Bateson imagines a kind of mutual rule changing called behavioral meshing in which relationships develop in step with decreases in preexisting differences in individual behavior. In rhesus macaques (*Macaca mulatta*), for example, there are predictable patterns of probabilities with which mothers and infants tend to leave one another. In polytokous carnivores, behavioral meshing could ensure that like-aged offspring, whatever their past experience, were simultaneously ready to encounter and to handle prey animals their mother brought to them. In both cases play may be a possible mechanism by which behavioral meshing could take place.

That genes and their environment interact in behavioral development, that "the environment" includes other genes, the animal's own body, and other organisms as well as physical objects, and that an animal's behavioral predispositions can themselves influence the way the animal perceives and manipulates its environment, so that self-selected and self-generated experience feeds back into developmental processes, are truths that have become obvious only at the cost of long and acrimonious controversy (Slater 1974). Although coherent theory is only now beginning to emerge from studies of behavioral development (P. P. G. Bateson 1976a), ethologists are justifiably excited about the glimpses that they have already caught.

According to Marler (1975), the most general principle emerging from studies of bird-song development is "that of the imposition of constraints on learning." Such constraints may be manifest at any one of a number of levels of organization of behavior and are by no means restricted to avian ontogeny (Hinde & Stevenson-Hinde 1973, Seligman & Hager 1972, Shettleworth 1972). The example of constrained learning offered by studies of bird song learning is sufficiently clear to suggest an informal model, the so-called template hypothesis (Marler 1970, 1975). Assume that certain information is present in certain elements of the nervous system from a very early age. The elements containing this information constitute the template. An organism responds to the consequences of its own behavior, comparing these consequences with the information provided by the template and modifying its behavior accordingly. The information present in the template is accessible to modification as well, for example, as a result of exposure to adult song. A template is therefore a "read-only" information element with respect to some environmental stimuli and developmental processes and a "write-only" element with respect to others. The result of this information structure is that behavior change can occur superficially and rapidly within given constraints, or slowly but more radically as the constraints themselves are modified by experience. Multilevel control systems of this sort are ubiquitous in systems engineering (Mesarovic, Macko, & Takahara 1971).

P. P. G. Bateson (1976b) argues that determinants of behavior form a continuum. These determining factors may affect a single pattern of behavior (specific factors) or may exert profound and general effects necessary for life itself. Moreover, behaviors for which specific developmental effects exist fall along a continuum whose extremes are defined by a complete lack of determinants and by multiple determinants. Because each of these determinants can be assigned to one of two separate dimensions that may be called inherited and environmental, classification of behavior patterns in terms of their developmental determinants requires four categories to be of any use. Some behavior owes its distinctiveness mainly to inherited determinants with specific effects and some mainly to environmental determinants with specific effects. Many behavior patterns are likely to be affected by both inherited and environmental determinants with specific outcomes. Or the distinctiveness of the behavior may arise "from the interaction of inherited and environmental determinants both having general effects on a wide variety of behavior patterns."

Speculation about play has tended to invoke developmental mechanisms requiring specific environmental stimulation, as in the hypothetical learning of communication signals or practice of fighting skills. But environmental determinants having more general effects exist and may be equally important. Whole-body endurance exercise, particularly running and swimming, trains the muscles and the circulatory and respiratory systems, enhancing physical capacity and thus affecting future performance of many different kinds of physically strenuous behavior patterns (Åstrand & Rodahl 1970, Keul, Doll, & Keppler 1972). Moreover, growth and connectivity of the central nervous system are affected by environmental stimulation, including that resulting from differences between activity in enriched, normal, and impoverished environments (Cummins, Livesey, Evans, & Walsh 1977, Ferchmin, Bennett, & Rosenzweig 1975, Greenough 1976, 1978, Levitsky & Barnes 1972, Rosenzweig 1971, Rosenzweig & Bennett 1977a,b, Volkmar & Greenough 1972). Here as well, the effects of increased brain size or connectivity on behavior must be manifold. Plasticity of bone, muscle, connective tissue, circulation, the sensory apparatus and the brain means that behavior such as play, which may result in significant increases in the amount of physical exercise and sensory or cognitive stimulation, can be interpreted using concepts of general as well as concepts of specific determination.

It is safe to assume that certain characteristics of the rearing environment will be present for all individuals who have a chance of surviving to reproduce. In group-living species, some individuals will be available as social companions. In mammals, the mother will be present at certain times to nurse and to groom her offspring. For this reason, behavioral elements involved in socialization and in parent-offspring relations may become increasingly dependent on experience for normal development (E. O. Wilson 1975 pp. 160–161). This process requires that features of the environment be sufficiently reliable or sufficiently general for evolution to incorporate them into developmental programs. If warmth, social contact, or physical movement has been present during the entire evolutionary history of a behavior pattern, it is difficult to imagine how developmental mechanisms could evolve independently of these conditions or could function in environments diametrically opposed to those to which they had become adapted.

When parental care is reliable, intense, and prolonged, the young organism's rearing environment is relatively buffered from fluctuations in the physical environment. (Of course, parent-offspring conflict may add considerable excitement and uncertainty to the situation.) Under parental care, many aspects of the phenotype need not be genetically programmed, but can develop as a result of reliably present environmental stimulation. If nature then pulls a dirty trick and changes the pattern of stimulation in the environment, there will be strong selection for behavior that buffers the phenotype. In fact, a behavioral response will probably be most likely (G. Bateson 1963, Mayr 1963 pp. 602–604). Could play first have evolved in this way? Suppose that in an ancestral reptilian species, somatic plasticity evolved because the young were highly motile upon hatching. Their bone and muscular development would come to depend on the level of physical exercise necessary to find food. Suppose further that evolution of mammalian lactation (Pond 1977) then meant that young need not be physically as active. It would then have been easier to evolve solo locomotor-rotational play (through neotenous prolongation of embryonic motility) than to change the rate constants of all the enzymes involved in synthesis and breakdown of body components.

Why plasticity? The question is too typological. Better to ask, For each trait separately, what is its position on a continuum ranging from resilience to plasticity of development? Should the effect of the environment be specific to that trait or relatively general?

Analyses of modifiability range widely, from the insights of ecological niche theory (Emlen 1975) to population genetic models (Feldman & Cavalli-Sforza 1976, 1977, Slatkin & Lande 1976) and theory on constrained learning, including biologically adaptive mistakes, stupidity, and even self-deception (Dawkins 1976b, Trivers 1974, E. O. Wilson 1978). Contributions from outside biology include the insights of Jean Piaget and powerful artificial intelligence models of skill (Sussman 1975) and cognition (Winston 1977). I return to these ideas in later chapters.

Genetic Constraints

Sociobiological theories of the family (Alexander 1974, Dawkins 1976b, MacNair & Parker 1978, Maynard Smith 1977, Parker & MacNair 1978, Trivers 1974) reveal how the social environment can constrain evolution of behavior development. As a result of less than perfect genetic identity of family members, conflict between parents and offspring and among offspring are inevitable, and stable coalitions may form (Hamilton 1975) in which learning becomes an important element (Nichols 1977). We must view developmental plasticity in the context of family conflict (Geist 1978b). If one young animal's social behavior can affect another's development and vice versa, neither the individuals directly involved nor the rest of the family are expected to wait passively while behavior-physiology takes over. Rather, developmental patterns and rates may themselves be subject to conflict, coalition, and manipulation, in ways that are only now becoming apparent. For example (Geist 1978b), what if play can be used by one littermate to physically exhaust another at a critical point in development? Will both individuals enter such a tournament if the outcome of such a match is sufficiently unpredictable and its value sufficiently great to make damaging competition effective? Geist's suggestion that play functions as a competitive mechanism requires critical scrutiny; I simply cite it here as an interesting sociobiological speculation.

Aggression, dominance, dispersal, and individual differences are all aspects of immature animals' social behavior that sociobiology can serve to clarify. Too frequently, discussions of these topics, especially about their relationship to play, have often been uncritically conceived and carelessly phrased. When young animals fight, what resource are they contesting, and why? Because dominance hierarchies and dominance relationships are results of competition for resources, we must not view them as independent entities (Symons 1978b). What is the nature of competition in families, and under what circumstances are young animals selected to perform behaviors that establish and maintain hierarchical relationships? What are the genetic interests of third parties, including parents and other siblings, in dyadic aggressive interactions of juveniles and dominance relationships of juveniles? Dispersal occurs because the disperser's future reproductive success (measured in units of inclusive fitness) is better served by dispersing than by remaining (R. R. Baker 1978, Parker & Stuart 1976), whether the proximate cause for dispersal be parental aggression, avoidance of parents by offspring, lack of playmates, or some combination of these and other factors. Sexually immature animals form complex and often-enduring societies marked by competition, cooperation, and adaptive developmental change. The evolutionary study of these societies amounts to sociobiology without sex. For some scientists and sociobiology-watchers, this constraint probably spoils all the fun. However, as I have argued in this chapter, the problems are sufficiently complex to reward biologists for years to come.

NOTES

1. I am fully aware that my discussion sidesteps the issue of units of selection and tends to amalgamate sociobiologists with some of their most eloquent critics. I cannot resolve the issue of levels of selection here. Evolutionary analyses of development come from human scientists who are expected to disagree sharply about the use, importance, and implications of their science. Many of these issues call for models rather than rhetoric.

2. In expanded form, the argument runs as follows. For P fixed, higher selection intensity at stage 1 (higher x^*) is always associated with higher stage 1 mortality [lower $p_1(x^*)$] if we compare life histories having different functions f. However, if f (and therefore x^*) is the same for all animals, and if we compare life histories for which P varies, stage 1 mortality will not depend on selection intensity at stage 1. If we compare life histories in which f and P both vary, higher x^* can actually accompany lower stage 1 mortality. One such case contrasts the life history for which $P = .9$ and $f = x^2$ with the life history for which $P = .5$ and $f = \sqrt{x}$. In the former life history, $x^* = \frac{2}{3}$, $p_1(x^*) = .4$; in the latter, $x^* = \frac{1}{3}$, $p_1(x^*) = .3$. (For simplicity, the above discussion is restricted to functions f having first derivatives monotonic in x. If f is sigmoid in form, multiple optima may exist, and the situation becomes even more complicated.)

3. My summary of Gould's argument (Gould 1977 pp. 343–345) does not address some underlying subtleties. Usually, specialized morphological adaptations, large size, and delayed sexual maturation evolve because they result in improved ability in survival under harsh physical conditions or in improved success at competition, predation, and/or predator avoidance. "The usual heterochronic result of such trends is probably hypermorphosis—an extension or extrapolation of ancestral allometries. With a delay of [sexual] maturation, differentiation can proceed beyond its ancestral level into the larger sizes of descendants" (Gould 1977 p. 341). However, such hypermorphic ontogenies produce large, specialized adults—not behaviorally flexible adults with hypertrophied brains for an animal their size. Retardation of somatic development must occur relative to sexual maturation in order to produce this result. Increased selective pressure for brain-mediated modifiability in sexually immature individuals would favor retention of the more rapid brain growth rates of earlier ontogenetic stages. At a given chronological age, the resulting brain would be more modifiable than that of the ancestral form. Of course, factors

favoring delayed sexual maturation could sometimes produce retardation of somatic development as well: arboreality, for example (Cartmill 1972, 1974, 1975, Eisenberg 1975, Fagen & Wiley 1978, W. J. Hamilton 1972). Delayed sexual maturation as such, general retardation of somatic development relative to sexual maturation, or retardation of specific features of somatic development could all, under appropriate circumstances, select for postnatal prolongation of fetal brain growth rates. Therefore, "the issue of causal direction and historical primacy" in brain evolution "is largely meaningless" (Gould 1977 pp. 371, 399, 401). To be sure, past debate on this issue has been bitter and prolonged. As usual, Karl Groos (1898) was among the first troublemakers, as illustrated by his much-quoted remark "Animals do not play because they are young, but they have their youth because they must play." I wonder whether the violence of debate about causal direction and historical primacy might stem from logical or political ties to issues raised by debates about the concept of intelligence. Is human intelligence a patchwork of specialized abilities, each subject to a particular set of biological constraints on learning ("intelligence constructed piece by piece as an enabling device to create the qualities," E. O. Wilson 1975 p. 313)? Or is it a general capacity for gathering and processing information, from whose gradual increase emerged "the capacity for language, the consciousness of personae, the long memories of personal relationships, and the explicit recognition of 'reciprocal altruism' through equal, long-term trade-offs" (E. O. Wilson 1975 p. 313)? If increased modifiability as such is primary and causal, then most uniquely human characteristics are the results of a single biological factor, namely, our hypertrophied brain, and human flexibility is expected to be general, pervasive, and overriding. If human intelligence is the result of separate ecological factors acting at different times in human evolution to enhance different specialized capacities, no overall generalization is expected as long as increased modifiability of one capability of the brain is independent of or tends to decrease modifiability of another capability. But if selection for one aspect of delayed somatic development yields generally enhanced growth of the brain as a by-product (because developmental constraints tie overall brain growth rate to rate of development of the characteristic in question), then specific environmental factors will necessarily produce a general increase in brain capacity, and fortuitous consequences of this increase may in turn affect the pattern of selective forces.

5

Biological Effects
of Play

Play seems aimless, capricious, and inconsequential. Play
sequences include behavioral acts or sequences of behavior also
occurring in high-risk adult activities requiring physical skill
and strength or endurance, but the products of these high-risk
activities are absent from play. Food is not ingested, risk of
death by predation is not avoided as a consequence of play,
conflict over a contested resource external to play is not settled,
and a zygote is not produced. Because play is risky and requires
time and energy, it is natural to ask just what beneficial effects
play could exert that might compensate for its apparent cost to
the player.

This chapter reviews evidence that play is costly behavior. It
also discusses six major hypotheses about beneficial effects of
play: that play develops physical strength, endurance, and skill,
particularly in those acts or combinations of acts used in social
interactions having potentially lethal consequences; that play
regulates developmental rates; that play experience yields specific information; that play develops cognitive skills necessary
for behavioral adaptability, flexibility, inventiveness, or versatility; that play is a set of damaging behavioral tactics used in intraspecific competition; or that play establishes or strengthens
social bonds in a dyad or social cohesion in a group. Additional
hypotheses about consequences of play will be considered.

Here and in succeeding chapters I present an evolutionary biological analysis of structure, function, and evolution of play

behavior. I consider the behavior's probable effects and estimate the costs and benefits of these effects in the currency of surviving offspring. I define circumstances under which benefits outweigh costs such that the behavior will evolve by natural selection. I consider structural and contextual evidence that the behavior has been shaped by natural selection to produce its presumed benefits. This mode of analysis yields new empirical questions and improves theory.

COSTS OF PLAY

Energetic costs of play have never actually been measured in the field. In view of the vigor and persistence with which young animals play, often to physical exhaustion, it would not be at all surprising if play were found to represent a significant fraction of a player's total daily energy expenditure. Indirect calculations of the instantaneous energy costs (cal/time) of play and other activities (Coelho 1974, Coelho, Bramblett, Quick, & Bramblett 1976) suggest that the four most energetically costly activities for a 5-kg primate would be swimming, playing, mounting, and "self-directed actions."

A field study of preschool children in Uganda (Rutishauer & Whitehead 1972) furnishes indirect evidence of energy trade offs between play and growth. Previously malnourished Ugandan children aged one and one-half to three years played less and rested more than well-nourished European children of the same age group. Height increases in both groups over the study period were equal despite a markedly lower caloric intake by the Ugandan children. As Read (1975) points out, these data suggest that the malnourished children preferentially used their limited caloric intake for growth rather than for energy-demanding play.

Time alone spent in play may be significant (Table 5-1). The highest figures reported to date are for confined animals, and these figures probably represent the maximum attainable in the wild.

Although the data of Table 5-1 are quantitative, they are not at all comparable. Overall averages conceal considerable age, sex, and diurnal and seasonal variations in frequencies of play. Moreover, sampling methods and techniques for calculating frequencies differ among investigators. For example, some frequencies are reported as minutes play/total observation minutes, others as percentage of total observation periods in which play occurred at least once. Despite these serious inconsistencies, the data agree that playtime represents 1 to 10% of the total time budget in nearly every species studied. Figures of this order of magnitude are a highly consistent feature of the data tabulated from field and confinement studies of 15 species of mammals.

Time and energy expenditure in play affects growth and survivorship indirectly. Resources allocated to play cannot be allocated to growth, fat storage, feeding, predator avoidance, or nonplay social behavior. Another possible cost of play is hypothesized by A. A. Gerall (1963, see also H. D. Gerall 1965), who claims that play experience retarded the sexual development of guinea pigs. Gerall's inference, based on an isolation experiment, is probably incorrect. However, play experience might conceivably delay sexual maturity if social factors, including pheromone production, could suppress maturation and reproduction.

In rodents (Batzli, Getz, & Hurley 1977, Lombardi & Vandenbergh 1977, E. O. Wilson 1975 p. 154), swine (Brooks & Cole 1970), and fish (Borowsky 1978, Sohn 1977), social factors affect age of sexual maturation. The inhibitory effects described to date for rodents do not appear to depend on contact play because growth and maturation of voles physically isolated from their siblings are still suppressed if olfactory contact is maintained via a common air supply (Batzli et al. 1977), and so far as is known, house mice do not play in contact (Chapter 3). However, vigorous leaping, rolling, jerking, and twitching movements can disperse airborne pheromones, just as contact play can result in exchange of high molecular-weight substances (S. Wilson & Kleiman 1974). Play may facilitate olfactory exchange in mammals (S. Wilson & Kleiman 1974), thus mediating such effects. Tactile or visual stimuli affecting developmental rates may also be frequent in play. Viewed solely as a delay in the age of first reproduction, play-induced suppression

Table 5-1 TIME SPENT IN PLAY BEHAVIOR

Species	Time Allocated to Play	Observer	Comments
Alouatta villosa, mantled howling monkey	3%	Bernstein 1964	Different ages
Alouatta villosa, mantled howling monkey	90–95% (locomotor) 0–60 sec/day (mother-infant) 2–15 min/day (infant-infant)	Baldwin & Baldwin 1973a	Infants, field
Ateles geoffroyi, red spider monkey	25%	Rondinelli & Klein 1976	Confined
Cebus albifrons, white-fronted capuchin	1%	Bernstein 1965	Confined
Saimiri sp., squirrel monkey	0–8% (field) 6–12% (confined)	Baldwin & Baldwin 1973b, 1974	
Cercocebus atys, sooty mangabey	9%	Bernstein 1976	Confined
Colobus badius, red colobus	3–14%	Clutton-Brock 1974	Field
Colobus badius, red colobus	3%	Struhsaker 1975	Field
Macaca mulatta, rhesus macaque	1–6%, max. 9% (1-yr-old males)	Levy 1979	Field
Macaca nemestrina, pigtail macaque	4–8%	Bernstein 1970, 1972a	Confined
Papio anubis, olive baboon	20%	Nash 1978	Infants, field
Papio ursinus, chacma baboon	3% (infants) 5% (juveniles and subadults)	Cheney 1978	Field
Papio anubis, olive baboon	3%	Rose 1977b	Field, all ages
Papio hamadryas, hamadryas baboon	50%	Leresche 1976	Confined

Hylobates lar, lar gibbon	10–20%	Bernstein & Schusterman 1964	Confined
Pan troglodytes, chimpanzee	0–300 play sessions/ 100 observation periods	van Lawick-Goodall 1968a	Field, various ages
Pan troglodytes, chimpanzee	4–10%	Merrick 1977	Confined
Rattus norvegicus, Norway rat	1–3%	Müller-Schwarze 1966	Confined, domestic strain BDIX
Panthera leo, African lion	1.5–6%	Schaller 1972	Field
Antilocapra americana, pronghorn	0.1–0.2%	Buechner 1950	Field

of reproduction represents a decrease in the Malthusian para-meter of population growth and therefore a decrease in fitness. Of course, beneficial consequences of delayed maturation (e.g., increased growth, outbreeding) would be expected to outweigh this cost. Such effects would need to be demonstrated if play were found to suppress reproduction.

Field observations demonstrate that play can result directly in injury or even in death. Playing juveniles have had serious falls (van Lawick-Goodall 1968a, Steinborn 1973). A study (Byers 1977) of confined Siberian ibex kids (*Capra ibex*) highlights the potential seriousness of these risks. Two kids fell vertical dis-tances of 4 to 5 m in play, and at least one-third of the 14 kids studied sustained, in play, "injuries bad enough to cause tempo-rary limps" (Byers 1977). Baldwin & Baldwin (1977), Budnitz & Dainis (1975), Hornaday (1934), van Lawick-Goodall (1968a) and Welles & Welles (1961) all emphasize the special risk of fall-ing in play. Play may be the only behavioral context in which these risks are ever encountered (van Lawick-Goodall 1968a).

Exposure to predation is another risk likely to occur in play (Hornaday 1934). Field observations (Hausfater 1976) of yellow baboon (*Papio cynocephalus*) predation on vervet monkeys (*Cercopithecus aethiops*) show that vervet play groups often formed at some distance from adults and that these juveniles were the ones caught by baboons. Playing vervets were vulnerable: of 19 chases of vervets by baboons, 9 (47.4%) were successful. We lack other field observations of predation on playing young, but this lack does not indicate that such events are infrequent. Both the activities of large predators and play itself are inhibited by the presence of human observers. Predation on nonhuman primates is notoriously difficult to observe in the wild. Is the risk of predation especially high in play? Hausfater's observations suggest that this may be the case.

Table 5-2 presents all reports known to me that suggest play may directly reduce the player's immediate probability of survival. This evidence, never previously summarized, indicates the costs that natural selection must overcome in order to main-

Table 5-2 OBSERVED RISKS OF PLAY

Fall down	Briedermann 1967, Budnitz & Dainis 1975, Byers 1977, Haas 1967, Kavanagh 1978, Kleiman 1977, Koford 1957, van Lawick-Goodall 1967, 1968a, Nance 1975, Petter, Albignac, & Rumpler 1977, Steinborn 1973
Separation from caregiver; alone for protracted period	Hausfater 1976, van Lawick-Goodall 1971, Sugiyama 1971, Welles & Welles 1961
Frighten prey	Wrogemann 1975
Elicit damaging or fatal aggression	Angst & Thommen 1977, Kurland 1977, F.L. Miller 1975, Mohr 1968, Vick & Conley 1976
Be caught in rocks or mud	Douglas-Hamilton & Douglas-Hamilton 1975, Marlow 1975
Danger from rolling rocks	Kummer 1968a
Painful contact with spines of cholla (*Opuntia* spp.)	Berger 1980 (and Fig. 5-1)

Fig. 5–1 Desert bighorn (*Ovis canadensis*) lamb and cholla (*Opuntia* sp.). In this environment, strewn with spiny cactus, lambs contact cactus spines during play. This contact, apparently painful to lambs, seems to decrease the frequency of the lambs' play. California (Joel Berger)

tain play in a population. Additional information on the actual frequency of such events in nature would be important.

Investigators concerned with child health are giving serious consideration to the risks of play. A U.S. Consumer Product Safety Commission study (Rutherford 1979) showed that nearly 100,000 injuries to American children each year are related to playground equipment; that the majority (66,000) of these injuries resulted from falls; and that most of those falls were from playground equipment "such as swings, slides, climbing bars and merry-go-rounds." Although proper equipment design and playground surfaces could reduce the frequency of such injuries, it is difficult to see how play could be zero-risk behavior even under ideal circumstances.

The frequency and severity of injuries occurring in human play are unknown in general. Could this "inconsequential" behavior prove to be a major source of injury? If so, are there biocultural reasons for taking such risks?

In view of these and other risks that players run, including the danger that social play will escalate into a damaging fight and the possibility of punitive intervention by third parties, we can no longer view play as innocuous behavior. We must recognize that play, like feeding, fighting, and sex, carries with it certain definite risks. Certain effects of play decrease the player's probability of surviving to reproduce. For play to evolve, this survivorship cost must be outweighed by other effects that benefit the player or the player's kin.

BENEFITS OF PLAY

All beneficial effects of behavior are not functions (Hinde 1975, Symons 1978b, Williams 1966). The study of the beneficial effects of behavior is not an end in itself (Symons 1978b), but a means to the study of function. If a behavior producing a beneficial effect in a certain environment shows evidence of evolutionary design for that effect and can evolve by natural selection to produce the effect in question in that environment, we may be reasonably certain that we have identified the function of that behavior.

Numerous beneficial effects hypothesized for play (Baldwin & Baldwin 1977 p. 383, E. O. Smith 1978 pp. 16–17, Symons 1978a p. 6) give this behavior the status of a wonder-working elixir. Apparently, play can do almost anything! Play is cited as the source of virtually any skill or information that can result from experience with conspecifics or with the physical environment during development. Why should play, rather than some other behavior having the same effect, be designed by natural selection to achieve this goal? To speculate freely on effects without asking hard evolutionary questions is a tendency correctly criticized by previous commentators on play. This habit results from approaching the search for function as if it were the search for the Philosopher's Stone. The function of apparently functionless behavior is treated as a great scientific mystery, and play research then becomes an effort to demonstrate the function of play. Different investigators then become zealous advocates of different functions.

A more practical goal for play research would be to formulate and to test biologically meaningful predictions about play. As discussed in Chapter 2, it is possible to base such predictions on very simple assumptions that do not completely specify beneficial effects of play. In addition, viewing play in terms of adaptation allows multiple working hypotheses about function to be refined by considering whether behavior having the hypothesized function could evolve by natural selection and whether the structure of play is consistent with the hypothesized function.

In recent years, the biological study of play has progressed significantly along both these paths. By making simple assumptions, Fagen (1977) and Konner (1975) developed predictive theory regarding life-history patterning and social structure of play. Similarly, the most successful design analyses of play have considered the simplest possible functions consistent with natural selection theory, namely, physical training (Brownlee 1954, Fagen 1976a) and physical skill development (Symons 1978a). Consideration of simple functions first does not preclude study of more complex consequences for play, such as development of cognitive skill or of social relationships. Rather, simple effects are treated as a set of easily manipulated null hypotheses. If these null hypotheses fail to explain a given aspect of play, then this aspect may be singled out for further analysis, and its functional implications may be considered as explained above: Can such an effect evolve by natural selection? Is the hypothesized function consistent with evidence on the design of the behavior? Why is behavior other than play not used to achieve the desired effect? In the absence of analyses of this kind, empirical observations and experiments, especially those seeking to demonstrate function, will almost always prove difficult to interpret.

Although progress to date has been satisfactory, as outlined above, it is important to recognize that basic understanding of behavioral development may change, and along with it criteria for simplicity of assumptions and effects. Perhaps, for example, as Austin (1974), Mead (1976), and Papoušek & Papoušek (1975) suggest, the distinction between physical and cognitive skill is not useful, and other more meaningful distinctions must be made. Behavioral development may well involve a few simple variables, but not those currently recognized as primitive (Simpson 1976, 1978). Play researchers should be prepared to consider the implications of these and other new ideas about development.

The possible beneficial effects to be considered below can all claim support of some kind. However, existing analyses of function are uneven both in scope and in depth. References cited in my discussion of each effect present relevant arguments and evidence in greater detail.

Physical Training: Strength, Endurance, and Skill Development

Play experience is likely to improve physical ability by enhancing strength, endurance, and skill (Brownlee 1954, Dobzhansky 1962, Fagen 1976a, Symons 1978a). Of course this athletic sense of the word "fitness" represents just one component of Darwinian fitness, but the contribution of physical ability to survival in nature is too obvious to require extended argument. Physical ability can even contribute directly to reproduction. For example, young male olive baboons (*Papio anubis*) were able to take over an estrous female from an older male competitor by leaping from branch to branch of sleeping trees (Packer

1979a,b). The young baboons outmaneuvered the older males and were able to separate a consorting male from his female consort.

The physical training hypothesis takes somatic, including behavioral plasticity, as given. However, a complete analysis of the problem would need to explain why animals must expend resources and risk injury or death in order to develop strength, endurance, and skill that in theory might merely mature to an optimal end point without any need for environmental feedback. Evolutionary questions of this type are the subject of modifiability theory (Chapter 4). A complete answer to this question lies beyond the scope of the current discussion and perhaps beyond the explanatory ability of current theory on development. In part, modifiability of physical ability should be adaptive because of the metabolic maintenance cost of excess physical capacity (and perhaps because of cybernetic costs of maintaining access to excess or unused skills) and because the optimal level of a physical ability for an individual in a population should depend on the physical ability level of competitors. Experience may be necessary in order to ensure resilience of developing ability. Physical ability may need to be modifiable by experience because developing animals change in size and in shape and in other ways (recalibration), because certain features of the environment are reliably present (economics of developmental programming, E. O. Wilson 1975 pp. 160–161), or because the amount of information needed to specify optimal levels of physical abilities (especially in changing or unpredictable environments) exceed an organism's available genetic capacity for information storage.

Somatic plasticity of plants (Bradshaw 1965, Hickman 1975), of locusts (Gould 1977 p. 312), and even of *Hydra* (Łomnicki & Slobodkin 1966) puts that of mammals and birds to shame. Most phenotypic modifiability in these large-brained vertebrates is, of course, behavioral modifiability. However, enduring adaptive physiological, histological, and biochemical responses to vigorous and repeated stimulation ("training responses" or "exercise adaptation") are known now to occur not only in skeletal muscle, connective tissue, bone, and the cardiopulmonary system (Åstrand & Rodahl 1970, Basset 1972,

Buller & Pope 1977, Burleigh 1974, Edgerton 1978, Goldspink 1970, Holloszy 1973, 1975, Holloszy & Booth 1976, Keul et al. 1972), but also in the endocrine and nervous systems (Bennett 1976, Greenough 1976, 1978, Rose 1969, Rosenzweig 1966, Rosenzweig & Bennett 1976, Tharp & Buuck 1974).

Regular performance of physical work by an animal tends to increase the organism's capacity to perform that work. Such "enrichment" of the animal's environment through regular work and exercise results in enduring adaptive physiological changes in mammals of several species, and particularly in young mammals.

The above meaning of the word "training" should be distinguished from a very different psychological use of the same word. In psychology, "training" is sometimes used to indicate a form of directed learning in which "responses are channeled and are followed by immediate reinforcement" (B. Beck 1975). I use "training" solely in the physiological sense indicated above.

Strength and endurance of particular muscle groups, maximum work rate, and endurance in whole-body work, such as running and swimming, have all been shown to adapt to a program of regular exercise. Muscle mass (Åstrand & Rodahl 1970), heart size (Poupa et al. 1970) and capillary density (Bloor & Leon 1970, Tomanek 1970), and vital capacity (Ekblom 1969a) can also adapt to exercise stress. Indeed, growth rate (Ekblom 1969a) and possibly even length of life (Arshavsky 1972) can respond adaptively and positively to a program of regular physical exercise, provided that the exercise program begins sufficiently early in ontogeny, before growth ceases in adulthood.

Physiological mechanisms that produce training responses, and indeed the training responses themselves, vary with type, frequency, intensity, and duration of exercise, and with age. "Work overload" exercises, such as weightlifting or static muscle contraction, which only stress certain muscle groups, will develop (increase the mass of) only those muscles exercised, whereas whole-body "endurance" exercise, such as running, cycling, jogging, or swimming, tends to increase the biochemical potential (but not the mass of) a wide variety of muscles and to improve cardiopulmonary system function (Åstrand & Rodahl 1970). Dynamic exercise may improve dynamic strength while leaving isometric strength unchanged (Hansen 1967). In humans, the immediate response to endurance exercise stress involves cardiopulmonary adjustments on a time scale of several weeks, but if exercise continues for a longer period of time, say several months, other physiological parameters will also adapt, further improving the organism's endurance (Åstrand & Rodahl 1970, Ekblom 1969b). A significant component of this second, long-term response to exercise is profound cellular and biochemical adaptation. Rats subjected to a program of endurance exercise show increased steady state amounts of enzymes and respiratory pigments. In addition, such trained animals exhibit greater use of fats during exercise than do untrained controls, and it is believed that this alteration in substrate source, which results in a reduced rate of glycogen breakdown, is related to the increased endurance capacity and delayed onset of fatigue seen in trained rats (Holloszy 1973, Keul et al. 1972).

Nature and magnitude of training responses can differ strikingly in organisms of different ages (Bloor & Leon 1970, Ekblom 1969a,b, Eriksson 1972, Hartley et al. 1969, Poupa et al. 1970, Rarick & Larsen 1959, Rowell 1974). For example, overloading of the heart in young rabbits leads to accelerated growth both of cardiac cells and of the cardiac capillary bed, but adults exhibit only changed cell volume (Poupa et al. 1970). Young men, but not older individuals, exhibit increased maximal oxygen extraction from blood (mixed arteriovenous oxygen difference) after exercising regularly for several months (Hartley et al. 1969). Cardiac hypertrophy in response to endurance exercise occurs both in young white rats and in adults, but adults require greater frequency and intensity of exercise than do juveniles in order to achieve the same magnitude of response (Bloor & Leon 1970). Adult humans and 11- to 13-year-old boys increased their maximum oxygen consumption by the same fractional amount after corresponding programs of training, but responses of different physiological systems produced this increase in boys and men, suggesting that "if physical training is begun as early as in prepuberty" greater physical capacity "may be attained in adulthood than if training were initiated later in life," since both systems rather than just one will be trained in the former case (Eriksson 1972). This physiological evidence suggests that training programs should begin early in life in order to be maximally effective, for the following three reasons:

1. Some training responses occur only in young animals.

2. Young animals often respond more rapidly or more strongly to training than do adults.

3. Even when a training response is not age-specific, the physical capacity of an animal who has trained since a very early age will at least temporarily exceed that of an age-mate who begins training in adulthood, if only because the first animal has trained for a longer total period of time.

Decreased muscular activity also causes physiological change, indeed physiological deterioration, in animals. After prolonged inactivity (bed rest, hospitalization, weightlessness of space flight, or simply a sedentary mode of life) endurance, work capacity (Åstrand & Rodahl 1970, Rowell 1974, Saltin et al. 1968), and muscular strength all tend to decrease. Histological and biochemical correlates of this "detraining" response are analogous, but opposite in sign to those observed after endurance exercise (Saltin et al. 1968).

Brain structure, like the structure of muscle, bone and other somatic components, may respond adaptively to experience. Brain components exhibit measurable anatomical, cytological, and biochemical training responses. These responses appear to be specific to certain types of experience. In fact, play and exploration are the only behaviors known to mediate this response (Ferchmin, Bennett, & Rosenzweig 1975, Ferchmin & Eterović 1977, Ferchmin et al. 1980, Greenough 1978).

Elegant experiments on environmentally induced growth of cerebral cortex indicate with increasing precision that the specific experience responsible for these changes is participation in playful social interaction, playful object manipulation, or performance of playful body movements. These experiments have ruled out many plausible alternatives to play experience, including observation by the subject of other rats housed in enriched environments that produce cortical growth (Ferchmin et al. 1975), nonplayful training of motor skills (Ferchmin & Eterović 1977), mere group living (Rosenzweig et al. 1968), or learning by operant conditioning (Rosenzweig et al. 1968). Recent reviews of these sophisticated experiments on play include Bennett (1976), Rosenzweig & Bennett (1977a,b), and Greenough (1978).

Evidence cited above suggests that young organisms, at least young mammals, must be viewed as possessing significant physiological plasticity. This evidence further suggests that behavior in young mammals should be viewed not only in terms of adaptation to the immediate environment or in terms of traditional ontogenetic and learning paradigms, but also in terms of possible impact on the young organism's developing physical capacity. Whenever a young animal exercises its muscles, it produces feedback, which will affect the growth and development of those muscles, perhaps at the expense of other muscles that are exercised less frequently or less intensely. One can even imagine different muscle groups competing for limited nutrients. In this physiological view of behavioral ontogeny, patterns of juvenile motor behavior are viewed as a mechanism for allocating exercise—and therefore "training" and nutrients—to particular muscle groups. Such plasticity would be extremely important if young animals developed individual behavioral strategies for vigorous activities, such as predation, predator escape, fighting, or courtship, since each individual would then tend to exhibit idiosyncratic sequences of motor behavior. As a result, each individual would levy a different set of physiological performance requirements on its muscles, and practice or repetition of idiosyncratic patterns of behavior would implement these requirements by inducing training responses in the particular muscle groups such motor sequences stressed most heavily. Note also that these plasticity mechanisms allow young organisms to influence each other's physical development, for better or worse.

How might such environmental molding of the phenotype occur in young animals? It is likely that a young animal deprived of all opportunities for regular vigorous exercise would not develop the physiological machinery necessary for later survival and reproduction. [That physical capacity itself, rather than prey abundance or time, may sometimes be a limiting factor for at least some animals, is suggested by an observation of

Eaton's (1969) on apparent physical exhaustion occurring in a cheetah following an unsuccessful hunt.] Brownlee (1954) and Geist (1971) argue that an animal that would not be expected to regularly exercise certain muscle groups (e.g., those needed for predation, escape, fighting, or reproduction) in a functional context until adulthood might well suffer from the physiological consequences of its inactivity, whereas an animal that regularly exercised such muscle groups in a nonfunctional context ("play") would train itself physically, in preparation for adulthood. Brownlee cites additional muscle groups that are regularly used by young animals in functional contexts, but that are not used in play. As argued above, current understanding of the physiology of exercise in mammals further supports Brownlee's hypothesis, suggesting that regular vigorous exercise must take place before adulthood to ensure adequate training. Once growth ceases, exercise is less effective and is therefore more costly in terms of time and energy. In adulthood, exercise will often occur in a functional context. Moreover, training of some physiological functions in adulthood appears to be impossible. For these reasons alone, young animals should play.

In the next section I discuss properties and characteristics of play behavior that may be explained in terms of physical training: "playful" form (including behavioral composition, interruption and repetition, rough-and-tumble, and approach-withdrawal); age-dependence; play deprivation and exercise deprivation; "warmup" play before foraging; play with familiar objects; and the phylogenetic distribution of play behavior. I then consider other aspects of play for which physical training does not appear to furnish a sufficient explanation.

Why playful rather than serious forms of activity? From a strictly physiological point of view, vigorous exercise in a non-play context does not differ from vigorous exercise in play. (Differences in muscle tone have been suggested, but such differences remain to be demonstrated electromyographically.) Functional forms of fighting, escape, and predation could be extremely costly in terms of time, energy, and survivorship, since they would tend to expose the young animal to very real dangers. In addition, the partner in such serious interactions should be selected to fight back, attack, or escape in ways that were not only functional, but that also produced the smallest possible physical training response in the animal initiating the interaction. In play, on the other hand, choice of training schedule and intensity is left to the individual, and natural selection should act on the behavioral composition (as argued by Brownlee), frequency, duration, and intensity of play bouts in order to produce optimal age-specific, species-specific, and sex-specific exercise programs. That is, play should train muscles or muscle groups that would otherwise only be used in such vigorous adult activities as prey-catching, fighting, predator avoidance, and mating (Ewer 1968, E. O. Wilson 1971b). Differential recruitment of muscle fibers and muscle groups (Basmajian 1972, 1974, Taylor 1978) is significant in this context because certain components of muscle are normally not recruited except in maximally vigorous physical activities. These activities are emergency behaviors and are generally accompanied by considerable stress or danger to the organism. Muscle components normally used only in emergencies could be recruited in play for training purposes. If recruited out of their "normal" sequence, they might impart an unusual, playful quality to the animal's movement.

Play is rich in brisk and lively body movements. These locomotor-rotational movements (S. Wilson & Kleiman 1974) fall into three general categories: locomotion, rotation of the whole body, and rotation of parts of the body. We know these characteristic behaviors as capers, jinks, and gambols. Animals at play cavort and frisk. (See Appendix III for a list of terms of reference denoting play-movements.) These behaviors also occur during predator avoidance (including parasite avoidance, Schaller 1977 p. 132) and during agonistic chases.

The single most frequent and phylogenetically widespread locomotor act of play must surely be a leap upward. Plato's *Laws* (Bury translation, 1926, pp. 91, 129, 159) cites this familiar tendency of young creatures to leap in play. A well-known photographer (Halsman 1959) who feels that jumping reveals a person's true character even photographed jumps performed by his human subjects.

The variety of play (Table 5-3) calls for systematic treatment,

Table 5-3 REPRESENTATIVE EXAMPLES OF LOCOMOTOR-ROTATIONAL MOVEMENT IN ANIMAL PLAY *

Vertical leap

PRIMATES	
Cercopithecus diana, diana monkey	Mörike 1973
Presbytis obscurus, spectacled langur	Horwich 1974
Macaca mulatta, rhesus macaque	Symons 1978a
LAGOMORPHA	
Lepus timidus, Irish hare	Webb 1955
RODENTIA	
Spermophilus columbianus, Columbian ground squirrel	Steiner 1971
Micromys minutus, harvest mouse	Frank 1957
Mus musculus, house mouse	Williams & Scott 1954
Rattus rattus, black rat	Mohr 1928
Microcavia australis, desert cavy	Rood 1970
Myoprocta pratti, green acouchi	Bell 1830, Morris 1962a
Dasyprocta aguti, agouti	Roth-Kolar 1957
CETACEA	
Lagenorhynchus obliquidens, Pacific whitesided dolphin	Brown & Norris 1956
CARNIVORA	
Canis aureus, golden jackal	Wandrey 1975
Vulpes zerda, fennec	Koenig 1970
Bassaricyon sp., olingo	Poglayen-Neuwall & Poglayen-Neuwall 1965
Mustela erminea, ermine	Müller 1970
Herpestes edwardsi, Indian grey mongoose	Rensch & Dücker 1959
Fossa fossana, fanaloka	Albignac 1973
PINNIPEDIA	
Callorhinus ursinus, Alaska fur seal	Rowley 1929
PERISSODACTYLA	
Tapirus sp., tapir	von Richter 1966
ARTIODACTYLA	
Hippopotamus amphibius, hippopotamus	Dittrich 1962

Vicugna vicugna, vicuna	Koford 1957
Muntiacus muntjac, muntjac	Dubost 1971
Cervus dama, fallow deer	Gradl-Grams 1977
Mazama gouazoubira, grey mazama	Frädrich 1974
Antilocapra americana, pronghorn	Autenrieth & Fichter 1975
Damaliscus dorcas, bontebok	David 1973, 1975
Antilope cervicapra, blackbuck	Schmied 1973
Ovis canadensis, mountain sheep	Geist 1971

Galumphing, lolloping

Macaca mulatta, rhesus macaque	Sade 1973
Papio anubis, olive baboon	Miller 1973

Running in circles

Symphalangus syndactylus, siamang	Chivers 1974
Otospermophilus beecheyi, California ground squirrel	Linsdale 1946
Dolichotis patagonum, mara	Dubost & Genest 1974
Tapirus sp., tapir	von Richter 1966
Antilocapra americana, pronghorn	Autenrieth & Fichter 1974
Bison bison, bison	Lumia 1972
Kobus leche, lechwe	Lent 1969

Chasing one's own tail

Erythrocebus patas, patas	Hall, Boelkins, & Goswell 1965
Canis familiaris, domestic dog	Fox & Stelzner 1966
Mustela vison, mink	Herter 1958

Bounce

Propithecus verreauxi, sifaka	Richard 1976
Presbytis johnii, Nilgiri langur	Poirier 1970b
Odocoileus hemionus, mule deer	Linsdale & Tomich 1953

Buck

Odocoileus hemionus, mule deer	Linsdale & Tomich 1953
Syncerus caffer, African buffalo	Sinclair 1977

Roll

Miopithecus talapoin, talapoin	Gautier-Hion & Gautier 1971

Table 5-3 REPRESENTATIVE EXAMPLES OF LOCOMOTOR-ROTATIONAL MOVEMENT IN ANIMAL PLAY (*Continued*)

Castor canadensis and *Castor fiber*, beaver	Kalas 1976, Schramm 1968, Wilsson 1968, 1971
Ondatra zibethicus, muskrat	Steiniger 1976
Mustela vison, mink	Poole 1978
Whirl	
Presbytis entellus, hanuman langur	Rahaman 1973
Macaca mulatta, rhesus macaque	Symons 1978a
Pirouette	
Pan troglodytes, chimpanzee	Aldis 1975
Dangle by one arm and turn in circles	
Pan troglodytes, chimpanzee	van Lawick-Goodall 1971
Tumble	
Hylobates lar, lar gibbon	Ibscher 1967
Gorilla gorilla, mountain gorilla	Schaller 1963
Handstand	
Presbytis entellus, hanuman langur	Blaffer Hrdy 1977
Back flip	
Macaca mulatta, rhesus macaque	Symons 1978a
Spin	
Cercopithecus aethiops, vervet monkey	Fedigan 1972b
Cervus canadensis, elk	McCullough 1969
Vault over another's back	
Cercopithecus campbelli, Lowe's guenon	Hunkeler, Bourlière, & Bertrand 1972
Macaca mulatta, rhesus macaque	Symons 1978a
Meles meles, European badger	Neal 1948
Somersault	
Blarina brevicauda, short-tailed shrew	Rood 1958
Presbytis entellus, hanuman langur	Blaffer Hrdy 1977
Pan troglodytes, chimpanzee	van Lawick-Goodall 1971
Spermophilus columbianus, Columbian ground squirrel	Steiner 1971
Meles meles, European badger	Eibl-Eibesfeldt 1950b

Ailurus fulgens, red panda	Keller 1977
Leopardus sp.	Wilson 1860
Cervus elaphus, red deer	Bubenik 1965
Hang upside down	
Cercopithecus diana, diana monkey	Mörike 1976
Hylobates lar, lar gibbon	Ibscher 1967
Ailurus fulgens, red panda	Keller 1977
Corvus corax, raven	Gwinner 1966
Nestor notabilis, kea	Keller 1975
Close eyes and roll sideways	
Cercopithecus diana, diana monkey	Mörike 1976
Theropithecus gelada, gelada	Emory 1975

*For additional references, see Aldis (1975) and S. Wilson & Kleiman (1974).

perhaps using movement notation techniques (as suggested by Golani 1976). Hops, springs, bounces, and bucks are variations on the basic vertical leap. Animals often run in play, using characteristic galumphing or lolloping gaits. They may run repeated courses—back and forth, circles, figure-eights, or semicircles—and may zigzag. Sudden starts and stops are frequent, as are sudden, sharp turns (jinks). Rotational movement is characteristic of play. Animals may somersault, roll, flip forward or backward, spin, whirl, pirouette, make handstands, chase their tails, rear, and kick up their heels. Rotations of body parts often occur in combination with whole-body movement. Often a vertical leap is decorated with body-twists, rear-kicks, or head-shakes. These acrobatics can be spectacular. Golden jackals (*Canis aureus*) are known to rear up on their hindlegs with such force that the hindlegs leave the ground; while in the air the animal twists its head and body, making a full turn in the air around the long axis of the body. Some individual animals prefer to make right aerial turns, others left turns (Wandrey 1975). Schaller (1977 p. 284) saw young wild goats (*Capra* sp.), urial (*Ovis ammon*), and bharal (*Pseudois nayaur*) jump and turn

180° in midair. In play, kids and lambs "run with huge bounds sometimes on a zigzag course," and buck "with hindlegs thrown up in the air and head waving from side to side" (Schaller 1977 p. 284).

Often these behaviors seem designed for sensorimotor stimulation as well as for physical exercise. Aldis (1975) correctly notes that play movements can effectively stimulate the vestibular apparatus, an inner ear sensory organ involved in maintaining the body's balance in space. In play, animals balance (Delacour 1933, Keller 1977), slide (Aldis 1975 p. 53, Emory 1975, Geist 1971, Nance 1975, Schaller 1963, Stevenson 1976, Vaz-Ferriera 1975), and fall, apparently on purpose (Aldis 1975 p. 55, Bourlière, Bertrand, & Hunkeler 1969, Donald 1948, Emory 1975, Gautier-Hion & Gautier 1971, Krämer 1969, Nolte 1955b, Sugiyama 1965a, Symons 1978a p. 47). Arboreal primates may play wild falling games, as observed by Leslie Milne (1924 pp. 8–9) during a journey in upper Burma:

As we came down into the valley, the air was oppressively hot, after the fresh air of the hills. We skirted long, terraced rice-fields, and saw two kinds of monkeys. Some of these seemed to be of a grey colour, with long tails and black faces; they were playing in a great tree near the path. They ran up the tree trunk, and when they arrived at the top threw themselves into the space below, catching at the branches as they fell, and then, having arrived at the bottom, they at once began the ascent again, exactly as is described by Rudyard Kipling in *The Jungle Book*. They seemed to express perpetual motion.

The monkeys described were evidently langurs (*Presbytis* sp., perhaps the still virtually unknown Phayre's leaf monkey *Presbytis phayrei*).

Motion as such can invite play (Aldis 1975). In addition to sliding, passive transport in play may include riding waves (Peterson & Bartholomew 1967), jets of water, or waterfalls (Knappen 1930, Roberts 1934, Stoner 1947). Young wild primates may even drop objects, apparently for the sole purpose of watching them fall (Baldwin & Baldwin 1977, Horr 1977, Hunkeler, Bourlière, & Bertrand 1972, Kummer 1968a).

Vestibular stimulation in play is familiar to students of human behavior. Aldis (1975) presents an excellent review of this aspect of human play, from the earliest games of parents with their infants to amusement-park rides. Clark, Kreutzberg, & Chee (1977) show that exposure to vestibular stimulation accelerates motor development in infants, an effect apparently recognized by Plato (1926 p. 5), who recommended that babies be rocked and carried. This regime is advisable, according to the writer, because when bodies receive most food, and when growth occurs most rapidly, then without plenty of suitable exercise countless evils are produced in the body. (For an experimental demonstration of this fact, see Hedhammar et al. 1974.) Therefore, according to Plato, young animals need the most exercise and need exercise the most. Our concept of play as physical training has not improved significantly in the two millennia since Plato. Questions about ultimate causation and evolutionary significance of play-induced modifications of the phenotype and of developmental rate remain to be formulated and answered. These contemporary questions about the effects of play are probably the only ones that an intelligent Athenian of Plato's time could not have asked.

Is there play before birth? Fetal mammals and birds of many species exhibit spontaneous, neurogenic motor behavior, including whole-body contractions and extensions, and movements of body parts (Bielański 1977, Fraser 1976, Fraser, Hastie, Callicott, & Brownlie 1975, Gottlieb 1976, Hamburger 1973, Narayanan, Fox, & Hamburger 1971). This spontaneous activity appears necessary for normal development (Gottlieb 1976).

In 1975 E. M. Banks first made me aware of the phenomenon of prenatal motility and of its general significance for the study of animal play. I later read Bekoff & Byers's (1979) thoughtful discussion of possible ontogenetic and phylogenetic relationships between prenatal motility and play. I wish to expand these authors' arguments by addressing some empirical results and evolutionary issues not cited in previous discussions of the topic.

Two functionally (and ethologically?) distinct forms of prenatal motor activity are known in some species of mammals. Just before birth in these species, the fetus actively changes its orientation *in utero* and attains the position in which it is born.

But long before this change in orientation, the fetus exhibits whole-body movements and movements of its body parts. The presumed function of these earlier movements is physical training—bodily exercise that facilitates neuromuscular development (Fraser 1976). The fetus' efforts to right itself before birth may be as strenuous, relatively speaking, as any physical work that the animal will do postnatally. That excellent physical condition may be necessary to achieve this change in position is suggested (but not demonstrated) by the fact that some fetuses seem to tire before attaining the correct position. These fetuses are actually at increased risk because being born in the wrong position can be dangerous.

The phenomenon of fetal exercise and reorientation *in utero* raises several evolutionary questions. If the animal can be born most safely in a given position, why does it not develop in that position? Presumably, because it is more efficient for the fetus to develop or to be carried in another position (but efficient from whose point of view—that of the mother or that of the fetus?). Furthermore, because in horses and in several other large ungulates more than one zygote can develop, but only one ordinarily survives to be born (e.g., Merkt & Günzel 1979), an additional individual's biological interests are sometimes involved. Prenatal motor activity in species whose females carry several fetuses at a time may not always represent innocent solo exercise. It could be a way to damage a competitor *in utero*. Is prenatal motor activity more frequent when more than one fetus is present? (It is certainly evident in single human fetuses during the second and third trimesters.)

The phenomenon of prenatal exercise raises the central question of the evolution and possible adaptive significance of phenotypic modifiability. Why should prenatal exercise be the mechanism of choice to increase physical capacity to a level adequate for performing necessary prenatal work? Would it not have been much easier for natural selection to program the required physical capacity into the animal, independently of experience, simply by fixing the values of synthesis rates of mitochondria, division rates of muscle cells, etc., as appropriate to develop an animal physically capable of righting itself before birth? Most biologists' answers to this question would probably invoke a hypothetical need for informational feedback of some kind in order to coordinate development among fetal systems or between the fetus and the mother, but the conflict hypothesized to occur between fetus and mother under certain conditions (Trivers 1974) is at least as relevant to this question as other hypothetical feedback mechanisms. Perhaps (and this is only a supposition) prenatal modifiability is natural selection's way of ensuring that only those fetuses capable of performing the necessary movements will survive to full term. A fetus that did not exercise would become increasingly weak as a result of detraining and might be resorbed or aborted. A developmental program, unlike an abstract computer algorithm, does not guarantee results. There should be some way to check whether the product actually works, and the best way to ensure verifiability is to build an organism whose development can be monitored through test programs and controlled based on the test results. Modifiability through self-generated movement ensures that organisms which are developing well will continue to do so, whereas organisms that genuinely fail to thrive will self-destruct at a rate proportional to their neuromotor inadequacy. Of course, the existence of mechanisms ensuring developmental resilience, the possibility of multiple optimal developmental pathways, and the fetus' theoretical conflict with the mother (and sibs, if present) over its developing physical capacity make the issue much more complicated than this simple model would suggest.

Although play includes a great variety of movements, some behaviors do not occur in play. Muscle groups exercised in frequently occurring activities performed in functional contexts by the young animal (nursing, mastication, vocalization, lying down or rising up, and traveling with the group) are not used specifically in play, at least in domestic cattle (Brownlee 1954). Nonstrenuous adult behavior patterns, including agonistic and sexual signals and displays, also tend to be absent from play. For example, the physical training hypothesis correctly predicts the absence from rhesus macaque play of specialized agonistic visual and auditory signals, as reported by Symons (1974). As Symons points out, hypotheses on play as a medium for developing communicative skills fail to account for this observation.

Exercise physiology cannot yet predict the exact frequency, duration, and intensity of optimal training programs, but certain physiological principles of optimal exercise may be cited. These principles lead to testable predictions about play, as the following discussion will indicate. It would be valuable to determine what sorts of behavior, experience, or exercise would be optimal for central nervous system training, but these responses are not yet understood sufficiently well to generate testable hypotheses.

INTERRUPTION AND REPETITION

To be most efficient, strenuous exercise should be both interrupted and repeated (Åstrand & Rodahl 1970). During the periods of interruption, light exercise should be performed (Åstrand & Rodahl 1970). As one would expect from these principles, interruption and repetition are commonly observed in play (Henry & Herrero 1974, Müller-Schwarze 1971) and have indeed been considered to be among its defining characteristics (Loizos 1966, 1967).

TYPES OF TRAINING

In humans, three distinct types of optimal exercise programs and three corresponding types of training responses have been identified by exercise physiologists (Keul et al. 1972). Brief (30-sec) periods of strenuous "overload" exercise involving high-intensity and even maximal muscle contractions mainly increase the strength and speed of the specific muscles exercised, through training effects that consist mainly of muscular hypertrophy. Slightly longer (1- to 3-min) periods of high-intensity exercise might be expected to increase aerobic capacity and local muscular endurance by acting on glycolytic and oxidative enzyme levels. Prolonged exercise periods (several minutes to an hour or more) at lower intensities ("endurance exercise") build endurance and aerobic capacity by training those cardiovascular, respiratory, and biochemical system components associated with oxygen transport and aerobic metabolism. Although absolute duration of these three types of exercise periods may de-

pend on body size or on some other characteristic of the organism, relative magnitudes of the three types of exercise period
should not change, since underlying biochemical mechanisms
are the same across species. Therefore, animals ought to exhibit
three distinct types of playful exercise corresponding to the
three types of training programs identified by exercise physiologists (e.g., Keul et al. 1972). The first type should involve
maximal, usually static (sometimes called "isometric") exercise:
pushing, pulling, shoving, biting, hanging by a limb, springing, jumping, squeezing, butting, pushing, pulling an immov-
able object, etc. This type of exercise should occur in brief, but
intense bouts that are frequently interrupted by rest periods.
The second type of exercise involves slightly lower intensity
and slightly greater duration than the first. Therefore, it would
be distinct from the first only under ideal conditions, since the
difference between the two types is established by the exact intensity and duration of exercise, whereas the sequences of
motor patterns used are the same in both cases (at least in humans). The third type should involve prolonged bouts of locomotion at submaximal intensity, including such motor patterns
as running, chasing and fleeing, swimming, or swinging from
tree to tree. This form of play has been termed "locomotor
play" by Müller-Schwarze (1971). I suggest that the first and
second types of play correspond to rhesus macaque "rough-
and-tumble" play (Harlow & Harlow 1965), the third to rhesus
"approach-withdrawal" play (Harlow & Harlow 1965), and
that these terms should also prove useful (for the physiological
reasons cited above) in describing play of animals other than
rhesus macaques.

As mentioned above, the first type of exercise has highly specific effects, training only those muscle groups stressed by the
exercise program. Given the variety of muscle groups that must
be exercised in young animals, and given the physiological fact
that only a few muscle groups can be exercised at any one time,
animals needing playful exercise must necessarily have developed a diverse set of exercise patterns; but the very diversity of
these patterns has often caused play to be viewed as a wastebasket, or catch-all, category of behavior. This view becomes
meaningless once the common physiological and evolutionary

basis of these behavior patterns is understood. The power of Brownlee's hypothesis is that it serves to unify and explain a collection of seemingly unrelated and seemingly inexplicable behavior patterns and to do so in physiologically meaningful terms.

Young animals tend to play more than do adult animals (Baldwin & Baldwin 1974, Dolhinow & Bishop 1970, Hinde 1971, Loizos 1966, 1967, Müller-Schwarze 1971, West 1974). This age-dependent scheduling of play may be explained as a consequence of natural selection, assuming that play results in immediate costs (e.g., diversion of time and energy from other activities) and eventual benefits (due to increased physical capacity caused by training effects of play; see Chapter 6). As argued above, exercise physiology supplies a well-defined mechanism by which play behavior can produce such eventual benefits. Indeed, the known age-specificity of many training responses suggests that play must occur during the juvenile period if it is to occur at all, since these responses often tend to be greatly curtailed in adulthood. In addition, as West (1974) points out, vigorous exercise in functional contexts, such as feeding, predator avoidance, social interaction, reproduction, and care of offspring, may be sufficient to maintain the adult organism's physical capacity. Play in adulthood would then be superfluous, and adult play must be explained in some other way. This argument is consistent with exercise physiology. The amount of exercise needed to enhance a given physical capacity level is considerably greater than that needed to maintain that enhanced level (Åstrand & Rodahl 1970). Once an animal has attained a given level of physical capacity after months or years of juvenile play, adult activities, though less vigorous or less frequent than juvenile play, may still constitute adequate exercise for maintenance of that level of physical condition. Adults might still be expected to play in order to maintain their physical condition during periods of food abundance or low predator pressure. And indeed, if adult mammals have been selected to exercise spontaneously in order to maintain physical condition, the phenomenon of spontaneous activity in caged individuals (e.g., wheel-running), which Collier (1970) considers "poorly understood" despite the fact that it "has been studied longer, more persistently, and with less profit than any other single phenomenon in animal psychology," becomes transparent and fully understandable. In fact, McClintock (1974) observed that adult rats in a naturalistic environment played more, and used their running wheel less, than caged rats with access to a running wheel.

Adults should also be selected to play with their offspring in order to increase the offspring's physical capacity and therefore the adults' own inclusive fitness. However, since the interests of parent and offspring in mammals and birds are not identical (Trivers 1974), one would predict parent-offspring conflict over the parent's participation in play with its offspring, as follows. Initially, parent and offspring should agree to play together. After a certain point in ontogeny, the offspring will continue to solicit play from its parent, but the parent will respond less frequently to these invitations. In confined domestic cats, waning of the mother-kitten bond appears to result from parent-offspring conflict over play participation and from disagreement over access to milk (Rosenblatt et al. 1962, Rosenblatt 1974).

Parent-offspring play is known in a number of species (e.g., rhesus macaques, Hinde & Simpson 1975; meerkats, Wemmer & Fleming 1974; chimpanzees, van Lawick-Goodall 1968a; certain caviomorph rodents, S. Wilson & Kleiman 1974). It is interesting that the extent to which the father is involved in such play can vary greatly from species to species within a given taxonomic group (S. Wilson & Kleiman 1974).

Play tends to begin shortly after walking begins, in late infancy in altricial mammals and during the first days of life in precocial mammals (Fagen 1976a). Man and the great apes seem to be the only clear exceptions to this rule. Absence of play in early infancy in altricial mammals is understandable if one considers the severe functional requirements placed on the infant. The time and energy demands of feeding (including possible costs of competition with siblings, as discussed below) and of maintaining body temperature through proper orientation to and contact with the parent, littermates, or nest material (particularly when young are not fully endothermic, Dawson 1972) would appear to preclude play at this stage of development. These demands (and therefore the cost in fitness due to the diversion of time and energy from feeding and temperature maintenance) should decrease with age, paralleling decreasing growth rate (grams weight increment per gram body weight per unit time). Exercise play will begin at that stage in ontogeny when the long-term benefits of exercise outweigh the short-term costs of such diversion of time and energy.

Arshavsky (1972) claims that heavy demands placed on skeletal muscle groups in early infancy by the need for temperature maintenance preclude the ongoing use of these muscles in other contexts and that when the animal attains a standing posture "the skeletal muscles are fully or partly relieved of the heat-regulating function they formerly performed." Slonim (1972) similarly relates onset of play to the development of thermoregulation. These arguments further support the evolutionary cost–benefit argument presented above. However, more data are needed on the amount and apparent functional significance of skeletal muscle activity in the early infancy of altricial mammals.

Intrinsic speed of contraction of skeletal muscle is fixed early in ontogeny for a given muscle group and a given adult body size (Close 1972). As a result, the intrinsic speed of the muscles of juvenile animals is not appropriate to current body size, but rather is suited to adult body size in that species. For example, an adult rat is not physiologically equivalent to a young carnivore whose current body size is the same as that of the rat. The adult rat's muscles contract at a speed appropriate to the functions they perform, whereas the like-sized young carnivore's muscles contract much more slowly under the same external load. The implications of this difference, both for day-to-day functional performance in the young and for future motor skill development, have not yet been spelled out in any detail. However, it should be possible to understand many previously unexplained features of juvenile movement patterns and motor function in the light of this difference. Might the apparent relaxed, loose body tone characteristic of juvenile play stem from the relatively slow contraction velocity of the muscles being used? And yet when adults play, their muscles appear to show the same relaxed, loose tone.

Although little is known of ontogeny of play behavior in birds, the above considerations suggest that young precocial birds should begin to play shortly after hatching (as in the case of the domestic chick, Guhl 1958), whereas young altricial birds should start playing soon after the acquisition of endothermy, or, if Arshavsky's argument is correct, soon after they begin flying. Young blackbirds (*Turdus merula*), belonging to an altricial species, learn to fly at 18 days and begin play-chases at 30 days (Messmer & Messmer 1956). Similar data for other bird species would be of interest.

During infancy in altricial homiotherms, the heavy demands of feeding, thermoregulation, and often, competition leave little time or energy for play and may also provide sufficient physical exercise. Play fails to occur in infants even when muscle groups later used in play are already functional, as in the case of piglets, who fight over access to nipples during their first days of life (McBride 1963, Hafez & Signoret 1969, Wesley 1967). Such early fighting also occurs to some extent in domestic cats (Rosenblatt 1971) and in hedgehogs (Herter 1965). Similarly feeding by nestling chaffinches (*Fringilla coelebs*) may involve vigorous struggles for position (Newton 1973). In the species cited, skeletal muscle has clearly developed to the point of being functional in early infancy, yet the animals cited do not play until they are older and have begun to walk or to fly (e.g., play-chases of fledgling chaffinches, Marler 1956).

In most altricial homiotherms the infant, like the adult, simply works and rests. In the great apes and man, however, an infant learning to walk has already been playing for several months. The relatively early onset of play in these primates suggests that their play behavior may indeed serve some function or functions other than physical training, since this behavior begins to appear at a time when skeletal muscle may require no additional exercise.

It has been suggested (Morris 1964) that the tendency for adult animals to play in captivity more frequently than in nature stems from a need for sensory stimulation or drive release. [Thorpe (1966) finds no evidence for this tendency; further studies are needed.] It may be hypothesized that increased play in captive adults results from regression to an immature ontogenetic stage. Such regression may involve psychological dependence of the animal on its human captors (a nonevolutionary explanation), or it may constitute manipulation of the humans by the animal: if humans use performance of play activities as a cue indicating an early stage of ontogeny and, therefore, a need for greater "parental" investment in the animal exhibiting play behavior, the animal may be selected to play in order to obtain increased benefits from its captors. (I imply an evolutionary explanation of regression here, see Trivers 1974.) But if certain animal species have been selected to keep in physical condition by playing when the environment itself does not provide adequate exercise, specialized psychological events need not be invoked in order to explain the phenomena described by Morris (1964) and others.

Close confinement prevents physical exercise. One would therefore predict play upon release from close confinement, as is indeed the case (Birdseye 1956, Brownlee 1954, Chepko 1971, Cooper 1942, Dalquest & Roberts 1951, Haas 1967, C. C. Hamilton 1967, McCabe & Hawkins 1946). Animals regularly shut in at a certain time may begin to play just before that time, or when they see their keeper coming to put them in (Kennion 1921, Walther 1965a).

The physical training hypothesis also correctly predicts that play will respond to bad weather just as it does to confinement. Animals tend not to play during periods of inclement weather (extreme temperatures, Bernstein 1972a, 1976, Feist & McCullough 1976, Ghobrial & Cloudsley-Thompson 1976, Müller-Schwarze 1966, Rasa 1971, Richard 1974, Schaller 1977, Walther 1973; prolonged heavy rain and wet grounds, Bernstein 1972a, 1976, Chivers 1974, Cooper 1942; high winds, Poirier 1972a, Rasa 1971). When inclement weather results in such play deprivation, play should rebound when good weather returns (rain check effect). Qualitative information indicates that play may occur frequently after such changes in the weather, especially on cool, crisp days (Haas 1967, Schmied 1973), but there is need for better data on this point. Because weather changes represent one of the most humane and obvious natural experiments on play deprivation, ethologists (e.g. Müller-Schwarze 1971) have often speculated on the possible behavioral and developmental effects of such environmental change, effects that could become the focus of an interesting field study if *differential* deprivation of play occurred (see below p. 317).

Evidence that animals generally tend to play most often in the morning and/or in the evening, but seldom at midday or early afternoon when the temperature is highest (Table 5-4), suggest that ambient temperature influences daily scheduling of play. This suggestion is supported by data showing that play occurs most frequently during cooler periods of the day in the dorcas gazelle (*Gazella dorcas*) (Ghobrial & Cloudsley-Thompson 1976) and by the fact that the morning play period of redtail monkeys (*Cercopithecus ascanius*) continues until a later hour on cooler days (Galat-Luong 1975).

Seasonal temperature or rainfall cycles also appear to influence occurrence of play. Play occurs more frequently during the cool, dry season in rhesus macaques on Cayo Santiago (Levy 1979), and play disappears completely during the hot, dry season in sifaka (*Propithecus verreauxi*) (Richard 1974). In a hot climate, scheduling of vigorous play during cooler hours makes good biological sense as energy conservation and thermoregulation. This result suggests that thermal stress may represent an additional cost of play. However, these arguments alone do not explain the apparent rarity of nighttime play. Existing observations generally cover hours when animals were visible to the unaided human eye. Data on nocturnal species and on occurrence of play at night are not often available except in captive animals under artificially reversed light-dark cycles. We must also recognize that the scheduling of play depends not only on its costs, but also on its benefits and on the costs and benefits of other activities that could take place at a given time (Chapter 5). Thermal stress is only one of many factors governing the allocation of time to play. Risk of predation, and such social factors as movement of the animal's caregiver are also important, especially for infants who must be carried (most primates) or who follow (Walther's "follower" ungulates, Lent 1974) or hide and remain motionless ("hider" ungulates, Lent 1974) when the caregiver moves. A similar caveat applies to tame animals, who may play most frequently at times when a human companion is present.

As suggested above, animals on a regular schedule of release and confinement may behave as if they anticipate the confinement phase, using the last part of their free period for play. Analogously, primates, lagomorphs, rodents, and ungulates

Table 5-4 TIMES OF DAY AT WHICH PLAY OCCURS MOST FREQUENTLY

Morning
Cercopithecus aethiops, vervet	Poirier 1972a
Equus zebra, mountain zebra	Joubert 1972a
Erythrocebus patas, patas	Hall 1965
Saimiri sp., squirrel monkey	Du Mond 1968

Morning and evening
Lemurs	Petter et al. 1977
Muntiacus muntjac, muntjac	Dubost 1971
Rangifer tarandus, caribou	DeVos 1960

Morning, midday, and evening
Papio hamadryas, hamadryas baboon	Kummer 1968a

Morning and middle to late afternoon
Cercopithecus ascanius, redtail monkey	Galat-Luong 1975

Early to midmorning and midafternoon to dusk
Odocoileus hemionus, mule deer	Linsdale & Tomich 1953

Dawn and early afternoon
Macaca sylvanus, Barbary ape	MacRoberts 1970

Late morning and early afternoon
Macaca fascicularis, kra macaque	Poirier & Smith 1974a

Morning, late afternoon, and evening
Meles meles, European badger	Schmid 1932

Midday
Presbytis cristatus, lutong	Bernstein 1969

All day
Presbytis entellus, hanuman langur	Sugiyama 1976a

Late afternoon
Vicugna vicugna, vicuna	Koford 1957

Late afternoon and early evening
Bison bison, bison	Lumia 1972

Late afternoon and dusk
Miopithecus talapoin, talapoin	Gautier-Hion 1970, 1971

Sunset
Theropithecus gelada, gelada	Bernstein 1975

Early evening
Equus caballus, domestic horse	Feist & McCullough 1976

Evening
Mungos mungo, banded mongoose   Neal 1970

Twilight
Cervus canadensis, elk   Trumler 1959a
Equus burchelli, plains zebra   Trumler 1959a

Evening and night
Vulpes zerda, fennec   Gauthier-Pilters 1966

During activity period after midnight
Hyemoschus aquaticus, water chevrotain   Dubost 1975

---

may also play just before or even during the onset of heavy rain (Gautier-Hion 1971, T. George pers. comm., Hladik 1973, Jaworowska 1976, van Lawick-Goodall 1968a, Michael 1968, Schaller 1967, S. Wilson & Kleiman 1974 and additional references cited therein), as if seizing the last available opportunity for exercise. With this information in mind, T. K. George and I observed pony foals during one summer (study animals and site described in Fagen & George 1977). We collected focal-animal data on play at all hours of the day including some night periods. During the study, carried out in the flat, open tornado country of east central Illinois (U.S.A.), there were eleven severe rainstorms accompanied by lightning and high winds as well as many smaller storms. Frequency of play was not significantly higher than the average for a given time of day when rainstorms occurred during that hour or when they occurred during the succeeding hour, whether play was measured by per cent time spent in play, by running steps taken, by turns during running, or by number of play bouts per hour (Fagen & George, unpub.). There are no obvious, even if non-significant trends in these data. However, these results may be inconclusive because of the difficulty of correcting for the amount of exercise taken by a given animal during the 24 or 48 hours preceding a storm. Moreover, the sole instance of group play observed during our study occurred before and at the onset of one of the heaviest storms. First one foal, then another began galloping, turning while galloping, rearing and body-twisting until the entire year-class of ten male and female foals, including some who seldom played otherwise, was actively exercising.

They did not appear to be chasing or fleeing one another. Each foal seemed to gallop independently in long arcs or in giant circles. The foals crossed each other's paths frequently at high speeds. Remarkably, they never collided.

Incidentally, this argument may help to explain the differences between the published results of Chepko's (1971) and Müller-Schwarze's (1968) play deprivation experiments. Chepko deprived domestic goats (*Capra hircus*) of play (as well as all other forms of vigorous physical activity) using physical restraint and found a substantial increase in play after the deprivation period, whereas Müller-Schwarze (1968) deprived black-tailed deer (*Odocoileus hemionus*) of play, but not of other opportunities for exercise, and found that their tendency to exercise, as expressed in "longer activity periods, more locomotion, and more exploration" (Müller-Schwarze 1971) increased as the result of his manipulation.

A more "biological" example of long-term confinement is the winter rest of certain mammals (e.g., bears and woodchucks). Do physiological detraining effects occur during such long periods of inactivity? If so, one would expect play to occur when hibernating mammals and other animals that are inactive, for the most part, during the winter emerge from rest. I have observed an instance of vigorous rolling and running in a woodchuck (*Marmota monax*) that had recently emerged from a burrow in early spring, but I cannot state whether this behavior is a general phenomenon during this season in this species. The prediction definitely holds for some bears. Krott & Krott (1963) describe vigorous approach–withdrawal play in young, free-living brown bears (*Ursus arctos*) in March following several months of inactivity during winter. Howard (1935) describes black bear play following a winter period of inactivity.

Additional instances of long-term play deprivation are known in mammals. Baldwin & Baldwin (1974) observed no play in two troops of squirrel monkeys (*Saimiri* sp.) in 261 hours of observations over 10 weeks. However, the young squirrel monkeys in these troops appeared to spend an unusually large amount of time foraging and traveling, and the physical exercise provided by these activities may have adequately trained the animals (though perhaps not optimally, as argued above).

On the other hand, as Müller-Schwarze (1971) notes, it is often the case that "young animals do not play for days or weeks because of disease or bad weather," and such natural play deprivation will produce physiological deterioration if the animals do not exercise adequately in some other context.

Attempts to demonstrate physiologically beneficial consequences of so-called "warmup" (moderate exercise preceding strenuous exercise) have produced both positive and negative results (Arshavsky 1972, Barger et al. 1956, Barnard et al. 1973, Elbel & Mikols 1972). In healthy humans, sudden exercise without warmup may produce cardiac ischemia (local blood deficiency) and abnormal electrocardiographic (ECG) responses, indicating potentially hazardous cardiovascular stress; but with prior warmup exercise, sudden exercise produced no ischemic changes, and ECG responses were normal in all subjects (Barnard et al. 1973). Pre-exercise of joints squeezes out synovial fluid and improves lubrication (C. Pond pers. comm.). Play behavior that can be interpreted as warmup is not unknown in animals. African wild dogs (*Lycaon pictus*) play before starting to forage (Estes & Goddard 1967, Kuhme 1965), as do timber wolves (Mech 1970, Murie 1944), and choz-choz (*Octodontomys gliroides*), South American caviomorph rodents (S. Wilson & Kleiman 1974). Preforaging play is also known in mountain hare (Flux 1970) and flying squirrel (Krishnan 1972). If preforaging play serves as warmup, other less physiologically meaningful functions need not be postulated for this behavior, although there is no evidence at present that such additional functions do not sometimes exist. [Warmup need not involve play behavior. Bumblebees (Heinrich 1975) and certain Lepidoptera (Kammer 1968) warm up by exercising their wings vigorously before flying, but such exercise cannot be termed "play."]

Physical training alone cannot account for all play behavior. In particular, manipulative and social play require further discussion. Well-known neurobiological training responses (e.g., differential environmental effects on brain growth and connectivity, pp. 19, 284) may aid in interpreting the behavioral and functional significance of many forms of play that the physical training hypothesis cannot explain.

Manipulative play, i.e., persistent, oriented, variable manipulation of objects, prey animals, or parts of the body, occurring without vigorous physical exertion and in a nonfunctional context, is found in young primates (Dolhinow & Bishop 1970, Hutt 1970, van Lawick-Goodall 1968a), carnivores (Leyhausen 1973b), cetaceans (Tavolga 1966, Tavolga & Essapian 1957), corvids (Gwinner 1966, Jones & Kamil 1973), and parrots (Derscheid 1947, Hick 1962, Jackson 1963b, Keller 1975, Low 1970). In man and in the great apes, manipulative play first occurs during early infancy (Hurlock 1972, Johnson & Medinnus 1976, van Lawick-Goodall 1968a, Schaller 1963). As already mentioned, early onset of play in these species suggests that such early play activity must serve some function other than physical training. Possible functions of manipulative play, including motor practice, development of tool use and problem-solving strategies, experimentation on the environment, and discovery or facilitation of novel behavior patterns, are discussed elsewhere (pp. 318–333, Fagen 1974a, 1976b, Fedigan 1972b, van Lawick-Goodall 1970, Leyhausen 1973b, Jolly 1966b, Sylva, Bruner, & Genova 1976, Tinbergen 1963).

Play behavior is often social. Playing animals typically orient to, solicit play from, and respond to play soliciting by conspecifics (Chapter 7, Bekoff 1974c, Dolhinow & Bishop 1970, Hinde 1971, Loizos 1966, 1967, 1969, S. Wilson & Kleiman 1974). Indeed, friendly, physically vigorous social interaction offers an ideal environment for physical training, since bone and muscle are trained "in context" without risk of serious injury. But there may be more to social play than strength and endurance training. Social play can furnish a context for developing fighting skill without the risks that would accompany serious fighting (Symons 1978a). Adult playfighting, and vigorous chasing courtship in adult animals (Bastock 1967, Marler 1956) could not serve this function, but could display the physical condition (strength and endurance) of the performer while assessing that of the partner.

Physically gentle social play involving prolonged, stable, exaggerated, repeated, and varied mutual visual, tactile, olfactory, and/or auditory contact is known in a variety of species including humans (Konner 1972), lions (Schaller 1972 p. 161), manatees (*Trichechus manatus*) (Hartman 1979), rhesus macaques (White 1977b), common (*Phoca vitulina vitulina*) and grey (*Halichoerus grypus*) seals (S. Wilson 1974a), the sifaka (*Propithecus verreauxi*) (Jolly 1966a), several species of voles (*Microtus*) (S. Wilson 1973, pers. comm.), and a bird, the kea (*Nestor notabilis*) (Keller 1975). Such gentle play is ethologically distinct from the brief, tense, inhibited play of young female primates and adolescent or adult primates of both sexes (Baldwin & Baldwin 1977, 1978, Breuggeman 1978, Kummer 1968a, Symons 1978a pp. 36, 73). Gentle play is probably very poor physical exercise and can hardly be considered effective training in fighting skill. As S. Wilson & Kleiman (1974) suggest, the most likely function of this play is development of a cooperative social relationship between the participants. Such play occurs most often between a mother and her young. It also occurs between infant rhesus macaques (White 1977b), human children (Konner 1972), juvenile manatees (Hartman 1979), and adolescent harbor seals (S. Wilson & Kleiman 1974). This plausible hypothesis deserves further refinement in the light of current social theory. Under what conditions should both partners be selected to behave in ways that make such exchanges mutually informative, and under what conditions should one partner be selected to deceive, manipulate, misinform, or withhold information from the other? Can natural selection produce asymmetry in such exchanges, so that one partner consistently takes the active role while the other remains passive? Might one partner repeatedly seek contact with the other, while the other resists?

In addition to these evolutionary questions, we must ask, from a mechanistic point of view, why behavior having the form of play has been selected to mediate these exchanges, rather than simpler, nonplayful forms of social investigation. In fact, young prairie voles (*Microtus ochrogaster*) actually exhibit two ethologically and contextually distinct forms of nonagonistic social contact, only one of which satisfies the structural criteria for play (S. Wilson pers. comm.). This result raises interesting questions about the differential effects of the two types of encounters and about their evolutionary implications.

Except for the particular sensory modalities involved, the instances of gentle play described above appear functionally equivalent to the nonvigorous interactive games that certain rhesus macaque mothers play with their infants (Hinde & Simpson 1975) and in part to human mother-infant contingency games (Aldis 1975, Dunn 1976, Stern 1974a,b, Watson 1972). It would be unfortunate, both from a scientific and from a political standpoint, if play research's current adult male-oriented (i.e., sexist) emphasis on development of skill for male-male competition caused gentle interactive play to be ignored.

TESTS OF THE PHYSICAL TRAINING HYPOTHESIS

The physical training hypothesis, like many other hypotheses in population biology and in animal behavior, is difficult to test by direct experiment. Although there is no substitute for a direct test, independent, indirect tests from several points of view in a variety of species can provide strong supporting evidence.

Four approaches to the problem of play as exercise have been used by ethologists. These approaches are quantitative analyses of duration distributions; time and energy budgets; contextual comparisons; and deprivation or substitution studies.

The first and easiest (but least direct) approach to the physical training hypothesis involves quantitative analysis of the empirical probability distribution of the duration of exercise in social play. A possible null hypothesis is that this duration is purely random, so that vigorous play is equally likely to end one second or five minutes (or any other time) after it begins. If this hypothesis is found to hold, it follows that physical fatigue cannot be a factor influencing the termination of vigorous play, for if fatigue were coercive on behavior then the probability of termination would necessarily increase as the animal became increasingly tired.

This null hypothesis is mathematically represented by the negative exponential probability distribution in continuous time (Metz 1974, Fagen & Young 1978) or by the geometric probability distribution when discrete acts such as running steps are counted. Interesting data relevant to these hypotheses are available. McDonald's (1977) analysis of California ground squirrel (*Otospermophilus beecheyi*) social play durations, like my analyses

(Table 5-5) of Symons's (1978a p. 67) data on duration of social play in young rhesus macaques, show that none of these distributions depart significantly from the negative exponential model. However, my analysis of solo locomotor (running) play of two domestic cats revealed significant departures from the negative exponential in the case of both individuals, and this seems to be the case also for bouts of solo running in domestic ponies (Fagen 1975). McDonald (1977) suggests that only the longer bouts he witnessed might serve as physical training. Levy (1979) similarly remarked that the longer bouts of rhesus play that she witnessed probably ended due to physical fatigue. If this is the case, departures from the negative exponential model would be difficult to detect, and they might be only qualitatively noticeable as a slight upward concavity in the log survivor function plot (Fagen & Young 1978) at the longest durations.

We may also obtain evidence bearing on the physical training hypothesis by considering young animals' resource budgets. Suppose a young animal devoted 95 or perhaps even 99% of its daily caloric expenditure (excluding thermoregulatory and digestive metabolism) to play. Or suppose that 95 or 99% of the total work done by a particular set of muscles, e.g., those used in running, in rearing, or in turning at high speed, occurred in play. Especially if the residual, nonplay exercise occurred in contexts that were rare, unreliable consequences of chance environmental events, such evidence would suggest that play is the major source of exercise in development. The physical training hypothesis could not be falsified based on such data,

Table 5-5 SOCIAL PLAY DURATION ANALYSES

	Median Duration (Sec)	Number of Data Points	Goodness of Fit (Chi Square) to Negative Exponential Probability Distribution
Rhesus macaque playfights (Symons 1978a)			
Infant–Infant	4	178	2.8 (3 degrees of freedom)
Juvenile ♀–♀	5	37	1.1 (2 d.f.)
Juvenile ♀–♂	5	128	3.5 (7 d.f.)
Juvenile ♂–♂	5	153	12.6 (10 d.f.)

although if opposite results were obtained, suggesting that play had little importance, it would be difficult to argue that play had been selected because of its physical training effects.

Biologists quantify animal life-styles by constructing activity (time-energy) budgets that specify the total or relative amounts of time and energy allotted to various activities (E. O. Wilson 1975 p. 142). Since incidence of human degenerative diseases depends in part on life-style (Abelson 1976), mammalian activity budgets are of particular interest because of this relation to health.

In what behavioral contexts do young animals obtain physical exercise? Observers of mammalian behavior have long claimed that play behavior accounts for most exercise of young mammals. Ethologists seem to have taken this claim, based solely on anecdotal reports, for granted. Quantitative data in activity-budget format, on the relative frequencies, in play and in all other contexts, of certain physically vigorous acts of young of a large mammal, would serve to furnish empirical evidence relevant to the previously speculative and untested exercise-play hypothesis.

T. K. George and I (Fagen & George 1977) evaluated the exercise-play hypothesis by measuring the occurrence of vigorous exercise in ten young Shetland and Welsh ponies (*Equus caballus*) from birth to age 6 weeks. We noted all instances of the two most frequent forms of obviously vigorous whole-body exercise in foals, namely galloping and turning while galloping. We chose these indirect measures of exercise because energy costs of mammalian running are well documented (Taylor 1973, Heglund, Taylor, & McMahon 1974) and because physical forces associated with rapid turns are likely to impose significant stress on limb bones and muscles. Other modes of vigorous exercise such as rearing and kicking often occurred during running bouts.

Solitary and social play accounted for approximately two-thirds of all galloping steps ("Steps," Table 5-6) and nearly all galloping turns ("Turns," Table 5-6). Since behavior of one highly active individual could in theory have influenced these values, the proportional contribution of play to exercise budgets was averaged over individuals by weighting each individual

Table 5-6 CONTEXTS OF VIGOROUS MOVEMENT IN
PONY FOALS
(Percentage of total exercise occurring in play)

| Measure of Exercise | Overall Average | Individuals * | | Ave |
		Min	Max	
Steps	68.3	54.5	84.6	65.2
Turns	95.1	85.4	100.0	94.6

*Ten separate values calculated, one per individual. The last column
displays the average (Ave) of these individual values

equally. Results were essentially unchanged (Table 5-6). The
frequency distribution of the number of galloping steps in an
instance of running (where an instance was defined as a straight
run demarcated by turns or pauses) did not differ significantly
between play and nonplay in any individual ($P \leq .05$, 2-way
contingency table analysis: variable 1, play/nonplay; variable 2,
short/long instances, where short instances were defined to contain up to twenty galloping steps and long instances over
twenty galloping steps). Long and potentially stressful bouts of
running were as likely to occur in play as in nonplay, but since
play-running was far more frequent than nonplay-running,
these long bouts occurred absolutely more often in play.

Results of this study and of a similar unpublished study of
young bison by T. George indicate that certain vigorous acts
are performed frequently or exclusively in play. Although
direct measurement of caloric expenditure in play remains to be
performed, this indirect evidence suggests that play is a major
source of physical exercise. Additional research is needed in
order to show that the exercise obtained in play is important
and that nonplay exercise, which may represent only a small
fraction of the total, is not in itself adequate to ensure normal
development.

In a valuable field study, Rose (1977b) measured frequencies
of certain physically vigorous acts in play of olive baboons
(*Papio anubis*) and in other behavioral contexts. The baboons ran
primarily in play and occasionally to keep up with movements
of their troop. One-half of all climbing took place in play. The
majority (73%) of leaps were associated with play; other leaps

occurred when baboons fed in trees. During play, leaps were often made from a running start, whereas during foraging most leaps were made from either a standing or running start.

Coelho (1974) and Coelho, Bramblett, Quick, & Bramblett (1976) calculated time-energy budgets for the activity of two primates, the red howling monkey (*Alouatta seniculus*) and the red spider monkey (*Ateles geoffroyi*), in the field. Their analyses were based on indirect, scaled measurements. Within the limitations of these calculations, play represents a major category of energetically costly behavior.

A third approach to physical training in play, one that demonstrates the value of ethological analyses of behavior, is contextual comparison. Suppose animals play significantly more when the environment permits exercise to be especially effective, at the cost of efficiency of attainment of other possible goals of the behaviors performed. Then, by showing that animals "go out of their way" to perform behaviors in a context that ought to maximize physical training effects even if no food or mates can be gained and social status remains as before, the ethologist has obtained data that again suggest that this behavior is being performed at least in part for its physical benefit to the individual.

Beck (1976) studied the shell-dropping behavior of two species of gulls (*Larus* spp.) on the Massachusetts coast (U.S.A.). When adult herring gulls (*L. argentatus*) dropped edible molluscs, they compensated for wind direction and appeared to compensate for wind velocity in a way that maximized their possibility of a successful drop (i.e., one that resulted in breakage of the shell, exposing the meat or making it more readily accessible). However, young herring and black-backed (*L. marinus*) gulls dropped both edible shells and inedible objects in a way inconsistent with foraging–efficiency considerations. For example, they would frequently drop inedible objects from great heights on very windy days. Maturation alone cannot explain this behavior. Older juvenile herring gulls are skilled at shell-dropping (though adults are more accurate).

Byers (1977) found that play of young Siberian ibex (*Capra ibex sibirica*) occurred preferentially on terrain optimally suited for physical training. The confined ibex band observed in this study inhabits a craggy "island" surrounded by a dry, flat-bottomed moat. The island itself offered a variety of terrain including some very steep slopes where footing was relatively difficult. The ibex played on sloped surfaces significantly more frequently than on flat surfaces, even though their enclosure contained equal amounts of flat and sloped area. This play was carried out despite clear physical risk: indeed, several falls were observed during the course of the study. The only exception to the predictions of the physical training hypothesis was social play of young males, which occurred most often on flat terrain. Social play of females, and locomotor play of both sexes, occurred on terrain that seemed maximally suited for training and conditioning. These data support the hypothesis that ibex locomotor and some ibex social play occur as preparation for life in a mountainous environment.

Byers' results on male social play focus on behaviors used in fighting, but not involved in skilled locomotor behavior. It was possible to show that in the case of these behaviors, for which there was no need to train on difficult terrain, ibex used flat surfaces significantly frequently in play. Here also, microhabitat selection suggested (but did not conclusively demonstrate) behavioral optimization of efficiency for play as physical training and skill development.

Deprivation of play ought to reduce physical capacity and skill. Substitution of equivalent physical activity for play should result in no reduction of these capabilities. If physical exercise is the only biologically important consequence of play, then behavioral compensation in the form of increased play should not follow deprivation if equivalent nonplay exercise occurs during play deprivation. Baldwin & Baldwin (1976a) and Oakley & Reynolds (1976) found that rebound effects followed work-induced play deprivation, but they had no controls for physical activity.

Although the role of play in physical training is not yet directly demonstrated, this concept has already entered popular culture. According to one authority, "we should have fun while we exercise, as children do in active play, and make it a natural part of our lives" (Beach 1978). Or, consider the following account (Kelly 1975):

As a girl and young woman I had been very active physically. I was slim, strong, and energetic. Then weight began creeping up on me at the rate of one or two pounds each year. At twenty-three I was wearing a size 10 and looking good. At thirty-three I was wearing a size 14 and looking bad.

One day, watching my little daughter at play, I caught myself feeling envious of her strong, young body. She was getting plenty of exercise. She was never still a minute, unless she was asleep. It hit me: Why was I wasting my time and energy watching her play when I could be playing too? I decided to spend one hour a day playing actively with her. She would be the boss. She would decide what we would play and how we would play. She was obviously an authority in that field. Since that day of decision I have climbed trees and fences, ridden my bicycle around the block beside her tricycle, jumped rather than walked through the living room, danced to *Yellow Submarine,* performed somersaults and donkey kicks on the lawn, etc.

The results have been wonderful. I'm a firmer, healthier size 12, and my daughter and I have both found new enjoyment.

It is gratifying to see one's intuition confirmed by such an enthusiastic correspondent. However, several questions remain about play as strength and endurance training. Do the effects of vigorous play in youth last throughout adult life, or does play simply prepare the body for some crucial period of development? One such crucial period is the period of group integration in many ungulates, during which the previously isolated young experience a virtual onslaught of society. They are sniffed, poked, nipped, and chased by older juveniles. In addition, play itself is frequent in these juvenile groups.

A serious question about play as exercise is raised by Symons (1978a). If, as Symons argues, the function of rhesus playfighting is to develop fighting skill, any effects on strength and endurance would be incidental. Moreover, because vigorous playfighting could reliably be expected to occur, physical development might even have become secondarily dependent on play experience (E. O. Wilson 1975, "vitamin effect"). One unfortunate consequence of the tendency of behavior to become secondarily dependent on experience is that advocates of each function can argue that everyone else's claimed function is merely an effect, an instance of secondary dependence. Here, deprivation experiments can prove nothing about function. As we have seen, much play, especially nonsocial play, seems designed as physical exercise, and its biological function is suggested by evidence of design if we take somatic plasticity as given. Furthermore, as Symons (1978a) argues, uniform deprivation of play, such as might occur during an extended period of food shortage or inclement weather, will have no adaptive significance if it affects all competitors equally. Geist (1971, 1978a) speculates that entire populations of mountain sheep may differ in just this way. The fundamental theoretical question about play as physical training appears to be "Why not calisthenics?" At least three answers to this question are available:

1. When one animal is playing, others will notice its behavior and begin to play in order to produce corresponding increases in *their* physical fitness for future or present competition. This argument is the simplest possible explanation of the contagiousness or infectiousness of play (a phenomenon reported by many authors including Espmark 1971, Hall 1962, Krämer 1969, Linsdale & Tomich 1953, Rensch & Dücker 1959, Rudge 1970, Schaller 1972, Ulrich, Ziswiler, & Bregulla 1972, Walther 1973).

2. One animal's play may stimulate competitors to interact playfully with it in a way that prevents its play from being effective as physical training. Can the structure of social play be simply explained in this way? If so, social play should exhibit two qualitatively different phenomenologies. The first is less likely to result in effective exercise than the second. We may predict which of these types of play will occur in particular dyads: where close kin are involved, play should be relatively good exercise unless special circumstances lead to sib competition. If there is just enough of a resource for two individuals, exclusive play dyads may form. That everyone else is also forming cooperative play dyads for the same reason means that there will be strong selection for effective exercise. The whole system is likely to collapse, however, since if resources are so limited play is unlikely. One dyad will defeat other dyads by more direct means, then play together later.

In this connection it is interesting to note some hidden costs of physical exercise. Too much exercise may be harmful, for it may cause growth of ectopic bone (C. Pond pers. comm.) or muscular hypertrophy leading to pinched nerves (K. Stayman pers. comm.). In humans, these problems are especially evident in dancers, athletes, and weavers and other artists.

3. The function of social play is not solely to obtain muscular exercise. Symons's (1978a) observations and analysis of skill development in rhesus play support this hypothesis.

Skill Development

Structure and content of playfights and play-chases point to training for future social interactions that involve direct conflicts of interest, physical danger, and high potential benefit to the performer, if successful, and require physical ability (Symons 1978a). The specific contexts indicated are escalated fighting, predation on large, dangerous prey, and predator avoidance. I regard skill development as one aspect of physical training. Historically, the study of skill development has been distinct from the study of strength and endurance training, and different sets of concepts have been proposed. But it would be difficult to separate the contributions of strength, endurance, motor skill, and cognitive skill to overall competence in these physically demanding activities. Moreover, other skills may develop as the result of play experience, particularly of play at early ages. Locomotor skill and postural control (Simpson 1976), tool skill (McGrew 1977), and skill in control or manipulation of items or other organisms (Konner 1975) are all implicated. Indeed, the ontogeny of play may be regarded as a progression through practice at controlling and manipulating "social objects of less and less contingent responsiveness" (Konner 1975). The preferred object of play becomes decreasingly controllable (McFarland 1973) as the animal develops. Because this view of play focuses on a single cognitive scheme for manipulating active objects whose behavior is contingent on the player's actions (Konner 1975), it represents a reductionist approach (Simpson 1978) centered on primitive abilities that can improve through practice. Therefore, skills analysis of play is potentially very powerful.

However, several limitations of the skills approach must be recognized. If play is dual to skilled action, following rather than preceding mastery and often continuing long into adult-hood (Bruner 1969, Bruner 1976 p. 14, van Lawick-Goodall 1968a, Miller 1973, Piaget 1962 p. 90, Schiller 1952), we can define play as one of two possible forms in which a given be-havior pattern, such as fighting or predator escape, can exist, but the developmental contribution of such dual behavior to its already-skilled counterpart is by no means clear. On the other hand, the view of play as a particular form of practice used in skill development suggests that the proper object of study is all of practice or perhaps even all of modifiability, and we must specify the structure, effects, costs, and benefits of dif-ferent forms of practice. We then attempt to understand what forms of practice are favored by natural selection under what circumstances, as well as why and when practice of any sort should be necessary. Finally, although the design (structure and context) of play definitely suggests that play serves to develop skill, there are numerous cases (Aldis 1975) in which animals apparently allow their playmates to attain the immediate goal of play (sniffing, mouthing, biting, or catching) or do not use their full physical capability to attain this goal. How would such behavior contribute to the development of skill? Is it an al-truistic act performed for the partner's benefit? Or is it simply an instance of self-handicapping (S. Altmann 1962b), one aspect of the compromises that play partners with different interests must reach in order to play at all? The skills approach to play can potentially make use of an enormous technical literature on skill development. This literature includes contributions from psychology (Bartlett 1958, Bilodeau 1966, Bruner 1969a, 1970, 1972, 1973b, Connolly 1970, 1971, 1973, 1977, Connolly & Bruner 1974, Connolly & Elliott 1972; Cratty 1964, Elliott & Connolly 1974, Fitts & Posner 1967, Holt 1975, Kay 1970, Paillard 1960, 1976, 1977, Schmidt 1975, 1976, White 1970), bi-ology (Blomfield & Marr 1970, Gregory 1969, Welford 1968), and artificial intelligence (Austin 1974, Goldstein & Papert 1976, Raibert 1977, Sussman 1975, Winston 1977). It is impossible here to do more than simply indicate the nature of some of these ideas. The following discussion is a brief introduction to the theory of skill.

A player controls its movements and the movements of its

partner in order to achieve some harmless goal, such as sniffing or mouthing, without allowing the partner to reach the same goal. The sequence of behavior used to attain this goal satisfies a series of simpler goals generated by a hierarchically nested stopping rule program (Dawkins 1976a, Sacerdoti 1977). Such programs freely combine short sequences of behavior called "chunks" (Austin 1974), "subroutines" (Bruner 1976, Mc-Farland 1973), or "melodies" (Fentress 1978) in order to execute plans (though not necessarily conscious ones).

Development of skill requires that appropriate subroutines be available and that they be orchestrated into an efficient and stable program. In developing these programs, animals make unavoidable mistakes—not necessarily because they are stupid, self-deceived, or manipulated by others, but because rational errors are always expected to result from powerful skill development strategies (Goldstein & Miller 1976, Sussman 1975). These errors, called "bugs" (a term used by computer scientists to denote a mistake in a computer program), result from unanticipated interactions between subroutines (Goldstein & Miller 1976, Sussman 1975). In learning juggling, a complex motor skill, several "bugs" become manifest simultaneously. Many of these "bugs" involve timing errors (Austin 1974).

Skill development through practice is like writing computer programs. Like programs, skills must be refined ("debugged" is the current jargon). In order to benefit from practice and to make effective use of the information contained in failures (interpreted as symptoms, or manifestations of "bugs"), developing organisms (or programs) should create their own training sequences and learn from their mistakes. The central question for interpretation of play then becomes to what extent these sequences should involve combinations of behaviors or contexts not associated with the perfected skill. A deductive theory of optimal practice is not yet available. The following principles have proved valuable in particular contexts:

1. In the development of physical skill, "bugs" come in bunches. Organisms should create learning environments in which the same "bugs" occur, but one at a time (Austin 1974). Self-generated training sequences should isolate "bugs" by attempting to solve subproblems individually, then in pairs, etc. (Sussman 1975).

2. Apparently, humans are able to advise themselves on certain motor skill tasks, but not on others. Sometimes, effective practice, even for skill not involving social interaction, must involve other organisms (Austin 1974).

3. It is often informative to practice slowed-down approximations of a new movement, then increase the speed (start slow, speed up paradigm of Raibert 1977).

4. The correct form or rhythm of movement can sometimes be established paying little attention to the details, which can be fine-tuned later (Raibert 1977).

5. Practice should be interrupted and repeated. Learning is more efficient when trials are distributed in time than when long practice sessions are employed (Cratty 1964, Raibert 1977, Welford 1968).

6. Movements should be broken down and practiced in parts. This procedure can redistribute the effects of practice that would normally generate data primarily for the initial sections of the practice movement (Cratty 1964, Raibert 1977, Welford 1968).

7. A good problem-solver creates related but smaller and easier problems and "fools around" with them before re-attacking the hard one head-on (Sussman 1975).

8. In developing their skills, organisms should actually encourage the occurrence of "bugs" by ruthless generalization. That is, existing routines should be applied to all situations, even those only vaguely resembling the ones for which they were first developed (Sussman 1975).

9. The more complex a skill to be learned, the lower the optimum motivational level required for fastest learning (Bruner 1976 p. 15). Only the simplest skills can be developed under pressure to produce outcomes external to practice itself.

10. Cerebral or spinal mechanisms may reflect habitual patterns of movement to such an extent that initial learning of new skills is physically blocked. It may be that some neuromuscular pathways acquire a habit of responding in certain ways, and it is not until that habit is broken that a new skill can be learned (Basmajian 1972). The animal must actually behave in antic and variable ways in order to loosen the grip of past habit on these neuromuscular pathways. Thus, play following mastery of a skill would allow the animal to "get unstuck" (Sussman 1975),

freeing the skill for generalization and recombination in new contexts.

11. The importance of the context in which a movement is first performed is stressed by numerous authors, including Blomfield & Marr (1970), Marr (1969), and Raibert (1977). A movement or skill mastered in one context must be systematically varied in practice. Generalization is not automatic, but requires new kinds of experience, including capricious variations having little or no obvious immediate utility. These variations would be particularly necessary for rarely performed behaviors. Simple locomotor and postural adjustments span a wide variety of contexts without playful practice. Rare emergency behaviors performed in play cannot possibly receive their full scope of exercise otherwise unless the animal is constantly in danger.

The current debate over the degree of discontinuity of behavioral development (Bateson 1978b, Goldberg 1978, Kagan 1978, Kagan, Kearsley, & Zelazo 1978) indicates that skill seldom appears *de novo*. Behavior becomes more general, more flexible, and is performed more smoothly and efficiently. In large part, this progression results from transfers of control within the brain itself. The timing of these transfers probably depends on experience, although details of the underlying processes are not known. The brain itself is an example of hierarchical design (Teitelbaum 1977). It can exert control over behavior at several different levels of integration. Developmental processes, including skill development, involve transfers of control from one stage of encephalization to another, and developing behaviors "go through successive levels of ontogenetic transformation of nervous control. Each stage of transformation in the hierarchy allows new controls to modify the action of the lower stage of integration" (Teitelbaum 1977 p. 21). For instance, in development of higher mammals, behavioral sequences initially achieve adequate if narrow competence. A period of experimentation follows during which precision is temporarily sacrificed for generality. The result of this experience may either be increased flexibility or long-term inhibition of skill (Bruner 1969a,b, 1970, 1972, 1973b, Leyhausen 1965, 1973b). Under stress, an animal may revert to using previously functional but crude techniques (such as indiscriminate biting in predation) (Leyhausen 1973b). This information suggests that skill development involves transfers of control in the brain that deepen the scope of and add flexibility to a core of fundamental abilities.

The above principles sound very much like a recipe for animal play behavior: short sequences repeated with variation, applied in a ludicrous variety of contexts, abstracted to focus on single subgoals or projects, frequently involving other animals, sometimes but not always temporally distorted, distributed in time, and motivationally independent of other causal systems. For additional structural and contextual parallels between play and optimal practice, see Konner (1975), Simpson (1976), Simpson (1978), and Symons (1978a).

The best available examples illustrating the logic behind playful practice involve development of skills that require simultaneous or sequential coordination of separate anatomical effectors. In the Australian magpie (*Gymnorhina tibicien,* family Cracticidae), manipulative play occurs most frequently when birds begin to integrate bill and head movements with foot and leg movements (S. Pellis pers. comm.). Successful predatory behavior involves coordinated use of these manipulative skills. Pole-tree swinging of orangutans involves manual-pedal coordination, whole-body control, and other more "cognitive" skills (Galdikas 1978):

Finally, much of orangutan locomotion very directly involved manipulation of materials with the hands and feet. An orangutan would frequently sway a pole tree back and forth until she could catch hold of an adjacent tree. Such locomotion actually required considerable expertise and wild orangutans seemed highly skilled at it. If the oscillations were not great enough, the orangutan would overshoot the target tree and even come falling down to the ground. On occasion we would see an orangutan testing the strength of a pole tree and then use two, rather than one, to carry her weight. Sometimes an orangutan would simultaneously manipulate several slender pole trees with a combination of hands and feet. That such forms of locomotion required learning is not so obvious from the behaviors of wild juveniles who usually relied on the superior weight of their mothers to sway pole trees over while travelling in the lower canopy. But it is clear from the behaviors of ex-captive orangutan infants and juveniles released at Tanjung Puting [an orangutan rehabilitation center in Central Indonesian Borneo where Dr. Galdikas and co-workers release former captives and rehabilitate them for life in the forest] how much learning was, in fact, involved. Such inexperienced animals would for months pick saplings for locomotion. These saplings would totally bend over bringing the orang crashing to the ground. Other times these ex-captives would pick pole trees so stiff that they could not be swayed over towards the next tree by their weight. Wild orangutans seemed to understand, at least in practise, rudimentary principles of mechanics.

A captive juvenile orangutan might discover the basic principle of pole-tree swinging in play if the youngster bounces around in a tree and reaches out to catch another tree as its own tree swings back and forth. The principle that a weight may be attached to a light support and given an impulse that propels the support toward a goal is also basic to a celebrated problem in experimental psychology, the Maier string problem (e.g., Battersby, Teuber, & Bender 1953). However, choice of appropriate weights and supports, as well as choice of the magnitude and direction of the impulse, requires considerable skill. Like most other primate locomotor and tool skills (B. Beck 1978, van Lawick-Goodall 1968a, McGrew 1977, Ripley 1967), orangutan pole-tree swinging appears to develop through some combination of playful and rote practice, observation of others, and physical-cognitive maturation. Development of this skill deserves thorough study. For comparative purposes, the link to the Maier string problem should really be exploited, since so much is known about cognitive aspects of human solutions to this problem.

During skill development, what sorts of mistakes arise from conflicts between components of complex skills such as the two just described? Are these mistakes given attention, isolated, explored, and corrected during playful and nonplayful practice? How are these goals accomplished? Do organisms ruthlessly generalize in each separate anatomical modality, then create simpler subproblems in order to "debug" their almost-right plans? These questions suggest opportunities for description and analysis of play behavior and of the ontogeny of skill in organisms ranging from birds to great apes.

The unit segments of play are simple structures resembling magic tricks, short-short suspense stories, or word problems in algebra. An element of uncertainty and expectancy is created, then resolved. This structure may represent attempts to isolate a single "bug," i.e., to focus on a particular difficulty (Simpson 1976). In one segment, strain builds and is then released when an initially uncertain event occurs to resolve the mounting tension. A play segment is thus seen to contain a single decision (Dawkins 1976a, Dawkins & Dawkins 1973, 1974). Alternatively, we may characterize a given play segment as a pair of events, the first under the player's control, the second contingent to some degree on the first, but also dependent on the surroundings and thus unpredictable (Simpson 1976). Sometimes the uncertainty may simply concern the timing of the resolving event, as in peekaboo or hide-and-seek. The chief quantitative feature of a unit segment of play is a particular time history of uncertainty about the resolvent event. This pattern might be quantified using information measures of the uncertainty of following acts as a function of the order of occurrence of the act in the behavioral sequence, or (in the case of time uncertainty) by the variance or coefficient of variation of the probability distribution of waiting time until the event, as a function of time since the beginning of the segment.

The communicative behavior of playing individuals often echoes the structure of play segments. Van Hooff (1972) remarked that children's play consists of an anticipatory phase followed by "the expected unexpected, the *pointe.*" This transition is accompanied by a change in communicative behavior: widemouth play-face accompanies the anticipatory phase and vocalized laughter accompanies the release. Marler & Tenaza (1977) describe progressive changes in chimpanzee laughter that occur as the vigor of tickling or biting play increases. Ripley (1967) refers to a "moment of suspense" in hanuman langur (*Presbytis entellus*) locomotor play. Immature langurs also tease adult males, running up to them, lunging in place, grimacing, squealing, and in some cases contacting them; pulling an alpha male's tail and even charging him were common feats of derring-do (Blaffer Hrdy 1977). The adult males so challenged steadfastly refrained from delivering the expected unexpected in the form of a punitive bite or slap. Peekaboo contains the same biphasic structure (disappear–reappear) (Aldis 1975, Bruner & Sherwood 1976, Darwin 1877), as do somersaults (put head down–tumble over), leaps (push off–land), sequences of climbing up and jumping off, the running games of ungulates in which they first travel away from their mothers or the herd and then back again, and play-wrestling segments in which one opponent or the other finds an opening after some efforts and succeeds in mouthing the opponent. This structure was also seen by Kirchshofer (1960) in play of a king colobus (*Colobus polykomos*) infant: the infant would climb up and look at its father, and when the father looked at the infant the infant would immediately turn and dash away. Loveridge's (1923) pet bat-eared fox (*Otocyon megalotis*) would crouch, then spring up and leap straight at wild baboon males, who would lope away.

The structure of a unit of play may be characterized in a number of equivalent ways. One formulation is so-called contingent responsiveness (Konner 1976a, Stern 1974a, Watson 1972). To what extent is the second phase of a play segment contingent on the player's behavior in the first? Contingent responses seem to be the goal of certain kinds of play in which members of two species interact. Young mammals, especially young ungulates, pursue birds and other animals that may suddenly hop away or take off. Plains zebra (*Equus burchelli*) foals playfully chase birds and other game in the wild (Klingel 1967). Once a young free-living chital (*Axis axis*) chased a myna bird (Schaller 1967). Blackbuck (*Antilope cervicapra*) calves playfully attack peacocks, pheasants, and ducks (Schmied 1973). Confined zebras chase birds and follow them as they hop along (Trumler 1959a). Confined oryx (*Oryx gazella*) sometimes playfully attacked birds in their enclosure or ran around, causing them to scatter (Walther 1965a). A five-month-old, free-living African elephant (*Loxodonta africana*) calf energetically chased and scattered egrets (Douglas-Hamilton & Douglas-Hamilton 1975 p. 91). In confinement, another young African elephant repeatedly chased a noisy fox terrier out of its enclosure, behavior described by Trumler (1959b) as play. Juvenile vicuña (*Vicugna vicugna*) in the wild chase geese and snipe (Koford 1957). The hornbill game of young Lowe's guenons and the repeated raven-chasing of free-living timber wolves, both of which also fit this pattern, were described earlier (Chapter 3).

Human parent-infant and child-infant games involve fine behavioral meshing in which the behavior of the two participants (Figs. 3-14, 3-15) seems designed to develop a relationship and/or the infant's ability to control its environment (Aldis 1975, Brazelton, Koslowski, & Main 1974, Brazelton, Tronick, Adamson, Als, & Wise 1975, Dunn 1976, Pawlby 1977, Stern 1974a,b, Watson 1972). The infant's experience with responses contingent on its own behavior is felt to generalize at later ages to increasingly less-contingent partners (Konner 1976). I feel it is important to identify two aspects of responsiveness: cooperativeness and contingency. For example, the behavior of a fleeing prey animal is strongly contingent on the predator's behavior, but the interaction is in no sense cooperative. Play seems to rehearse skills later used in contexts of maximum contingency and minimum cooperation. These contexts correspond to the sociobiologist's "conflict of interest."

In humans' finger games or toe games (e.g., "This little pig went to market") played with infants (Aldis 1975 p. 249), the parent recites a rhyme with rising inflection, then at the peak of the game pokes the infant and says "THERE." The infant responds with gales of laughter. Darwin (1877) and Bruner (1976) speak of the moment of "surprise," Koestler (1976) calls it "when the penny drops," and a movement notation specialist (Golani 1976) would describe the structure of play in terms of joints (stable positional configurations from which unilateral departure did not occur) and extrications (actions by one partner that break the impasse). In many cases these mini-interactions occurring in play are so cryptically simple that the only uncertainty concerns the timing of an exciting event. When will the hidden face reappear? When will the playmate hiding behind the tree jump out and yell "Boo"? When will the kitten hiding behind the chair jump out and catch your foot, hugging and kicking it? For those students of play who enjoy Bruner's (1976) and Dawkins' (1976a) linguistic analogies, these little suspense stories recall human riddles and jokes. They lead up to a climax released by the punch line. Even human magic tricks have this structure. The operator (commonly known as the magician) creates an atmosphere of tension as he prepares his trick, then at the instant of greatest possible suspense reveals the surprise.

A given play segment will be repeated again and again, often with slight variations, or even in different contexts, until the animal becomes physically tired or is interrupted. Again, this repetitive (though not blindly repetitive, Austin 1974) structure is expected, based on skill development theory.

Bruner (1976) might consider each segment an attempt to master a given subroutine. According to this view, manipulative subroutines are "practiced, perfected and varied in play" (Bruner 1976 p. 43). It is difficult to compare different hierarchical metaphors about behavior. What is the relationship of Bruner's subroutines to Sussman's almost-right plans and subgoals? See McFarland (1973) for additional discussion of distinctions between plans and subroutines. These metaphors and verbal models, however suggestive, must give way to rigorous formulations involving model programs, block diagrams, and computer simulations as suggested by observations and analyses of play itself.

THE NATURAL HISTORY OF PRACTICE

Skill can develop without dual play. Female langurs refine infant handling techniques by handling infants (Blaffer Hrdy 1976, 1977, Lancaster 1971). Loggerhead shrikes refine the orientation of their prey-handling movements through experience (S. Smith 1972, 1973). Young reed warblers (*Acrocephalus scirpaceus*) improve their techniques of handling and swallowing flies simply by attempting to capture actual flies (Davies & Green 1976). These authors speculate that young birds try a wide variety of methods of mandibulation (mandibulative movements of older birds are relatively stereotyped) and so learn more rapidly, but data on this point are not available. Once a bird swallows a fly, its fly-handling techniques become rapid and very efficient, and the previous forms of mandibulative behavior are no longer seen (Davies & Green 1976). Similarly, male rhesus monkeys and male baboons mount females with increasing skill as they mature (Goy & Wallen 1979,

Hanby 1976, Hanby & Brown 1974, Mason 1965a, Owens 1976). Young oyster-catchers (*Haematopus ostralegus*) use a variety of movements to peck at small inanimate objects until they finally develop the complex skills needed to open mussels (Norton-Griffiths 1969). As described in Chapter 3, young oyster-catchers learning to open mussel shells perform variable sequences of motor patterns, similar in form to the adult patterns, that are directed at many types of edible and non-edible objects and that may meet some of the criteria for play, although sufficiently detailed descriptions are not available (Tinbergen & Norton-Griffiths 1964).

Chimpanzees at Gombe Stream manufacture and use tools to dip for driver ants. This activity, unlike termiting, is risky because an ant can inflict a ferocious and painful bite. The skill of dipping for driver ants is mastered slowly, without play or extended practice, primarily through observation of skilled performers (McGrew 1977). Learning of bird song requires a period of variational practice that might even be considered play (Ficken 1977, Marler 1977). Three criteria might be used to differentiate practice from play. Playful forms of skilled behavior exhibit variation that increases with time as the animal repeats the behavior in successively different contexts, whereas practice reduces variation as mastery is approached. Play behavior is structurally and contextually distinct from the adult behavior it rehearses, and play continues after mastery. We might say that play develops certain primitive, general skills that can be used in particular contexts in earnest after a certain level of mastery (not necessarily the final one) is achieved. However, the skill then continues to be further developed, elaborated, and generalized in play even after it is put to use in the service of particular functional goals. Accordingly, we tend not to call the variational practice of young birds "play" because it converges on a narrow band of adult behavior (mussel opening or fly handling). In contrast, animal playfighting and play-chasing develop skills used in all contexts involving disputed control of the performer's body. These contexts include escalated fighting, predator avoidance, and conflicts between sexes over the opportunity to mate. It would be difficult to argue on this basis that young chimpanzees' play at sponging was any more playful

330 Animal Play Behavior

than the spare time pecking activities of young oyster-catchers. However, the two skills are quite different, and as the protocol of object play reported by McGrew (1977, see Chapter 3, p. 111) indicates, the behavior has considerable structure. It seems to consist of hierarchically organized subroutines, each exercised separately in play. Sponging involves tool use and tool construction. It may also be more complex in other ways, such as the number of primitive skills involved or the number and difficulty of "bugs" manifested. Would it be more difficult to develop a computer simulation of oyster-catcher mussel opening or of chimpanzee sponging? What types of "bugs" would manifest themselves in each case? What sorts of training sequences and practice activities would be optimal in each case?

Obviously, quantitative comparisons of play with simple practice are yet to be carried out. Formal models that define meaningful statistics for use in such comparisons are already available. Dawkins (1976a) suggests some ways in which hierarchically organized behavior can be quantified. In addition, the theory of piecewise Markov processes (Kuczura 1973) furnishes a versatile model of hierarchically organized behavior. A piecewise Markov process is a probabilistic model of a sequence of events (such as behavioral acts) organized into segments. Within each segment, the sequence of acts is generated by a Markov process. Changes from one segment to another are determined by a separate Markov process. As a result, the model animal's behavior is governed by one transition probability matrix for a period of time, then by another transition probability matrix. The model is an excellent one for sequences of behavior incorporating changes in mood or in motivational state.

Mistakes are interesting, empirically observable behavioral events. What kinds of mistakes do animals make while practicing skills? Mistakes can be of several types. (1) In the laboratory, an animal may not learn certain discriminations because it was never selected to do so in nature; the task is biologically irrelevant. (2) "Mistakes" can in theory result from adaptive deception or from self-deception as predicted by sociobiology (Alexander 1974, Dawkins 1976b, Popp & DeVore 1979). (3) In play, a "mistake" can also mean biting too hard or playing too roughly as a result of imperfect information, even when it is in both animals' interest to maintain play (as argued in Chapter 4). Unlike (2), these errors are "honest" mistakes, and natural selection will act to minimize their probability. (Play may also break down as a result of irreconcilable conflicts of interest over play itself; unlike those preceding, these transgressions will not be forgiven by the partner.) (4) Of most interest in the present discussion are tactical mistakes made during playfighting that allow the partner to reach its goal of mouthing or sniffing, as well as mistakes occurring during solo play. An infant may misjudge distance or force when jumping; a foal or fawn may stumble or fall accidentally during locomotor play; playful tool practice may manifest certain "bugs." The most important empirical contribution of skills theory is that it calls for ethological observations on the tactical mistakes animals make in play. Can we describe and classify these "bugs," as Austin (1974) did for mistakes of human subjects learning to juggle? Can we observe the development and generalization of subroutines, and the "debugging" of almost-right plans? Are mistakes in timing, in sequencing, in subroutine choice, or in subroutine performance responsible for the opponent's success in social play?

The skills approach to play is promising. It is only fair to add that the underlying theory is often frustratingly vague and metaphorical. To my knowledge students of artificial intelligence have not yet attempted to simulate animal play. Sussman's and Austin's discussions make it quite clear that something like play is precisely what is needed to produce optimal training sequences and to "debug" almost-right plans in a flexible, generalizable, and efficient manner.

Could play facilitate development of certain cognitive skills useful in fighting, predation, or predator avoidance? Social theory suggests a number of possible abilities that could contribute to success in these activities and that could be practiced. The ability to assess the ability or determination of one's opponent is crucial in animal conflict resolution behavior (Parker 1974, Popp & DeVore 1979). In social primates, coalitions are important mechanisms for gaining resources (Cheney 1977, Hamilton 1975), and coalitional skill, including the ability to estimate the likelihood of being chosen by another individual as a partner (Nichols 1977), may contribute to reproductive success in these species. Because optimal strategies for aggression depend on the costs and benefits associated with different fighting tactics, it may be important to develop the ability to estimate the relative magnitude of these values from experience (Treisman 1977).

Adaptive modification of the probability of using a given behavioral tactic to resolve conflict over a resource is just one example of behavioral scaling (Chapter 2; E. O. Wilson 1975), a widespread phenomenon of major biological importance. What factors cause timber wolves, chimpanzees, domestic cats, wildebeest, equids, and individuals of many other species to modify their use of space and social behavior in response to the abundance, distribution, and variability of resources and of competitors in their environment? Are special abilities involved? Could such abilities develop through play? Should they, in an evolutionary sense?

Harassment, threat, bluff, and deception are important in much animal conflict (Blaffer Hrdy 1977, Maynard Smith & Parker 1976, Popp & DeVore 1979). Skills for producing these effects, for resisting harassment and threat, and for seeing through another's bluff and deception might be practiced in play. What kinds of experience would be most effective in developing these skills? Could and should such experience occur in play?

Negotiation requires social skills for which play may constitute effective preparation. A playfight between two animals is always a compromise between different, often diametrically opposed interests. Unless animals are perfectly matched and share the same developmental history, they may need different kinds of exercise or practice. For instance, older individuals will seek to play longer and more vigorously than younger individuals. These same conflicts will occur between males and females. For play to occur at all, some sort of initial compromise must be reached, and play will only continue as long as it satisfies the interests of both parties. The preliminaries to play, involving signaling of various kinds, may be interpreted in terms of such negotiation. One skill that play experience perhaps teaches is the ability to resolve conflicts by negotiation. Does this skill generalize to other kinds of conflicts over other resources in adulthood? What, behaviorally, is negotiation: Adjustment of an individual's direction of movement as a group moves? Non-damaging fighting, including boundary displays? Subtle changes in relationships between parents and offspring, between sibs, or between mates? When does it pay an animal to negotiate?

These points await attention. The important theoretical exercise of specifying what sorts of experience would serve to develop the skills cited must still be carried out. Whether play of some sort should be used to generate the required experience is an equally important question that remains open.

Regulation of Developmental Rates

In theory, environmental cues, including behavior of other individuals, result in switches from one developmental pattern to another. The timing of developmental transitions might also be under behavioral control (e.g., age at sexual maturity, p. 366). Rosenblatt, Turkewitz, & Schneirla (1962) suggest that domestic cat kittens' rough and boisterous play, in which the mother is often a passive participant, helps accelerate the weaning process. This phenomenon is not an artifact of domestication; I have seen the same playful harassment of the mother by kittens of three species of wild cat raised at Brookfield Zoo (*Felis manul, F. margarita, F. silvestris*). Similarly, vestibular stimulation (Clark, Kreutzberg, & Chee 1977) and other forms of environmental intervention can accelerate motor development in human infants and in young of other species, including the laboratory rat (Geist 1978a, Konner 1975, 1977a). The evolutionary status of these effects is unclear. Why should kittens be selected to behave so as to decrease their access to milk? One speculation is that they are ready to begin learning to handle prey, but it is easier for the mother to lactate than to catch and bring back prey, so that conflict occurs over the time at which the mother begins to hunt for her kittens. By pestering their mother, they increase her cost of staying around the home area and make hunting energetically preferable from her point of view. A more mundane explanation of the phenomenon cited by Rosenblatt et al. is that the kittens' vigorous play gives the mother the information that they are behaviorally ready to begin learning predatory skill. How else would she obtain this information? If there is any conflict of interest over the time at which she initiates hunting for the kittens, there would be pressure on both sides to send deceptive messages and to resist deception.

Developmental rate effects mediated by social behavior or by environmental complexity must be explained by citing the biological interests and information requirements of each party affected by the potential change in timing.

Play as Aggressive Competition

Biologists from Darwin to W. D. Hamilton have successfully argued that animal behavior (including social relationships) evolves to benefit genes and individuals, not groups or species. This insight suggests a need for closer scrutiny of behavior that appears to violate the modern canon of genetic self-interest. Altruism, communication, ritualized or conventional fighting, mating, parental care, and sib relations have all been reinterpreted in this manner. By analogy, play itself may have been viewed naively and uncritically in the past. As argued in Chapter 7, even if social play is truly cooperative behavior from which each partner gains benefits greater than those possible when each plays alone, we expect that each partner will still behave so as to maximize the benefits (measured in reproductive success) that such cooperation affords the particular gene(s) involved. The major constraint on expression of such selfishness is that the partner must also behave to maintain the relationship as long as continued play is to that partner's benefit. These statements follow directly from social theory (Hamilton 1975, Popp & DeVore 1979, Trivers 1971). Partners' interests in play may sometimes agree sufficiently to rule out conflict [just as in Trivers's (1974) model of parent-offspring relations, which predicts that parent-offspring conflict over parental investment may be slight or non-existent at times]. Gentle social play between close kin (p. 308) may represent an example of such extreme cooperation. However, it would be naive to assume perfect agreement between all play partners at all times.

Even a mother and her offspring may be selected to disagree over their play in particular circumstances (p. 299; Chapter 7).

Though this concept of play as selfish cooperation may annoy or even offend some behavioral scientists, it seems timid indeed compared to the proposal (Geist 1978b) that even the most mutual, stable play interactions actually represent damaging, aggressive competition:

An individual enhances his reproductive fitness, however, not only by successfully competing for resources, but also by directly reducing the reproductive fitness of other individuals. . . . Granted a high probability that individuals born together will also live together, it pays not simply to defeat prospective competitors early during ontogeny, but to somewhat impair their body growth and competitive ability as adults. This can be done by interfering with their acquisition of resources needed for growth, or alternatively to interfere with the assimilation of these resources. . . . It is obvious from the foregoing that "play" in juveniles takes on a somewhat less innocent image than it previously had. If a juvenile can reduce the body growth of his "playmates" by gaining dominance, he inflicts on them a permanent disadvantage. This could conceivably be done by (1) imposing stress that causes a reduction in circulation in tissues of low growth priority through hypertension, with concomitant reduction of the size of blood vessels in these tissues by a mechanism proposed by Ooshima, Fuller, Cardinale, Spector, and Udenfriend (1975), resulting in less than maximum possible growth; and by (2) increasing his playmates' cost of living by costly excitation. (Geist 1978b pp. 3–4)

Geist's central insight is that developmental plasticity makes an individual vulnerable to competitors. Exercise adaptation, learning and practice of skill, behavioral meshing, and all other forms of dependence of developmental paths and rates on social experience allow an individual to be manipulated by others whose biological interests may not be the same as its own. Competitors may withhold or falsify information. They may manipulate another's experience in order to serve their selfish ends. Under such circumstances, natural selection should favor developmental resilience. The impressive resilience of human development (Dunn 1976, Kagan et al. 1978) may have resulted from selection for the ability to resist selfish attempts to influence the development of infant and child (compare Trivers 1971), if these arguments are correct. In animals (such as *Homo sapiens*) necessarily dependent on environmental information for survival at virtually all postnatal life-history stages, the evolutionary developmental balance between plasticity and resilience must be quite sophisticated.

Although Geist's proposal opens new directions for speculation about play, it suffers from several weaknesses. It stresses the delayed benefit of access to resources in adulthood, but damaging play (if it exists) could well secure immediate access to resources. After all, dominance hierarchies should not only be resource-specific (Popp & DeVore 1979), but also age-specific, both because the value of a given type and amount of resource will differ for organisms of different developmental ages and because competitive abilities vary with age. Furthermore, such competitive behaviors as infanticide, nest and habitat destruction, and interference with mating and offspring rearing (Geist 1978b) force the victim either to submit or to defend itself. But if play is damaging, an animal can simply choose not to play at all. Therefore, a strategy of damaging play is not evolutionarily stable against one of no play (Chapter 9). If each animal can potentially damage the other in play, but the outcome of any given bout is unpredictable, then damaging play may still occur in theory, though it is not clear how the "winner" of such play is to be identified in practice. In a vigorous and mutual playfight between like-sized, equally healthy individuals of the same sex, each player might expend roughly equal amounts of energy, occupy available roles at the same frequency, and score the same number of positional "wins." If the damage hypothesis refers to one-sided play of a strong with a weak partner, as in the bullying, teasing, or intimidation reported in play of primates (Bertrand 1969, Itani 1959, Mörike 1973, Symons 1978a pp. 75–76, 198) and ungulates (Linsdale & Tomich 1953), the damage argument may be valid, but it is not clear what the weaker partner stands to gain from such play or why it should have risked damage in such an obviously unequal contest.

In summary, the idea that all play represents a damaging competitive tactic is interesting, but not compelling. Based on the important theoretical insight that developmentally plastic organisms are vulnerable to damaging social manipulation, it fails to pass the test of evolutionary stability; neither does it address the question of design. Is social play actually designed by natural selection as a mechanism for harming other organisms' development? Could behaviors other than play serve this purpose?

The concept of genetic self-interest suggests a third hypothesis distinct from the two (selfish cooperation and damaging competition) already proposed. This "cheating" hypothesis [proposed and criticized by Bekoff (1978b) and Fagen (1978a)] suggests that animals may exploit opportunities to change play into agonistic fighting. An animal would begin play, then make an escalated attack on its playmate. By doing so it would damage or intimidate its opponent, thus gaining access to a current or a future resource. An initial play attempt could also serve to deceive watching adults, who might then disregard the pair long enough for the cheater to harm its partner. How could we recognize such cheating if it actually occurred? Bekoff (1978b) suggests that play-signals could themselves be used as means of deception. Animal A invites B to play. Animal B, who wants to injure A, deceptively accepts A's play invitation. With its guard down, A approaches B playfully and B then attacks A using damaging fighting tactics.

Failure to desist following a mistake may also be interpreted as cheating. Animal A cries out in pain when B accidentally bites too hard in a playfight, but instead of biting more gently or facing away B bites even harder. When A tries to escape, B pursues it, aiming blows at it and ignoring its submissive gestures.

Increasingly rough play that transcends the other's tolerance could follow or accompany failure to desist. One-sided mauling and consistent adoption of a single role when partners are equally matched might also be viewed as cheating.

The behavior of playing animals could allow us to infer motivation, including cheating. Cheaters might act guilty. For example, some juvenile rhesus macaques playing roughly with an infant ran away immediately when it squealed, "anticipating" its mother's intervention (S. Altmann 1962b). Data on modes of termination of play interactions, on stimuli for and responses to pain vocalizations and signals of threat and submission in play, and on play bouts that change to escalated fighting are of great interest if we wish to test the cheating hypothesis. What behavior actually constitutes a breach of fair play, and how do participants respond to transgressions?

Breaches of the rules do occur. Social conventions notwithstanding, a playfight may occasionally end with threat displays or even with an escalated fight. The terms applied to describe such transitions are as varied as the possible explanations of these rare events (Hinde 1974 p. 253, Kurland 1977, Levy 1979, Owens 1975a, Symons 1978a). They have been called "mistakes" and "misunderstandings," and play has been said to "breakdown," or to "escalate into," "merge into," "spark over to," or "go over to" agonistic behavior. The immediate cause of such events appears to be that an animal who behaves roughly to another fails to heed the other's responses, does not restrain itself, and continues its rough behavior despite the other's fear or pain vocalizations or attempts to flee. Although the initial misunderstanding can be explained either by invoking imperfect information (we say that a "mistake" occurred) or by assuming more sinister motives on the part of the transgressor (Konner 1975), an animal who persists even when its partner cries out and attempts to escape cannot be unaware of the change in its partner's mood. It may prove valuable to distinguish between cases in which the initial invitation or solicitation to play elicits an agonistic response, cases in which momentary transgressions occur, but the animals quickly calm down and resume playing, and cases in which an ongoing play interaction ends in mutually agonistic behavior. The first type of event may simply represent the only way in which a tired or an ill animal can repulse the enthusiastic advances of a conspecific unaware of its condition or of a conspecific with whom play offers little benefit. The second situation is best explained as an honest mistake that might occur most frequently in very young animals who have not yet learned to restrain themselves or to recognize play-signals (e.g., Blurton Jones 1967) or in rapidly developing animals who may not know their own increasing strength and who may lack control and coordination (e.g., adolescents). An animal will persist in a fight resulting from play for the same reasons that it will persist in any other fight, namely that if it wins it will gain undisputed and immediate access to a limiting resource, valuable object, or place, or it will cause the loser to avoid disputes with it in the future, thereby reducing the likelihood of additional conflict; whereas if it loses it will not be penalized to an extent sufficient to warrant consideration of other means to these ends. The central question remains: When play turns to fighting, what are the animals fighting about, and what benefit does the winner stand to gain? Is one animal simply conditioning the other not to be quite so rough? If so, the two will play again, perhaps even more frequently, in the future. If access to some resource in the immediate or distant future is at issue, play will decrease sharply after the interaction. Young animals' access to current and future resources is actually settled, so far as is known, by fights that start as fights (Bekoff 1974c, Symons 1978a) or by maternal intervention in agonistic interactions between young (Cheney 1977). The benefit of escalating a playfight may be that the partner learns not to play as roughly with the transgressor, that the partner no longer solicits play from an animal who no longer wishes it as a playmate, or that the transgressor's mother intervenes in the transgressor's behalf before the victim's mother can respond to its cries.

Some cases of escalation may represent a conflict between one animal who wishes to keep playing with a second, and the second animal who has nothing further to gain from play with the first, may have something to lose, and therefore seeks to terminate the play relationship once and for all. Stuart Altmann (1962a), for example, suggests that the increasing frequency with which play breaks down in older rhesus macaques results from older juvenile males' unwillingness to accept an equal or subordinate role, even in play, for such acceptance might jeopardize their still-uncertain status.

In theory, who should be decreasingly willing to play with whom? Do possible changes in play frequencies and partnerships follow serious breakdowns of play? Evidence bearing on these questions would add weight to the above speculations. Cases in which play erupts into an escalated fight must be examined more closely. Are these breakdowns honest mistakes resulting from participants' imperfect information, or do they actually represent cases of cheating? Because cheating behavior could result from natural selection on learning itself, making animals unable to gather or to process the information necessary to prevent such mistakes, or even producing self-deception regarding playfulness, it is important to consider possible mechanisms very carefully if these opposing insights are to be transformed into testable, exclusive working hypotheses. The rules of social play are cooperative. Unilateral deviation from these rules results in immediate cessation of play. If inhibitions temporarily break down, the game is typically suspended and partners may face away from each other or groom themselves until tempers cool (Sparks 1967). If one participant has imperfect information about its strength or about the sensitivity of its opponent, and physical pain occurs, the other animal often ignores its discomfort and continues to play, or it utters a pain vocalization that usually makes the attacker desist (p. 398).

Bekoff (1978b) incorrectly states that no instances of deceptive use of play-signals can be found in the ethological literature. To be sure, in all the years that animal play has been observed, no animal has yet been seen to invite play and then immediately attack its partner. However, a chimpanzee was observed to initiate play with another chimpanzee who had food. During their play the initiator sneakily grabbed the food (Menzel 1975).

Thus, despite the general view among social theorists (Dawkins 1976b, Otte 1975, Trivers 1971, 1974, Wallace 1973) that all animal communication is expected to be exploited for selfish interests, we lack evidence that play-signals are regularly used for this purpose.

The argument that young animals might refrain from cheating in play for the benefit of the group as a whole, although possibly valid under very stringent constraints and within narrow parameter values, is not likely to explain most cases of play-fighting in animals, for this behavior appears to have evolved in nearly all mammalian species and in many species of birds (Chapter 3). It is difficult to imagine how the extremely restrictive conditions for the occurrence of group selection could have been satisfied during the evolution of so many different kinds of animals.

The cheating hypothesis will be further discussed below (p. 435) and related to ontogenetic changes in quality of social play.

It is correct to suppose that social interactions between individuals who are not genetically identical will often reveal evidence of conflicts of interest and that these conflicts will be resolved by some competitive behavioral tactic. As I argue in the next section of this chapter, play has incorrectly been viewed as behavior selected for effects conferring an advantage on the social group rather than on the individual performer and its kin. I believe (Chapter 7) that conflicts of interest over play are real and that behavioral observations reveal evidence for such conflict as well as evidence of negotiated settlements. But this conflict occurs over initiation, termination, form, pattern, and content of play itself, not over some other resource that the "winner" of a playfight gains or the "loser" loses. Social theorists have yet to propose a plausible behavioral mechanism by which play, in and of itself, could function covertly as a competitive tactic. If such a mechanism exists, it remains to be argued how play is designed to this end and under what circumstances behavior so designed could evolve by natural selection.

Bonding and Cohesion

The metaphor of social group as superorganism is among the most pervasive in behavioral science (Symons 1978b). Dyads or larger social groups are viewed as functional units having internal structure that serves the unit's homeostatic requirements. This structure is described using physical metaphors. Individuals form bonds, groups cohere, bonding and cohesion are reinforced. Bonds can be strong and tight or weak and loose. Bonds can be cemented. Each such social unit is considered to exhibit an equilibrium between forces of attraction and repulsion, or alternatively, between cohesive and dispersive elements. In an unfortunate conflation of topological and physical metaphors, the strength of bonds is also described using an imaginary social distance metric according to which dyadic relationships are "close" or "distant." These terms are convenient and come so readily to mind that it is almost impossible, though undesirable to discuss social behavior without them.

The above concepts require that social behavior be either cohesive or disruptive. Social behaviors must then be classified as cohesive (communication, grooming) and disruptive (aggression, territorial behavior) according to their effects on strength of bonds or spacing of individuals, to cite a second, Euclidean distance metric that further confuses the issue.

Play has a special role in this belief system. Like grooming, play exerts a cohesive force that keeps individuals and societies together (in Euclidean and social space) and strengthens bonding and cohesion. In a word, it is social glue.

The metaphor of play as glue has two far-reaching implications. It defines play as a bonding force in dyads. Therefore, when play is absent, individuals are no longer bonded and must disperse. It defines play as a bonding force in societies. Therefore, members of socially cohesive species will exhibit more frequent, more intense or more stable play than members of socially dispersed species. These ideas have long dominated behavioral scientists' views of play. They have significantly influenced both data and interpretation, and their influence has been less than salutary.

The concepts of bonding and cohesion, and all hypotheses and interpretations based on them, are unsatisfactory in the study of social behavior. The implicit social theory contained in these concepts is fundamentally incorrect. A social group (dyad or society) is not an integrated, functioning unit. "The individual members of an animal group are imperfectly related genetically, achieve reproductive success at one another's expense, and compete for the same resources" (Symons 1978b p. 201). When animals maintain physical proximity, cooperate, perform altruistic acts, disperse, space themselves out, groom, communicate, fight, or play, they do so to promote social cohesion or to strengthen a bond only if these proximate effects are likely to ensure the ultimate survival of their genes. I illustrate this point by citing examples of the influence of cohesion theory on recent social play research.

Play supposedly keeps individuals together in Euclidean or social space. West (1974) suggests that play "helps keep littermates on friendly terms" and is adapted to "promoting physical contact, thus keeping kittens together." Similarly, Poole (1978) suggests that play functions to "reinforce amicable relationships between juvenile members of a social group." Again, "the early ontogeny of social play and a delay in the appearance of rank-related aggression may be responsible for the development of strong social bonds and a coordinated social group" (Bekoff 1977a). The logical conclusion drawn from this erroneous premise is, not surprisingly, that individuals who do not interact socially with littermates "will not develop strong ties to their group and will be the most likely individuals to leave their natal site first. Other sibs, who have interacted more with each other, will show a delay in dispersing until their first potential reproductive season" (Bekoff 1977a).

There is no evidence for causal relationships between play and physical or social distance. In chimpanzees, sleeping dyads and play dyads do not coincide (Riss & Goodall 1976). The same principle, namely that partner choice patterns depend on the activity in question, holds in many other species as well (Mori 1974, Rasa 1977, Rhine 1972, Rhine & Kronenwetter 1972, Rosenblum & Lowe 1971, Soczka 1974, Yamada 1963). Social groups may be playful, but noncohesive (Mc Guire 1974) or cohesive, but nonplayful (Baldwin & Baldwin 1974). Chimpanzee dyads may be cohesive without ever playing (Menzel 1975). During the dry season, no play occurs in groups of the sifaka (*Propithecus verreauxi*), but troop cohesion is maintained (Richard & Heimbuch 1975). In a multivariate analysis of squirrel monkey social behavior, play behavior did not contribute strongly to either the "dominance" or the "affiliation" factor (Strayer et al. 1975). Apparent lack of association between characteristics of felid play and characteristics of adult social organization has already been discussed (p. 168). In the square-lipped rhinoceros (*Ceratotherium simum*), social play behavior only rarely occurs between individuals who are companions in the same group and, in one field study, was never seen between close companions; it generally develops from nasonasal meetings between two individuals from different groups (Owen-Smith 1973 p. 599). In the euro (*Macropus robustus*) and in the red kangaroo (*M. rufus*), young-at-foot frequently playfight with their mothers, and evidence to date suggests that this behavior is significantly more common in male than in female young (D. B. Croft 1980 and pers. comm.). In these kangaroo species, adult males seem to fight more often and for greater benefits than do adult females, but there is no evidence that males form stronger social bonds at any age. In the absence of any indication of greater cohesion between males and their conspecifics than between females and their conspecifics, evidence on kangaroo behavior certainly seems to support a training rather than a bonding function for mother-offspring play in the two species cited.

Biologists who study aggregation (E. O. Wilson 1975 Chapter 12) and dispersal (R. R. Baker 1978, Packer 1979a, Parker & Stuart 1976) from an evolutionary point of view provide a very different perspective on social cohesion and dispersion. Do young animals play in order to stay together as juveniles and/or throughout their lifetime, or perhaps even in order to learn behavior maintaining species-specific social structure (a misleading superorganismic concept in itself, given the ubiquity of ecological scaling of social behavior)? More likely, they stay together to play (or for other equally important reasons). Do they disperse because they are no longer bonded? Or, do they disperse because it is best for their genes that they do so? The distinction between proximate and ultimate causation is important here, of course, but even the proximate chain of causation must be more tightly argued. Why are animals selected to avoid each other? If avoidance is the proximate stimulus for dispersal, animals can use it to manipulate each other's behavior for selfish ends. No proximate social stimulus can be taken as given, since it can always be controlled by the performer to achieve a desired effect.

All of this is not to say, of course, that behavior differentially promoting interindividual recognition, familiarity, meshing, negotiation, behavioral prediction, and cooperation is not fascinating. On the contrary, it is an evolutionarily significant biological mechanism (p. 490). Why is it performed? How does it achieve these goals? Consider grooming, for example. Why should a monkey have to spend hours every day grooming another monkey in order to ensure its future cooperation? Why not just sleep in contact or sit nearby? There are several answers to this question. If different animals must compete for the groomee's cooperation, only one or two can groom at a time. Grooming represents an exclusive time investment. A monkey cannot forage or monitor for predators while grooming as well as it can when not doing so. The groomer thus indicates its desire and ability to give up resources to the groomee. By doing so, it says, in effect, "I am ready, willing, and able to co-operate with you." Finally, given a high overall level of groom-ing behavior in a troop, unilateral deviation will be penalized. Therefore, grooming is evolutionarily stable and could remain at a high level even if originally brought there for hygienic, par-asite avoidance reasons.

The biology of friendship is important in its own right (S. Altmann 1962a p. 409, Cheney 1977, Packer 1979a, Trivers 1972). Klingel (1967, 1974a) observed that young plains zebra (*Equus burchelli*) stallions would leave their natal band at an ear-lier age if they had no playmates. They then found playmates from other bands. That these play relationships might be bene-ficial if they endured sufficiently long is suggested by Darwin's (1874 p. 571) observations of equid social behavior. Two wild horse stallions came in together and took females from a band led by an English stallion. Had these individuals previously met, become acquainted, and learned to cooperate through play? If so, how? Mechanisms other than play might also lead to future cooperation. It is necessary to specify just how play fa-cilitates such cooperation. In the equid case, is it genetically best for a stallion X to join his half-brother to steal females? If no half-brothers of X are in his band, perhaps an older full brother or half-brother Y has already dispersed. Will Y prefer to join his inexperienced brother or to remain in his current dyad? By as-suming appropriate coefficients of relationship, probabilities of stealing given numbers of females as a function of the size and experience of the stallion pair, and probabilities of insemination given that a female is successfully stolen we can compute op-timal dyad composition from each individual's point of view. These calculations are not the central issue here. I merely wish to suggest that there is much more to bonding and dispersal than glue.

The concept of play as glue has also shaped the collection, in-terpretation, and very nature of comparative phylogenetic data.

For example, Moynihan (1964) asserts that play wrestling "seems to be almost or completely lacking in night monkeys (*Aotus trivirgatus*), presumably as another consequence of their slight degree of gregariousness." (See p. 92 for a description of possible play-wrestling in night monkeys.) A positive correlation between juvenile playfulness and adult group size has been claimed for several other groups of species. In at least nine mammalian taxa (Table 5-7, see also Chapter 3) the tendency of young to play *together,* with body contact, and in closely-interacting groups of more than two animals, is apparently associated with the sociality (group size, group permanence, average physical closeness, "cohesion," or degree of cooperation) of adults. In interpreting this information, it is essential to recognize that the tendency to play socially may be entirely independent of the overall tendency to play with or without social interaction and body contact. For example, one animal could play ten times more frequently and/or readily than another animal, but all of the first animal's play might be solo play whereas all of the second animal's play might be playfighting. Moreover, the degree of social interaction in play is relative and perhaps even multifactorial, as demonstrated by the phenomena of socially facilitated solo play, companion-oriented play, dyadic play, and group play (Wilson & Kleiman 1974). It is important to note that the mother or father can be a passive or even active companion for social play in species offering few other playmates (e.g., Madagascar viverrids, caviomorph rodents, monotokous felids, polar bear, pygmy hippopotamus; see Chapter 3). Degree of parental involvement in play is an additional variable that must be considered whenever correlations between juvenile social play tendencies and adult sociality are sought. It is also entirely possible that Table 5-7 simply reflects unconscious culling of the natural history literature. In how many species, genera, and families would no such correlations be found? Perhaps the species of Table 5-7 are isolated exceptions rather than examples of a general trend. Finally, as discussed in Chapter 3 (p. 78), these associations may well reflect observer effects, differing degrees of habituation, relative brain size, greater physical proximity of potential playmates in physically cohesive species, demographic or developmental constraints,

Table 5-7 JUVENILE PLAY AND ADULT SOCIALITY

Species	Reference	Characteristic of Juvenile Play	Characteristic of Adult Sociality
Canids	Bekoff 1974c, Fox & Clark 1971, Fox et al. 1976	Frequency, stability	Adult group size
Cervids	Chapter 3	Solo vs. dyadic vs. group	Adult group size
Conilurine rodents	Happold 1976	Solo vs. social	Adult group size
Damaliscus dorcas, Kobus defassa	David 1973, 1975, Spinage 1969, Verheyen 1955	Solo vs. social (nursery herds only)	Adult group size
Damaliscus korrigum	Monfort-Braham 1975	Frequency of group play	Adult group size
Macaca nemestrina, Macaca radiata	Kaufman & Rosenblum 1965	Solo vs. social	Adult physical proximity
Marmota spp.	Barash 1973a,b, 1974a,b, 1976	Frequency	Adult group size and tolerance of young
Microtus agrestis	Wilson 1973	Frequency	Adult tolerance of each other and of young
Tree squirrels (*Tamiasciurus hudsonicus, Sciurus* spp.)	Ferron 1975, Horwich 1972	Solo vs. social	Adult group size

and perhaps even the subconscious influence of the cohesion hypothesis on housing conditions, management practices, and study design. (Many of the studies cited were carried out on confined and even hand-reared animals.) Nevertheless, I believe that the information cited in Table 5-7 is more than a giant artifact of superorganismic thought. Why, then, have young of social species apparently been selected to play *together?*

John Terborgh (pers. comm.) suggests the following plausible explanation. Any change in the environment that increases the availability of playmates may select for an increased tendency to play socially if playmates continue to remain available for a sufficiently long time. This hypothesis is completely independent of cohesion or bonding effects, for a playmate can potentially offer improved physical training, brain stimulation, developmental rate adjustment, or virtually any other hypothesized effect of play. Thus an increase over evolutionary time in adult group size or permanence could result in greater abundance and reliable availability of playmates, and therefore in selection for increased tendencies to interact in play (subject, of course, to genetic constraints on *cooperativeness* of play). This increase in adult group size is necessary but not sufficient to ensure that a young animal will have playmates. Adult group size variance, mean brood size, birth rates, interbirth intervals, survivorship, birth synchrony, and age at dispersal will also affect the outcome, as will the temporal correlations of these parameters.

This hypothesis might seem not to explain the apparent correlation of juvenile play sociality with adult sociality in animals (e.g. canids) giving birth to multiple offspring. However, if mean litter size at birth is constant but survivorship to the playful age is higher in more social species, or if variability in the size of the playful age-class is lower, as would be the case if "extra" adults or subadult helpers assisted in feeding younger pups, then like-aged and older playmates would be more *reliably* available in the more social species. Tendencies to play socially would then be more likely to evolve in those species, according to the mechanism proposed by Terborgh. This argument suggests that the most important additional variables to measure in comparative studies of this problem would be demographic: age distributions, life tables, and their variation over time. Cohesion metaphors could scarcely have indicated the theoretical importance of the juvenile age distribution, or of demographic variation as such, in selecting for tendencies toward sociality of play. Moreover, as indicated by information on orangutan and on margay social play (Chapter 3), the tendency to play socially is not a fixed characteristic of a species or even of an individual. It is a biological dependent variable that will change, within limits set by natural selection, as a function of individual requirements for play and of demographic opportunities. If such variation can be demonstrated in nature, it will represent yet another case in which social behavior exhibits ecological scaling, with the slope and intercept of the scale function determined by natural selection. In mammals and birds, there is no reason to assume fixity of these slope and intercept parameters. Individual tendencies to engage in social play as a function of population density of potential playmates can themselves be affected by experience. Natural selection may produce experiential modifiability of an individual's tendencies to play with others in preference to playing alone. This phenomenon has already been hypothesized. For example, it is claimed that domestic dogs and other mammals have "critical periods of socialization" (Fox 1972a). Of course (Bateson 1979), this hypothesis in its most extreme version could not be true even in theory. But sensitive periods of social development do occur and may be interpreted biologically (Bateson 1979). Are there also sensitive periods for development of preferences for play with others versus play alone? If so, to paraphrase the title of the Bateson paper just cited, how do they arise and what are they for? Are they "adaptive" in the strict biological sense of the word, or are they simply selectively neutral (or perhaps even costly!) by-products of a more general adaptive tendency toward experiential modifiability of social behavior? (Such tendencies cannot always be adaptive; as discussed in Chapter 4, plasticity is a two-edged sword.)

Learning Specific Information

This category of effects includes practically every type of biologically relevant information that an imaginative ethologist might consider. It has been suggested that play teaches animals to identify their kin (S. Wilson & Kleiman 1974 pp. 362–363) and species (Poirier 1972a), to learn the properties of objects (Eibl-Eibesfeldt 1967), to assess abilities of others relative to one's own abilities (Fagen 1974a), and to learn cultural behavior of their troop (Baldwin 1969). To this list we might add simply learning the value of all environment-dependent parameters in all models of social behavior. These hypotheses beg the related questions of design and mechanism. Why should such information be gathered in play rather than in some other less costly behavior? Do structure and context of play suggest that play is designed specifically for gathering the given information, and not for some other purpose? In what situations (species, environments) is this information most important for reproductive success? Is play most frequent, intense, and stable in these situations? Using arguments of this sort, Symons (1978a) shows that specific learning may well occur as a fortuitous effect of play, but that no stronger statements about specific learning in play can be made. Hutt's (1966, 1970) experiments on exploration and play in human children suggest the same conclusion. Weiss-Bürger (1975) showed that play and exploration by ferret-polecat hybrids in an artificial tunnel system initially improved their maze-running ability, but that this improvement did not persist when the animals were retested. Again, this study may merely have demonstrated incidental learning.

Play and Behavioral Flexibility

The audacious suggestion that play experience develops competence not in specific motor or cognitive skills, but in the very art of being alive, of being a vital, authentic person, has had a compelling influence on nonbiologists. One of the world's most respected students of play, the psychiatrist Erik Erikson, told a Harvard audience in 1973 (Erikson 1977) that, of the children he had studied thirty years before, those with the "most interesting and fulfilling lives were the ones who had managed to keep a sense of playfulness at the centre of things" (Bruner 1976 p. 17). That play experience contributes to human creativity, flexibility, imagination, innovativeness, productivity, or versatility is frequently suggested (Bruner 1976 p. 17, Einstein in Hadamard 1945 p. 142, Hutt & Bhavnani 1972, Rumbaugh 1974, Sutton-Smith & Sutton-Smith 1974 p. 5). Nonhuman primates may soon be subject to similar speculation. Play looms particularly large in the lives of three outstanding females: the chimpanzee Flo (van Lawick-Goodall 1971), the Japanese macaque Imo (Kawai 1965), and the rhesus macaque Sarah (Hinde & Simpson 1975). Flo was a good mother with an especially playful and close relationship with her children. Imo discovered several novel behaviors that other young monkeys apparently learned while playing with her children. Sarah, a very warm and sensitive mother, was one of the few mothers in her rhesus colony to play approach-leave games with her infant. The few rhesus mothers who do play this game all have well-meshed relationships with their infants, and in at least one case (Eliane), the infant of such a mother played the same kinds of games with her own infant. These animals may define a new biological image of parental competence: the nurturing, responsive parent as creator and player.

Humans are absurdly easy to typecast using distinctions based on experimental paradigms for nonhuman behavior. Simple dichotomies are visible, compact classifications that can always be aligned to mean us versus them. Human (or nonhuman) flexibility and the like can hardly be considered one-dimensional variables, and we have no idea of the skills involved, much less of the ways in which they develop. Rats reared in unstimulating environments where they seldom if ever play develop a classic behavioral syndrome of narrowness, impulsivity, emotionality, and lack of flexibility. Rats reared in enriched environments where play occurs are behaviorally plastic, flexible, and versatile; they respond effectively to novel situations (Hinde 1966 p. 385, Levine 1962, Marler & Hamilton 1966). In a series of elegant experiments, Dorothy Einon and Michael Morgan have demonstrated that play interaction, or something very much like it, is necessary to produce these effects and that it is impossible to explain the results either in terms of arousal or in terms of inhibition (Einon & Morgan 1976, 1977, 1978a,b, Einon, Morgan, & Kibbler 1978, Morgan 1973, Morgan, Einon, & Morris 1977). A brief discussion cannot do justice to these ingenious experiments on play. They include such features as manipulation of the quality of social interaction by allowing rats contact either with undrugged or with drugged companions (Einon, Morgan, & Kibbler 1978). This manipulation allowed rats belonging to different test groups to interact with different kinds of partners: chlorpromazine-injected animals who did not initiate social interactions, but sometimes responded to a normal partner, amphetamine-injected animals who "spent considerable time in stereotyped sniffing and following but frequently failed to respond to their partner," or undrugged rats. Partners in the undrugged condition "spent much of the hour in chasing and wrestling." Behavioral flexibility in certain tests only developed in those rats whose partners could play with them. Also, differences in flexibility between rats whose partners had been injected with different drugs showed that the experimental animals were reacting to drug-specific changes induced in partner behavior, rather than to nonspecific changes in partner characteristics that resulted from receiving an injection as such (e.g. smell of alcohol used to clean injection site). These experiments suggest that play experience aided rats to develop the ability to respond flexibly in a test situation requiring rapid switches between different patterns of behavior.

Geist (1978a) carries the argument considerably further. He speculates that, when environmental conditions favor dispersal, individual experience under these conditions should produce individuals "preadapted to confront and solve diverse problems . . . equipped to handle the unexpected." This dispersal phenotype should be behaviorally adaptable, self-disciplined, and tolerant of diversity, with highly developed abilities to control its environment. The contrasting "maintenance" phenotype is rigid, narrow, emotional, fearful, overreactive, and withdrawn. Geist implicates play as one factor contributing to this difference. Geist proposes that play facilitates flexibility and thereby prepares an animal to be a successful disperser. Bekoff (1977a) takes the opposite view. He views play as social glue and argues that individual animals who do not play sufficiently with other members of their natal group "will not develop strong social ties" with their siblings (or, more generally, groupmates) and will be the most likely individuals to disperse of their own accord. Flexibility hypotheses and cohesion hypotheses make opposing predictions about dispersal. However, as argued above, the premises on which these hypotheses are based are highly questionable. Although these predictions are testable, their theoretical basis will require additional refinement. Existing evidence on animal dispersal reviewed by Baker (1978) and by Packer (1979a) suggests that males, the more playful sex in most species considered, are also the dispersing sex.

In chacma baboons (*Papio ursinus*) and rhesus macaques (*Macaca mulatta*), some males that play together leave (disperse) together at an early age, whereas other males that play together remain for a longer time in their natal troop (Cheney 1978, Levy 1979). In gorillas (*Gorilla gorilla*), chimpanzees (*Pan troglodytes*), and African wild dogs (*Lycaon pictus*), females disperse (degree of sexual dimorphism in play is unknown for these species), whereas both sexes disperse in hamadryas baboons (*Papio hamadryas*), lar gibbons (*Hylobates lar*), and plains zebra (*Equus burchelli*). In domestic cats, males always disperse, females sometimes disperse (Laundré 1977), and females play more with objects but not with littermates when raised with males (Bateson & Young 1979). Proximate causes of dispersal include differential access to natal group and non-natal group females (Packer 1979a), changes in maternal activity [West (1974) suggests that dispersal contributes to a decline in play observed in older juvenile domestic cats, contra the prediction of the cohesion hypothesis], and numerous other factors (Baker 1978, Packer 1979a), including eviction of males by females and differential availability of food. The question of causal relationships between play and dispersal is more complex than existing simplistic hypotheses suggest. This question needs to be analyzed from a formal theoretical point of view, using evolutionary models based on those of Baker (1978), Packer (1979a), and Parker & Stuart (1976). These models address both proximate and ultimate causation of dispersal.

Flexibility, creativity, or innovativeness hypotheses (Chapter 8) are not new and enjoy some empirical support. As currently formulated, they are also naive. Might not biological constraints on creativity exist, analogous to biological constraints on learning, so that creativity itself could only be selected to function in certain spheres of activity, and there to only a limited extent dictated by natural selection? What are the evolutionary limits placed on creativity? Also, is creativity a necessary absolute, or does it only pay an animal to develop its creative ability to the extent that every other animal does? What are the costs of excessive creativity or flexibility? Are optimal and maximal values of this complex of phenotypic characters necessarily synonymous? These questions, as well as many more (see Symons 1978a), stand between the flexibility hypothesis and biological legitimacy.

I consider most evidence to date on behavioral flexibility effects of play suggestive at best. The sole exception to this generalization, Einon and Morgan's work, is very important. These authors convincingly demonstrate that differential rearing experience affects cognitive skill and basic dimensions of personality. Their experiments further indicate the contributions of play to such experience.

Because possible links between play and versatility represent one of the more humanistic aspects of play research, this discussion will end with a literary and historical footnote. The contrast between two famous seventeenth-century Englishmen, George III and Samuel Johnson, perfectly exemplifies the concepts discussed above. According to Sir Lewis Namier (1961 p. 90), the unfortunate George III was reared in a narrow and restrictive environment. He was "sensitive and brought up under abnormal conditions . . . a youth who had never been allowed to be young and who had not managed to grow up; not diverted from serious work by passions or pleasures, but sinking helplessly into sad indolence." A very different sort of man was Samuel Johnson (Bate 1977). As an Oxford undergraduate, Johnson went sliding in Christchurch Meadow (and afterwards nonchalantly informed his tutor where he had been) (Bate 1977 p. 90). When in a holiday mood, Johnson kicked off his shoes, raced a lady across the lawn, leaving her far behind, and then returned "leading her by the hand, with looks of high exultation and delight" (p. 359). Johnson loved racing, climbing, and jumping (p. 485), once performed a convincing imitation of a kangaroo (p. 466), was always "the first to join in childish amusements, and hated to be left out of any innocent merriment that was going forward" (p. 434). In the course of a country footrace with his friend, John Payne, "Johnson caught up the diminutive Payne in his arms and placed him on the branch of a tree they were passing. He then continued running as if he had met with a hard match, releasing Payne from the tree with much exaltation on his way back" (p. 485). An incident (p. 367) that took place when Johnson was 55 years old is particularly illustrative. During a walk with his friends, the

Langtons, Johnson came to the top of a very steep hill and announced that he wanted to take a roll down. The Langtons tried to stop him. But he said he had not had a roll for a long time, and taking out of his pockets his keys, a pencil, a purse, and other objects, lay down parallel at the edge of the hill and rolled down its full length, turning himself over and over till he came to the bottom.

SUMMARY

Play has measurable costs and apparent benefits. Structure and context of play suggest that it functions to develop physical ability, including strength, skill, and endurance. Play may also be very important for cognitive skill development. Hypotheses about cognitive functions will definitely be worth pursuing when the biology of cognition is better understood. Because play functions in development, its adaptive effects are delayed in time. This time delay may be on the order of days or weeks, as in physical training; or, as in cognitive skill development, adaptive effects may be delayed until adulthood. It is possible to consider short-term beneficial effects of play, but these effects may be fortuitous. Theoretical analyses of the biological effects of play suggest new kinds of empirical questions. What are the direct survivorship costs of play? How do time and energy spent in play affect growth, fat deposition, and survival? Do the predictions of skill development theory agree with the structure and content of play and with observable age, sex, population, and species differences?

Play has baffled students of function. The dilemma is as follows. If play is an alternative rather than a precursor to skilled behavior, such as escalated fighting, we assume that the players can already perform skillfully and well. Then what skill and what strength remain to be developed? If there are still skills to learn or muscles to train, playfighting is by definition a developmental precursor of agonistic fighting, and the validity of play as a behavioral category becomes questionable. Symons (1978a) considers this problem; he agrees both that (1) the behavior of young organisms can be understood as a developmental process and that (2) some behavior patterns such as fighting occur in two distinct modes in the same animal at the same age, one of which is play.

This problem can be resolved. Playfighting and escalated fighting develop in parallel, and interactively. An animal at any given level of fighting ability is always potentially capable of improving its ability by playfighting, although the absolute amount of such possible refinement decreases with age. As Symons (1978a) argues, the issue is proficiency. When young monkeys of any age actually fight, they certainly do not do so with adult skill and effectiveness, but their bites are far from inhibited. Fur flies, cries of pain are uttered but disregarded, and the loser avoids the winner thereafter. Adult proficiency, Symons argues, results in part from repeated use of skill and strength in play.

This point suggests another observation on duality. In allowing the animals themselves to tell us by their overt behavior whether they are playing or fighting, we admit the possibility that the animals themselves recognize this duality and that even a very hard blow delivered in play will elicit a playful response if the recipient views the interaction as playful. The animal's own attitudes and intentions are crucial, and fortunately for the study of social play, they can at least sometimes be inferred from overt behavior. If an animal can respond to the same stimulus either playfully or agonistically, then existence of this duality at any age is the basic condition for validity of playfighting as a separate category of behavior.

The criterion of duality may also be applied to solo and item-oriented forms of play. If a young monkey can climb a bush, walk away from its mother, and return, or run from a disturbance in its environment, but only sometimes does so repetitively with slight variation and loose body tone while making a play-face and possibly exhibiting other locomotor-rotational movements, then it can be said to perform locomotor play. The same criterion may be applied to play with objects and with non-conspecifics. This approach to item-oriented play may suggest that some forms of repetitive and diverse manipulative practice resemble locomotor and social play to a lesser extent than these two types of play resemble one another. However, quantitative comparisons are needed.

Three general classes of functional hypotheses about play appear to have current standing in biology. Training hypotheses are supported by data on design, on sex differences, and on context. Their major theoretical weakness is that they take modifiability as given. Their major empirical weakness is that they fail to explain adult play, courtship play, and gentle interactive play. Developmental rate hypotheses (including aspects of behavioral meshing) are wholly mechanistic and have a growing empirical basis, but their adaptive significance in terms of inclusive fitness and evolutionary stability is seldom clear. Their chief theoretical weakness is that they take modifiability as given, ignore constraints imposed by biological self-interest, and fail to distinguish between adaptation and effect. Cohesion hypotheses, including short-term recognition, explain adult social play, courtship play, and gentle interactive play. They fail to address play alone. They do not explain why play, rather than some other form of behavior, is necessary in order to produce the hypothesized behavioral effects. Their chief theoretical weakness is their fundamentally superorganismic point of view. Their chief empirical weaknesses are that they are partly based on flawed correlations between observed frequency or quality of play and adult social structure, that they ignore bonds and cohesive societies that form and endure without play, that they disregard families (e.g., hippopotamids, felids) containing species that differ in sociality of adults but not in playfulness of young, and that they fail to explain why females should play less frequently than males in such species as equids, red kangaroo, euro, baboons, and rhesus macaques in which intensity and duration of female-female bonds may match or even exceed that of male-male bonds.

Evolutionary reviews of animal play can easily avoid the pitfall of advocacy. At least three working hypotheses (training, developmental rate, and cohesion) should be considered. Biologists might formulate and test competing predictions about sex differences, age differences, individual differences, and population differences in quantity and quality (including stability) of play and in solicitation, acceptance, refusal, prolongation, and termination rates, based on these three hypotheses. I consider training, developmental rate, and cohesion hypotheses equally orthodox. Without doubt, each hypothesis will prove valid in particular instances.

6

Play and
Life-History
Strategies

Play is primarily behavior of young animals, although free-ranging adults of some species [e.g., human, chimpanzee (van Lawick-Goodall 1968a), African lion (Schaller 1972), timber wolf (Mech 1966), and spotted hyena (Kruuk 1972)], as well as adult animals in zoos and adults of domesticated species, often play. Yet play appears to be virtually absent from the behavioral repertoires of animal species other than birds and mammals (Chapter 3, Meyer-Holzapfel 1956b, E. O. Wilson 1975). And, still more curiously, play may frequently occur in one population of a species, but may be completely absent in another population (Baldwin & Baldwin 1974). Moreover, as Susan Wilson (1973) has shown for the short-tailed vole (*Microtus agrestis*), play may only occur in particular generations of multivoltine species. In this chapter, I attempt to explain this puzzling variation in the occurrence and scheduling of play behavior by adopting an evolutionary approach to play within the framework of the ecological theory of life-history strategies. This population biological theory seeks to explain widespread variation among animals in number of offspring, age at sexual maturity, survivorship, longevity, and growth rate. The theory of life-history strategies (Cole 1954, Gadgil & Bossert 1970, Ricklefs 1977, Schaffer 1974, Schaffer & Gadgil 1975, Schaffer & Rosenzweig 1977, Stearns 1976, 1977, Wilbur, Tinkle, & Collins 1974) attempts to explain why different organisms allocate time and energy to maintenance, growth, and reproduction in such a variety of ways. The central assumption of life-history theory is the following: since "any organism has limited resources of time and energy at its disposal" (Gadgil & Bossert 1970), it is necessary to allocate or budget these resources carefully among survivorship, growth, and reproduction, the component processes of the life history, in order to maximize fitness.

The theory of life-history strategies provides a firm theoretical basis for evolutionary analyses of the development of behavior. It is potentially applicable to the study of play behavior because, for many species of mammals and birds, play is as much a part of the life history and the resource budget as survivorship, growth, or reproduction (Table 5-1).

My intent is to view play in life-historical terms in order to relate its age-specific effects to fitness. This approach makes it possible to model evolution of optimal age-dependent schedules of play behavior. These schedules are quantitative prescriptions of the fraction of total resources that must be allocated to play at each age in order to maximize fitness. It is conceivable that under some conditions the optimal schedule may not include play at all (zero resource allocation at all ages). Under such conditions the propensity to play is selected against and will tend to disappear from the behavioral repertoire of the animals in question. In this model, play behavior will be analyzed in the life-historical cost-benefit framework just described in order to obtain a theory that relates play to fitness and that leads to predictions about the amount of play during different stages of an animal's development, in environments that differ in various ways. The model to be derived is a formal restatement of the following simple and relatively uncontroversial, though still speculative, assumptions about play in animals. Play requires time and energy. An animal at play may risk physical or social injury. It will therefore be assumed, as suggested by my earlier discussion (Chapter 5) of costs of play, that play tends to affect immediate fitness negatively. For example, energy expended in play might instead have been allocated to feeding, growth, or maintenance. Time spent in play is time that cannot be devoted to nursing. Playing juveniles may fail to notice the approach of a predator. In a life-historical context, such negative effects would be expressed in terms of possible harmful direct influence on current survivorship, growth, and reproduction. This assumption rests on evidence that play is usually interrupted in the presence of immediate danger or when continued play would seriously threaten the animal. Factors that threaten immediate well-being also tend to depress play (Berlyne 1969): biting insects (Espmark 1971), inclement weather (p. 302), including extremes of wind and temperature (Rasa 1971), inadequate maternal care (Goodall 1965), introduction of an aggressive adult male (Bernstein & Draper 1964), overcrowding (Leyhausen 1973b), starvation (see below, p. 370), treatment with drugs that disrupt energy metabolism (Slonim 1972), and unfortunately for the comparative study of play, presence of a human observer (Flux 1970, M. Fox 1972b, Gartlan 1970, Ibscher 1967, Sugiyama 1965a). Sick, injured, or handicapped animals (including humans) are less likely to play than healthy animals (Autenrieth & Fichter 1975, Fedigan & Fedigan 1977, Garvey 1977, Hodl-Rohn 1974, Kritzler 1952, van Lawick-Goodall 1968a, 1971, Mason & Berkson 1975, Toepfer 1971). In the experience of veterinarians and pediatricians of my acquaintance, behavioral change, particularly failure to play, is the first and most obvious sign of illness in an animal that may otherwise appear very healthy. (Although I was unable to find scientific evidence for this widespread belief of clinicians, it agrees with my own experience and might merit formal study.) These and other observations suggest that quantity or quality of play may measure an animal's physical and emotional well-being (Cummins & Suomi 1976, Darwin 1874 p. 77, Geist 1978a, Harlow 1974, Hughes 1977, Mason 1965b, Poirier & Smith 1974b).

The converse of the above relationship between play and health has also been proposed. Play reappears as an animal recovers from stress or disease, and this correlation is consistent with observational (Fedigan & Fedigan 1977, van Lawick-Goodall 1971), clinical (Grantham-McGregor et al. 1979), and experimental (Cummins & Suomi 1976, Mason & Berkson 1975) evidence suggesting that social contact, by offering particular types of stimulation, actually aids recovery from forms of trauma ranging from severe malnutrition (due to protein and energy deprivation) to bereavement and social deprivation. In each study cited, play was one obvious component of the subjects' beneficial social relationship. Careful experimental work and systematic observations will be required in order to demonstrate that play experience itself actually contributed to recovery.

If play ever affects fitness positively, benefits resulting from play are likely to be delayed or cumulative rather than immediate. This statement is more complex than it sounds because the effect of any behavior on the expected number of surviving offspring involves many chains of causation and a combination of different time scales. Moreover, some play may actually lead to immediate benefits, although these effects are probably secondary. Rhesus macaques (Breuggeman 1978) and pronghorn fawns (Autenrieth & Fichter 1975) apparently use play to displace conspecifics. Mothers often use play as a distraction technique (Bruner & Sherwood 1976, van Lawick-Goodall 1967, 1971). A hungry chimpanzee played with another chimpanzee who had some food, distracted the playmate in this way, and grabbed the food (Menzel 1975). Mothers also apparently use play as a consolation prize when weaning their offspring (Clark 1977, S. Wilson 1974b). Actually, the basic delayed-benefit structure is still present for one of the two participants in each case cited above, and these accounts do not pose a serious threat to the delayed-benefits assumption. Each account has the form "X plays with Y and thus gains immediate benefit b." However, Y's gains from play are still delayed.

If in play the partner is bullied or intimidated, the effects of bullying or intimidation may still be long term. In fact, the only known case in which play directly resulted in the death of one of the players (Angst & Thommen 1977) occurred in the kra macaques (*Macaca fascicularis*) of the Basel Zoo. Captive adult males play with male infants. They do not play with female infants, but they sometimes kill them. Of the many cases of infant-killing reviewed, only one involved play. Atypically, the male played with one female infant, which he killed one day when several females attacked him as he played with it.

The delayed-benefits assumption approximates the effects of play that serves to modify developmental rate or to switch an animal from one pattern of development to an alternative evolutionary optimum. In these cases, the benefits begin immediately and last for the animal's entire lifetime. However, because future benefits are still present, optimal allocation schedules for these cases should be close to those predicted by a delayed-benefits model.

Even if benefits and costs of play occur at the same age, analysis of play needs to consider the entire life-span. Effects of play on immediate survivorship, growth, and reproduction may vary with age, either because the animal's play itself changes with age or because the impact of a given change in survivorship or fecundity is very different in animals of different ages. For illustrations of this point in the contexts of parental care and of aggression, see Trivers (1971) and Popp & DeVore (1979), respectively.

Under what circumstances can delayed benefits of play outweigh its immediate costs? These benefits must compensate not only for immediate costs, but also for the delay of reproduction to later ages that such life histories often entail. For this reason alone we would expect less investment in play in rapidly growing populations, since costs of delayed reproduction are greatest when population growth is rapid.

Two life-history models are presented in this chapter. The first is a simple model relating immediate costs of play and delayed benefits to fitness. This model gives some idea of the magnitude of benefits required if play is to evolve at all. The second model differs from the first by explicitly representing functional relationships among play, growth, survivorship, and reproduction.

Consider the economics of life-history trade offs in a sexually immature animal. With all other factors held constant, expected surviving offspring will decrease if juvenile survivorship decreases or if age of first reproduction increases. These effects are two possible costs of play. The second cost is not immediate, but would seem to make sense if, for example, play only becomes possible if the juvenile period is extended. Delayed benefits could include an increase in adult survivorship, an increase in adult fecundity, or a decrease in the age of first reproduction. A theoretical framework for this analysis requires precise formulation of questions like the following: If I decrease my juvenile survivorship by some amount as a result of play, how many additional future offspring must my play yield in order that my contribution to the next generation exceed that of a nonplaying animal whose chance of surviving to reproduce is greater and whose fecundity is less? Such questions are easily analyzed using life-history models (Caswell 1978, Cole 1954, Demetrius 1969, Emlen 1970, Hamilton 1966, Michod 1979).

In the simplest formulation, an animal has probability c of surviving to age at first reproduction j. During its reproductive life it gives birth to B offspring per unit time. Its probability of surviving to age $x > j$, given that it survives to age j, is given by the exponential $\exp[-\rho(x-j)]$, where parameter ρ measures the force of adult mortality. Fitness is measured by r, the intrinsic rate of population growth, defined by the equation (Wilson & Bossert 1971)

$$\int_0^\infty \ell_x \, m_x \, \exp(-rx)\,dx = 1 \qquad (6\text{-}1)$$

Substituting for ℓ_x and m_x,

$$\int_j^\infty c \, \exp[-\rho(x-j)] \, B \, \exp(-rx)\,dx = 1 \qquad (6\text{-}2)$$

This integral equation yields an implicit algebraic expression for r:

$$\frac{Bc \, \exp(-rj)}{r+\rho} = 1 \qquad (6\text{-}3)$$

Feasibility of various life-history trade offs can be measured by substituting appropriate values for the quantities in Eq. 6-3. For example, a large, long-lived vertebrate, such as elephants or humans, might have $r = 0.03$ year^{-1}, $j = 15$ years, $c = .5$, $B = 0.4$ year^{-1}, and $\rho = 0.1$. In this case, a 10% reduction in c due to play calls for an over 11% increase in B in order for play to evolve. Since an 11% increase in B increases total newborn offspring produced by an adult,

$$\int_j^\infty B \, \exp[-\rho(x-j)]\,dx = \frac{B}{\rho} \qquad (6\text{-}4)$$

from 4 to 4.4, play that decreases juvenile survivorship by 10% must only increase an adult's total expected newborns by less than one-half of an individual in order to evolve. This trade off does not appear unreasonable.

Because in theory play accelerates development, it is interesting to consider trading juvenile survivorship for decreased age at first reproduction. With $r = 0.03$, $j = 15$, $c = .5$, $B = 0.4$ and $\rho = 0.1$, a 10% decrease in c must be traded for an over 4-year decrease in j. It is difficult to imagine how play could single-handedly decrease age at sexual maturity from 15 years to under 11 years. An effect of this magnitude might be possible in a small mammal if social play facilitated contact with a phero-mone that accelerated development.

The trade off between juvenile and adult survivorship can also be analyzed using this model. With the parameter values specified above, ρ must be under .085 to compensate for a 10% decrease in c. This change means that yearly per cent adult survival must increase from 90 to 92%—not an unreasonable modification.

The model just discussed demonstrates that the hypothesized delayed benefits of play are probably feasible in an evolutionary sense. Physical training and skill development effects provide the necessary mechanism for such delayed benefits. Analysis of the model also suggests that a play-induced increase in developmental rate, expressed as a decrease in age at sexual maturity, must be substantial in order to evolve. No mechanism producing effects of this magnitude is known in large mammals or in birds. However, we might expect rodent or insectivore play to function at least partly in this way.

We may further explore the implications of this model by specifying functional relationships between play and its effects. Under the assumption that play is risky, juvenile survivorship c will be increasingly reduced as more and more resources are devoted to play, and we write $c = C(y)$, where y is the absolute amount of resources devoted to play during the juvenile period and $C(y)$ is some decreasing function of y. This function probably falls off more rapidly when resources are scarce than when they are abundant. For example, the impact of 10 resource units spent on play on total budgets of 1000 and 11 units would be totally different, even fatal in the latter case if successful predator avoidance cost two or more units.

Birth rate B will increase with y under the assumption that play training improves escalated fighting ability used in securing territory, achieving dominance, or protecting young. It will also increase with y if play improves physical ability used in feeding young. Adult mortality ρ will decrease with y under the assumption that play improves physical ability for feeding and/or predator avoidance in adulthood.

For the purpose of this analysis the value of j is fixed. Since $\partial r/\partial j = -r/[(r+\rho)^{-1}+j]$ by the implicit function theorem applied to Eq. 6–3, decreasing age at first reproduction will always increase r. From the expression for $\partial r/\partial j$ we see that if r is small, adult mortality low, and age at first reproduction high, a decrease in j will improve fitness very little. Such a decrease can even be expected to have a negative impact on fitness by decreasing available playtime during the abbreviated juvenile stage.

The implicit function theorem applied to Eq. 6–3 also yields $\partial r/\partial y = [\partial(Bc)/\partial y - \exp(rj) \quad \partial\rho/\partial y] \ / \ (\exp(rj) + Bcj)$. If a value of y can be found such that $\partial(Bc)/\partial y = \exp (rj)\partial\rho/\partial y$, fitness r will be maximized by that y (minima are ruled out by the functional forms chosen). These equations yield information about the optimal value of y.

As play becomes more effective in reducing adult mortality, $\partial\rho/\partial y$ decreases and the optimal y increases (Fig. 6–1). A decrease in total available resource or a decrease in juvenile survivorship will have an opposite effect, decreasing $\partial(Bc)/\partial y$ and decreasing the optimal y (Fig. 6–2). In sexually dimorphic species in which females can reproduce without play experience, but in which males require play to develop skills used in fighting for the opportunity to reproduce, $\partial(Bc)/\partial y$ will be less for females, and young males will allocate more resources to play than will young females (Fig. 6–3).

Each of these predictions is amply supported by existing information on play. Play will be most effective as training in adult survival skills when animals can play socially (Chapter 5), make informative mistakes, and learn from them. Such play requires that the animal have the capability of behaving in playful as well as nonplayful modes, of generating its own training sequences and of attending to them, of putting together and

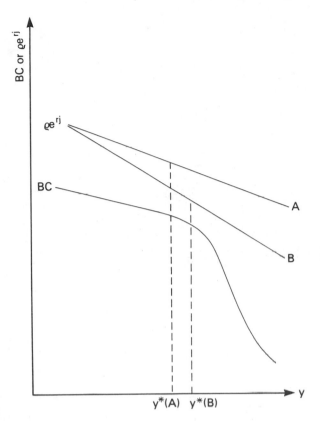

Fig. 6–1 Increase in play predicted with increase in effectiveness of play for adult survivorship.

"debugging" almost-right plans, and of varying its behavior in informative ways. Furthermore, social play requires that participants can compete successfully for play opportunities (Chapter 7). An increase in play with relative brain size is therefore expected.

If the leap from these requirements to relative brain size seems hard to accept, consider the storage and time requirements of computer programs that actually accomplish these tasks in artificial intelligence and in strategic and tactical systems analysis. I challenge any skeptic to implement such a simulation using a small programmable calculator.

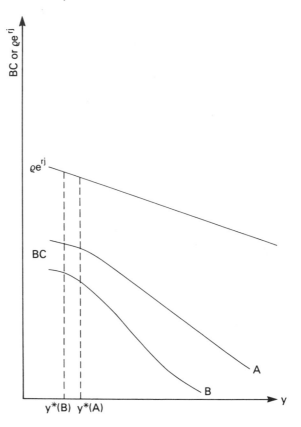

Fig. 6–2 Decrease in play predicted following decrease in resource availability or in juvenile survivorship.

The only valid information on this point concerns gross phylogenetic differences. Mammals and birds play, but individuals of virtually all other species do not. The reptiles and fish span the enormous brain-size gap between mammals, birds, and invertebrates (Jerison 1973). Occasional reports of play in these two groups are available (Chapter 3). Interactiveness, stability, elaborateness, frequency, and intensity of play apparently increase within the orders Mammalia and Aves in parallel to the size of the cerebral hemispheres, or surface area of cerebral cortex, relative to body size or brainstem mass. I stress that this increase may only be apparent. The required quantitative com-

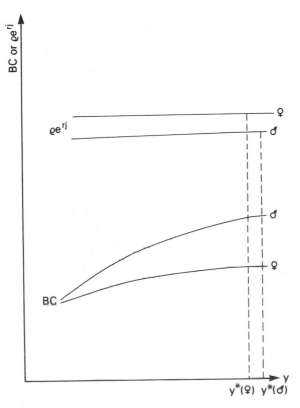

Fig. 6-3 Differences in play predicted in sexually dimorphic species when males benefit more from play.

parisons have not been made. Species (rather than environmental) differences in these parameters may only be found when grossly different species are compared. For example, contrast human and chimpanzee play with the play of rabbits, house mice, or meadow voles (Chapter 3). This effect is particularly striking in birds. Portmann & Stingelin (1961) define a hemispheric index in order to analyze relative mass development in the avian brain. Most birds for which this index can be calculated yield values of 2 to 6. The macaw *Ara ararauna* has massively hypertrophied cerebral hemispheres (index = 28.02). Next on the list is the common raven (*Corvus corax*) (18.95). Most other birds known to have highly developed cerebral hemispheres, according to Portmann & Stingelin's index, are either corvids or raptors. This information precisely mirrors the known prominence of play in different bird species. Quiring (1941) presents indices of brain development in birds based on brain weight–body weight relationships. Of the small number of bird species measured by Quiring, the common crow (*Corvus brachyrhynchos*) was by far the most highly cerebralized.

Marsupial-placental differences in play also reflect differences in relative brain size. The grazing macropodids of Australia show the lowest brain-body weight ratios of any terrestrial herbivore (Eisenberg 1975). So far as is known, play of these marsupials is simpler and less frequent than play of grazing rodents, ungulates, or primates (Chapter 3). The carnivorous Dasyuridae, of Australia, show the highest level of brain development (Eisenberg 1975) and the most elaborate play (Chapter 3) known in any marsupial, but dasyurid carnivore brains are not as developed as those of eutherian carnivores in comparable niches in the main continental areas (Eisenberg 1975). Dasyurid-eutherian carnivore differences in play are not nearly so obvious as apparent differences between play of grazing macropodids and grazing ungulates.

The model predicts that reduced total resource availability will decrease the proportion of resources allocated to play. Primates and ungulates deprived naturally or experimentally of food, or forced to work harder for a fixed food ration, respond with an immediate decrease in play (Baldwin & Baldwin 1972, 1974, 1976a, Dasmann & Taber 1956, Geist 1971, 1978a, Hall 1963, Loy 1970, Marsden 1972, Richard & Heimbuch 1975, Rosenblum, Kaufman, & Stynes 1969, Schaller 1972, 1977, Southwick 1967, Zimmermann, Geist, & Ackles 1975). A result that the model cannot explain is that growing domestic dogs, fed ad lib in the laboratory, overate and appeared to play less than dogs on a regular meal schedule whose caloric and nutrient content was adequate to ensure normal growth (Hedhammar et al. 1974). Overnutrition in the experimental dogs resulted in rapid growth and abnormal skeletal development. The overfed dogs could not move or be touched without exhibiting behavioral signs of pain. Aside from this experiment, whose biological (as opposed to clinical) relevance is unclear, I am not aware of any data suggesting that animals play less when resources are abundant.

Play also decreases when juvenile survivorship decreases or when threats to juveniles are present, as predicted by the model (Fagen 1977). As mentioned earlier in this chapter, one such threat, the presence of a human observer, represents a potential biasing effect in phylogenetic comparisons of play in different species. For example, because non-group-living, forest-dwelling species, unlike group-living species in open country, do not have social predator defenses or the opportunity to detect predators at great distances, a human at a given distance will represent more of a threat to survival in the former species, whose play will be depressed more. [However, animals can be conditioned to play, as in the case of zoo animals fed by the public as a reward for their play. One result of such conditioning is that certain individuals may play *only* when a human observer is present! Gehring (1976) gives a possible example of play conditioning in a confined brown capuchin (*Cebus apella*).]

Sex differences are evident in play of many species. In many group-living primates, for example, male play is more frequent or more vigorous or both than female play (Baldwin & Baldwin 1977, Symons 1978a). In Siberian ibex (*Capra ibex*) (Byers 1977), in black-tailed deer (*Odocoileus hemionus*) (Linsdale & Tomich 1953), in Norway rats (*Rattus norvegicus*) (Olioff & Stewart 1978), and in golden hamsters (*Mesocricetus auratus*) (Goldman & Swanson 1975), females play as if training for predator avoidance, males for intraspecific escalated fighting. It is important to stress that these differences are only quantitative. Females often wrestle and males often chase. Female rhesus chase as frequently, but not as vigorously, as male rhesus (Symons 1978a). Certain individuals exhibit play styles typical of the opposite sex. The model suggests that these sex differences in play should be present only when fighting skill is differentially important in males and in females. The model predicts relatively few sex differences in carnivores. Both male and female carnivores must display physical skill in feeding. Females, as well as males, hold and defend territories, and defend and train offspring. Play is not less frequent or less intense in females than in males in many carnivore species: domestic cats (*Felis catus*) (Barrett & Bateson 1978) and other felids (Lindemann 1955), golden jackals (*Canis aureus*) (Roberts 1977), domestic dogs (*Canis familiaris*) (Aldis 1975 p. 106), dwarf mongoose (*Helogale parvula*) (Rasa 1977), and four species of North American canids (Bekoff 1974c, Hill & Bekoff 1977). Zimen (1972 p. 228) suggests that play of male timber wolf cubs (*Canis lupus*) may be less frequent than that of females. Implications of the model for pair-bonding or group-living carnivores, such as African lions (*Panthera leo*), African wild dogs (*Lycaon pictus*), spotted hyena (*Crocuta crocuta*), timber wolf, and red fox (*Vulpes fulva*), depend on the details of male-male and female-female competition in these species. The weasels would be a particularly interesting group in which to study sexual dimorphism in play, since physical dimorphism is marked in weasels, but functional requirements placed on skill in the two sexes are similar to those of other carnivores, not to those of sexually dimorphic primates.

Human social playfighting and play-chasing appear to exhibit the same pattern of sex differences found in other primates (Aldis 1975, Blurton Jones 1967, 1972 p. 109, Blurton Jones & Konner 1973, P. K. Smith 1977), as expected from the current sociobiological hypothesis that male humans have been selected for strength and skill in hunting large game and as intraspecific fighters (E. O. Wilson 1975). However, this difference appears valid only for wrestling and chasing. Human sex differences in sociodramatic and fantasy play appear to be minimal in young children (P. K. Smith 1977). Chimpanzees, unlike humans, are not yet reported to exhibit significant sexual dimorphism in play. Chimpanzee fighting ability is important in male-male resource competition, especially in competition between troops. Will juvenile male chimpanzees be found to playfight and play-chase more often or more vigorously than juvenile female chimpanzees, as these observations and the model suggest?

Differences in frequency and, to a lesser degree, in style between play of juvenile male and female rhesus macaques led Symons (1978a) to suggest that play served as physical training and that males played more because skill in escalated fighting was more important for a male's reproductive success than for a female's. Levy (1979), in an independent and equally detailed field study of rhesus play, replicated virtually all of Symons's results, observed the same patterns of sex differences, and concluded that since females also playfought as juveniles and fought damagingly as adults, whereas within the male sex play was most frequent and stable among those males who appeared to form long-term social bonds, rhesus play evolved not to develop fighting skill but to develop cooperative social relationships.

Additional information on primate sex differences in play further illustrates the complexity of this issue and the multiplicity of possible interpretations. Infant female Japanese macaques may actually play-fight more frequently than infant males (Mori 1974). The higher frequency of play-fighting in juvenile male macaques may merely reflect differential development. Perhaps female primates of many species must concentrate their play (and train their fighting skill) at an earlier age than do males because they become sexually mature at an earlier age and because while they are still immature they must devote large amounts of time to interactions with infants in order to develop mothering skills. Indeed, perhaps female primates of some species playfight less often as immatures because they fight in earnest more frequently as immatures (as is the case in rhesus macaques, Levy 1979) and therefore require less training.

In captive talapoins (*Miopithecus talapoin*), adult females socially dominate adult males but juvenile males were observed to play more frequently than juvenile females (Wolfheim 1977). Analogously, in captive family groups, female lion tamarins (*Leontopithecus rosalia*) exhibit greater intrasexual aggression, while juvenile males "appear to be more assertive in play bouts, doing more wrestling and chasing than female siblings" (Kleiman 1979). In both cases, juvenile males are the playful sex in species in which females appear to have been differentially selected to exhibit skill in aggressive competition, and in which long-term social cooperation occurs within each sex and between sexes. Therefore, although the topic of sex differences in play is undeniably fascinating and evolutionarily interesting, and although it offers a seductive and fashionable blend of sex, aggression, play and hormonal control of behavior, I find no reason to give this topic special status in the scientific study of play. Sex differences in play are as complex and multifactorial, and as difficult to explain simply, as any other intrapopulation difference in play.

I formulated a more elaborate life-history model (Fagen 1977) in order to answer some additional questions about evolution of play behavior. These questions were:

1. Can alternative patterns of development, one including play and one without play, evolve in the same environment?
2. How does evolution of play in the model depend on the functional relationships by which resources allocated to play are translated into changes in growth, survivorship and fecundity?
3. What forms do the model's optimal play schedules take as a function of age? Are there "optimal periods" (Lancaster 1971) for play?

Three series of experiments were carried out. In the first series, a two-age life-history was used and the functions of the model were systematically varied over a wide range of functional forms. Values of fitness maxima were found by exhaustive visual inspection of tables of fitness values for candidate life histories. Results of this series of experiments shed considerable light on questions (1) and (2) above, but did not illuminate question (3), since in these two-age simulations, play could occur during, at most, one age of the life history. Using this procedure it was, however, possible to carry out numerous experiments rapidly and inexpensively. In order to investigate question (3), longer life histories were generated using various parameter and function values. The computational cost of these more elaborate analyses was considerable, and a second series of experiments, in which all optima were found for a set of three-age life histories, had to be truncated, as it would have been prohibitively expensive to carry out the entire series. In an earlier series of experiments, the computer program was stopped after finding a single (not necessarily unique), optimal five- or six-age life history. This third series of experiments serves to extend and confirm the results of the second series with respect to question (3).

Results of numerical experiments performed with the model (Fagen 1977) are as follows.

1. Alternative developmental patterns are extremely common. Of 216 experiments carried out, 166 (77%) resulted in several such alternative stable equilibria. These multiple optima are also common in the three-age life histories. Multiple optima are frequent both when play evolves and when it does not evolve. In a total of 432 experiments with two-age life histories, play evolved a total of 169 times, and of these 169 experiments 112 (67%) resulted in multiple optima. These experiments typically revealed two alternative optima, one having play at age 0, but no reproductive effort at age 0, the other having reproductive effort at age 0, but no play at age 0. In some cases three distinct optima were found.

2. Natural selection may act to remove play completely in an optimal life history: 77 of 216 two-age experiments in which play was relatively safe yielded no optima containing play; 186 of 216 experiments assuming high risk from play saw play eliminated from the life history. Three-age and longer optimal life histories also gave this result in a number of cases.

3. Both in the two-age and in the three-age life histories, concavity of the survivorship cost function for play tended to favor the evolution of play much more strongly than concavity of the birth cost function or concavity of the play profit function. Results with sigmoidal functional forms tended to fall between these extremes.

4. Play may be present only at age 0, present only in juvenile stages, or present at all ages except the last age. Its schedule may be monotone decreasing, but is never monotone increasing. If play is present at all, it is present at age 0. In general, play tends to occur more frequently and at a higher level in young animals in the model life histories.

The model has four biological implications: optimality and nonuniqueness of the play strategy, possible elimination of play by natural selection, existence of "optimal periods" for play, and appearance of play early in ontogeny.

In the model systems considered, play evolves as an optimal but not necessarily unique adaptive strategy in certain environments. Play cannot be dismissed simply as an artifact or an epiphenomenon. In the same model environment and species, two or more alternative optimal life-history strategies may evolve. Some of these include significant amounts of play. Others include none at all. Thus we cannot always expect play to evolve in all local populations of a given species in nature. For mammals, play may effectively solve certain adaptive problems, but in some environments play will not be the only possible solution. That certain populations within a species will exhibit play, whereas others will not, is expected from the theory presented above. It would be incorrect to argue that the playful animals comprising one such population were any less fit than the nonplayful animals making up another. Nor does the absence of play in certain populations of a species mean that play is not highly adaptive in other populations of the same species inhabiting the same environment. In general, it will be impossible to predict which of a set of multiple optima will evolve in a given population. Random or historical causes will often determine the outcome.

Schaffer & Rosenzweig (1977) discuss the problem of multiple optimal life-history strategies. They argue that in many species there are mechanisms that permit populations to switch directly between adaptive peaks. Play may simlarly be switched on and off in response to environmental variation.

Play does not always evolve in model life histories, even though it is always assumed to have some (delayed) beneficial effect. When costs of play are excessive, natural selection eliminates play behavior completely from the repertoire of model species. In some cases moderate changes in a single environmental parameter will again permit play to evolve, suggesting that, in nature, temporal variation (for example) could alternately favor playful and nonplayful strategies. Such variation on a seasonal basis amounts to the sort of predictable environmental change that, according to Levins (1968), would select for animals having a switch mechanism, i.e., the capacity to express a playful or a nonplayful phenotype depending on conditions. A possible example is the short-tailed vole (*Microtus agrestis*), in which play of the young appears to be a seasonal phenomenon, occurring in spring litters but not in fall litters (S. Wilson 1973).

The results reported here clearly demonstrate the existence of "optimal periods" for play in my model of ontogeny. In a typical life-history strategy obtained in the numerical experiments, the amount of time and energy devoted to play in the youngest age-class was nine times that allocated to play at any other age. For all numerical experiments conducted with five-age life histories, play attained its maximum value either at the youngest age (age 0) or at the next age (age 1). This result is entirely consistent with existing data on age specificity of play in animals (Baldwin 1969, Bernstein 1975, 1976, Bramblett 1978, Cheney 1978, Geist 1971, van Lawick-Goodall 1968a, Merrick 1977, Nash 1978, Owens 1975a, Poole 1966, Rosenblum 1968, Symons 1978a, S. Wilson 1973, West 1974, Zimen 1972). These data generally exhibit an initial steep rise in frequency of play, followed by a gradual decline with age. Some interesting exceptions to this rule result from stressful changes in mother-infant relationships. Female mammals may reject their offspring during a temporally limited period of parent-offspring conflict over amount or type of parental investment in the young (Trivers 1974). Often access to milk is contested, but maternal rejection may represent conflict over other resources, such as the opportunity to be carried on the mother's ventrum (J. Altmann 1980). In chimpanzees (Pan troglodytes) (Clark 1977), anubis baboons (Papio anubis) (Nash 1978, Owens 1975a), chacma baboons (Papio ursinus) (Cheney 1978), and domestic sheep (Ovis aries) (Sachs & Harris 1978), play decreases in frequency and/or intensity during the period of parent-offspring conflict over parental investment, then increases to roughly its original level after the period of conflict ends. [Demographic factors may be responsible in part for the observed decrease; see Cheney (1978) for an example.] These bimodal schedules are predicted by the model. Temporary decreases in play occur in model life-history schedules at ages when play would represent a particularly heavy drain on the resource budget (Fagen 1977).

Play, if present at all, always occurs at the earliest age of the model life histories. This property of my model is consistent with data on precocial mammals, but it is not the case in nature that altricial mammals begin to play immediately following birth. Rather, play begins when the young animal acquires the strength and coordination to exhibit adult locomotor patterns (Fagen 1976a). This process involves developmental changes in fiber types in striated muscle (Burleigh 1974, Close 1972). To predict the optimal scheduling of this developmental event would require an evolutionary theory of behavioral ontogeny that included not only play, but also other aspects of development, such as nursing, perception and cognition, and physiological change.

Perhaps the earliest known age at onset of play is that of wildebeest (*Connochaetes taurinus*) and caribou (*Rangifer tarandus*). These migratory, group-living ungulates give birth to highly precocial young that often run and frisk around their mothers within an hour or so of birth (Estes 1976 and pers. comm., Müller-Schwarze pers. comm.). Apparently, precocity and early play form part of an antipredator ontogenetic strategy. This strategy also involves synchronized births and rapid early development (Estes 1976).

Age at termination of play is a highly variable phenotypic characteristic. In the second model in this chapter, play ceases around the time an individual first reproduces because play at successively later ages yields successively fewer cumulative benefits and because resources devoted to reproduction are more effective for producing surviving offspring than are resources devoted to adult play. Many proximate factors and various mechanisms are involved in these changes. Increases in individual size, strength, and weaponry may make social play breakdowns more costly (S. Altmann 1962a). The impact of falling, making sharp turns, leaping in the air, and similar high-acceleration behaviors may result in injury to a large, heavy animal in play. The same activities would be less likely to injure a younger, lighter animal. As they mature, female primates of many species devote increasing amounts of resources to grooming others or to observing and trying to handle other females' infants. Both of these cases probably represent substitution of one delayed-benefit behavior (grooming, infant-oriented activities) for another (play).

The models in this chapter were not designed to predict differential allocation of resources among competing delayed-benefit components of the phenotype. For example, energy used in play could instead be used for growth or for fat deposition (Fagen 1977, Müller-Schwarze 1978a p. 382). The observation that a given calorie of fat can only be burned once, whereas a given calorie devoted to play that develops skill in capturing prey may save calories every day, suggests that fat storage is a specialized adaptation to predictable future food shortages occurring during resource crashes or during periods in which no time can be devoted to feeding. [Cost of fat deposition must similarly be considered, for each calorie of fat stored produces at least one additional calorie of heat loss (Blaxter 1960). See Pond (1978) for further information on vertebrate fat deposition and its consequences.] Play prepares the animal for temporally unpredictable events predictably requiring skilled behavior.

Ontogeny of play is not completely specified by the model. An increase in resources devoted to play at a given age could produce more vigorous play without leading to any change in frequency or duration. Play might even become more interactive or more elaborate as a result of the additional resources invested in it. Players could change their probabilities of play solicitation and acceptance or their willingness to self-handicap. These qualities of play are reflected only implicitly in the shapes of the cost and benefit functions.

What about such gross generalizations as "social species play more"? Ambiguous as phrased, this statement could be taken to imply that individual young of species forming large permanent adult groups play more frequently or more vigorously or invest a greater percentage of their total resource budget in play in some other way than do individual young of species whose adults do not form such groups. No aspect of the model would be influenced in obvious fashion by adult group size or permanence. One remote possibility is that selection for greater encephalization is a characteristic of particular types of social evolution. If so, these differences in encephalization would be present from the earliest age through prolongation of fetal brain growth (Gould 1977). More complex, stable or efficient play would then be possible, but this result would be a consequence of encephalization, not of social organization as such. It would be expected under whatever circumstances a more highly encephalized phenotype proved more adaptive.

The life-history model of play is potentially important in many areas of play research. Perhaps its most important application is in design and interpretation of manipulative experiments on play. Each simulated life history generated by the model is, in effect, an experiment. Effects of environmental manipulations can be simulated by changing appropriate aspects of the model. For example, an experimenter could double the risk of play at selected ages and measure the resulting change in amount of play in the optimal life history corresponding to the increased risk. This type of manipulation represents the typical play deprivation experiment, in which the experimenter seeks to alter the amount of play through environmental manipulation. However, computer models also permit a second kind of experimental manipulation that has, so far at least, proved as impossible in practice as it is desirable in theory. In this paradigm, the simulated animal is deprived of play without depriving it of anything else. The model allows this procedure to be carried out very straightforwardly, as follows. Suppose the optimal amounts of play at ages 1, 2, and 3 in a given environment are P_1, P_2, and P_3, respectively. We can artificially set the quantity of play that actually occurs at age 1 to 0 by appropriate programming and then see how the simulated animal responds to this play deprivation by asking the computer to calculate optimal quantities of play at ages 2 and 3, given the new quantity (0) of play at age 1.

Computer models make it possible to compare different experimental paradigms, to predict the results of experiments scientists cannot yet perform and to compare these results with those of simulated experiments based on paradigms actually in use, and to clarify underlying conceptual issues. I am confident that the results of simulations of play deprivation will provide logical explanations for the empirical fact that results of experimental play deprivation depend on the paradigm, manipulation, species, and environment used (Baldwin & Baldwin 1976a, Chepko 1971, Müller-Schwarze 1968, Oakley & Reynolds 1976, Rosenblum, Kaufman, & Stynes 1969).

Models discussed in this chapter suggest that many different factors can affect the amount of resources devoted to play at a given age. Accordingly, many explanations of age, sex, envi-

ronmental, and species differences in the amount of play observed will always be possible in theory. Benefits of a given
amount of resource invested in play depend on the relationships
between resources invested in play and increments in ability and
on the relationships between increments in ability and increments in components of fitness. For example, some abilities,
such as innovative skill (Chapter 8) or skills related to dispersal
(Geist 1978a), might be beneficial in certain environments and
harmful in others. The importance of components of fitness
further depends on the animal's age and on its overall life-history pattern. Costs of a given amount of investment in play
depend on the total amount of resource available, on the negative effects of play on growth, survivorship, and reproduction,
and on the importance of these components of fitness. The
forms of the cost functions measuring negative effects potentially depend on total resource availability, on availability of opportunities for play, on the activity of predators and competitors, on the riskiness of the physical environment, on the
playing animal's state of health and well-being, and on the animal's age. Furthermore, both fitness effects of ability and importance of components of fitness will depend on the social environment, including other animals' life-history tactics and the
player's distribution of genetic relatedness to the animals in its
environment. Because of this multiplicity of factors, any model
of optimal qualities of play involves judicious simplification.
Sensitivity analysis may be useful in identifying a few key variables expected to be important in particular situations. For example, numerical sensitivity analysis of one model discussed
above (Fagen 1977) suggested that the most important determinant of the amount of resources devoted to play was the survivorship cost of resources allocated to play behavior.

To restate a difficult "why" question by asking "why not"
may seem playful or even flippant, but this intellectual gambit
furnishes novel perspectives on many problems, including the
problem of play. Why do certain animals or kinds of animals
consistently fail to play? Three instances of this phenomenon
merit special attention. The first two cases, discussed previously
in this chapter, are (1) a decrease in play frequency with increasing age leading to relative rarity of play in adults of most

species (see also p. 377 and Levy 1979) and (2) virtual absence of play in species other than mammals and birds (p. 218). The third case, fully as puzzling as the two previously cited, is that some healthy, vigorous, and even socially dominant juveniles seldom if ever play, have difficulty soliciting play, and are rarely asked to play (Bekoff 1974c, C. C. Hamilton 1968, Eaton 1974 p. 149, Rasa 1977, Symons 1978a p. 64). These cases seemed unlikely, both in terms of the physical training hypothesis and in terms of the cohesion hypothesis, and they appeared to be isolated, if interesting aberrations (Fagen 1978c). However, an informal survey of field mammalian ethologists at the XVIth International Ethological Conference in Vancouver (1979) suggested otherwise. In nearly every case, the ethologists I questioned had seen this phenomenon in "their" species, even though their research did not emphasize play and quantitative data were not collected or not yet analyzed. This surprising result suggested to me that evolutionary biologists might not always have asked the right questions about ultimate causation of play. Instead of merely investigating the reasons why animals play, we might also ask ultimate and proximate questions about healthy immature individuals who do not play even when littermates, peers, or group-mates do.

The number of possible explanations for failure to play is surely as great as the number of possible explanations for play itself. Such complexity calls for formal analysis. The life-history models of this chapter and the simplified causal network of Fig. 6-4 suggest the kinds of analysis required. An animal may fail to play because beneficial effects from play are impossible in a given environment, because the benefit of play is too small, because the cost of play is too large, because other beneficial effects are preferable at that age (resource budget constraints), because resources required for play are preempted by other ages of the animal's life history (life-historical constraints), or because of historical constraints. Furthermore, the pattern of phenotypic effects of behavior and the signs and magnitudes of these effects, as well as their diverse consequences for survivorship and fecundity, will vary from environment to environment. If an animal inhabits a patchy environment, the arrows, signs, and implicit magnitudes in Fig. 6-4 might differ in

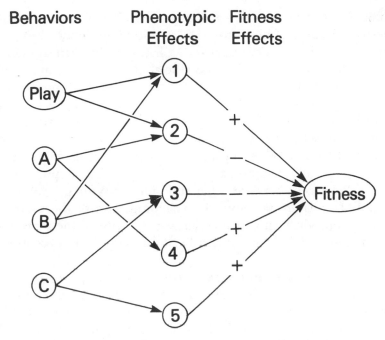

Fig. 6–4 Diagram of factors influencing occurrence of play.

different types of patches. They are also likely to change with age.

1. Effect Impossible. At a particular age, physical (including phylogenetic) constraints may prevent play from having a given type of beneficial effect. Benefits may be impossible if the animal has not yet mastered essential information or if essential features of the physical environment are absent. Demographic constraints, such as inadequate quantity or quality of partners, may preclude benefits from play.

2. Effect Possible, but Too Small (Fig. 6-4, effect 1 negligible). Benefits of play may exist, but they may be so small that they are unimportant. All physical training programs (including skill development), for example, seem ultimately to reach a point of diminishing returns. No phenotype is infinitely plastic. Levy (1979) discusses this law of diminishing returns as a possible explanation for a decrease in play in older juvenile rhesus macaques. Life-historical considerations may ultimately explain the above phenomenon, as has already been argued in general for senescence. Any beneficial effect of play may yield diminishing returns with age, at least in expanding populations, even when the effect itself (measured, for example, as per cent increase in maximal aerobic capacity or in isometric strength of a given muscle) is the same at all ages. The reason for this demographic law of diminishing returns is that an individual's greatest reproductive impact on an expanding population is more likely to occur when that individual is young. However, all populations are not always expanding, and furthermore, as delayed reproduction and survival past the end of the reproductive period both indicate, this rule admits significant exceptions.

3. Effect Possible, but Cost Too Large   (Fig. 6-4, magnitude of effect 2 exceeds magnitude of effect 1). This reason for absence of play is probably the most frequently cited in the literature. It explains the sudden decline in play in bighorn lambs (*Ovis canadensis*) following a one-week period during which lambs were observed to collide eight times with spiny cacti (*Opuntia* spp.) (Fig. 5-1), and during which effects of heat stress and nutritional stress were insufficient to account for the subsequent decrease in play (Berger 1980). It explains short-term absence of energetic, vigorous play in food-deprived or temperature-stressed animals. It could also explain a decline in play frequency with increasing age. Levy (1979), for example, argues that the cost of play in juvenile male rhesus macaques increases with age because play comes to resemble fighting in older animals. Through this resemblance, play is more easily mistaken for fighting, and the danger of physical injury through escalation also increases.

If a beneficial effect produced by play can also be produced in larger amounts or at lower cost by an alternate behavior (e.g., B in Fig. 6-4), play may be temporarily or permanently replaced by a behavior that is a means to the same end.

Adaptation to a changing environment involves costs as well as benefits. If play is beneficial in one type of environment, but not in another, play may fail to evolve at all because of these costs, including (1) the cost of ascertaining whether play's benefits are likely to exceed its costs in a particular environment, (2) the costs of incorrect decisions (to play or not to play), (3) the costs of switching to and from a playful phenotype, (4) the cost of switching to play when the environment will change again before play's benefits can be realized, and (5) the analogous cost of switching to alternative, nonplay behavior in an environment that changes too rapidly for the animal to track.

4. Other Effect at Same Age Preferable (Fig. 6-4, behavior C). If play is beneficial, but not necessary, another beneficial behavioral effect may take precedence over beneficial effects of play in the animal's resource budget, and the animal would then use its time, energy, or other limiting resource for behaviors other than play. Levy (1979) indicates that older juvenile female rhesus macaques may play less frequently than their male peers, due in part to these females' attraction to infants. Time spent in nonplayful practice of the skills of infant interaction may be more beneficial to females of this age than time spent in play with other juveniles.

5. Other Ages Have Priority. Life-historical constraints may result in absence of play at a given age if ontogenetic programs do not allocate sufficient resources to play at that age, as discussed earlier in this chapter. As a consequence, costs of play may be particularly high at a given age (see 3, above).

6. Historical Constraints: Multiple Optima. As discussed above, multiple life-history strategies can, in theory, evolve by natural selection in different populations in the same environment. Some of these strategies may lack play at all ages, whereas others include it. However, if individuals can switch to the highest peak on the fitness surface of life-history strategies (conditional on their life history to date, of course) whenever the environment changes, this explanation for absence of play in entire species or higher taxa becomes less compelling, since at least some portions of the taxon's range will sometimes favor play.

The foregoing discussion suggests a number of reasons for

386 | Animal Play Behavior

animals' failure to play. These arguments explain why an animal that can play does not do so and how play may be potentially beneficial to an animal and yet fail to occur. Because animals evolve in temporally and spatially heterogeneous environments, each environment associated with a particular pattern of qualitative and quantitative effects on the phenotype and on fitness, and because adaptation to a changing environment suggests a number of non-intuitive costs and constraints (Levins 1968), the question "Why do animals fail to play?" involves interesting complexities.

The problem of play's ontogeny resembles problems of population and community ecology in that both have resisted unaided intuition and unsophisticated theory. Simple verbal explanations of differences in quantity of play have enjoyed legitimate success in the past due to a combination of biological intuition, intelligent choice of problems in which a few significant biological relationships were evident from the outset, and great good fortune. Because it is interesting to predict age, sex, and environmental differences in quality (Chapter 7) as well as in quantity of play, the models discussed in this chapter are by no means the only possible models of ontogenetic changes in play behavior. Future formal theoretical analyses will surely address the problem of ontogenetic changes in directly measurable characteristics of play (e.g., stability, elaborateness, reciprocity, intensity) in a more straightforward manner than do the models presented here.

7

Biology of
Social Play

Social play at its cooperative best is a biological showpiece. Evenly matched and closely related partners cooperate in apparent mutual physical training and skill development. Their play is non-injurious. It does not harm their social relationship and may even strengthen long-term prospects for their cooperation. An older individual may play altruistically with its younger sibling. Special communicative signals and stabilizing techniques ensure that play is fair to both participants.

Play between individuals is less idyllic when partners are not evenly matched, are not close genetic relatives, and can exploit play interactions for ulterior purposes. Social manipulation, cheating, bullying, and intimidation may then become the norm.

Social play reflects biological adaptation. Like other aspects of social behavior, it has been selected to adjust to a broad range of environmental conditions in the service of inclusive fitness. Therefore, it is pointless to claim that play is either essentially cooperative or essentially competitive. To a subjective human observer, the natural history of play has its bright side and its dark side. This fact would be of no scientific importance except that observers whose subjectivity derives from their assumptions about social behavior tend to see only one side of the phenomenon.

A sociobiological approach to animal social play assumes that each individual behaves in its own genetic self-interest. It follows that an individual's tendencies to initiate, maintain, and terminate play are all products of natural selection. Accordingly, individuals will behave so as to play in those ways that, and with those partners who, contribute most strongly to that individual's inclusive fitness. Because individual preferences for play styles and play partners may not always coincide, a certain amount of conflict over play is theoretically possible. This chapter will explore the concept of an individual's biological interest in play and the social consequences of similarities and differences in individual interests.

The study of social play reflects diverse theoretical perspectives. Past hypotheses viewed play as pure cooperation for practice and skill development, as a mechanism for informing individuals of their relative fighting abilities and dominance ranks, or as a superorganismic mechanism for establishing and strengthening social bonds and social cohesion. These hypotheses were generally based on concepts of group selection. As previously argued (Chapter 5), none of these arguments is entirely consistent with natural selection theory or with observed characteristics of play. With the advent of sociobiology, attempts to interpret play as damaging competitive behavior have similarly been unsuccessful. No hypothesis yet proposed explains how social play behavior could in itself represent an evolutionarily stable strategy of damaging competition for current or future resources (Chapter 5).

Group-selection, superorganism, and damage hypotheses all fail to recognize basic evolutionary theory. Analysis of social play begins by considering some implications of these insights.

COOPERATION AND COMPETITION IN SOCIAL PLAY

Current understanding of the functions of animal play suggests that individuals play in order to obtain physical training, to train cognitive strategies, and to develop social relationships (Chapter 5). (No conscious purpose is implied, of course.) Play with another individual may be as effective or more effective for physical and cognitive training than play alone. However,

no matter what function of play is at issue, the interests of any
two individuals in play will rarely coincide. Each animal should
have its own requirements for experience in play. The nature of
optimal play interaction from the individual's point of view
depends on that individual's size, sex, developmental level, and
past experience. For example, a training program optimal for
developing juvenile monkeys' chasing and fleeing ability would
be far too strenuous, difficult, and dangerous for infants. The
same program would not stress or challenge adolescents adequately. Furthermore, two individuals might require qualitatively different kinds of training. One might need to practice
chasing and fleeing, the other wrestling. Finally, animal A
might benefit from a cooperative social relationship with B, but
B might not similarly benefit from A's cooperation. Thus A
would seek to develop its relationship with B through play, but
B would resist A's initiatives.

Different individuals' needs and requirements in play do not
inevitably agree. When such disagreement exists, a "conflict of
interest" is said to occur over play. (Note that even identical
twins' interests in play might conflict in this sense.) Conflict
stems from the fact of individual genetic and phenotypic differences. Social theory predicts that the ways in which animals
manage this conflict will differ depending on the animals' genetic relatedness, on the magnitude of the differences between
individual optima, and on the social environment.

Conflicts over social play may also arise when potential
partners disagree about type of play, or time or place at which
play occurs. For example, two animals of the same age, size,
and sex may both benefit maximally from moderate intensity
wrestling, but if one animal has just fed or has not played for a
long time, and the other is about to feed or has just played, then
the second may resist the play solicitations of the first. Or, since
mothers may intervene in play if their offspring vocalize in distress, each individual will prefer to play at a closer distance to
its own mother than to its playmate's mother (Fig. 7-1).

Just as in the case of sexual reproduction (Davies & Halliday
1977), any social play interaction necessarily involves a compromise between partners' differing optima. It is this need to compromise that makes play a challenging form of social coopera-

Fig. 7–1 Is the location of a playfight potentially subject to conflict? These Himalayan langur infants (*Presbytis entellus*), aged 1 year and 6 months, respectively, are playing near the younger infant's mother, an advantageous position for the six-month-old. Melemchi, Nepal. (John M. Bishop)

tion. For example, suppose that A and B differ in preferred intensity of play. For purposes of illustration, assume that at A's preferred intensity, A gains 5 units of individual fitness, B 2. If B's preferred intensity prevails, A gains 3 units, B 4; A and B must compete for the resource of play itself. They may settle their conflict by escalated or non-damaging fighting. They may compromise by playing at some intermediate intensity value or by playing at each individual's preferred intensity part of the time. Or they may not play at all, particularly if the cost of resolving their conflict exceeds the benefit expected from play. But many additional factors must be considered. If A and B are genetically related, each should consider the effect of play on its own inclusive fitness. A third animal, C, whose interests in play coincide with those of A, may be temporarily unavailable: should A wait, and should B concede more to A? How much is it worth not to lose a playmate to a third animal? Other animals in the population may already have reached a compromise that yields benefit 2 to each partner, in which case A and B are unwise to refuse to play together and any compromise at all would be preferable to not playing. Suppose B's older sibling can be enlisted by B at minimal cost to attack A if A does not play at B's preferred intensity. Thus A must consider costs of such third-party intervention as well as benefits of play with B. If A and B communicate about their play, A could misinform B about A's requirements or willingness to compromise in order to lure B into play. Or A could attempt to manipulate B's assessment of its own requirements. If B's costs of withdrawal from play with A, search for another partner, and negotiation with this new partner are sufficiently high, it might pay B to play with A even on A's terms. If play only exerts beneficial effects after many interactions occur, negotiation of single interactions on a case-by-case basis is problematic. It may not be worth bargaining for a single interaction if the cost of bargaining exceeds the maximum possible gain per interaction. And yet the magnitude of the cumulative benefit from many such interactions could vary greatly depending on whose interests prevailed in any given interaction. These considerations suggest a possible mutual advantage in long-term play relationships. Experience with a given partner might serve to reduce the cost of negotiation. Could such relationships be stable despite minor amounts of cheating? What factors would favor desertion and what factors would contribute to stability?

The strategic questions posed above have parallels in many areas of evolutionary biology. The problem of optimal partner choice is a two-player game for which the one-player optimization analog is the ecological theory of food choice and habitat selection (MacArthur & Pianka 1966, Schoener 1974a,b). Equally important are parallels with sociobiological theory: mate selection and parental care (Maynard Smith 1977, Trivers 1972, 1974), grooming (Seyfarth 1977), parent-offspring conflict and sib competition (MacNair & Parker 1978,

O'Connor 1978, Parker & MacNair 1978, Trivers 1974), reciprocal altruism (Trivers 1971), cooperation (Hamilton 1975), and conflict resolution (Maynard Smith & Parker 1976, Maynard Smith & Price 1973, Parker 1974, Popp & DeVore 1979, Treisman 1977). It is not unreasonable to suggest that a basis for formal evolutionary analysis of social play is now available.

As argued above, social play poses many problems from an individual's point of view. It would be difficult even for an adult human to make correct decisions about play using game theory, mathematical optimization techniques, and a computer. Could an intellectually naive six-month-old rhesus monkey do any better? The surprising answer is that young animals can and do behave as if they are making just such sophisticated decisions (Dawkins 1976b). Young animals are behaviorally sophisticated (Trivers 1974) and can be expected to show considerable competence in gathering and processing information in those few limited areas where such competence is biologically advantageous (E. O. Wilson 1975 p. 156). The minimum competence needed to make decisions about social play should be specified. Although this problem requires further investigation, basic requirements are (1) the ability to recognize individuals, (2) a tendency to find different kinds of play exciting, boring or uncomfortable depending on past experience, on the nature of the interaction, and on the identity of the partner, and (3) the ability to associate play experiences, pleasant and unpleasant, with the partner involved.

Crucial to the arguments above is evidence that young animals actually compete for and disagree about play, as they do regarding access to other resources. This evidence exists mainly, but not solely as a result of studies on large groups of nonhuman primates in which the number and variety of potential play partners is greatest. Play solicitation may be refused (Bekoff 1974c, Bertrand 1969, Jolly 1966a), displays may serve to communicate refusal (Drüwa 1977, Schifter 1968, Tembrock 1960), refusal may be aggressive (Ehrlich & Musicant 1977, Fedigan 1972b, Gauthier-Pilters 1966, Hodl-Rohn 1974, van Lawick-Goodall 1971 p. 229) or submissive (Drüwa 1977), and refusal to play may even elicit aggression from the unsuccessful solicitor (Ehrlich & Musicant 1977, Kurland 1977). An unsuc-

cessful solicitor may try to force another animal to play by bullying it (Bertrand 1969) or by attempting to prevent it from
leaving the solicitor's vicinity (G. Fox 1972). During solo running play, one animal may direct aggression at another playing
animal who runs too close to it (Walther 1964, 1965a). Two
young primates may compete aggressively for the opportunity
to play with a third (Rivero pers. comm.). One player may try
to prevent its partner from leaving a rough play bout (Bertrand
1969, Steiner 1971, Symons 1978a p. 198). Play may become
too rough, intense, or threatening for one partner, who vocalizes (primates, S. Altmann 1962b, DeVore 1963, Du Mond
1968, Hall & Mayer 1967, van Lawick-Goodall 1971 pp. 81,
129, Mörike 1973, Moynihan 1966; rodents, Horwich 1972,
Steiner 1971; carnivores, Dalquest & Roberts 1951, Gauthier-
Pilters 1966, Herter & Herter 1955, Kaufmann 1962; pinnipeds,
Rand 1967) or attempts to struggle free (Symons 1978a p. 198).
Transgressor and/or victim may threaten (Baldwin & Baldwin
1976b, Drüwa 1977, Fedigan 1972b, Gauthier-Pilters 1966,
Symons 1978a p. 64) or act submissively (Drüwa 1977, Melchior
1976, Nolte 1955a) (Fig. 2-3). The transgressor may adopt ap-
peasing behavior in an apparent attempt to placate or to reassure
its partner (Drüwa 1977, van Lawick-Goodall 1971 p. 129,
Sparks 1967, Wasser 1978, White 1977). Play may even escalate
into damaging fighting (M. Altmann 1952, Fox 1969a, Kurland
1977, Symons 1978a p. 64).

The tendency of third parties to intervene in play, always (as
far as is known) in behalf of a player who is a close relative, furnishes additional evidence of the biological importance of play
in general, of the relevance of inclusive fitness considerations to
the study of social behavior, and of the resulting importance to
animals themselves of conflict over play. Just as mothers and
other close kin may intervene to protect their offspring from
possible or actual physical danger in play (Douglas-Hamilton &
Douglas-Hamilton 1975, Kinzey et al. 1977, Kummer 1968a,
van Lawick-Goodall 1967 p. 310, Nance 1975 p. 132, Pfeffer
1967, 1972), parents act to ensure that their offspring play with
like-aged partners rather than with older juveniles. Parents
achieve this effect by threatening or attacking older individuals
who attempt to play with the youngsters [mountain sheep (*Ovis*

canadensis), Spencer 1943] or by breaking up mixed-age play [feral horses (*Equus caballus*), Feist & McCullough 1976]. Whose interest might such behavior serve? Older animals may intervene in rough playfights on behalf of their kin if their young relatives begin to vocalize (S. Altmann 1962b, Baldwin & Baldwin 1974, Bernstein 1965, 1967, Bolwig 1959, Chivers 1974, Douglas-Hamilton & Douglas-Hamilton 1975 p. 96, Du Mond 1968, Fedigan 1972b, Hansen 1966, Hinde, Rowell, & Spencer-Booth 1964, Kummer 1967, Limbaugh 1961, Mörike 1973, Owens 1975b pp. 392, 397, Spencer-Booth 1970). A young animal for whom play is going badly may even solicit aid from a third individual (Hanby & Brown 1974, Owens 1976). In contrast, individuals not closely related to any of the players may act to break up stable play (Farentinos 1971, Graf 1966, Linsdale & Tomich 1953, McKay 1973, Mizuhara 1964, Rudge 1970, Steiner 1971, Varley & Symmes 1966, Verheyen 1955).

Although examples cited above stress the uncooperative side of play, cooperation is to be expected when players' interests coincide. Just as sociobiological tradition serves to focus attention on conflict and competition in play, an older ethological tradition based on superorganismic perspectives has naively, but effectively highlighted play's cooperative aspects. Again, I wish to stress that the fundamental evolutionary logic of social play predicts extremes of cooperation as well as competition, depending on the particular situation. If play interactions and relationships appear more stable and more cooperative than would be expected from simple calculations assuming ruthless self-interest, it then becomes necessary to specify how such stability or cooperation could evolve by natural (including kin) selection.

The idea that social play represents an extraordinary form of cooperative social behavior is an uncritically accepted tenet of the biological study of play. Extraordinary cooperative social conventions, a social "playground fence," seem to ensure play's stability and safety. A requirement that play shall not harm either the players themselves or their social relationship is said to be enforced both by participants' own restraint and by individuals' actions outside the play interaction, including parents, other juveniles, and adults. That these stabilizing conventions exist is clear. It remains to demonstrate how they serve the genetic interests of those individuals whose behavior maintains them.

Although earlier observers, including Charles Darwin (1898) and Gregory Bateson (1955, 1956), identified extraordinary conventions governing social play, Stuart Altmann (1962b) first presented an actual outline of these rules. He noted that rhesus macaque play interactions appeared to terminate quickly "unless the 'games' that were played were 'fair' games," i.e., unless each individual had about the same chance of attaining its tactical goal (of "winning"). Fairness resulted if like-aged monkeys played together or if a monkey whose strength or dominance status exceeded that of its partner actually held back ("self-handicapping") by matching the intensity of its acts to that of its partner's acts or even by taking a defensive or subordinate role, fleeing from a smaller partner or falling down near it. Moreover, monkeys at play used special signals to communicate readiness to play, to solicit play from a partner, and to maintain play.

Subsequent studies of animal social play confirmed that these and other mechanisms served to reduce risks associated with play and to stabilize play interactions. By virtue of these social conventions animals at play avoid injuring each other and reduce the likelihood of misunderstandings that would cause play to escalate into potentially dangerous fighting. Of course, this is the case only when such stability is in both individuals' interests.

Restraint

Playing animals seldom if ever use damaging tactics. Bites are inhibited and are delivered with less than full force. Such bites are more correctly described as nips. They do not break the skin. When one animal grasps another it does not tear the other's fur or hair out. Animals with sharp claws, including Felidae and Ursidae, do not use them to injure their play partners. During play, felids usually keep their claws retracted. Hoofed and horned mammals at play do not inflict injury with their anatomical weapons. These conventions are in force even during vigorous and physically exhausting play-chases and playfights. There is some evidence that young animals play more roughly and that restraint develops gradually (Aldis 1975 p. 24, Cooper 1942, Fox 1970, Schaller 1972 p. 156). An animal at play often seems to ignore physical discomfort. It may jump or even fall heavily to the ground or bump its head without interrupting play. The same stimulus that elicits defensive threat or withdrawal in nonplay situations fails to produce any interruption in an ongoing play bout. It is as if the player's threshold for pain had increased (Konner 1975). Or does the animal sense the pain but ignore it?

Self-handicapping

Larger, older, or more socially dominant animals do not use their full strength in play with smaller, younger, or more socially subordinate animals, and they seem to modify their strength and skill to match that of their partner (Figs. 7-1, 7-2). Meerkat (*Suricata suricatta*) fathers rarely chase their offspring in play (Wemmer & Fleming 1974), but are often chased by them. When lion cubs (*Panthera leo*) of unequal sizes play together, "the larger of the two animals contains its strength" (Schaller 1972 p. 162). Olive baboons (*Papio anubis*) (Owens 1975b), hamadryas baboons (*Papio hamadryas*) (Leresche 1976), rhesus macaques (Symons 1978a), chimpanzees (Loizos 1969), and other species (Aldis 1975) also self-handicap and thereby make play interactions less unfair. One consequence of self-handicapping is that dominance distinctions are less evident in play than in other contexts.

The term "self-handicapping" denotes that an animal's behavior reduces its probability of achieving its tactical objective in play and thereby prolongs the play interaction. Such behavior appears inconsistent with the statement that animals seek to win playfights (Symons 1978a). Loizos's (1969) dominant chimpanzee initiated a playfight by fleeing from its subordinate; an adult rhesus may flop down in front of an infant (Symons 1978a); here, self-handicapping may be the only way to initiate a playfight if a less capable individual will refuse another's invi-

Fig. 7–2 The evident difference in size and speed between these two domestic dogs (*Canis familiaris*) suggests a need for self-handicapping by the larger dog if this play-chase is to remain "fair." California. (John M. Bishop)

tation to join in a playfight that it would quickly lose. Similarly, prolongation of a playfight may depend on the willingness of the weaker or less capable individual to persist in a bout with a stronger opponent, making immediate termination the penalty for lack of restraint.

Control or Modulation

Restraint and self-handicapping in play comprise the first stage in a system of controls serving to reduce the risk of misunderstanding or injury. This first stage of the system cannot be expected to be absolutely foolproof (Fig. 7-3). If one animal unexpectedly turns the wrong way it may receive a bite on a tender region of its body, but the bite may actually have been aimed at another, less tender region. One partner may misjudge the severity of its actions. It may not be aware that another's recent experience has frightened it temporarily or that the other has a minor cut or bruise that should not be contacted in play. Ani-

Fig. 7–3 Domestic dogs (*Canis familiaris*). Play-chase turning into a playfight. Same dogs as in Fig. 7–2. Note the facial expression of the dachshund: a wide-eyed direct stare, covered teeth. California. (John M. Bishop)

mals' information is imperfect; they cannot use words to tell each other about such conditions in advance, and the affected party may itself have forgotten its minor discomfort until its playmate inadvertently mouths or paws a tender spot. Moreover, it is difficult to imagine ways in which animals could inform each other before the fact of the exact amount of roughness or intensity that each is willing to tolerate. How can a stronger animal know, before the fact, to what precise extent it must restrain itself with a given partner, except through experience? How could animals know each other's tolerance beforehand? This task would be impossible even for humans. We would need to issue tables specifying the maximum tolerable pressure or torque that could be applied to each part of our bodies as a function of our nutritional condition, level of fatigue, recent experience, and so on. Instead, the response to these unavoidable fluctuations that exceed tolerance levels is behavioral and communicative: the affected animal vocalizes (see above, p. 340). Such vocalization may bring about third-party intervention. Older juvenile rhesus macaques seem to know that if an infant playmate screams, the infant's mother is likely to intervene, for they run away as soon as the infant vocalizes (S. Altmann 1962b). A human analog is the use of the word "uncle" by a bested player to request interruption of a playfight. Usually the partner desists or pauses, and both animals may temporarily withdraw, facing away from each other or using their time-out for lower-priority behaviors such as autogrooming (Baldwin 1969, Dalquest & Roberts 1951, Levy 1979, Steiner 1971, Tembrock 1960). The animals' attention is thus temporarily deflected, the one from the other. When both animals face away or when social grooming occurs, both transgressor and victim act in a way antithetical to that occurring in an escalated fight. The aggressor desists and both animals calm down. Play frequently resumes only seconds after these events occur, whereas an escalated fight can affect the subsequent behavior not only of the participants, but also of their nearby conspecifics for hours, as in rhesus macaques (Symons 1978a). A brief pause in vigorous play suffices to permit animals to calm down immediately, even when one partner has momentarily violated the rules. It would be interesting to analyze circumstances under which *both* partners should be selected to respect such pauses.

After repeated unsuccessful attempts at maintaining play, partners may switch to a different style of play in which postures or physical relationships used are less likely to elicit serious fights. They design a game. For instance, young female rhesus macaques, in whom play-wrestling breaks down more frequently than in young males, tend to play-chase rather than to playfight (Symons 1978a). One sand cat (*Felis margarita*) kitten, aged 12 weeks, at Brookfield Zoo frequently mounted its littermate and bit its neck soon after play bouts began; the littermate repeatedly vocalized, hissed, and tried to get away. After several asymmetric bouts of this type and after several unsuccessful solicitations by each kitten, the mountee jumped playfully into a hollow under a rock, of its own accord (it was not chased). The two kittens then playfully and silently angled at each other with their paws for several minutes. They became so involved in their game that they reversed roles four times (one

Fig. 7–4 Domestic cat (*Felis catus*) orienting to partner's tail in play. Same interaction as in Fig. 7–13. Nong Khai, Thailand. (Robert Fagen)

kitten charged out and the other fled in) while playing for thirteen continuous minutes. As soon as one kitten reached the safety of the crevice, pursuit stopped and the angling duel resumed. No other play bout during the one-hour observation period lasted longer than a minute.

Use of environmental features (Figs. 7-4, 7-5) allows play to be rougher or more vigorous while maintaining its stability. Animals may hide in and playfully defend burrows, bags, tunnels, or crevices (Christen 1974, Dücker 1968, Herter & Herter 1955, Loveridge 1923, Rowe-Rowe 1971, Weiss-Bürger 1975, Yate 1898). They may chase or spar around a tree trunk or other obstacle (Darling 1937, Gentry 1974, van Lawick & van Lawick-Goodall 1971 p. 113, van Lawick-Goodall 1967, Neal 1962, Ripley 1967, Rudnai 1973). Increased vigor of stable play results from use of environmental features by young African elephants (*Loxodonta africana*): "Should an obstacle such as a fallen tree come between the two contestants during their play-fight then they will suddenly threaten with redoubled fury, each knowing that in fact it cannot get to grips with the other" (Douglas-Hamilton & Douglas-Hamilton 1975 p. 98). Orienta-

Fig. 7-5 Use of an object in social play can permit increased vigor and intensity while ensuring the stability of the interaction. The object may be contested, bitten, thrown, or pulled more roughly than could a conspecific, without causing pain or physical damage to the partner. Might such interactions represent a relatively safe way to display or to assess physical ability? The tug-of-war "game" between these two domestic dogs (*Canis familiaris*) is a frequent form of object-oriented social play in many species of mammals. Connecticut. (Phillip Parker)

tion to tails in play may produce the same effect (Figs. 3-34, 7-4).

Differences in time spent in social play or in time spent playing with particular classes of partners result from individual behavioral tendencies. These tendencies may, of course, depend on the partner's characteristics and on aspects of the environment. In order to see how different kinds of behavioral tendencies might affect gross measures of social play, Nicholas Mankovich and I formulated a simple flow-chart model of dyadic social play (Fig. 7-6). As suggested above and by the model, differences in amount of social play performed can result from

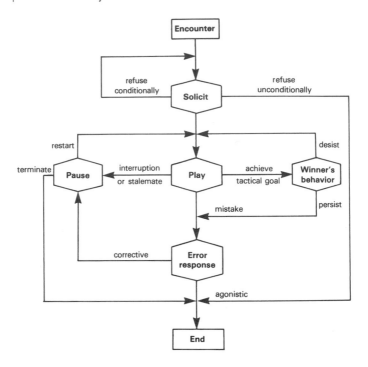

Fig. 7–6 Kinematic graph diagram of dyadic play. Normal end of play: the left branch (after a pause of varying length). See text for explanation. Simplified from a model developed by N. Mankovich and Robert Fagen.

differences in ten or more separate parameters of behavior, including encounter, solicitation, refusal, acceptance, prolongation and termination, pauses within play bouts, and intervals between play bouts. Therefore, although solicitation and acceptance are probably the most obvious components of this system (as well as the easiest to measure), differences in solicitation and refusal rates alone do not fully explain differences in frequencies of social play. Animals that play more overall may successfully initiate play at the same rate as do less playful animals, but their play bouts may be longer. Or, more playful animals may differ from less playful animals only in the tendency to perform fewer nonplay activities between play bouts. Two animals may play at equal frequencies even if the first successfully initiates more play. This result can occur when the first plays for shorter periods than the second.

Processes of differential solicitation, refusal, prolongation, and termination of play discussed above result in nonrandom composition of play dyads. As a general rule, similar (like-sexed, like-aged, like-sized, or closely related) partners play together with greater than chance probability (S. Altmann 1962b, Cheney 1978, Dunbar & Dunbar 1975, Fady 1969, Fagen 1974a, Levy 1979, Owens 1975b, Rasa 1977, Schaller 1972 p. 162, Soczka 1974, Symons 1978a pp. 56–60). Data quality, sampling techniques, and analytical methods used to demonstrate these results vary among authors. For the most part, however, these differences appear to be real and do not result merely from differential physical proximity or from differential playfulness. Play partnerships do not merely reflect general preferences for feeding, grooming, resting, sleeping, companionship, or physical proximity; they are specific to play (C. C. Hamilton 1967, Mori 1974, Rasa 1977, Rhine 1972, Rhine & Kronenwetter 1972, Riss & Goodall 1976, Rosenblum & Lowe 1971, Yamada 1963). Play partnerships, although highly nonrandom, seem to be less so than feeding or grooming relationships, at least in Japanese macaques (Mori 1974, Yamada 1963). We cannot infer preferences directly from dyad composition data because some animals have no choice but to play with a less preferred partner. Even process data—rates of solicitation, acceptance, refusal, prolongation, and termination—may reflect individuals' assessment of the probability that play will occur with that partner as well as their assessment of the benefit to be derived from such play. In fact, these behavioral rates should reflect all the complexities discussed at the beginning of this section, particularly inclusive fitness constraints and resulting behavior of third parties.

Play is an exciting and difficult social phenomenon because of the factors cited above. The full complexity of these social constraints on play is best realized in large, multi-age social groups of immature individuals in close proximity to adults. As Altmann, Altmann, Hausfater, & McCuskey (1977), Baldwin & Baldwin (1971), Blaffer Hrdy (1977 p. 294), Jolly (1972), and Konner (1975) point out, the demographic environment determines the number and characteristics of potential play partners. Play has so long been viewed as a fundamentally social phenomenon that we are genuinely surprised to discover play in moose (*Alces alces*) (M. Altmann 1958, 1963, Geist 1963, Stringham 1974), margay (*Leopardus wiedi*) (Leyhausen 1963), pangolin (*Manis tricuspis*) (Pagès 1972, 1975), two-toed sloth (*Choloepus didactylus*) (McCrane 1966), and other mammals that do not live in groups of adults and that give birth to only one offspring at a time. Non-group-living adult mammals that give birth to several offspring at a time incidentally produce a social environment in which genetic relatedness is uniform and in which littermates differ primarily in sex, size, developmental level, and experience. Some mammals [e.g., beavers (*Castor canadensis*), orangutans (*Pongo pygmaeus*)] live in families consisting of one or both parents, with offspring of mixed ages. Others (e.g., equids) form assemblages in which several females simultaneously produce single offspring that grow up with like-aged companions—half-siblings, cousins, nieces, and nephews. In large assemblages of hoofed and marine mammals, unrelated immatures play, fight, and travel together in like-aged groups. Young of other social mammals, such as African elephants (*Loxodonta africana*) and some humans, form mixed-age groups of close kin in which degree of relationship may vary. This pattern reaches its extreme in large, expanding primate populations in which any juvenile has several potential companions of the same age, sex, and parental social status. Within a given species [e.g., yellow baboons (*Papio cynocephalus*), J. Altmann et al. 1977)] demography of immatures may vary from small, mixed-age groups to large, same-age groups. Despite great variation in the number and phenotypic distribution of potential play partners, three primary play-group patterns are most common in mammals: like-age sib groups, like-age groups of unrelated individuals, and mixed-age kin groups (Konner 1975). The missing logical possibility is that of mixed-age groups of unrelated individuals. Such groups could only occur in small populations in which same-age partners are unavailable. A possible example is the Cape fur seal (*Arctocephalus pusillus*), in which yearlings often play with black pups-of-the-year (Rand 1967). Generally, however, weaned young of marine and hoofed mammals form spatially distinct, year-class play groups.

SOCIAL PLAY STRATEGIES

An animal's tendency to play socially should depend on the costs and benefits immediately associated with each interaction, on the ways in which that interaction affects future play interactions with the same partner, and on the current availability of other partners. Although the same fundamental strategic problems are always present, details differ greatly as the context of play ranges from single or twin offspring with one or two parents (margays, young pronghorn fawns, marmosets) to a mother and her offspring of several ages (gibbons, orangutans) to litters of like-aged animals of different sex and developmental level (tigers, Norway rats) or to primate troops with different numbers and types of potential companions (e.g., certain populations of *Macaca mulatta*). An animal seeks the best available opportunities for play. Demographic circumstances may limit such opportunities to those furnished by play without partners. Demographic constraints may also bring an individual to play with unlike-aged companions, with companions of the opposite sex, or with companions belonging to a strange kinship or social group, even though the same individual would tend to restrict its play to others similar to itself if a sufficient number of such others were available. In evolutionary time, play may adapt more deeply to its social environment. If a particular demographic distribution of partners is a constant feature of the environment for many generations, the effects of play may become dependent on that particular distribution. Thus, for example, human infants seem curiously inept at peer play (Konner 1975), although not as totally so as was once supposed; see Eckerman, Whatley, & Kutz (1975) for an example of peer play competence in two-year-olds. If an individual must play simply to equal the level of competence attained by current or future competitors, then it is only necessary to play as effectively as everyone else does. Thus, to the extent that lack of companions equally affects everyone's opportunities for play, individual requirements for play will parallel play opportunities. If an animal later emigrates from its natal group and attempts to join a larger group whose members had played frequently as juveniles, the strategy of playing only as much as current potential companions play is ineffective, but an individual's chance of joining a new group may be extremely low unless that individual already has kin or friends in that group.

Simple models of social play strategies will be defined in order to address these and other questions, including optimal degrees of self-handicapping and partners' relative responsibility for play interactions (initiation, maintenance, termination).

Optimization and Game Models

The tendency to play, like the tendency to feed (McFarland & Sibly 1975, Sibly 1975), should depend both on incentive (the availability and quality of opportunities for play) and on deficit (the animal's requirements for play, which may depend on the social environment as argued above). This homeostatic formulation may prove useful for experiments that compare tendencies to perform various types of play. Incentive reflects probability of access, cost, and skill, strength, or endurance benefit; deficit reflects relations between time, energy, survivorship, skill, and surviving offspring. Play is not homeostasis-maintaining behavior, nor is it analogous in any sense to such behavior unless the arousal metaphor proves biologically meaningful after all. However, this formulation predicts that an animal's tendency to initiate, accept or reject, maintain, and terminate play will depend on probability and cost of access to play opportunities, on the changes in physical and cognitive ability that result from the interaction if one occurs, and on the fitness effects of increments in physical and cognitive ability.

How will probability of access and costs and benefits of play affect play tendencies? The quality of a play interaction, including its intensity, degree of physical contact, degree of contingent responsiveness, and degree of reciprocity may be represented by a vector-valued variable x. The probability of a subject achieving play with individual i is given by p_i, the benefit of such play to the subject by $B_i(x)$, and the cost of play to the subject by $C_i(x)$. At any given time, play will be solicited from that individual i who maximizes the product $p_i[B_i(x)-C_i(x)]$, provided this quantity is positive. However, this formulation of play is not complete. The individual solicited will only agree to play if the subject maximizes the same quantity from the solicited individual's point of view. Furthermore, as argued above, individuals' optimal x values may differ. The subject's probability of access to individual i for play may even depend on i's assessment of the value of x at which the subject is prepared to play. Past experience, future prospects for play, possible genetic relatedness of the potential partners, and possible intervention of third parties are additional factors to be considered. Individuals may misinform potential partners about their own benefits, costs, preferred play quality, and access probabilities, and they may try to influence those of the partner. Or they may even try to deceive the partner into a false assessment of the parameters of its own benefit-cost equation. A closer look at these parameters is therefore necessary.

KINSHIP

The benefit-cost equation above may be easily rewritten to accommodate possible kinship. In terms of inclusive fitness, with r_1 the regression coefficient of genetic relatedness between the subject and individual i, the subject will be selected to solicit play from the individual who maximizes

$$p_i\{[B_i(x) - C_i(x)] + r_i [B(x) - C(x)]\} \qquad (7\text{-}1)$$

This equation suggests that when potential playmates differ primarily in relatedness to the solicitor, kin will be preferentially invited to play.

Benefit, Cost, and Quality

How do increments in ability depend on the qualities of a play interaction, and how do these increments influence the expected number of surviving offspring? How does play experience produce time, energy, and survivorship costs, and how do these costs affect individual fitness? Answers to these distinct questions ultimately depend on the dynamic structure of social play.

Physical roughness, measured by frequency, duration, and rate of exertion, appears to be an important quality of social play from a participant's point of view. If an infant entered

Fig. 7–7 Clasping and pushing.

Fig. 7–8 The juvenile stands as the two grapple.

Fig. 7–9 The subadult uses his foot in an attempt to push away the juvenile.

Fig. 7–10 The subadult grasps the juvenile's head.

Fig. 7–11 The juvenile sits back, evading the subadult's grasp.

Figs. 7–7 through 7–11 Himalayan langur (*Presbytis entellus*) playfight between juvenile and subadult males on a tree branch. Photographic sequence of a single playfight, which began with the juvenile (right) approaching and embracing the subadult and ended with the animals, seated, facing each other. Shortly thereafter, they playfought again. Melemchi, Nepal. (John M. Bishop)

409

juvenile-juvenile play on an equal basis it would be badly hurt. Apparent distress vocalizations usually occur in play as a result of actual or anticipated physical discomfort (see above, p. 398). By the same token, the amount of physical training achieved will depend on the amount of exercise that occurs.

Subjectively, animals appear to participate in play for the experience of striving to attain control over the partner's movements or over some part of the partner's body, while at the same time preventing the partner from attaining such control (Figs. 7-7, 7-8, 7-9, 7-10, 7-11). Tactics used are subject to certain rules or constraints on behavior that prohibit or minimize damage to the partners' bodies or to their social relationship. Control means control of movement (including restraint) and implies access to the animal or to the body part controlled, which may therefore be freely sniffed, mouthed, or manipulated by the animal in control. The importance of these tactics is apparently that they result in efficient, low-risk training of physical and cognitive abilities used for escalated fighting, predation and predator avoidance, and perhaps cooperation.

Biting harmlessly without being bitten is the apparent tactical goal of play in rhesus macaques (*Macaca mulatta*) (Symons 1978a), in Steller sea lions (*Eumetopias jubatus*) (Gentry 1974), in mountain gorillas (*Gorilla gorilla*) (Carpenter 1937, Schaller 1963), and in Shetland and Welsh ponies (*Equus caballus*) (Schoen, Banks, & Curtis 1976). Chimpanzees at play may also strive to tickle without being tickled (van Lawick-Goodall 1968a). In other species the objective may be to sniff, to otherwise sense, or to touch some part of the partner's body: nosing the nape of the neck in the caviomorph rodents *Octodon degus* and *Octodontomys gliroides* (S. Wilson & Kleiman 1974), vicuña (*Vicugna vicugna*) putting the neck over the partner's neck (Koford 1957) (perhaps as a controlling position for biting), and touching the partner in the "tag" games of red deer (*Cervus elaphus*) (Darling 1937), of humans (Aldis 1975 p. 210), and of other mammals (Aldis 1975 p. 17). Sometimes the target of social play is to gain control of an object in the partner's possession (as in rhesus macaques, Southwick, Beg, & Siddiqi 1965, Symons 1978a) or over a particular physical location ("King of the castle," Bertrand 1969, Darling 1937). In both cases possession alternates and the object or place is not monopolized or used exclusively by one animal. The tactical goal of a play-chase in white-tailed deer (*Odocoileus virginianus*) may be to get in front of the partner (Michael 1968). In human playfighting the objective is often simply to throw the partner to the ground and to maintain position on top of him or to pin him down (Aldis 1975 p. 178).

The tactical objective of social play is thus to achieve control of the opponent without being controlled, analogous to a checkmating position in the game of chess. In chess, the victor never actually captures the opponent's king, but rather achieves a position in which the next move would achieve this capture.

When the outcome of a sequence of maneuvers is no longer in doubt, when successful defense produces a stalemate, or after an interruption, animals may continue their interaction from another starting position or may reverse roles. For example, the partner who succeeds in escaping in a play-chase may turn and chase the other animal. These observations, the fact that interactions continue after the tactical goal is reached, and the phenomenon of self-restraint or self-handicapping (discussed below) all strongly suggest that this final position serves to constrain behavior along certain lines of pursuit, eliciting behavioral tactics the performance of which is of value to the participants. The apparent purpose of play is some form or forms of experience occurring while attempting to achieve a positional objective against a partner's defensive moves or while defending against a partner's attempts to achieve it. In Symons' (1978a) words, "the striving or competition" (rather than mere achievement or defense of the goal) "appears to be its own reward."

Because players have tactical goals, we may refer to the winner and loser of a playfight, or we can say that play is going badly for one animal if it consistently fails to achieve control while its partner consistently succeeds. Two indices of these outcomes are the total number of such wins achieved by both participants and the ratio of one partner's wins to those of the other.

Degree of body contact is another significant aspect of social play. Maximal body contact occurs in wrestling. Wrestling animals roll and twist on the substrate, using all four limbs, their whole bodies, and their mouths to control each other's position and velocity. Less contact occurs in mutual upright grappling. Here, at least one limb serves as attachment to the substrate (ground or tree branch), while one or more other limbs and the mouth are used to push, pull, and hold the partner. In grasping, animals contact and hold each other with one or both forelimbs only. Sparring or slapping is similar to grasping, but contact only involves hitting, slapping, or kicking at, not holding. If animals run to and from each other while sparring or slapping, the interaction resembles a game of tag. If no physical contact occurs while an animal pursues or flees its playmate, the interaction is the familiar play-chasing. This body contact spectrum encompasses most known social play interactions (Aldis 1975, Symons 1978a). Logically, any degree of physical roughness may occur together with any degree of body contact: compare vigorous play-chasing (during which animals may run hundreds of meters at top speed while making rapid turns) with the gentle contactual holding and rolling of infant rhesus macaques (White 1977b) or aquatic and marine mammals (S. Wilson & Kleiman 1974).

In ungulates and in other animals that lack grasping limbs, maximal contact occurs in play as animals ram, push, ride, rest their bodies on, shoulder, and mouth one another. In sparring and butting matches, only head-to-head contact occurs. Young elephants may chase each other, slapping at each other with their trunks (Sikes 1971 p. 270). Tagging games and play-chases are well known in ungulates (Chapter 3).

An important parameter of social play is its degree of role reversal, mutuality, reciprocity, or symmetry. For example, rhesus macaque wrestling includes identifiable on-top (Fig. 1-2) and on-bottom positions (Symons 1978a p. 29). The on-top monkey leans over or on and holds its partner. The partner lies on its back. Similarly, a rhesus macaque "behind" its partner places its ventral surface against or near the partner's dorsal surface, grasps the partner, and prevents it from turning (Symons 1978a p. 31). One macaque may rest its body on, ride, or mount another. Similarly, at a given instant of a play-chase one partner chases the other who flees. For any degree of contact and intensity, the occupation frequency of a given role by one partner can range from 0 to 100%. In other play patterns, such as grappling and sparring, separate roles are not defined. The relative amounts of this type of play and of the amount of play involving well-defined roles define a second index of mutuality or symmetry, conceptually distinct from the first (Levy 1979). Mutuality measures are probably not independent of win frequencies and win-loss ratios (see above, p. 395).

Degree of contingent responsiveness in play, defined as the degree to which one animal's behavior may be predicted from its partner's immediately preceding action or actions, may be measured using information theory (S. Altmann 1965, Losey 1978). Another quality of play is its diversity, which is defined in various ways including the order of the simplest Markov chain model that adequately describes the data; the average number of statistically significant transitions between the behavioral acts of one participant and the immediately following acts of its partner; or the number of different kinds of acts occurring in play. Information-theoretic measures of contingent responsiveness are mathematically related to information-theoretic measures of diversity, variability, and predictability. These final two qualities are the best available measures of the brilliance and artistry that delight human observers of play.

Stability of play was discussed separately (p. 393) because it results from individuals' tendencies to prolong or to terminate play in accordance with each participant's assessment of its interests based on inclusive fitness.

Qualities of play discussed above are surely not all independent. How many independent variables are required to describe the phenomena outlined?

In theory, what would the optimal qualities of play be for a given animal? Except possibly for strength and endurance training, this question cannot yet be answered. However, given such answers, it is possible to rank partners in order of relative desirability. Konner (1975) speculated that a slightly older partner would be preferred "since the things that need to be learned are learned more easily from those a little ahead of oneself than from those, so to speak, in the same quagmire." Symons (1978a p. 60) failed to find age-asymmetric patterns of play initiation in rhesus macaques and suggested (Symons 1978b) that "play with peers and younger animals may, in fact, provide certain kinds of practice benefits not provided by play with older animals. . . . Perhaps for each animal there is an ideal profile of play partners, a profile that varies with age, sex, and proficiency." These data actually do not contradict Konner's prediction, because, as argued above (p. 390), the probability of an initiation attempt will depend both on the likelihood of successful initiation and on the benefit of play, given such success. Therefore, if older animals were selected to resist, younger animals would not attempt to initiate play with older animals as frequently as the potential benefit from such play might indicate. Similarly, older animals would initiate play with younger ones more often than expected because their expected success rate for such solicitation was high.

Play-signals

Communicative behavior serving to solicit, invite, and maintain social play is prominent and widely recognized in animals. Darwin (1898) discussed his pet dog's play-signals, including the play-face and the play-bow. W. J. Smith (1977 pp. 146–147) reviews communication and discusses displays associated only with play, including the play-face.

Play-signals appear in all sensory modalities used by mammals. The relaxed, open-mouth grin or play-face (Figs. 1-2, 3-1, 3-2, 3-20, 3-24, 3-25, 3-26, 3-30, 3-31, 7-12) is near-universal. Some species add particular communicative features to the play-face. One example of this phenomenon is the unique eyelid display of the douc langur (*Pygathrix nemaeus*). This Southeast Asian primate, one of the most strikingly colored of all mammals, is marked with sharply contrasting patches and areas: the face (bright yellow with pale blue eyelids and white whiskers), the head (brown with a bright chestnut band below the ears), the body (mottled grayish), and the rump and tail (white) (Walker 1975 p. 464). Behavior of *P. nemaeus* has been studied at the Köln (Hick 1972) and San Diego (Kavanagh 1978) Zoos. Douc langurs accentuate their play-face with "a vivid display of the very pale blue eyelids, with the eyes closed"

Fig. 7–12 Himalayan langur (*Presbytis entellus*) subadult male approaches a juvenile male for playfight. Note the subadult's facial expression. Melemchi, Nepal. (John M. Bishop)

(Kavanagh 1978). Prolonged eyelid lowering in a social context occurs only in play. The eyelids may be lowered momentarily in sexual displays (Kavanagh 1978), presumably as part of the social play that often precedes copulation in these primate harlequins (Hick 1972). Sometimes the eyelid display alone suffices to initiate or to maintain play (Kavanagh 1978).

Particular postures and movements used to elicit or to maintain play include gamboling play-gaits (Blauvelt 1956, Darling

1937, van Lawick-Goodall 1968a, Sade 1973, Schaller 1972, Simonds 1965, Symons 1978a, Struhsaker 1975 p. 171), rolling (Bertrand 1969, Owens 1975a, Schaller 1972, Schramm 1968, S. Wilson & Kleiman 1974), specific tail positions in bovids (Kiley-Worthington 1976, Schloeth 1961), and especially in primates, looking between the legs (Baldwin 1969, Du Mond 1968, Owens 1975b, Ploog, Hopf & Winter 1968, Reynolds 1961, Symons 1978a) and arm extension (Hess 1973). The playbow, in which an animal lowers its forequarters and places its chin near or on the substrate, is a characteristic play-signal of canids (Bekoff 1974c, 1977b, Fentress 1967, Fox 1972a, Tembrock 1957) and of lions (Schaller 1972), but, curiously, not of other felids (in which the same position, often accompanied by noisy scratching with the claws, is used when stretching and may represent a low-intensity threat) or other mammals. Felids other than lions begin play by approaching a conspecific broadside, arching the back and dorsiflexing the neck, and sometimes even leaping vertically (Barrett & Bateson 1978, West 1974), or by pouncing (West 1974). Quiet vocalizations, a sort of "all's well" signal, often accompany play (Table 7-1). Olfaction, a sensory modality crucially important to mammals, but whose potential use in play has only recently been recognized (Fagen 1974a, S. Wilson 1973, S. Wilson & Kleiman 1974), may be used by canids (Fig. 2-2, Fox 1972a, Tembrock 1960), mongooses (Hinton & Dunn 1967 p. 45), rodents (S. Wilson 1973, S. Wilson & Kleiman 1974), and pinnipeds (S. Wilson 1974a). Because play-smells, unlike facial expressions, postures and gaits, or vocalizations, are not perceived by human observers, they represent a challenge to students of mammalian communication. An exemplary approach is that of Susan Wilson (1973), who took ether extracts from portions of the body frequently sniffed in play by short-tailed voles (*Microtus agrestis*). She applied these extracts to other voles that normally did not elicit play from conspecifics, and these treated voles then elicited play, whereas ether-treated controls did not.

Avian and mammalian play-signals also exploit tactile modes of communication. A gentle nip serves to initiate play in certain parrots (Hick 1962, Ulrich, Ziswiler, & Bregulla 1972) and in Steller sea lions (*Eumetopias jubatus*) (Farentinos 1971). Touching,

Table 7-1 VOCALIZATIONS SIGNALING PLAY (REPRESENTATIVE EXAMPLES)

Primates: Chivers 1974, 1976, Epple 1968, Gautier-Hion 1971, Hall 1965, Hopf et al. 1974, Jolly 1972, van Lawick-Goodall 1968a, Leresche 1976, MacKinnon 1974a, Marler & Tenaza 1977, Schaller 1963, Struhsaker 1967b, Symons 1978a

Rodentia: Ewer 1966, Steiner 1971, S. Wilson & Kleiman 1974

Carnivora: Bates 1944, Ewer & Wemmer 1974, Fox 1970, Rasa 1973, Schaller 1972

Artiodactyla: Schloeth 1961

---

slapping, pawing at, or pushing elicit play in a variety of animals (D. Altmann 1972, Bopp 1968, Chepko 1971, G. Fox 1972, Gautier-Hion 1971, Haas 1967, Hassenberg 1977, Redshaw & Locke 1976, M. D. Rose 1977b, Savage, Temerlin, & Lemmon 1973, Weiss-Bürger 1975).

This diversity of play-signals raises an interesting but unsolved theoretical question. If the message to be communicated is simply "This is play" (Bateson 1956), why should most species have from ten to a hundred distinct play-signals? Just as in the case of bird song repertoire diversity (Krebs 1978) and species diversity (Hutchinson 1965 pp. 26–78), many plausible proximate and ultimate answers to this question are available. I now discuss a particular hypothesis based directly on the evolutionary considerations outlined above.

What might play-signals communicate other than readiness to play? If different individuals have different play requirements, play-signals might serve to inform or misinform another individual about these requirements. At the very least, play-signals might communicate the strength of an individual's interest in play in addition to an all-or-none ready/not ready message. This goal might be accomplished by controlling the intensity of a graded play-signal, by varying the frequency of a highly stereotyped play-signal, or by using different signals to communicate different levels of motivation.

The strength of an animal's interest in play can be communicated by play-signals that follow refusal of that animal's initial solicitation. Coyotes (*Canis latrans*) may chase their own tails immediately following an unsuccessful play-solicit (Bekoff 1974c). A captive hyacinthine macaw (*Anodorhynchus hyacinthinus*) first solicited play by inclining its head obliquely and springing on its partner, but if this solicitation failed the macaw would then bite its partner lightly (Hick 1962). An individual initially reluctant to play with a partner whose interests in play were known to differ from its own might first refuse that partner's solicitations. If the partner still persisted, such persistence could indicate the partner's willingness to play at values other than its optimum. I observed a possible instance of such negotiation in two female margay (*Leopardus wiedi*) kittens who differed in age by roughly one year. The older kitten was twice the size of the younger and played much more roughly with humans. Play would be expected to be less important as physical training for the older kitten, who was almost an adult. Because of her age, size, and previous play experience it is understandable that she preferred infrequent, but strenuous play. The younger kitten was capable of playing stably with the older kitten. Her smaller size often put her at an advantage in arboreal play-chases, but she was at a disadvantage when they wrestled on the ground. On the ground, the older kitten would repeatedly refuse the younger kitten's solicitations until the younger kitten became extremely vigorous and active. An intense wrestling match then began.

Conflict of Interest Over Social Play: Evolutionary Stability of Compromise

If an individual is willing to play at values other than its personal optimum, it increases its potential number of play companions, but decreases its expected return from a play interaction. (It may also reduce negotiation costs.) A strong and experienced individual may compromise by playing less roughly, by using less than maximum skill, or by agreeing to play in ways, in places, and at times that increase its partner's chances of winning and/or decrease the partner's chances of discomfort or injury. A weak, inexperienced individual may agree to the additional risks and decreased returns of rough play. In a population of animals that vary both in preferred quality of play and in willingness to compromise, what strategy of compromise is evolutionarily stable, in the sense that a rare variant willing to compromise either less or more than the population norm will not increase in frequency?

The natural history of social play suggests that dissimilar play partners differ along a dimension that might be termed "toughness." Large animals are tougher than small animals, offspring of socially high-ranking individuals (rank means that the same preferential access relation between individuals exists for several resources or that access priorities are transitive for a single limiting resource; see Hinde 1978, Popp & DeVore 1979) are tougher than those of low-ranking individuals, and juvenile male rhesus macaques are tougher than juvenile female rhesus macaques. "Toughness" may be interpreted as resource-holding potential (Parker 1974), as potential for inflicting damage, or as probability of victory in a conflict. In rhesus macaque play (Levy 1979, Symons 1978a), this difference is expressed in several ways. Tough animals solicit play at higher rates, are less likely to terminate play by leaving, and tend to take the attacker's role in role-type play. They also tend to play less roughly with non-tough partners than with tough partners. The result of these differences is that those play dyads that occur most frequently are those that minimize the difference in partners' toughness. For example, play between one-year-old males and two-year-old females is unusually frequent, apparently because contributions of age and sex to toughness balance each other. Dyads of partners differing in toughness are unstable, even though the tougher partner solicits, initiates, and self-handicaps, apparently because the tougher partner takes the "attacker" role, while the less tough partner remains tense or fearful. Relative contributions of the two partners to this unstable situation have not been measured. It is important to point out that this pattern may be specific to large populations of rhesus macaques in which peers are abundantly available. In mixed-age sib groups, animals may tend to compromise more in play because they have no partners similar to themselves. Play between littermates may reflect size, developmental and sex differences, and possibly also differences in social rank if a hierarchy of priorities of access to resources is defined among littermates. However, the problem of social rank in developing animals is more complex than generally recognized. Littermates or broodmates may compete for food, warmth, grooming, protection, play, and other resources, including life itself (MacNair & Parker 1978, O'Connor 1978, Parker & MacNair 1978, Trivers 1974). Such competition may take various forms (E. O. Wilson 1971a), including that of aggressive behavior. Furthermore, hierarchies of social dominance, in which priority of resource access is transitively defined among individuals and in which conflicts are settled by the lower-ranking individual's withdrawal, are only expected under certain conditions, and may be defined for some resources, but not for others (Popp & DeVore 1979). Indeed, differential development of littermates makes it virtually inevitable that individuals of different sizes or at different developmental levels will differ in resource priorities, so that, for example, pup A will consistently give way to its larger littermate B for several weeks when their mother regurgitates food, but A will preferentially obtain access to milk or warmth.

How, in theory, should conflict over play be resolved? This question is answerable, at least in a few simple cases, by use of game theory models [see Dawkins (1976b) and Maynard Smith & Parker (1976) for background information on such models]. The first model to be discussed will assume two types of preferred play qualities (Tough and Gentle), two strategies (Stingy and Generous), random mixing of strategy types, single encounters, no third-party intervention, and no genetic relatedness of participants. The assumption of no genetic relatedness will be altered for a later model of sib-sib play. In my models, I assume that when like players (e.g., two Tough or two Gentle players) meet, they play at the optimum for their type. When a Generous animal meets a Stingy animal, they play at the optimum preferred by the Stingy animal. When Generous animals of different types meet, they play at some intermediate value. When Stingy animals of different types meet, play does not occur. The payoff in surviving offspring is greatest at the optimum and decreases monotonically away from the optimum. Tough animals get V_A at their optimum, αV_A when they play generously with animals who are both Gentle and Generous, and

βV_A when they play at the optimum for Gentle animals. Gentle animals gain V_B at their optimum, $\alpha' V_B$ in Generous–Generous matches with the Tough, and $\beta' V_B$ when they play at the optimum for Tough animals; $\beta < \alpha < 1, \beta' < \alpha' < 1$. In the special case of physical roughness we can expect the cost of self-handicapping to be far less than the danger of rough play for a Gentle animal, and therefore $\alpha' << \alpha$, $\beta' << \beta$ in this special case. We must consider two possibilities: either each animal can occupy both Tough and Gentle roles (as would be the case for an animal growing up in the midst of a variety of partners), or an animal's role is fixed (e.g., when successive year-classes are of different sizes and most immatures encountered by a yearling are two-year-olds rather than infants). In the first case, consider four strategies: G always Generous, SG Stingy when Tough and Generous when Gentle, GS Generous when Tough and Stingy when Gentle, S always Stingy. If a given individual is Tough one-half the time and Gentle one-half the time, the payoff matrix for this game is Table 7-2, and the only two evolutionarily stable strategies are SG and GS (Appendix IV). If fraction p of the population is Tough and fraction $(1-p)$ is Gentle, two evolutionarily stable populations are (1) Tough animals who are Stingy and Gentle animals who are Generous and (2) Tough animals who are Generous and Gentle animals who are Stingy (Appendix IV). Populations composed entirely of Generous or entirely of Stingy animals are not evolutionarily stable. Evolution is most likely to produce the second of these two possibilities when generosity costs the Gentle more than it does the Tough. As argued above, this is likely to be the case for physically rough play. In the second model, Stingy animals always receive their theoretical optimum return from play no matter what their partner's identity because they either play with their own type or with Generous animals of the opposite type. However, Generous animals gain less than their optimum from the average interaction because they must play at the optimum for the opposite type whenever they encounter an individual of that type. (Remember that in the only evolutionarily stable populations in this model, the two different types play opposite strategies.)

These simple models yield several predictions about play. In-

Table 7-2 PAYOFF MATRIX FOR MODEL I

		Against:		
	G	GS	SG	S
To: G	$\frac{1}{4}[V_A(1+\alpha) + V_B(1+\alpha')]$	$\frac{1}{4}[V_A(1+\beta) + \alpha'V_B]$	$\frac{1}{4}[V_A(1+\alpha) + V_B(1+\beta')]$	$\frac{1}{4}[V_A(1+\beta) + V_B(1+\beta')]$
GS	$\frac{1}{4}[V_A(1+\alpha) + 2V_B]$	$\frac{1}{4}[V_A(1+\beta) + 2V_B]$	$\frac{1}{4}[V_A(1+\alpha) + V_B]$	$\frac{1}{4}[V_A(1+\beta) + V_B]$
SG	$\frac{1}{4}[2V_A + V_B(1+\alpha')]$	$\frac{1}{4}[V_A + V_B(1+\alpha')]$	$\frac{1}{4}[2V_A + V_B(1+\alpha')]$	$\frac{1}{4}[V_A + V_B(1+\alpha')]$
S	$\frac{1}{2}(V_A + V_B)$	$\frac{1}{4}(V_A + 2V_B)$	$\frac{1}{4}(2V_A + V_B)$	$\frac{1}{4}(V_A + V_B)$

terests in physical roughness of play will differ between different-sized animals, between males and females, and between offspring of dominant and offspring of subordinate females (because a dominant mother can always rescue her offspring from rough play). We would expect greater concessions (including self-handicapping) by larger animals, by the larger and stronger sex, and by dominant individuals or their offspring. Observers of play very frequently report this outcome (see above, Self-Handicapping; also, dominance, Loizos 1969, Zimen 1972). We would occasionally expect to find the opposite pattern, in which younger, weaker, or less powerful individuals submit willingly to rough treatment and even return to rough play with stronger individuals again and again (evidence, Andrew 1962, Bertrand 1969, Itani 1959, Symons 1978a p. 76).

The model predicts that when the biologically important quality of play is its physical roughness, rougher animals should self-handicap in play with gentler animals, whereas Gentle animals should play as they would with other Gentle animals. When inevitable mistakes occur, Tough animals should be highly tolerant, whereas Gentle animals should protest violently. This combination of tolerance by larger or older animals, including adults, with vociferous protestation by smaller or younger animals, including infants, is extremely common in nature. Examples of such protestation were reviewed earlier in this chapter (p. 398). (For examples of tolerance, see Bourlière, Hunkeler, & Bertrand 1970, Box 1975a, Douglas-Hamilton & Douglas-Hamilton 1975 p. 91, East & Lockie 1964, Ewer & Wemmer 1974, Fedigan 1972b, Fellner 1968, Gundlach 1968, Hill 1966, Neville 1972b, Poirier & Smith 1974a, Rowell, Hinde, & Spencer-Booth 1964, Schaller 1963 p. 254, Struhsaker 1975 p. 66.) As kinship theory would suggest, adult males, whose genetic relatedness to young is always less certain than that of the young's mother, may be less tolerant than mothers (Hill 1966, Verheyen 1955), but strange females can also be highly intolerant (e.g., Linsdale & Tomich 1953). Dominant male Columbian ground squirrels (*Spermophilus columbianus*) disrupt juvenile play, but males from an area other than that of the juveniles were most violently disruptive (Steiner 1971). Bekoff (1978b) correctly suggests that this difference probably reflects a smaller degree of relatedness of the juveniles to invading males than to resident males.

In New Zealand sea lion (*Neophoca hookeri*), older juvenile males tolerate the play of pups and even join in it, whereas in the Australian sea lion (*Neophoca cinerea*), older juvenile males bite and toss small pups without provocation (Marlow 1975). This difference may result from shorter male tenure in the latter species, producing weaker genetic relationships between age-groups; it may be a consequence of more intense competition for space on the rocky beaches where *N. hookeri* breeds (*N. cinerea* breeds on sandy beaches); or it may reflect the two alternative evolutionarily stable strategies discussed above.

Because the Stingy are indifferent to the identity of their companions, whereas the Generous can gain by associating preferentially with other Generous animals, natural selection will act to produce positive assortment by strategy type. Assortment will become more positive until the demand for playmates of like type exceeds the supply. Thus we expect like to prefer like as playmates, but less strongly in small populations.

The implications of kinship for the model are difficult to spell out in detail without formal analysis paralleling that by Mac-Nair & Parker (1978) and Parker & MacNair (1978). However, analysis of a simpler model (Model II) suggests that some interesting phenomena may occur. Consider two siblings, A and B, related by r. Sibling A is bigger, stronger, or older ("tougher") than sibling B (sex differences may translate into differences in size and/or strength). Because of their differences, A and B have different optima for play, resulting in V_A and V_B, respectively. If A and B both generously compromise to play, they gain V_A/k and V_B/k, respectively, where $k > 1$. If one sibling is Stingy and the other Generous, play occurs at the Stingy sibling's optimum and the other sibling reaps no individual gain (rather than the fractional gain assumed in previous models). If both are Stingy, no play occurs and neither gets anything. Table 7-3 gives the expected gain in inclusive fitness for each sibling as a function of the possible strategies employed by both.

To solve this game we seek a so-called Nash equilibrium solution, a choice of tactics for each player that does not permit unilateral deviation by either player (Ho 1970). Clearly, mutual

Table 7-3 PAYOFF MATRICES FOR MODEL II

| | **Against B when B is:** | |
	Stingy	**Generous**
To A if: Stingy	0	V_A
Generous	rV_B	$\dfrac{V_A + rV_B}{k}$

| | **Against A when A is:** | |
	Stingy	**Generous**
To B if: Stingy	0	V_B
Generous	rV_A	$\dfrac{V_B + rV_A}{k}$

stinginess is never a solution. However, mutual stinginess and therefore failure to play can evolve in a slightly more realistic model in which an individual that plays at its partner's optimum actually suffers a finite cost C. In this model, A's inclusive fitness when Generous against a Stingy B is given by $rV_B - C$, and B's inclusive fitness when Generous against a Stingy A is given by $rV_A - C$. Then for sufficiently large values of C, individual animals are selected not to play at all, even though their populations would benefit if they compromised and played at an intermediate value—a classic sociobiological dilemma. Each of the other three combinations of tactics is stable under particular conditions, as the following analysis will demonstrate.

Case 1: $k - 1 > r$. Here, relatedness is low relative to the cost of compromise. When it has relatively little to gain from play, $V_A/V_B < r/(k-1)$, the Tough sib should self-handicap and its partner should be Stingy. Conversely, when the Gentle sib has significantly little to gain, $V_A/V_B > (k-1)/r$, play will be one-sided in favor of the Tough sib, and its partner will submit to mauling and bullying. When both sibs stand to gain comparable amounts, $r/(k-1) < V_A/V_B < (k-1)/r$, both strategies (bullying and self-handicapping) are stable. However, sibs may prefer different stable strategies. If $V_A/V_B > 1/r$, then both sibs agree on bullying; if $V_A/V_B < r$ then both agree the Tough sib should self-handicap; if $r<V_A/V_B<1/r$ then A wants to bully B and B wants A to self-handicap. In particular, if compromise is not very costly ($k<2$) the sibs will always disagree. Here, the participant that first announces its intentions can make its partner play at the initiator's preferred point. Or, each could attempt to misinform the other about the value of V_A/V_B. Each sib will attempt to exaggerate its own benefit from play and deemphasize the other's benefit in order to give the situation the appearance of one in which both would agree on style of play. This type of misinformation could be communicated by an individual's exaggerated displays of enjoyment (laughter, antic or manic acts, including unusual types of movement) during its preferred style of play and by its exaggerated protests (pain vocalization, struggling, threatening, or behaving submissively) during the partner's preferred style of play. Note that such deception is only expected over a narrow band of V_A/V_B values. When sibs' relative benefits are truly very different, the model predicts no such deception. Occasionally, an animal squealing in play is actually attacked by nearby immatures who were not previously playing [e.g., African wild dog (*Lycaon pictus*), van Lawick & van Lawick-Goodall 1971 p. 75; baboons (*Papio* sp.), Rivero pers. comm.]. One possible interpretation of this behavior is that the squeals were exaggerated and that intervention occurred as a form of punishment.

Case 2: $k-1<r$. Here, relatedness is high relative to the cost of compromise. As in case 1, self-handicapping by the Tough sib is stable for V_A/V_B sufficiently small (namely, for $V_A/V_B<(k-1)/r$), whereas a bullying play relationship is stable for V_A/V_B sufficiently large (for $V_A/V_B>r/(k-1)$). The predictions of the model in this case for $(k-1)/r<V_A/V_B<r/(k-1)$ differ sharply from those obtained in the previous case for intermediate V_A/V_B values. Here, compromise is the only stable strategy.

Would it ever pay one sibling to misinform the other regarding their cost of compromise k? Consider the simple case $V_A=V_B=V$. If $k-1>r$, in truth, then the Tough sib is better off gaining V as a bully than gaining $V(1+r)/k$ in a compromise. However, if its sib refuses to be bullied, and $k-1<1/r$, then by deemphasizing the value of k the Tough sib would be better off compromising than self-handicapping, since $V(1+r)/k > rV$. The Tough sib might accomplish this form of misinformation by failing to communicate decreased enjoyment when bullying play changed to mutual play or by failing to communicate increased enjoyment when ceasing to self-handicap. Under the same conditions ($k - 1 < 1/r$), the Gentle sib may find it advantageous to attempt to misinform its sib about the cost of compromise if the sib refuses to self-handicap. It could perhaps achieve this by failing to communicate changes in its enjoyment when play style changed. The other parameter subject to possible misinformation is degree of relatedness. It is difficult to imagine how nonhuman animals could misinform one another about their relatedness (Popp & DeVore 1979), but if some behavioral characteristic, such as a play-signal, particular locomotor-rotational movement, or style of play, was associated with one genealogy other animals might rapidly imitate it in order to simulate a closer degree of relatedness. Such play-signal "fads" are well known in chimpanzees (van Lawick-Goodall 1973), but their relationship to misinformation needs to be explored further.

If A is B's older sibling, V_A/V_B and k should both depend on the age difference between A and B. Plotted as a function of B's age, V_A/V_B will decrease monotonically, at least in the simplest case (V_A/V_B a ratio of two Gaussian distributions with equal variances). According to the model, play between kin who are closely related or who are close in age should follow a different ontogenetic course from play between kin who are distantly related or who are far apart in age. An older individual should bully a younger relative when they first begin to play together. As the younger playmate ages and the benefits of the two from play become more nearly equal, closely related or similar-aged kin should begin to play mutually, whereas distantly related or distant-aged kin should enter a phase marked by reversals between bullying and self-handicapping, misinformation of various sorts, and conflict over play. At later ages, in both cases, the older animal should self-handicap.

A natural extension of the model to include a third related playmate (Fig. 7-13) or a parent might be difficult, but very in-

Fig. 7–13 Domestic cat (*Felis catus*) play involving three kittens. Two kittens in the background spar with their paws, the kitten in the foreground makes paw contact with one hindleg of the kitten in the middle. Social play involving three or more participants may or may not be stable. Nong Khai, Thailand. (Robert Fagen)

teresting (compare the three-person game discussed by Hamilton 1975). There is need to consider strategies for sequences of contests, possibly by using finite state game models (Ho 1970). We might also consider a third possible tactic, intermediate between stinginess and generosity. However, the major prediction of this simple model is that qualities of play relationships between individuals should undergo predictable ontogenetic changes and that the pattern of these changes will vary predictably with relative age and genetic relatedness. Longitudinal data on play between known individuals are rare indeed. Bekoff (1974c) on canids, Cheney (1978) on chacma baboons, van Lawick-Goodall (1971) on chimpanzees, and White (1977b) on rhesus macaques are among the few ethologists to report such information, but not in sufficient detail to test the model's predictions.

In nature, play between individuals belonging to different categories (male-female play or juvenile-infant play, for example) presents an interesting opportunity to test sociobiological theories of conflict of individual interests in play. In rhesus macaques, the most frequent form of male-male play is wrestling and the most frequent form of female-female play is chasing (although both forms of play occur commonly in both sexes) (Levy 1979, Suomi & Harlow 1977). If males of a species usually wrestle together, whereas females of that species usually chase together, what will happen when a male and a female play together? In this battle between the sexes, whose interest will prevail? The simplest analysis of this question assumes that males benefit more from wrestling and females from chasing, leading to a conflict of interest over mixed-sex play.

An evolutionary model discussed above (p. 426) predicts that under certain conditions the partner who declares its intentions first has the advantage in determining the style of play. Therefore, the form of inter-sex play should often depend on the identity of the initiator. Conclusive data are not available on this point, but Suomi & Harlow (1977) suggest, consistent with this prediction, that the likelihood that male rhesus will wrestle (as opposed to chase) with females depends on the sex of the individual initiating play.

Field studies of play between sexes in rhesus yield interesting if unexpected results. The models of social play strategies discussed above all make the prediction that play between members of different classes should either take the same form as intra-class play or should exhibit intermediate ("compromise") characteristics. However, this is not always the case in practice.

In young rhesus macaques, the relative amounts of wrestling and chasing in inter-sex play seldom fall between the corresponding intra-class values and rarely even coincide with either value. Instead, completely counter to the predictions of the simple sociobiological models discussed earlier, the inter-class values generally represent extremes! For example, the proportion of total playtime spent in physical contact (and also the proportion of contact time spent in "gentle," or, more accurately in this case, tense, Symons 1978a p. 64, contact) is greater in inter-sex play, in play of yearlings with two-year-olds, and in play of yearlings with three-year-olds than in play within the given age-classes or sex-classes (Levy 1979). Play between animals of different ages or sexes often seems to a human observer to be potentially threatening and anxiety-producing to the smaller or weaker participant, despite the tougher partner's obvious self-handicapping. Interestingly, when an older female and a younger male play the age and sex components of toughness approximately cancel each other, leading to play in which qualitative characteristics resemble those of intra-class play (Levy 1979).

A closer look at the nature of play-chasing in rhesus helps explain the apparent paradoxes just discussed. Chasing may sometimes occur, not because it results in a greater gross benefit to both partners, but as a less costly (if less beneficial) substitute for wrestling. Females' tendency to escalate in wrestling actually exceeds that of males (Symons 1978a p. 64; no data on escalation of chases in either sex). Therefore, although females might prefer to wrestle all the time if no costs were involved, the high cost of wrestling with other females might force them to chase with other females instead if the *net* (benefit minus cost) payoff from chasing with females exceeded the net payoff from wrestling with them. If this is the case in nature, then females might chase with females, but wrestle with males, and females would reap a greater payoff from inter-sex play by playing at the male "optimum" than by playing at that characterizing female-female play. This result could be incorrectly interpreted as male victory in male-female conflict over the style of inter-sex play.

The foregoing argument still fails to explain extreme values in inter-sex play. Intuitively, the reason for the extreme frequency of chasing in inter-class play appears to be the gentler partner's fear or anxiety about wrestling, even in play and even if the partner self-handicaps, with a larger and/or stronger ("tougher") playmate (Levy 1979). In biological terms, this fear would result from the gentler partner's assessment of its expected cost of injury in inter-class play. The models discussed above failed to predict the observed extreme properties of inter-class play because of their failure to recognize this sociobiologically significant component of the cost of play.

Analysis of inter-class play is only one empirical approach to sociobiological questions raised by social play. Analysis of play involving asymmetry, identifiable roles, and possible reciprocation represents a second empirical approach to the sociobiology of play between animals. For example, Wasser (1978) suggested that distinct roles of "predator" and "prey" occurred in one form of play between tiger (*Panthera tigris*) cubs. If predatory skill is refined by this type of play, Wasser argues, then perhaps the animal playing "predator" actually benefits, not the animal playing "prey." Given this supposition, one can imagine a selfish animal who would only play if it took the role of "predator." This is countered, according to Wasser, by recipients' influence on the initiator's behavior. Only certain responses by the "prey" will serve to train the predator. Therefore, "a partner withholding responses may suffer from similar treatment later on," and this ability of recipients could "prevent one individual from persistently playing predator while refusing to play prey."

The concept of reciprocal altruism also proves useful in analyzing data on interactions between asymmetry, short-term stability, and long-term stability of play in rhesus macaques. In rhesus, play encounters that escalate are not rougher than play in general, but they appear to be unusually asymmetric: an encounter that escalates into a fight is virtually one-sided even before the real fight begins, with one partner (generally the older monkey or a male with a female partner) performing the majority of mock attacks while the other remains passive (Levy 1979). Both asymmetry and escalation are especially frequent in older rhesus (Levy 1979). Suppose that failure to reciprocate in play by reversing roles resulted in one partner reaping more than its share of the benefits of play (from the other partner's point of view). Then, asymmetric play would be especially likely to break down. The particular "games" to be played and the intensity of play are not the only issue over which conflict and negotiation can occur when animals play. The data and inferences just discussed make it clear that role reversal and reciprocity are also important and merit further investigation.

Finally, what are the implications of the fact that social play may involve repeated interactions between individuals? Animals may be concerned with each other's short-term and long-term attitudes and intentions as well as with overt behavior. Successful play requires recognition, prediction, and cooperation, i.e., "trust." Agonistic fights, especially those resulting from misunderstandings that occurred in play, which the partners could not resolve other than by escalated fighting, would damage this trust. Because an animal at play may be vulnerable to injury, an untrustworthy partner (one lacking the desire or ability to inhibit its own acts and its responses to the other's mistakes) would be shunned in favor of partners with whom the individual could play well. Because skill is refined in small increments over a long period of time, it is important to find play partners that one can trust in the long run. Players could even be selected to behave so as to develop their partners' skills only by small increments, as a safeguard against cheating. As a result, the most skilled animals would be those that succeeded in cooperating on a long-term basis in play. One cost of dominance to dominant juveniles may be lack of participation in play and lack of opportunities to refine skill. Thus the ability to win a dominance fight early in life may not persist, and an early hierarchy may later be overturned as a result of training or bonding experienced in play by the subordinate animals. Dominance reversals do indeed occur in coyote pups (Knight 1976).

AGE DIFFERENCES IN SOCIAL PLAY

Frequency of social play changes with age, declining in adulthood (Chapter 5) except in particular cases discussed below ("Social play of adults," p. 438). Qualities of social play, including stability, symmetry, and mutuality, also change with age (S. Altmann 1962b, van Lawick & van Lawick-Goodall 1971 pp. 124–125, Levy 1979, Symons 1978a). Older animals, especially certain individuals, appear to become increasingly conscious of status, increasingly unwilling to lose playfights, and increasingly unwilling to accept a subordinate position in play, whereas younger playmates appear tense or anxious when older individuals solicit play from them. In a litter of golden jackal (*Canis aureus*) cubs, around the time of weaning and ingestion of large pieces of meat, Rufus, the largest, strongest, and most aggressive cub, "was most likely to bully, transforming a game into fighting at the least provocation" (van Lawick & van Lawick-Goodall 1971 p. 125). At this time, gentler animals appear to approach play with increased tension, anxiety, and fear. Female rhesus macaques exhibit this attitude toward play even as yearlings, and especially toward female playmates (Symons 1978a pp. 64–65, 153–154). These changes may suggest that as physical ability improves through play, play must be more like true fighting in order to further improve ability, behavior of immatures becomes a better predictor of adult behavior (including competitive ability), animals begin to assess each other, to misinform each other, and to cheat in playfights, the cost of play increases and the benefits of play decrease, trust breaks down, and play decreases in frequency. In this sense, play may be said to contain the seeds of its own destruction. The facts are not quite this simple, however. As adults, the golden jackal Rufus and his three sibs returned to their mother's den when her next litter was born four months later. At this time the four grown sibs played together mutually, boisterously, and without evident aggression or bullying, and they even solicited play from their tiny new siblings (van Lawick & van Lawick-Goodall 1971 p. 140). This anecdote suggests that qualities of play are adaptive aspects of individual behavior and that unidirectional ontogenetic trends in stability or mutuality of play have no special status. This is precisely the message of the game models of this chapter, but better models of this kind are needed.

As Janet Levy (1979) has argued, ontogenetic changes in quality of social play represent an unsolved theoretical problem. In theory, play could decline in frequency with age in the absence of any change in quality, merely through decreasing tendencies to solicit play or to accept others' solicitations. Instead, the whole character of social play appears to shift. Stable cooperation gives way to provocative self-assertion and status seeking, as exemplified by human adolescent bullying and teasing in play.

In rhesus macaques, a decline in frequency of social play during development is accompanied by certain changes in the quality of play (Levy 1979). Play of older juveniles, especially males, is less mutual and more asymmetric, one partner performing mock attacks while the other partner adopts a defensive role. In these older animals, play becomes one-sided. Apparently, less "tough" (subordinate, younger, or smaller) individuals become anxious, tense, or fearful about the intentions of a "tougher" partner. They do not solicit play from it and reject its play solicitations even if its behavior suggests every intention of self-handicapping.

When partners must play either at one individual's optimum or at the other's, they could take turns being Generous. The evolution of long-term play relationships embodying this rule may be seen as a special case of the evolution of reciprocal altruism. If behavior ensuring long-term reciprocal altruism requires more brain power than most nonhuman animals can muster, then such altruistic play interactions are unlikely to be stable against cheating in nonhumans, and natural selection can at best produce two simple rules of conduct.

(1) Play with animals whose requirements for play are as much like yours as possible, and prefer other individuals for play according to the extent to which their preferred quality of play matches yours. A corollary of this rule is that animals should readily forgive the transgressions of long-time play companions, but should strongly avoid strange animals with whom play has not gone well. A second corollary is that if companions having similar requirements are not available, kin whose age is closest to one's own should be preferred.

(2) Behave so that the benefit obtained from any single play interaction is sufficiently small to make cheating of little use.

These rules make selection of play partners very easy because the most similar individuals will always prefer each other. However, because individual differences will always exist, some conflict is expected. Under conditions where self-handicapping is likely to evolve according to the models discussed earlier, most conflict over play should represent contests for partners, especially in small social groups. Two animals might fight for a third partner. Disagreements over termination would occur when one animal attempted to leave a second in order to play with a third.

Data bearing on these predictions are generally not available.

Levy (1979) and Symons (1978a) report that play between individuals of different sexes is relatively rare and unstable in rhesus except for play of one-year-old males with two-year-old females. In this combination, could preferred quality of play be more nearly equal than in the complementary combination of older males with younger females? Rivero (pers. comm.) reports 43 instances of play escalating to fighting in a mixed group of yellow and hamadryas baboons at Madrid Zoo. Four frequent apparent causes of this escalation were a solicited individual refusing to play, a third individual attempting to play with one participant in an ongoing play bout, excessive roughness, and intervention by an older individual.

One factor that distorts the simple preference of like for like is apparent preference by chacma baboon infants for play with offspring of high-ranking individuals (Cheney 1978). Older individuals do not usually show such preferences in chacma (Cheney 1978) or in rhesus macaques (Levy 1979). It is therefore likely that these are not true play preferences, but rather consequences of differential availability and physical proximity. Females seek to groom high-ranking females and to remain close to them. Therefore, the infants of these females are brought into most frequent contact with the infant of the highest-ranking female (Cheney 1978).

As discussed above, periods of ontogeny during which previously stable play breaks down with increasing frequency represent the best available evidence that animals may sometimes cheat in play, exploiting play interactions for ulterior, selfish ends. To date, cheating of this sort appears to occur in two types of animals—young carnivores beginning to feed on meat supplied by adults, and adolescent primates. At these times, the society of juveniles is in a state of unrest because of new resources and/or rapid development of the members. We may consider this situation from two separate points of view: that of an animal seeking to establish or reaffirm its status with respect to new resources or a changing pattern of competitive abilities and that of an animal who had developed rapidly, secured access to resources important at a young age, and who now faces a potential challenge with respect to the new resource or as others begin to catch up in competitive ability.

1. Animals "on the way up." As Popp & DeVore (1974) argue, these individuals may compete

for the sake of status *per se*. That is, they are competing for the considerable future benefits that will accrue if they are able to win in initial encounters, i.e. begin to establish a higher status *vis-à-vis* their opponents. By competing at an intensity that exceeds the value of C_a [their expected permissible cost of aggressive competition] for the immediate situation, the opponent will be misled with respect to the cost-benefit function for the actor's competitive behavior, i.e. the opponent overestimates the actor's intrinsic competitive ability, or the value of the resource to the actor, or underestimates the cost of competition per unit time to the actor. By doing so, an actor can increase the rate at which the same opponent avoids, or terminates quickly, aggressive encounters in similar circumstances (since the opponent believes he can not gain in future competition against such an "outstanding" or "determined" individual). (Popp & DeVore 1974)

An individual using this aggressive strategy will take all possible steps to assert itself against possible competitors. It is unlikely to accept a temporary reduction in status even in play, and it should not disregard the opportunities that play offers to emphasize its determination to win. In play, such animals, like Rufus the golden jackal, will provoke quarrels, stage disagreements, and escalate on the slightest provocation, since such behavior tends to devaluate their own competitive costs. It is usually assumed that the target of such intimidation is the partner, but the actual target may in fact be a third individual whom the actor is unwilling to challenge directly. Possible evidence for this statement is supplied by observations that two individuals may play stably and apparently cooperatively for several minutes, then suddenly escalate when a third individual approaches (domestic cats, pers. obs.; baboons, H. Rivero pers. comm.; rhesus macaques, C. Berman pers. comm.).

2. Animals "on the way down." Developing animals' benefits from a given resource, their competitive abilities, and their agonistic alliances all change with age. Therefore, an immature animal who dominates its potential competitors at an early age may face increasing challenges at a developmental stage transition (e.g., onset of feeding on meat, adolescence), even if its own competitive abilities are still increasing (at a rate less than

that of its competitors). The "prestige effect" hypothesized by Popp & DeVore (1974) for aging dominant individuals is equally valid for these juveniles. There are (at least) three possible strategies for such a dominant juvenile.

(1) It may act as a cheater, teasing, bullying, and escalating in play situations while avoiding conflict over actual resources. At the same time, it may play selectively with animals much younger than itself, especially with younger kin of its competitors. These younger animals need play as training and may be unable to resist the attraction of playing with a larger individual. The cheater would then bully these younger animals, either in order to take resources from them, in order to impress its peers, or in order to increase its peers' cost of competition with the bully who injures their youngest kin. The bully's victim then finds itself in a classic sociobiological trap. The bully may injure it, then immediately solicit play and even adopt an inferior fighting position, as if to say "I was only joking." The bully may even diabolically succeed in deceiving its victim into believing that their interactions really are play and into coming back for more.

(2) "Cheater" is not the only possible strategy for a dominant bully under increasing pressure from below. A second strategy is to be an "aloof rejector." Competitors on their way up may seek to play with the bully in order to cheat against it (as argued above) or simply in order to test their growing competitive ability against the bully's declining power. As Popp & DeVore (1974) argue in their discussion of the "prestige effect" in adults, fading individuals "are expected to conceal their declining competitive abilities" and should behave so as to give "the impression of complete disinterest" in tests of relative strength. Thus, the bully will be an aloof rejector, avoiding play and rejecting solicitations even though it remains in perfect health and good physical condition. However, this strategy, like cheating, can only be a stopgap, since sooner or later the bully's bluff will be called. A third strategy is then appropriate.

(3) The third strategy for the dominant individual, now fading rapidly, is to use play to seek alliances with other individuals. The bully suddenly becomes a "compulsive solicitor," desperately seeking play (and presumably friendship, if the co-

hesion hypothesis has any meaning) from animals it previously dominated and ignored. Bekoff's (1974c) dominant coyote, who had such trouble soliciting play, may have been on the way to becoming such a compulsive solicitor. If the bully forms friendships, it may do so with other former bullies now in the same state, and they may form a peripheral group (Levy 1979) or may emigrate together (Cheney 1978, Packer 1979a). If no other individuals find it in their interest to accept the former bully's advances (a situation evoked by the stock saying "Where were you when I needed you"), the former overlord becomes a social isolate and may eventually emigrate alone. This scenario may furnish an evolutionary basis for Bekoff's (1977a) speculations on play, dominance, and dispersal in young social mammals.

Cheaters, aloof rejectors, and compulsive solicitors—what could be farther from the classic view of play as innocent, selfless cooperation? The dark side of play has received little attention in the past. It seems destined to receive ample emphasis in the future, given the current seductiveness of social theory. However, the dark side of play is not a perplexing mystery. The most exciting aspects of play awaiting evolutionary analysis are those baffling cases in which play currently appears to be wholly cooperative, without any obvious indications of selfish cooperation or cheating (e.g., gentle play, p. 308). Here, sociobiological analysis of play faces its greatest challenge.

SOCIAL PLAY OF ADULTS

Adults of many mammalian species play socially. This play is of interest because of the adult ability to seriously injure other animals and because hypotheses about developmental roles for play that are specific to juvenile or infant stages require modification in order to explain play in a skilled and trained adult.

Adult social play appears to fall into four major demographic categories: play between adults of the same sex; mixed-sex adult play, often preliminary to mating; group play; and play of adults with immature conspecifics, including infants. The variety of likely functions that such play serves, and the variety of contexts in which it occurs, both point out the opportunistic nature of selection for playfulness.

Like-sex Adult Play

Playfighting and play-chasing by adults within their own sex is generally rare compared to corresponding activities of juveniles, but it does occur, notably in spotted hyenas (Kruuk 1972). Adult female Steller sea lions chase and playfight with each other in the evening when they visit tide pools (Farentinos 1971), female howling monkeys play with each other as adults (Bernstein 1964), and African lionesses "readily play into old age" (Schaller 1972 p. 164). Male-male play is rare in adult lions (Schaller 1972 p. 164) but frequent in bonnet macaques (Simonds 1965).

Mixed-sex Adult Play

The courtship ceremonies beloved by ethologists are as infrequent in mammals as they are common in birds and fishes. With a few interesting exceptions [e.g., the green acouchi (*Myoprocta pratti*), Kleiman 1971], mammals eschew the formal, ritualized display sequences for which such animals as great crested grebes and three-spined sticklebacks are justly famous. Instead, they may sniff and threaten each other, the male may simply chase the female, or in some instances, vigorous interactive play may occur, which, except for the size of the animals, is virtually indistinguishable from the play of juveniles (Aldis 1975 p. 111, Bekoff 1974c, Ewer 1963, Kaufmann 1962, Mech 1970, Tembrock 1958, Woolpy 1968). It is difficult to see how play in this particular context would be especially suited to skill development, physical training, or central nervous system stimulation; one of its consequences may be to bring the animals into readiness for mating, but this hypothesis fails to explain why some species (like many primates) are ready to mate without play. One function of courtship play may be to display one's own prowess at hunting, fighting, and predator escape and to test that of the partner, or simply to ascertain whether the partner is sufficiently healthy and has enough energy to wrestle and run. This assessment hypothesis can be indirectly tested by comparing species having different systems of parental care. If this assessment is involved in premating play, one would predict that the most interactive and egalitarian play would occur in species in which both parents cared equally for the young, since each parent would need to test the other. Judging from the preponderance of canid examples cited above, this may well be the case. Relative interactiveness of courtship play is an excellent topic for canid-felid comparisons, since paternal care of offspring is as common in the Canidae as it is rare in the Felidae (Kleiman & Eisenberg 1973). Schaller (1972) describes playful avoidance of the male lion by the female; this play is far less balanced than that of the canids cited and is not reported to exhibit role reversal. The courtship play of free-living and confined cheetahs involves several males and a female and is apparently more similar to canid premating play than to that of lions (Eaton 1974, Florio & Spinelli 1967, 1968). The premating interactions of the confined group of male cheetah observed by Benzon & Smith (1975) were unrestrainedly aggressive and appeared far from playful. Here, as in the case of the lions, felid courtship behavior seems far less playful than that reported for canids. In confinement, male and female felids often live together in relative peace and may frequently play with each other, phenomena seldom recorded in the wild. I observed behavioral preliminaries and three copulations between margay (*Leopardus wiedi*), two between sand cats (*Felis margarita*), and one between jaguarundi (*Herpailurus yagouaroundi*) in indoor enclosures. The only pair to exhibit play was the jaguarundi, and their behavior was rather like that of the lions Schaller observed. The female rolled on her back and playfully cuffed the male, who directed inhibited bites at her head and chest. This interaction was followed by a chase with the female in the lead. When the female was approached by the male, she crouched, the male mounted, and mating occurred. Because confined adult carnivores housed in compatible pairs often obtain much of the stimulation they need by playing with each other, this set of observations is by no means conclusive. Existing evidence suggests that canid premating play is in fact more interactively social than felid premating play, as predicted by the assessment hypothesis.

Group Play

Except in the spotted hyena (*Crocuta crocuta*) (Kruuk 1972), in lagomorphs (Flux 1970, Lockley 1974), and in humans, play in mixed-sex groups of adults appears to be rare in nature unless it occurs as a preliminary to mating. However, one exception to this rule is known and more can be expected. African hunting dogs (*Lycaon pictus*) play socially before going out to hunt (Estes & Goddard 1967, Kühme 1965) as do timber wolves (*Canis lupus*) (Mech 1970) and choz–choz (*Octodontomys gliroides*) (S. Wilson & Kleiman 1974). Traditionally this play was viewed as arousing the pack for the hunt and as coordinating group activity. Estes & Goddard (1967) call it a "pep rally." Exactly what behavioral systems are aroused, what effects are produced, and just how play optimally coordinates group activity are questions for additional investigation; the biology of pep rallies is not well studied. I suggested in Chapter 5 that this play serves as physical warmup, but there may be more to the story. Do these animals need to know which individuals are rested and well fed and in good physical condition in order to coordinate their hunting behavior more effectively? They cannot ask each other how they feel, but they could conceivably obtain such information from play. It could also be the case that play originally served as warmup and was so predictable that it gradually became a stimulus necessary for inducing behavioral readiness to hunt. A tired or ill animal would have a higher threshold for the stimulus. The play of one or two individuals would spread infectiously through a pack of healthy and well-rested animals like an epidemic, but if only a few animals were ready and the rest were tired or unwilling the "infection" would be confined to these few foci and would not spread. As a result, no hunt would occur. This mechanism is a logical possibility for coordinating many kinds of group activity that require animals to be in a prepared or excited state and to know that group-mates are also in this state. The wild dogs could, in principle, of course, have evolved a less energetically costly "ready to hunt" signal, but if they must play anyway in order to warmup, this play can then serve a dual function. Although it was perhaps originally selected as a warmup adaptation, any change in its present properties that resulted in less efficient group arousal is likely to be counterselected.

We might expect warmup play in other socially hunting carnivores whose cooperative techniques are similar to those of the wild dog or of the wolf. For instance, a "bout of greeting," which may include play, precedes pride lions' departure from their resting site (Schaller 1972 p. 120). According to Schaller, pride lions may hunt cooperatively, although they do not always do so.

Play of Adults with Immature Conspecifics

Why should a physically fit, fully skilled adult, whose daily feeding and predator avoidance activities give it abundant and perhaps even excessive stimulation, throw itself on its back in front of an infant and wave its legs in the air while the infant flails clumsily, but perhaps painfully, at it? We could argue that this is the adult's way of relaxing after a hard day, but "relaxing" is not defined nor is the mechanism by which play is optimal for producing this (undefined) result. In any case, appeal to human experience suggests that after a hard day at work, the last thing that a parent needs is a boisterous romp with an overenthusiastic child. Zahavi (1977) ingeniously suggests that this play is an assessment mechanism, a behavioral way to test the parent's willingness to continue to contribute to the offspring in the future. Mechanistic and evolutionary details of this speculation deserve further clarification. Zahavi's argument may apply either to formation of a new relationship or to maintenance of an existing one. It seems more likely that the parent will not be able to know just how much play or what type of play the offspring needs unless the offspring solicits play in this graded manner. Of course, the offspring may ask for more than it really needs, because the parent may not be willing to offer its offspring the requested amount of play. However, since the offspring's needs will enter into the parent's calculations, it is in the interests of both to be able to communicate about the off-spring's state of need.

A parent can facilitate the persistence of its own genes by par-ticipating in its offspring's (adaptive) play. In fact, play of adults with infants and juveniles is by far the most frequently cited of the four types of adult play. Such play often takes place be-tween a mother and her offspring (chimpanzee, van Lawick-Goodall 1967, 1968a; fossa, Albignac 1969a, 1975; margay, K. Wiley pers. comm.; pygmy hippopotamus, S. Wilson & Klei-man 1974; pinnipeds, Farentinos 1971, Paulian 1964, S. Wilson 1974b) or between a father and his offspring (meerkat, Wem-mer & Fleming 1974). The adult, and especially the male, may allow the young animal to climb, hit, bite, tug the adult's tail, and the like, treatment that would hardly be tolerated if at-tempted by another adult (Bourlière, Hunkeler, & Bertrand 1970, Schaller 1972). In many reported cases of adult-immature play, there is no information on the genetic relatedness of the pair (Bernstein 1964, Hall 1962, Schaller 1972). However, in the Serengeti, for lions belonging to a pride, Bertram (1976) has es-timated that the average degree of genetic relatedness between a male and a cub is 0.31 and that average relatedness between a female and a cub is 0.50 if the cub is her own and 0.15 if it is not hers. The adults that Schaller observed playing with cubs were very likely to have been close kin of the cubs.

Adult female rhesus monkeys in free-living groups play with infants and with juveniles, but they were not observed to play with other adults. In 9 of the 14 cases observed, the individuals were closely related (mother–daughter or sister–sister play) (Symons 1978a).

Adult play may benefit adults in ways other than those men-tioned above. An adult gains status and protection when it holds an infant. Playing with an infant could effectively lure it away from its mother. In the most sinister version of this scen-ario, the (unrelated) adult would then proceed to kill the infant, but although infanticide occurs in nature, it is never preceded by play, and the selfish intent of adults who play with infants is probably status and protection, not infanticide. (In captivity, play may lead to infanticide; see Angst & Thommen 1977 for an example.) Play in adult interactions almost seems to carry a

Fig. 7–14   Human invites play by showing a toy to the dog.

Fig. 7–15   The dog solicits play with a play-bow.

444

Fig. 7-16 The dog leaps playfully and seizes the toy in its mouth.

Figs. 7-14 through 7-16 Domestic dog-human play. Photographic sequence of a single interaction. Pennsylvania. (Dave Dorn)

paradoxical message of long-term goodwill when it precedes a displacement or other competitive interaction; the initiator achives its immediate goal by playing (Breuggeman 1978), and the other individual both benefits from the play and is reassured that this event should not be taken to indicate that future contact between the two will be competitive. Similar behavior is seen in common seal (*Phoca vitulina vitulina*) mothers, who sometimes play with their offspring before leaving while weaning them (S. Wilson 1974b).

INTERSPECIFIC PLAY

The possibility of play between members of different species (Figs. 3-7, 7-14, 7-15, 7-16) is interesting for several reasons. First, individuals must recognize and respond to another species' play-signals, so that interspecific play would offer examples of interspecific communication, a phenomenon rarely demonstrated in nature. Second, differences in preferred content of play, and differences in body size and shape, ought to make interspecific bouts unstable or require extreme self-handicapping if they are to be maintained. Third, individuals of different species must respond correctly to each other's control mechanisms in order to maintain play. Fourth, because members of different species have particularly few genes that are identical by descent, interspecific play should be especially subject to cheating except for cases in which reciprocal altruism has evolved.

Table 7-4 lists examples of interspecific play reported from wild populations. Many more instances are known from animals in confinement (e.g., Maple & Zucker 1978). Instances of non-mutual play, e.g., a predator repeatedly tossing and catching a prey animal without killing it, are omitted. The phenomenon is rare, but, as Rose's (1977a) observations fully confirm, interspecific play does exist, even in the wild.

Play occurs between species that interact in other ways as well. Baboons were reported both to play with and to hunt and kill vervet monkeys in Amboseli National Park, Kenya (Hausfater 1976). However, following a decrease in the quality of the Amboseli habitat (mid-1960's), all baboon-vervet play ceased and has not yet resumed, whereas baboon predation on vervets continues (S. Altmann pers. comm.). Young baboons in Amboseli still play by themselves and with one another, but they threaten young vervets.

Chimpanzees both play with and hunt and kill baboons (van Lawick-Goodall 1971). Several species of West African forest primates form multi-species assemblages (Struhsaker 1975). Young of these monkeys play interspecifically as well as intraspecifically (Gautier-Hion & Gautier 1974).

Table 7-4 OBSERVATIONS OF INTERSPECIFIC PLAY IN WILD ANIMALS

PRIMATE–PRIMATE

Cercopithecus aethiops–Papio cynocephalus	Altmann & Altmann 1970
Papio anubis–Theropithecus gelada	Dunbar & Dunbar 1974
Cercopithecus nictitans–Cercopithecus pogonias	Gautier-Hion & Gautier 1974

Papio anubis–Pan troglodytes	van Lawick-Goodall 1971
(Hylobates sp. unsuccessfully solicits play from infant *Pongo pygmaeus)*	MacKinnon 1974b
Presbytis obscurus–Symphalangus syndactylus	McClure 1964
Cercopithecus aethiops–Colobus guereza	Rose 1977a
Cercopithecus ascanius–Colobus badius	Struhsaker 1975
Cercopithecus mitis–Colobus badius	Struhsaker 1975

PRIMATE–RODENT

Cercopithecus campbelli–Heliosciurus gambianus	Bourlière, Bertrand, & Hunkeler 1969 (see also Bourlière, Hunkeler, & Bertrand 1970; Hunkeler, Boulière, & Bertrand 1972)

PRIMATE–CARNIVORE

Canis aureus–Papio ursinus	Saayman 1970
Cercopithecus campbelli–tame *Crossarchus obscurus*	Boulière, Bertrand, & Hunkeler 1969

PRIMATE–ARTIODACTYL

Papio anubis–Tragelaphus scriptus	Douglas-Hamilton & Douglas-Hamilton 1975
Papio anubis–Aepyceros melampus	Grzimek & Grzimek 1960
Papio anubis–Tragelaphus scriptus	Grzimek & Grzimek 1960
(Macaca fascicularis unsuccessfully solicits play from adult *Sus barbatus)*	Kurland 1973

CARNIVORE–CARNIVORE

Canis familiaris–Vulpes vulpes	Hurrell 1962

CARNIVORE–PINNIPED

Canis familiaris–Neophoca hookeri	Walker 1975 p. 1292

CARNIVORE–ARTIODACTYL

Gazella thomsoni–Otocyon megalotis	van Lawick & van Lawick-Goodall 1971
Alopex lagopus–Rangifer tarandus	Stefansson 1944

CARNIVORE–BIRD

Canis lupus–Corvus corax	Mech 1966, 1970
Corvus albicollis–Felis catus	Schlater & Moreau 1933

BIRD–BIRD

Corvus sp.–*Haliastur indus*	Neelakantan 1952

CONCLUSIONS

In view of the constraints on play imposed by individual self-interest, it is remarkable that play between individuals occurs at all, especially considering its hypothesized function in developing physical ability. Play with others, like any cooperative social interaction, reflects compromise between individuals' differing interests. When compromise is impossible, or when play interactions are influenced by other interactions in the relationship between two individuals, social play becomes unstable or ceases entirely. Not surprisingly, from this point of view, the most stable, cooperative, and enduring play occurs between close kin in resource-rich environments—especially between parents and offspring or between siblings.

Social play offers excellent opportunities for the study of conflict and compromise in relationships. It raises fundamental questions about animal communication and about sociobiological constraints on behavioral development. Under what circumstances should animals be selected to choose play with others? Why are there so many different kinds of play-signals? Why do some apparently healthy individuals seldom if ever play? Can competing predictions based on the skill and cohesion hypotheses be formulated, quantified, and tested, and can the hypothesized role of play in social cohesion be outlined more precisely by distinguishing between behavioral adaptations and behavioral effects? Do qualities of an individual's play interactions with others mirror or even determine the quality of their overall relationship?

The study of social play is a curious *macédoine* of processes, results, biologically opaque terms like "cohesion," "trust," and "pep rally," and loaded words like "mistake" and "cheating." Sociobiological studies of social play that acknowledge developmental mechanisms can be immensely valuable. Ethological methods make it possible to measure relevant variables without understanding their interactions. Such methods rely on imprecise theoretical ideas that require expression in the form of explicit statements, empirical justification, and in some cases, clarification and testing with the aid of mathematical models.

8

Play, Innovation, Invention, and Tradition

In Vladimir Nabokov's *Look at the Harlequins* (1974), Baroness Bredow, the narrator's extraordinary great-aunt, advises her difficult young grandnephew:

"Stop moping!" she would cry. "Look at the harlequins!"
"What harlequins? Where?"
"Oh, everywhere. All around you. Trees are harlequins, words are harlequins. So are situations and sums. Put two things together—jokes, images—and you get a triple harlequin. Come on! Play! Invent the world! Invent reality!"

Does play "invent the world"? Ethologists believe that an animal's world is created not only by its specific sensory capabilities (in the restricted sense of von Uexküll's *Umwelt*), but also by its behavior. Behavior *qua* experience creates later behavior epigenetically. What an animal does determines the experiences it has and therefore the way it perceives and perhaps even understands its world. In particular, a novel skill can add new functional capabilities to an animal's behavioral repertoire. Behavioral novelty furnishes new food sources (Kawai 1965, Konner 1977a, Mayr 1963), changes an individual's social position (van Lawick-Goodall 1971), extends a species' geographical range (Kawai 1965), and even opens a new ecological niche (Mayr 1963).

The biological study of animal inventiveness exploits both mechanistic and evolutionary modes of analysis. What behav-

ioral processes can yield novel skills? Under what environmental circumstances will these processes evolve, and by selection at what level of population structure? B. Beck (1975, 1978)
and Fagen (1974a), like earlier commentators on the problem
(e.g., Jolly 1966b), recognized that these questions of origin
could not be answered without considering social factors, in
particular the mechanisms of dissemination of novel skills and
the evolution of social channels for the spread of novel behavior. Play has been viewed both as a mechanism of origin for behavioral novelty and as a mechanism of dissemination to other
members of the society in which the novel skill originates.

Play could produce innovation directly or developmentally.
According to direct-effects hypotheses (Einstein 1945, Fagen
1974a, Fedigan 1972b, Jolly 1966b, Miller 1973), during its play
the animal puts together a new plan or treats an object in a new
way, and the resulting behavioral effect has some utility to the
performer or its kin. As Symons (1978a) points out, this hypothesis is naive. Even though blind trial and error can sometimes lead to breakthroughs in skill development (e.g., juggling, Austin 1974), these hypothetical effects of play may
simply be by-products of behavior designed to develop a given
behavior pattern through self-generated feedback.

The second mechanism is that envisaged by flexibility hypotheses about play (Chapter 5). According to these hypotheses, play experience with an object, with a skill, with response-
contingent interactions, or with another individual develops the
player's ability to be flexible and versatile in those contexts, or
perhaps even in general (Mason 1978), in the future. How these
abilities develop, and how experience should be designed to foster these abilities, is not clear. However, if play develops skill
through generalization and subroutinization, it is not a very big
jump from skill development to flexibility. Answers to these
questions would be useful in all kinds of applications, and of
course many suggestions have been made (e.g., Bruner, Goodnow, & Austin 1956, Holt 1967), but these suggestions shed no
light on the basic biology of the problems considered.

The models to follow make no explicit distinction between
the two mechanisms hypothesized above.

Acquisition of invented behavior patterns by conspecifics in
any species results in a kind of cultural transmission. The un-

iqueness of human culture may be said to reside both in processes of origin and in processes of dissemination, as the following discussion should indicate.

The study of play and behavioral innovation is problematic. Nowhere in play research has observational and experimental confirmation of theory in human behavior been more satisfying, or the extension of this relationship to nonhuman species more frustrating. Available theory and evidence suggest that human and chimpanzee play, including both specific and diversive exploration, function as a learning mechanism and facilitate tool use as well as other forms of skill and cognitive development (Dansky & Silverman 1975, Golomb & Cornelius 1977, Lieberman 1977, McGrew 1977, Sylva 1977, Sylva, Bruner, & Genova 1976, Sutton-Smith 1979). But nonhuman parallels to human playful discovery are rare. Manipulation of inanimate objects is itself uncommon in nature, even in "higher" primates like baboons and macaques (Symons 1978a). Among primates, only the great apes (Menzel 1974, Rumbaugh 1974) appear to share the human propensity for exploratory manipulation of inanimate objects and that to a relatively limited extent. B. Beck (1975, 1978) discusses interspecific variation in ability to invent and use tools and in relative use of alternative behavioral mechanisms conferring these capabilities.

Only an inventive animal can discover a new skill. But after this skill is first displayed by some member of a population of animals capable of observation learning, any other member may potentially become a performer. Whether this potential is realized depends on the events, deterministic or probabilistic, that shape that animal's individual history: on the chance, in the first instance, of contacting performers, and, subsequent to that, on the chance of learning the skill from those contacted.

Questions of origin and of dissemination are independent at the level of mechanism, but not at that of evolution, since "theft" of useful discoveries from the inventor by genetically unrelated conspecific observers would tend to counterselect genetically based individual tendencies toward innovation. Although selection above the individual level is unlikely here, it cannot be excluded *a priori* (Harper 1970, Poirier & Smith 1974b, Wade 1978a).

Genetic predispositions toward innovative behavior face a

continuing evolutionary challenge from unrelated noninventive observers. How, therefore, can the ability to make discoveries be selected? The models presented in this chapter address the simplest case of this problem. How can innovators benefit sufficiently from their genetic predisposition towards inventiveness? Will their genes persist over time, or will theft resulting from behavior of observers neutralize any temporary selective advantage and counterselect genetic tendencies toward innovation?

The model to be presented considers a hypothetical single diallelic autosomal locus (D, d) that determines presence or absence of the ability to invent new behavioral acts or skills. This random-mating, large-population, individual-selection model neglects mutation, migration, and drift. Invented behaviors are assumed transmitted from performer to nonperformer in two possible ways:

(1) Transgenerationally, from parent to offspring. Feldman & Cavalli-Sforza (1976) base models of cultural evolution on this transmission mechanism alone.

(2) Intragenerationally, from performer to peer. This mechanism should be as important in social species as that preceding. One may cite evidence for its operation in the field (Kawai 1965) and in the laboratory (Jones & Kamil 1973).

I view transgenerational and intragenerational transmission of cultural traits as equally orthodox and as equally appropriate in general models of cultural evolution. Indeed, interesting biological questions may be posed regarding the vivid, stormy, and sometimes painful interdigitation of transgenerational and intragenerational learning processes—of socialization by parents and socialization by peers. This phenomenon, in the guise of "the family versus the peer group," has already been approached from many points of view. The approach taken in this chapter is to define explicit models of two learning mechanisms, one transgenerational and one intragenerational. The parameters of these models determine, in part, whether or not innovative tendencies will evolve. It would be beyond the scope of this chapter to attempt evolutionary analyses of the processes that in turn determine the values of these parameters. However, such analyses are probably important in understanding possible differences between human and nonhuman playful discovery.

Concern with the possible fate of playful, creative individuals in variously structured social systems suggests equal emphasis on innovation and tradition and on parents and peers. Feldman & Cavalli-Sforza (1976), in contrast, considered only the roles of tradition and of parental influence; in so restricting the scope of their inquiry, they chose to disregard several biologically important factors. On the other hand, their theoretical exploration of those phenomena of particular interest to them is thorough and diverse. Their formulations assume a single autosomal locus at which two alleles may affect behavior. Their alleles, however, affect learning of a preexisting tradition. I wish to consider alleles that affect the ability to create novel traditions or skills, but I will assume no genetic differences in the ability to acquire this cultural phenotype from others for the sake of modeling simplicity. In both the models below and in those of Cavalli-Sforza & Feldman (1973a,b) and Feldman & Cavalli-Sforza (1976), a large, closed, randomly mixing population of diploid animals is assumed. In both cases, individual selection occurs and depends on acquisition of the skill. Feldman & Cavalli-Sforza (1976) vary other assumptions of their model in order to conduct a systematic exploration of selected aspects of biocultural evolution. As modeled by these authors, transmission of the parental skill may be complete or incomplete; genotypes may vary in learning ability; and either or both parents may act as teachers, or alternatively one parent may be arbitrarily defined as the teaching parent. The learned skill may either be advantageous or disadvantageous. However, these authors view learning that occurs outside the family environment as an extrinsic factor akin to mutation in genetics; they discuss possible complications due to this factor, but they present no formal models that embody it. Furthermore, they consider only special cases of their general model in which fitness is determined solely by the presence or absence of the skill. My model, in contrast, includes possible genetic differences that influence selection on the trait.

The question of possible modes of interaction between genetic and cultural evolution is a central problem of sociobiology (May 1977, Wilson 1976). The model below, initially formulated to answer specific questions about the evolution of innovative play behavior, incidentally addresses the question "What are the modes of interaction of genetic and cultural evolution, and under what conditions do they accelerate or slow one another?"

The selective advantage of an allele that affects behavioral tendencies depends on the social environment in model systems. For example, in Boorman & Levitt's (1973) social carnivore altruism (reciprocal fitness transfer) model, rare altruist alleles were counterselected. Counterselection occurred because most altruists never met other altruists in the population and perished before encountering a cooperative partner. But in a model population consisting mainly of altruists, selfishness was actually counterselected. Similarly, Maynard Smith & Price (1973) and Maynard Smith & Parker (1976) demonstrated theoretical frequency-dependent selection of alleles that affect tendencies to use non-damaging as opposed to escalated fighting tactics. Clearly, frequency-dependent selection is important in social evolution because altruism, cooperation, learning ability, and even play (see below) will thrive only in particular social environments defined by present and past gene frequencies. Feldman & Cavalli–Sforza (1976) observe that dependence on past environments implies a complex delay phenomenon that in turn confers various nontrivial kinds of dynamic behavior on genetically based models of biocultural evolution. These dynamic complexities, familiar (albeit in another context) to ecologists (see May 1974), have arisen in several areas of theoretical population biology. This circumstance seems destined to enhance biologists' consciousness of dynamical systems theory, while diminishing their confidence that all equilibria must be stable and unique, and all results intuitive.

MODELS OF THE EVOLUTION OF INNOVATIVE PLAY

This chapter presents two families of formal mathematical models of evolution of innovative play by individual selection in a large, randomly mixing population with discrete generations. The first model assumes no cultural dissemination of the innovative act (either intragenerational or intergenerational), the second assumes both types of dissemination.

The basic elements of the model are discoveries made by playful individuals. An individual has probability f of discovering a useful performance. The effect of the performance is fitness benefit σ. The cost of being a playful individual is τ fitness units. Innovative performances are transmitted horizontally with probability c (to peers, kin, other youngsters, or adults) and vertically with probability h (to a performer's offspring). Discoveries are made in a different new context every g generations.

Probability of discovery f is likely to be influenced by several factors, among them the animal's degree of physical and cognitive readiness and the extent of its previous experience in the context in which innovation is to occur (Beck 1978). An additional factor that may affect the value of f is the animal's rearing environment, including its demographic environment. Early rearing in an environment rich in response-contingent stimulation may be necessary to develop the ability to apply manipulative abilities and experience to novel situations (Konner 1975, Mason 1978, Menzel, Davenport, & Rogers 1970). Such adaptability may be a necessary prerequisite for innovation (Beck 1978).

Another factor influencing f is the degree of enrichment in the animal's recent environment. To maximize f, a period of deprivation of response-contingent stimulation or a period in which play was prevented by other environmental pressures (Chapter 6) should be followed by environmental enrichment or by an increase in opportunities for response-contingent stimulation. [Response-contingent stimulation strikes one as an unlikely resource from an ecological point of view, but Konner (1975) discusses it quite forthrightly in his analysis of sociobiology of immatures.] This ontogenetic pattern of response-contingent stimulation—an enriched rearing environment to develop versatility, then a period of deprivation and subsequent enrichment—should produce a strong rebound in play (Baldwin & Baldwin 1976a, Oakley & Reynolds 1976).

The demographic environment is also crucial to innovation.

The potential innovator must be isolated from conspecific play companions, and especially from unrelated conspecific thieves, in a resource-rich environment. If Symons (1978a) is correct in suggesting that the design of play for skill development will normally yield social play and acquisition of the same complex species-typical skill by all individuals, we expect that a resource bloom will produce a large cohort of socially playing animals and no innovation, whereas lack of resources will yield small cohorts and inter-age play (primarily among kin) or, under severe stress, no play at all.

If play and immaturity lasted several years in ontogeny, an individual born a few years before a resource bloom would be a demographic isolate. However, it might not have grown up in a stimulating environment. Or, if a cognitively precocious individual found the games of its age-class boring, but was not physically able to succeed in soliciting play from older individuals, this individual would be effectively isolated. Demographic variations in small groups might also isolate an individual.

The hypothesis of demographic constraints on innovation receives support from recent observations of play in captive animals. Object play is more frequent and more innovative when a playmate is not available. Pluta & Beck (1979) found that object play was more frequent in one singleton polar bear cub than in one pair of twins, who playfought and play-chased together. This object play included skilled throwing of large objects, behavior hypothesized to function in polar bear predation, but not yet actually observed in the wild. Dolgin (1978) found that juvenile chimpanzees provided with novel objects in both solitary and social settings used the objects more creatively when alone, as indicated by the greater number of behaviors invented and larger variety of actions performed during the solitary trials. The objects were actively manipulated for equal amounts of time under both conditions.

Benefit σ will depend on the type of discovery and on the individual's ability to exploit the opportunities provided by the novel act. Most novel behaviors known seem to involve feeding (Beck 1975, 1978, Kawai 1965, McGrew 1977), but chimpanzees also invent patterns of social communication (invented agonistic displays, van Lawick–Goodall 1971; invented play-signals, van Lawick-Goodall 1973), and birds invent new songs (Jenkins 1978). Cost τ depends on the risks taken in play and on the immediate growth and survivorship consequences of using resources in play. An animal in good health and enjoying abundant resources along with protection from predators (including parasites) and from accidents is best suited to absorb the cost of play.

Rate g of successive discoveries is probably subject to the same factors that affect f.

Transmission rates c and h depend on the animals' ability to attend, observe, learn, and perhaps teach (Beck 1975, 1978, Feldman & Cavalli-Sforza 1976, 1977, McGrew 1977). The quality of the parent-infant relationship may be an important factor in successful vertical transmission (van Lawick-Goodall 1968a).

The two families of models (Appendix V, Appendix VI) have been analyzed in some detail. Several principles emerge. In the absence of cultural transmission of behavior, innovative tendencies are selected when an individual's discovery probability f exceeds its cost-benefit ratio a (ratio of the fitness cost of playing to the fitness gained from discovery). A genetically based play ability, represented by the d allele, goes to fixation under these circumstances. But under the opposite set of conditions ($f < a$), the d allele is eliminated.

Cultural transmission changes this picture completely. Now stable polymorphism is the rule, and playfulness can be selected even when $f < a$. These results will now be discussed in greater detail.

Results of the Innovation Model
with Cultural Transmission

The model of Appendix VI has the following properties. Frequency of the play allele increases when this allele is rare, since initially most performers are animals who have discovered the novel behavior themselves rather than learning it from others. As the discovery begins spreading in the population at large, playful animals continue to suffer the cost of their innovative tendencies, but no longer enjoy exclusive possession of

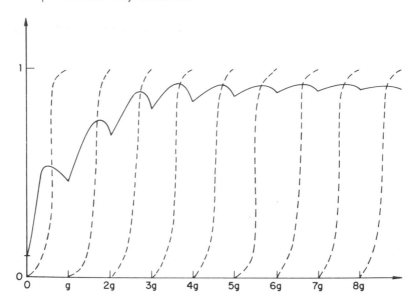

Fig. 8–1 Qualitative behavior of model of evolution of innovative play: play (*d*) allele frequency (solid line) and frequency of performers (broken line) as a function of time. Abcissa labeled in increments of *g,* the number of generations that separates successive new types of discovered performances. Explanation in text.

the new act, and their fitness is now lower than that of the average non–inventive individual. The resulting decrease in frequency of the *d* allele continues until a new type of discovery is underway. If during this cycle the frequency of the *d* allele has increased (decreased), more (fewer) performers are produced, the new behavioral act spreads more (less) rapidly, and as a result the frequency of the *d* allele soon begins to decline (increase). This negative feedback produces a small-amplitude equilibrium cycle in *d* frequency of period *g* generations (Fig. 8-1). Qualitative properties of the model do not change if the allele conferring innovative tendencies is assumed to be dominant. Numerical results reported below apply to a recessive *d* allele.

Numerical experiments show that equilibrium *d* frequency always increases with increase in benefit σ and with decreases in risk τ (Fig. 8-2) and cycle length *g* (Fig. 8-3). Higher discovery

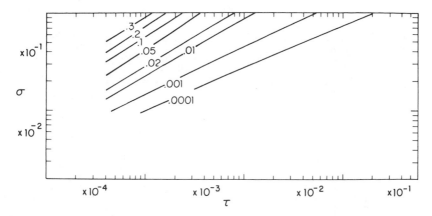

Fig. 8–2 Equilibrium frequency of d allele as a function of benefit σ and cost τ for $f = 0.005$, $g = 100$. Explanation in text.

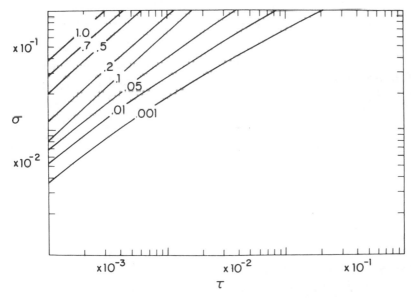

Fig. 8–3 Equilibrium frequency of d allele as a function of σ and τ for $f = 0.005$, $g = 10$. Explanation in text.

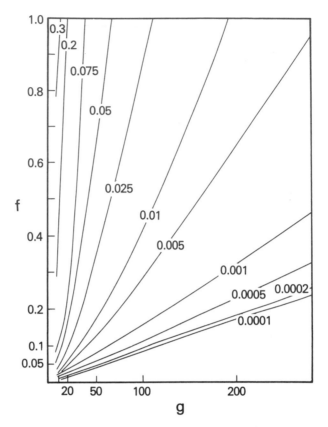

Fig. 8-4 Equilibrium frequency of d allele as a function of f and g for $\sigma = 0.04$, $\tau = 0.004$. Explanation in text.

probabilities f may either increase (Fig. 8-4) or decrease (Fig. 8-5) equilibrium frequency of d. When f is large and g small, an increase in f tends to produce a decrease in equilibrium d frequency (Fig. 8-5, upper left-hand region). In this case the most adaptive behavior facilitating invention is actually relatively inefficient, presumably because higher efficiencies produce too many performers, resulting in an outbreak of new behaviors, which then spread too rapidly for the innovators to realize any lasting advantage.

I calculated rates of genetic and cultural evolution for all runs, but found no consistent relationship between these two state derivatives (Figs. 8-6, 8-7, 8-8).

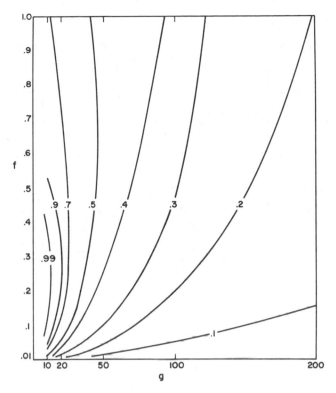

Fig. 8–5 Equilibrium frequency of *d* allele as a function of *f* and *g* for $\sigma = 1$, $\tau = 0.01$. Explanation in text.

Complex short- and long-term relationships between the rates of genetic and cultural evolution are manifest in the above results. Short-term effects (those occurring during a single cycle of discovery) are expressed by calculating correlations between the rate of gene-frequency change and the rate of change of the frequency of performers. Because of lag effects mentioned above, there is no simple rule relating the instantaneous rates of short-term gene-frequency change and of short-term cultural evolution in this model.

The rate of cultural evolution can be conceptualized in a different way that may serve to shed additional light on this problem. If parameter *g* itself is viewed as a rate of long-term cultural change, it is clear from the foregoing that rapid cultural

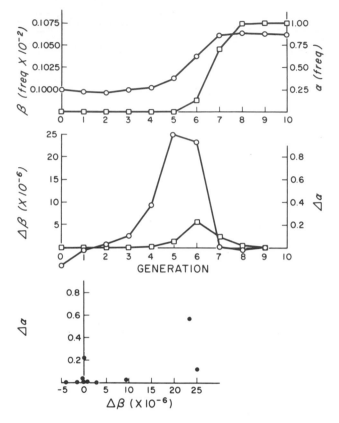

Fig. 8–6 Top to bottom: state-variable histories for d allele frequency β (circles) and frequency of performers α (squares); rates of genetic evolution $\Delta\beta$ (circles) and cultural evolution $\Delta\alpha$ (squares), calculated from the above by taking derivatives; scatter diagram showing no obvious relationship between these rates. Parameters $f = 0.05$, $g = 10$, $\sigma = 1$, $\tau = 0.05$. Explanation in text.

change facilitates rapid genetic evolution and specifically favors innovative play.

Even this very simple model illustrates two important biological principles of cultural evolution:

(1) The value of a single parameter of individual behavior can determine whether rapid cultural evolution will accelerate, retard, or have no influence on genetic evolution. All three possibilities are to be expected in nature. In turn, any environmental state variable, such as level of resource availability or population

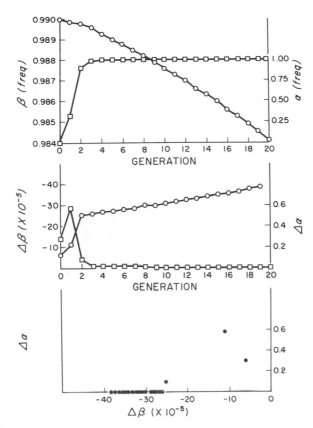

Fig. 8–7 Plots as in Fig. 8–6, for parameters $f = 0.05$, $g = 20$, $\sigma = 1$, $\tau = 0.05$. Explanation in text.

density, that affects behavioral tendencies will automatically affect the sign and magnitude of the genetic-cultural interaction. This problem might merit additional modeling that explicitly incorporated an environmental state variable.

(2) Gene-culture interactions occur simultaneously on several time scales, and short-term retardation is in no way inconsistent with long-term acceleration.

If the capacity for observation learning is absent, inventiveness will only evolve provided $f\sigma > \tau$ (Appendix V). Observation learning may be said to provide a mechanism for evolution of inefficient (low f) tendencies toward discovery of novel

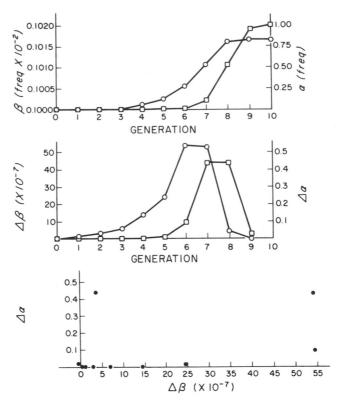

Fig. 8–8 Plots as in Fig. 8–6, for parameters $f = 0.05$, $g = 10$, $\sigma = 2$, $\tau = 0.5$. Explanation in text.

acts having relatively small (low σ) benefits and occurring in severe (high τ) environments. But inventiveness may disappear when observation learning abilities are present, and strength of counterselection will increase with efficiency of observation learning. This property of the model must be reconciled with the empirical fact that observation learning can occur at rates that result in fixation of new behaviors within a single generation. Multiple types of discoveries within a single generation could produce high d frequencies even in a population of animals capable of observation learning, and such multiplicity is known in nature (Kawai 1965). Formation of play-groups including only dd animals would also favor evolution of inventiveness. Secrecy, communication of misinformation concerning discoveries previously made (including advertisement and display of nonfunctional novelty and concealment of beneficial discoveries), as well as free exchange of discoveries among close kin coupled with restricted nonkin interactions (as might occur in many mammalian and avian societies), would also favor innovators. Finally, new discoveries could be shared with noninnovators in return for protection. Protective behavior of noninventive individuals would produce a decrease in the value of τ for those innovators toward whom this behavior was directed. These guardians of creativity would tend to maintain physical proximity to playing animals, in order to learn from them and to protect them. The guardians would keep nonplayful animals away from the players and would come to the players' assistance if necessary. Within a coalition of unrelated guardian and player, cheating is expected within the limits imposed by the theory of reciprocal altruism. Probably, therefore, the best guardian from the standpoint of natural selection is a nonplayful heterozygote (Dd). This prediction assumes that the play allele is recessive, but parallel arguments hold if inheritance is polygenic or if penetrance is incomplete.

Consider a small social group in an environment of highly dispersed, rich resource patches. If the group includes an innovator lacking play companions and having the necessary combination of early experience and maturation, behaviors ordinarily used in social play will sometimes produce innovations instead. Newly provisioned populations of mammals and many captive animals satisfy these conditions admirably. In this anomalous environment dissemination to peers is maximally unlikely. The most likely beneficiaries of such innovation would be younger sibs and cousins, if present, or the innovator's own offspring, at a later age. Might this unlikely set of circumstances have preceded such periods of human innovation as the so-called Pleistocene explosion of symbolic behavior (Pfeiffer 1978 p. 227)?

Failure of human observers to recognize the possible innovative significance of play in nature, and these observers' characterization of animal play as inefficient, uneconomical, and/or apparently functionless (Berlyne 1969, Loizos 1966) may be ascribed in part to historical factors beyond the scope of this discussion. Personal, political, and cultural biases sometimes make play very difficult to appreciate on its own merits. Why study "useless" behavior? Subjectivity is one of several reasons for human failure to identify innovative play. Total observation time may be inadequate to observe a very rare event. A study based on thousands of hours of observation is admirable and might be perfectly acceptable for most purposes, but even this extended observation period might fail to include discovery of a novel behavior (Appendix VII). Furthermore, total observation time is not the same as time spent watching juveniles play. A 1000-hour study of social behavior might include 20 hours or less in which a juvenile at play was in sight and was the focus of the observer's attention.

The present model suggests that additional factors might contribute to the failure of human observers to recognize innovative play or its results. Innovative play can be relatively inefficient and still be selected; the most adaptive play may, in fact, be inefficient; innovators may have been selected to conceal useful inventions; and an extended period of time (Beck 1975) may separate initial manipulation and subsequent insightful use of an object.

It must nevertheless be conceded that innovative play behavior is phylogenetically restricted. Despite many potentially mitigating circumstances (see also the factors discussed above), behavioral novelty has only been observed, *in the field,* in a few species of primates, even though some local populations have now been studied continuously for a decade or more. Moreover, most of these cases can be explained by citing contact with humans and access to human artifacts (Symons 1978a). A few taxa—primates, carnivores, cetaceans, elephants, equids, corvids, psittacids—exhibit innovative potential in confinement. In phylogenetic perspective, innovative behavior is an isolated and aberrant phenomenon, which Susan Wilson has justly characterized as the Sirenia of animal play. However, we humans, who have made innovative play into a way of life, may possibly be excused for our parochial interest in this topic.

ART, DECEPTION, AND PLAY

The idea of possible parallels between animal play and human art pervades the history of play research. Commentators from Groos (1898) to Klopfer (1970) view human dance, music, literature, painting, and sculpture as direct extensions of animal play. Even Plato (*Laws,* Bury tr. 1926, p. 159) traced the origin of dance to "the habitual tendency of every living creature to leap."

E. O. Wilson (1978 p. 206) argues that art, as a means of direct human communication of personal experience, will survive sociobiology's anticipated explanation of "artists, and artistic genius, and even art." Still, I would be willing to wager that a randomly sampled dancer or sculptor, for example (there is no "average" artist), would view sociobiologically based aesthetics with the same amused distaste that properly greeted Freudian, Jungian, and Marxian aesthetics.

The profound impact that social ideas have on creators and on creativity is illustrated by the position of artists in a totalitarian state and by that of black jazz musicians in a racist society. This would be the case even if humanists, critics, and artists themselves did not have political and professional interests vested in the ways society views and interprets art.

Biological interpretations of art as a form of human play are only vague metaphors and will remain so until significant advances are made in the biology of cognition. I discuss these suggestions in order to present the current unsatisfactory state of understanding of this topic and in order to demonstrate the extent of work that remains to be done.

One model (McCulloch 1945, Minsky 1977, Winston 1975) proposes that the brain functions via interactions between intercommunicating agents within it, not all of which share the same information or goals. These agents exchange messages that "in human discourse would be called advice, suggestions, remarks, complaints, criticism, questions, answers, lies and conjectures" (Winston 1975 p. 10). This heterarchical framework furnishes a possible mechanism for self-deception. In sociobiology, where lies and deceit are viewed as natural and normal forms of behavior in which subtlety and complexity are limited only by evolutionary constraints on the liar's cognitive ability, the most effective deception of others requires deception of self. This conclusion is based on the argument that self-deception renders "some facts and motives unconscious so as not to betray—by the subtle signs of self-knowledge—the deception being practiced" (Trivers 1976).

Deception, including "gestures, feints, and ruses designed to mislead" (Thorpe 1972) is practiced in play (Hamilton 1975, Thorpe 1972). Play behavior may even be viewed as primitive self-deception in which the animal simulates damaging fighting, fleeing (Kirchshofer 1960), or predatory behavior (Groos 1898, 1899, Rensch 1973, Reynolds 1976). This aspect of play is emphasized in Gregory Bateson's discussions (1955, 1956) of play in logical terms as an improper not. Play is not fighting, but the negative in this comparison is an improper not because play differs from fighting in a sense different from that in which other behavior, e.g., grooming, differs from fighting. According to Bateson (1955, 1956) and Groos (1898, 1899), this self-deception in play is even conscious. To the player, "the illusion is perfect, while on the other[hand] there is full knowledge that it is an illusion" (Groos 1899, see Bruner, Jolly, & Sylva 1976 p. 80).

How these fuzzy concepts map into heterarchical brain models must be left to philosophers, sociobiologists, and artificial intelligence experts to determine. I hypothesize that one consequence of developing a successful, but crude behavioral subroutine, skill, or coping strategy is that an animal "gets stuck" (G. Bateson 1978, Sussman 1975) in the given paradigm (G. Bateson 1978, Kuhn 1962), in the logical rules governing that class of actions. See Nabokov's *The Defense* for an example of a chess player whose entire world became a chess game. Scientific colleagues of mine have told me that sometimes, when they concentrate too hard on a problem, their dreams are phrased totally in terms of the problem, and all elements of the world outside science become elements of an experimental protocol or variables in a mathematical model being developed; they lose all independent existence. The closest scientific parallel to this phenomenon is a theory that can explain all its exceptions (see Gould 1977 on the triumph of recapitulation) or a single working hypothesis that its proposer adopts to the exclusion of all others (Chamberlin 1965). A heterarchical brain model will show analogous behavior if one agent becomes so powerful that it monopolizes its local domain and perhaps neighboring domains. All goes well until the agent fails (perhaps when one of its plans is extended to a new problem). At this point, classical learning theory would claim that the behavior was negatively reinforced and would decline in frequency. Actually, the situation is that of the levels problem, as discussed by G. Bateson (1978). The dominating agent places the blame on its subordinates, and they accept it! You cannot teach a rat not to explore by giving him an electric shock when he puts his nose into boxes; the negative reinforcement simply teaches the rat that exploration is an excellent strategy because it reveals which boxes are not safe. The same type of problem occurs in criminal justice, in treatment of alcoholism, and in psychotherapy (G. Bateson 1978). It is a case of fatal self-deception. What is required is to treat self-deception with a taste of its own medicine and obtain a loosening of the rules via play (G. Bateson 1956 p. 216). Play confuses the logical types and illogically negates the dominant agent in a way that reveals new possibilities. At this point, depending on which metaphor one prefers, the subgoal conflict is resolved, the "bug" is found, the dictator is overthrown, the revolution is successful, the cognitive or motor paradigm shifts, a new instance of self-deception replaces an ontogenetically older one, or the developing organism pulls itself up by its bootstraps.

The natural history of scientific creativity provides examples of this process. Play has generated insights into major scientific problems at times when individual scientists found themselves hopelessly stuck. Albert Einstein hints at essential relationships between play and creative thought. In a letter to Jacques Hadamard (Hadamard 1945, p. 142) Einstein stated:

The words of the language, as they are written or spoken, do not seem to play any role in my mechanism of thought. The psychical entities which serve as elements in thought are certain signs and more or less clear images which can be "voluntarily" reproduced and combined. There is, of course, a certain connection between those elements and relevant logical concepts. It is also clear that the desire to arrive finally at logically connected concepts is the emotional basis of this rather vague play with the above mentioned elements. But taken from a psychological viewpoint, this combinatory play appears to be the essential feature in productive thought.

Kekulé describes the benzene ring first taking shape in the playing, dancing flames of his fireplace. The idea for the fluctuation test in microbiology is said to have occurred to Salvador Luria as he watched slot machines in Nevada. At the University of Michigan in Ann Arbor, a story told of the invention of the bubble chamber describes D. Glaser, the inventor, sitting in the Pretzel Bell, a local bar, watching bubbles form on the side of a beer mug. The biologist Arthur Hasler, seeking the mechanism by which migratory salmon find their way back to their home stream, developed his home odor hypothesis during exploratory play. The emphasis on chemical sensation is almost Proustian:

I once went with my family to the West where we climbed a mountain, one that I had climbed as a child. We approached a waterfall on a trail, covered with moss and lichens, where a fragrance came toward me that produced a rather surprising reaction. My mind became filled with youthful memories. First, the names of long-forgotten school chums flashed into my mind, and I recalled the sight of the place above that waterfall where we had played. I thought, "Here you are, Art, coming back home and it *smells* like home." My thoughts continued, "If you were a salmon, how would you do it?" (Bateson 1956, p. 235)

Compare Baudelaire's *Le Flacon* or Marcel's reaction to the almost-forgotten taste of tea and madeleines in Proust's *Du côté de chez Swann*. [See also E. O. Wilson 1975 (pp. 155, 231–235) and Wilson 1978 (p. 69).]

These unsatisfactory metaphors are the best currently available links between play and human creation. By some unspecified mechanism the cognitive processes of transformation, abstraction, and innovation involved in Kuhnian fashion in animal play are hypothesized to propel human artistic activity, paradoxically recombining experience and creating novel forms of expression (Bates 1960 p. 227, Burgers 1966, Erikson 1977, Groos 1898, 1899, Huizinga 1950, Klopfer 1970, Koestler 1976, Morris 1962b, Thorpe 1962, 1974). Play has been termed an essential element in the contributions of artists as diverse as the Beatles (Mellers 1973), Edward Gibbon (Clive 1976), and Vladimir Nabokov (Nabokov 1970).

I suspect that the basic variational, recombinational, self-deceptive procedures apparently used in play to develop physical abilities, cognitive strategies and social relationships (see Chapter 5 and Mason 1978) achieve the same immediate goals of recombination and paradigm shifting at a more abstract level in human creative activity. But if play is, as this proposal implies, a set of bootstrapping skills used in developing other abilities, then we had better start by investigating these skills in their evolutionary context, not by spinning incomprehensible techno-poetry from fashionable sociobiological, cybernetic, and psychiatric jargon. I do not deny that these basic skills, retained in adulthood through human neoteny, are available for use in the functional context of behavioral innovation and in additional contexts ranging from religion (E. O. Wilson 1978) and sex (Simpson 1978) to art and science. Animal play offers a relatively uncontroversial and unpoliticized framework for scientific study of certain basic skills having important cognitive implications. Because scientists are still very much at a loss to explain creativity of any sort, including that occurring in art and in science, such a framework is valuable. It should prove at least as intellectually productive as earlier attempts to understand art by giving chimpanzees drawing materials (Morris 1962b; apparently an early experiment of this sort provided the initial stimulus for Nabokov's *Lolita,* Nabokov 1970 p. 313!).

9

Epilogue

Play, as Shakespeare had his character remark, needs no excuse. Why a final chapter?

The best final chapters of biological books are glistening codas (Darwin's *Origin*), apotheoses, oracles. They point out the intellectual and social importance of the work, suggest applications of the ideas presented, and identify promising future research directions and unsolved problems. They place the preceding work in its most general philosophical context. The worst final chapters, as all Frank Merriwell readers know, exist to market someone or something—the author, the author's ideas, the current book, and especially, the book to come.

Many excellent biological books (e.g., R. R. Baker 1978, Maynard Smith 1978) do not need and do not include philosophical concluding chapters. Others (e.g., Schaller 1972, 1977) philosophize briefly, but effectively. In this final chapter I summarize controversies introduced in earlier chapters. A section entitled "Warnings, Limitations, and Present Needs" evaluates the book's own approach to play. I discuss research opportunities and unsolved problems. I omit an accepted biological finale, a "people chapter." Authors of biological books often conclude by arguing that their ideas represent a novel approach to understanding humans or to solving human problems. Such may also be the case for biological ideas about play presented in this book. But this general argument has already been pursued sufficiently well, pro and con, by sociobiologists and their opponents and by educators, psychiatrists, and psychologists, including Owen Aldis, Gregory Bateson, Jerome Bruner, Erik Erikson, Catherine Garvey, Harry Harlow, Dorothy Singer, and Jerome Singer. I recognize that most arguments for human importance of play research lack scientific polish. However, I am neither an educator nor a clinical psychologist. If this book succeeds in improving old ideas about play and in suggesting a few new but vague ideas, it will also succeed in furnishing additional material to support theoretical studies of human evolution and of human culture. Educational and child-rearing applications of biological theory on play should be pursued by qualified educators and psychologists who are prepared to use biology without abusing it.

This chapter concludes, as did W. D. Hamilton's (1977) review of Dawkins's *The Selfish Gene,* by striking a different chord. There are two sorts of unsolved problems in play research: those that can be formulated in relatively concrete terms at this time and those that cannot. The only available approach to the second type of problem is essentially literary. I therefore conclude with nonscientific perspectives on play offered by Hemingway, Nabokov, Rilke, and others. Each passage evokes associations among play, time, and human development. I hope that this material will stimulate biological questions about becoming and continuity.

CONTROVERSIES

Current and future controversies about play behavior are the topic of this section. After stating my conclusions about current controversies, I attempt to identify controversial aspects of this book and to anticipate criticisms that may be directed against it. These arguments also serve to introduce my own evaluation of the work's limitations.

Social Play: Cooperative or Competitive?

Biologists propose several paradigms for fitness effects of play with others. According to the traditional view of social play as

Table 9-1 EVOLUTIONARY INSTABILITY OF
SELFISH (DAMAGING) PLAY

	No Play	Selfish Play
No play	0	$-b$
Selfish play	$-a$	$-(C_1 + rC_2)/2$

cooperative behavior, social play increases both participants' expected reproductive success (or the welfare of the group in the superorganismic formulation of play as a socialization and cohesion mechanism). Bekoff (1978b), Fagen (1978a), and Geist (1978b) propose and attempt to evaluate the hypothesis that play, like escalated fighting, is a mechanism of aggressive competition. According to their proposals, play is mutually harmful, but the "loser's" reproductive success decreases more than does the "winner's." Because these authors identify no resource over which competition occurs in play, they do not explain why animals should play at all. In fact, these proposals are invalid as stated because play behavior of this type is evolutionarily unstable against a strategy of not playing at all (Table 9-1). In Chapter 7, I propose that an animal plays at times, in places, with partners, and in ways that maximize its own inclusive fitness, that conflicts of interest may occur over play, and that fitness effects of play with others can range from cooperative patterns ($+$, $+$) to competitive (bullying) patterns ($+$, $-$).

Play Behavior or Play Behaviors?

Because I consider play to be a behavioral mechanism for developing skilled behavior, I consider play alone, play with animate and inanimate non-conspecific items, and play with others to be different aspects of the same phenomenon (Chapters 2, 4, 5, 7). The closest analogous behaviors are those used in feeding. Feeding involves different kinds of items, different kinds of acts, and compromises between energy, nutrients, water, time costs, survivorship risks, and dietary poisons (S. Altmann 1974, Schoener 1971). Like other functional behavior, play may acquire new functions in evolution and in ontogeny. Play, for example, may perhaps be used to manipulate the behavior of others in displacement, in weaning, and (like grooming) in securing future social cooperation.

Practice or Innovation?

I conclude that the case for evolution of play as an immediate source of novel, innovative behavior in nonhumans was previously overstated by myself and by others. However, I am equally convinced that evolutionary developmental arguments for play as biologically adaptive preparation for later behavioral flexibility and versatility are valid subject to evolutionary constraints on learning. "Flexibility" hypotheses may also need to recognize the distinction between plasticity and resilience in development. This controversy has some interesting philosophical implications to be considered below (p. 492).

Metacommunication

I have not previously addressed the current controversy in play research over the utility of Gregory Bateson's concept of metacommunication (G. Bateson 1955, 1956). According to Bateson, animal play is important in the study of animal communication and cognition because it suggests that nonhuman animals can communicate on more than one conceptual level, that these animals can communicate about communication itself, and perhaps even that they can recognize and use the concept of an improper not. W. J. Smith (1977 pp. 236–238) and Symons (1978a pp. 92–98) offer several criticisms of the idea of metacommunication. These ethologists justifiably criticize Bateson's cognitive inferences. I accept Bateson's fundamental idea that an animal's play-signals serve to communicate contextual information about its subsequent behavior. This proposal is supported by quantitative comparisons of behavioral sequences with and without play-signals (e.g., Bekoff 1975) and can be further analyzed using the theory of piecewise Markov processes (Kuczura 1973; see discussion above, p. 59).

Training, Bonding, Developmental Rate, or ?

Why do animals play? This question is not easily answered. The difference between the conclusions of two excellent recent and independent quantitative studies of play in free-ranging provisioned rhesus macaques inhabiting two separate islands off the Puerto Rico coast illustrates this point. Levy (1979) and Symons (1978a) collected the same kinds of data on rhesus play and demonstrated the same patterns of age and sex differences. Yet these two studies, representing the most thorough empirical field research conducted on play to date, resulted in diametrically opposed conclusions regarding function. Symons concluded that rhesus playfighting evolved to train fighting skill, but Levy concluded that play developed social bonds.

Although Symons and Levy differ over the relative importance of training and bonding effects of rhesus play, they both reject the hypothesis that play facilitates innovative learning. Paradoxically, the best experimental evidence to date indicates that play is a mechanism by which animals can influence their developmental rate, brain weight, and the flexibility of their behavior as well as certain aspects of learning ability (Chapter 5). It remains to be demonstrated whether such effects occur in nature, and despite Geist's (1978a) inspired speculations on this topic, their evolutionary significance needs to be further clarified.

Only tentative conclusions may be drawn at this time regarding function and evolutionary significance of play. The clear correlation of play with level of brain development (especially cortical development) and play experience's specific effects on cortex suggest that play mediates an adaptive balance between cortical and subcortical control of such behaviors as fighting, predation, and predator avoidance that are represented at multiple levels of brain organization. The fundamental trade off is one familiar to engineers and computer scientists, since it involves issues in hierarchical system design. Small units, such as analog computers in process control, are fast, inexpensive, but inflexible, since they are unable to analyze past experience or to respond appropriately to novel situations. Large units, such as digital computers, can integrate past experience and devise appropriate responses to novelty, but they are expensive and slow. In theory, then, different environments should call for different allocations of control of particular behaviors to cortex and to lower centers. When the environment changes in a way that favors long-term integration and restructuring of information, the cortex should be built up. Play should respond to those features of the environment that reliably predict the need for greater cortical control. Such "enriched" environments are precisely those in which play occurs. Of course (Bruner 1976), stimulation is not analogous to a vitamin and play may only be necessary for development in certain environments.

There is crucial need for more precise specification, in ecological terms, of environments in which play is expected to be most beneficial. Experimental data on relative frequencies of play in different environments are not always relevant to this point, since failure to yield beneficial effects in a certain environment is only one of several possible reasons for lack of play. As argued in Chapter 6, absence of play can also be explained by an increase in the cost of play, by performance of nonplay behavior yielding the same beneficial effect, or by preemption of resources by higher-priority behavior (resource budget constraints). In addition, as argued above, life-history strategic trade offs can result in a decrease or in a total absence of play at one age in order to conserve resources for a later age or as a result of resource depletion occurring at an earlier age. Therefore, experiments on environmental enrichment that control for these factors (e.g., by ensuring that adequate food is present) are most relevant to these questions.

Preliminary conclusions drawn from such studies are fascinating, if difficult to draw precisely. Novelty, complexity, and new resources that cannot be defended are only parts of the total picture, since not all novel stimuli elicit play and too much novelty can be intimidating. Nor is arousal the answer, since an animal's optimal level of arousal should itself be subject to evolutionary constraints: it will vary with age, sex, social status, and the environment. When animals are limited by their own skills and physical capacities rather than by constraints imposed by their social or physical environment, play appears to be necessary, as in a supportive environment that allows an individual to develop its potential to the maximum. But how can the individual reliably determine that its environment is of this type and will remain so for a sufficient time to justify the risk of an investment in play? It would be highly disadvantageous to the animal to commit resources to development through play in a hostile or hopeless environment in which increased strength, skill, or behavioral flexibility are irrelevant. It would also be costly to begin playing immediately in an environment about to return to a state in which play was wholly disadvantageous. These strategic considerations suggest why play occurs only after an initial period of investigation and exploration of a novel environment (Hutt 1970). Certain features of the environment reliably indicate what can only be termed an expanded perceptual and cognitive world: a balloon given by a scientist to two bored zoo otters (Bateson 1956); protein-rich sprouting vegetation (Geist 1966) or the combination of dietary enrichment and social stimulation (Grantham-McGregor et al. 1979); contact with immature conspecifics who are no longer strangers; a change to fair weather; cessation of stress; the end of a long, difficult project; escape from danger; release from confinement. According to this argument, play was selected to develop flexible cortical control of particular behaviors along with the improved physical machinery without which such control would be pointless: An improved physical plant requires better integrated, more flexible, and more insightful management. Because mind and body are inseparable in this view (compare Mead 1976) and because the particular behaviors and relationships involved may vary, whether we label these effects training, bonding, or developmental flexibility will depend on species, environment, and individual and may chiefly be a matter of taste. At least two questions remain, however. Why does play seem not to be involved in development of certain kinds of behavior? Is it sufficient to cite associated risks and alternate mechanisms (e.g., chimpanzees fishing for driver ants) or probable lack of multilevel brain control (e.g., human skill at chess)? And in what sorts of environments should such mechanisms evolve? In particular, how do animals guard against other animals' possible tendencies to manipulate their competitors' development, damaging the competitor by giving its cortex a suboptimal degree of control of the behavior involved?

The above discussion of function has at least three nonspecific implications, some of which I discuss below (p. 487).

First, the concept of an animal in an expanded world suggests some lines of Shakespeare (Brownlee 1954):

> And when the mind is quicken'd out of doubt
> The organs though defunct and dead before
> Break up their drowsy grave and newly move
> With casted slough and fresh legerity.
> (*Henry V,* IV.i. 20–23)

Second, the idea of an expanding consciousness is interesting in relationship to brain evolution, play and drug use (see below p. 490).

Third, although the speculations on play and brain biology above surely appear grossly naive when evaluated in the light of contemporary brain science, it is possible that they may suggest particularly horrible manipulative experiments on the brain and play, as well as ill-considered social programs. In taking responsibility for these speculations, I hasten to add that not enough is known, either about the brain or about play, to justify proposing actual intervention at this time, either in the context of brain research or in the context of social policy. Computer models such as Marr's (1970, 1971) of cortex, Sussman's (1975) of skill development, and the life–history model of play deprivation suggested in Chapter 6 will aid in clarifying these suggestions and in ensuring optimal use of research resources. A sound conceptual basis for elegant experimental research on play similar to that of Einon, Ferchmin, Morgan, Rosenzweig, and others can lead to better environments for children and for other young animals. These better environments could include research laboratories in which excellent research and humane policies are inseparable.

New Controversies

Play is a sensitive area of human discourse. For this reason and for others (Chapter 2), the study of play is as controversial as any area of contemporary behavioral research. Currently, many sophisticated and entrenched approaches to behavior coexist. Some, such as the Piagetian formulation of cognitive development, recognize play, but give it epiphenomenal status, whereas others neglect it entirely. Although I do not consciously seek controversy in this book, I recognize that it is impossible, in a full-length biological work in a relatively uncharted area of behavior, not to make naive scientific and political statements, especially when the behavioral phenomena analyzed have clear social and political implications. Some aspects of this book that could potentially lead to controversy are as follows.

Within Play Research: Analysis of play using hypothetico-deductive evolutionary theory may offend ethologists and comparative psychologists who prefer data to mathematical models and physiological metaphors to evolutionary logic. I would be quite willing to admit the validity and desirability of multiple approaches to play behavior, but I feel, for reasons stated in Chapters 1 and 2, that at this time an evolutionary approach like the one presented here can lead to fundamental, novel insights and to exciting, rigorous empirical research on play. Some ethologists may be upset at my stubborn insistence that play is a single meaningful behavioral category. I admit that this question is still open, but I do not feel that it is the most important question facing students of play.

Within Sociobiology: Many aspects of this book are not sociobiologically orthodox. I stress development, even developmental mechanisms. Play is not a central sociobiological topic because its contributions to reproductive success are not nearly as obvious as those of sex, fighting, or feeding. I am much less ready to propose simple biological explanations of human behavior than are most sociobiologists, even though such simple explanations are audacious, usually have solid grounding in evolutionary theory, and may even be right. Nevertheless, sociobiological analysis of human art and aesthetics appears feasible if limited in scope (Wilson 1978 p. 206). I would welcome serious attempts, however incomplete, to apply biological ideas on animal play to the analysis of these phenomena.

Within the Study of Behavior: Developmental processes and phenomena can be described and analyzed from a bewildering number of points of view. I would applaud an approach to play that emphasized individuals and employed phenomenological, epigenetic, or even frankly literary perspectives. The phenomena described and the resulting phenomenological models are of interest in their own right. If evolutionary developmental theory cannot explain these phenomena or predict how the parameters of these models will differ as environments change, this failure represents a challenge to theory, not a weakness of the epigenetic, developmental, or individual ontogenetic approaches to play. Non-evolutionary developmental studies, such as those by Robert Hinde on rhesus macaque social behavior and by Patrick Bateson and Jay Rosenblatt on domestic animals, are paradigmatic. This work has fundamentally altered our understanding of humans' relationships with their animals and with their own children, as well as our understanding of behavioral development. There is every reason to expect that studies such as these will continue to be of value as long as humans ask questions about the "what" and "how" of behavior.

Within Biology: Is play adaptive? Charles Darwin, a pioneering observer of animal play in his own right (e.g., Darwin 1874, 1877, 1898), expressed some doubts (Romanes 1884). In a posthumous essay on instinct, Darwin discussed small and trifling instincts which, "if really of no considerable importance in the struggle for life, could not be modified or formed through natural selection." Some instincts, according to Darwin, prove far less trifling upon closer inspection than was first apparent. "But some instincts one can hardly avoid looking at as mere tricks, or sometimes as play" (Darwin, in Romanes 1884 pp. 378–379).

Current-day students of play might give Darwin's words serious attention. Recent critiques and redefinitions of the biological concept of adaptation (Lewontin 1978, Oster & Wilson 1978, Schulman 1978), and Symons's (1978b) critical review of previous attempts to study biological functions of play, fully confirm Williams's (1966) contention that the term "adaptation" must be used carefully, and they underline E. O. Wilson's (1971b p. 458) warning that evolutionary optimization works with unknown precision. Play research based on evolutionary developmental biology, on sociobiology, and on optimization models is sure to be a prime target for such criticism. Play, unlike sex or aggression, makes no obvious contribution to reproductive success. If a prominent and easily defined anatomical feature like the human chin (Gould 1977) can be shown to be a compound feature rather than a unitary trait, what about an elusive behavior like play, which may be defined most easily using purposive conceptual schemes derived from everyday human discourse? And what are we to make of development, uncritically viewed as adaptation? Developmental processes may indeed be profoundly influenced by natural selection to produce adaptive results (Gould 1977), but other results of these changes in developmental patterns may have nothing whatsoever to do with adaptation, whether or not they owe their existence to natural selection. Development can be influenced by random environmental noise, even though the results of such a random process might still appear purposive or adaptive to a human observer (Cohen 1976).

To an evolutionary biologist aware of these contemporary ideas on adaptation, play offers stimulating possibilities. Perhaps play will truly prove inexplicable within an adaptationist framework. Should this be the case, play would become a uniquely important biological phenomenon—the axolotl, so to speak, of a new evolutionary developmental biology. Such unique status ought to prove vastly more exciting than an experimental or comparative study that satisfactorily revealed adaptive differences in play between ages, sexes, populations, or species. After all, what is so wonderful about one more adaptation, however difficult to demonstrate?

Available evidence to date points to some adaptive significance for play (though it is wholly incorrect to view play behaviors as a whole, or differences between particular types or sequences of play, as necessarily adaptive). This evidence includes age, population, species, and sex differences cited in the text; playground, plaything and playmate choices; specific motivation for play; and existence of play-signals. The fact that play still occurs despite measurable time, energy, and survivorship costs is also important in this context. It would be difficult to explain the eyelid display play-signal of *Pygathrix nemaeus,* the canid play-bow, or the change that came over Miro the hanuman langur when he finally found someone to play with, as developmental noise or as epiphenomena.

At this time, two characteristics of play might especially interest pluralists. The theoretical result (Chapter 6) that optimization models of play in the life-history commonly have several optima suggests that play might indeed represent something less than "a biological imperative." Empirically, the apparent correlation of play with relative brain size suggests the possible interpretation that play will become more elaborate, frequent, stable, and/or interactive whenever natural selection increases the degree of encephalization in a population, although the reason that brain size increased in the first place might have had nothing to do with effects of play as such. Reinterpretation of play in the light of new ideas on adaptation and natural selection may become a central issue of play research in the 1980's.

Beyond Behavior: The humanities. I suppose that some humanists might be offended, in ways and for reasons about which I do not presume to speculate, at my approach to human art and aesthetics by way of the study of animal play. Literature and literary criticism are difficult and sophisticated areas of human activity. A sociobiological school of literary criticism, if one ever evolves, should prove roughly as successful and about as interesting as Freudian criticism. The great problems of time, mind and development, human purpose, and consciousness are far too large and complex for any one group of scholars to monopolize; in the language of behavioral ecology, they are not economically defendable.

Beyond Research: The field of politics. Social and political implications of the act of conducting biological analyses of human behavior, particularly human development, should be obvious to anyone who has followed the *Sociobiology* controversy. Socially, the concept of biologically based self-deception might be viewed as especially pernicious, because it enables us to explain our opponents' perverse behavior by pointing out how natural selection has constrained them to hold erroneous opinions and to attack others. But what of play? Is the topic of this book sufficiently important in a political sense to warrant socially oriented criticism? Marx and Engels, for example, were concerned with labor, work, and production, not with play. Radical social critics and biologists who choose to comment on politically sensitive biological topics would be very naive to join battle over play without acknowledging the political implications of biological approaches to play behavior.

If play is held to be important to health and to development, then challenges to this tenet of faith will be seen to threaten children themselves. Is the child's "right to play," stated so effectively during the International Year of the Child, compromised by research demonstrating that play is dangerous or that play has no beneficial effect in certain environments? Biological research on play is relevant to virtually every aspect of child care and early education. For example, at the beginning of this century, biological interpretations of play in the light of recapitulation theory deeply affected educational and social policy (Gould 1977). This book should demonstrate the tentative nature of current conclusions on the causation and function of play as well as the need for more sophisticated and more cautious application of these conclusions.

WARNINGS, LIMITATIONS, AND PRESENT NEEDS

The analysis of play presented in this book answers old questions by raising new ones. There is need for much more thorough biological analysis of human play, and especially of those aspects of human play that appear to be unique. In addition, we need better formal mathematical models of play. In general, these models should be simple (or else they cannot be tested), should focus on a few essential variables, and should make nontrivial, testable predictions. Specific needs include three-player game models of social play, life-history models that predict ontogenetic changes in qualities of play (especially social play), and models of play with others that account for individual differences and repeated interaction. There is also need for artificial-intelligence models of playful skill development, even though these models are in general highly complex and often can only be simulated piece by piece.

The work bases many crucial arguments on second-hand, verbal descriptions of animal play, on studies of confined and laboratory animals by myself and by others, and on a few highly visible field studies. This data base is, however, shared by all ethologists and sociobiologists. In view of current threats to animals and to their environment, a case could be made for the proposal that the most crucial future need in play research will be for systematic observations and film records of play in animals in undisturbed environments under naturalistic conditions. Although we can hope that data relevant to the next generation of theory on play will be collected serendipitously in the course of current projects, preservation of natural diversity is the only tactic that will ensure that we will be able to try to answer questions not yet formulated. The familiar philosophical argument (e.g., Fagen 1978a, Gould 1977) suggesting essential subjectivity of scientific description therefore represents one of many arguments for preserving natural diversity.

There is need for more and better visual documentation of animal (including human) play. In part, the current lack of such material stems from the basic difficulty that play, like dance, is dynamic and can seldom be represented adequately by single or sequential illustrations. Only a truly gifted photographer of animals can capture the essential characteristics of play on film. Cine films and videotapes of play, and use of movement notation systems to describe play, would all be valuable. I have cited existing films of play in this work wherever possible and appropriate. Increased efforts to obtain visual records of play would benefit science, conservation, dance, and the general public.

Several interesting questions about animal play may be answered by future research. For example, some years ago David Barash (pers. comm.) suggested to me that social play could be viewed as a mechanism for assessing, or for learning to assess, other individuals' competitive skills, in the context of G. Parker's (1974) evolutionary analysis of assessment strategies. The ancestry of this idea in superorganismic concepts of socialization and dominance hierarchy definition is somewhat questionable, and my own application of the psychological theory of social comparisons to predict play partner choice (Fagen 1974a) was naive, although not incorrect. As I briefly suggest in this work (Chapters 5, 7) animals may assess or learn to assess their own abilities or those of others in play, and this system is therefore vulnerable to communication of misinformation. However, basic theoretical questions must be answered. Is any information preferable to no information at all? What aspects of a young developing animal's behavior and environment predict future behavior and future environments, especially if animals' abilities are constantly changing (in part, because of play itself)? Worse yet, of what use is assessment in play if development is resilient and early variation among individuals thus tends to disappear (or to be swamped by new sources of variation among individuals)? This problem suggests a basic question asked by modifiability theory. If play is preparation for the future, what time scale of prediction is implied, and what aspects of the current environment are sufficiently predictive of future environments to make such adaptation beneficial? For example, in the simple case of physical training, we may argue that physical ability is not genetically predetermined because the optimal level of physical ability varies from generation to generation. But what factors ensure that the amount of exercise obtained by a young animal in play is appropriate to the physical ability required by that same animal as an older juvenile or as an adult? It would be difficult to assess current resource-holding potential (RHP) from probability of victory in a trivial dispute (Popp & DeVore 1979), much less future RHP. In sexually dimorphic animals, the larger sex is only slightly larger during youth (Trivers 1974), and adult size is achieved as a result of accelerated growth at adolescence (see Fagen 1972a, Geist 1978a, Laws 1959). Moreover, fighting weapons seldom develop until adolescence. Therefore, size and skill of an immature animal can only weakly predict adult size and fighting ability in such species. I speculate that play can only indicate how well development is going—whether the animal is well-fed, uninjured, receiving excellent parental care, etc. Play may integrate experience over a certain period (whose length and fading function would be subject to natural selection) of past ontogeny, so that past trauma or stress would be manifest in an individual's frequency or quality of play (including solicitation and refusal rates, willingness to take a given role, intensity and elaborateness, probability of uttering pain vocalizations, and so on). Play indicates current and past well-being, and, to an extent determined by the feasibility of developmental resilience, future well-being as well, as suggested by accounts of defective, fragmentary, regressive, or infrequent play in orphaned, handicapped, injured, or seriously ill primate infants [e.g., Fedigan & Fedigan 1977, van Lawick-Goodall 1968a, 1971]. Therefore, again within the limitations imposed by the possibility of future developmental resilience, frequency and quality of play could indicate current and future competitive and perhaps cooperative ability to an even greater extent than did actual fighting or cooperation. An evolutionary biology of aesthetics might consider this hypothesis in addition to others previously proposed. At the very least, assessment suggests why we find play aesthetically pleasing, though not why we sometimes find it threatening, baffling, or impenetrable.

Is some play, like new music and new art, too much for the participant/observer, whereas other play is tiresome and boring? This scale linking boredom, aesthetic pleasure, and intimidation echoes game-theory models presented in Chapter 7 and represents a biological alternative to arousal metaphors.

Modifiability theory, the evolutionary analysis of developmental and phenotypic plasticity and resilience, analyzes somatic plasticity, learning, and developmental canalization. This body of knowledge proved crucial for understanding the evolution of play and for criticizing existing hypotheses. However, perhaps because I could not research all the literature on evolution of phenotypic plasticity in plants, I found existing modifiability theory exciting, but incomplete, sometimes difficult to apply, and occasionally group-selectionist. Future theoretical progress in play research will almost surely depend on future advances in the evolutionary theory of optimal modifiability. My reading in this area and my attempts to apply the theory to play suggested need for further clarification of several issues. What is the nature of the trade off between optimality and stability in development? What factors determine the balance between the cost of resilience and the cost of suboptimality? In the development of behavior, what homeostatic mechanisms function in order to control what variables? What behaviors or behavioral mechanisms are allowed to vary in order to hold what behaviors or mechanisms constant? Are personalities or Mason's (1978) "coping strategies" the fundamental variable conserved by resilient behavioral development, or are they results of variation that occurred in order to stabilize some other component of the behavioral phenotype?

Phenotypic change that results from environmental change can be interpreted either as adaptation or as damage. When the environment changed from X to X' and the phenotype changed from Y to Y', did this change occur because Y' is superior to Y in environment X' or because Y is still optimal, but the environmental change stressed the phenotype so deeply that resilience mechanisms could only bring it back as far as Y'? For example, if age at first reproduction increases when nutritional level decreases, is delayed reproduction optimal in a life-historical sense in an environment of low resource availability, or is the phenotype simply damaged by its low level of nutrition? The basic questions here are those of precision of natural selection and limits of developmental canalization. They ramify throughout the substructure of theoretical ideas about play. In free-living rats, for example, is the environment ever sufficiently impoverished to produce the behavioral syndrome observed by Einon, Morgan, Rosenzweig, and others in the laboratory? If so, is the deprived rats' impulsive, inflexible behavior to be interpreted as damage sustained in order to maintain some other component of the phenotype, as a coping strategy or as an adaptation? Similarly, when resources are preferentially allocated to high-growth priority tissue (Geist 1978a), we do not identify the reduced development of the low-priority tissue as an adaptation in itself, but view it as an expression of the cost side of the cost-benefit equation. Processes and results have been confused here. Improved modifiability theory would serve to clarify the issues involved. At the very least, we should be very careful about identifying isolated results of developmental processes as adaptations. They may simply reflect costs of canalization of some higher-priority component of the phenotype.

My analyses revealed a need for tighter sociobiological analyses of social behavior among immature animals. Although much of this information appears implicitly or even explicitly in recent papers by Alexander, Konner, Trivers, and others, the implications need to be pointed out in more detail. When immature animals fight, what do they fight about and why? What does "dominance" mean in immature animals, and who benefits from the underlying behavior? How do tactics of competition among sib, peer, kin, and mixed-age immature individuals change with age, and what are the essential environmental variables that determine this change? What is the sociobiological significance of phenotypic modifiability in development, and what are the behavioral strategic consequences of the fact that developmental rates and pathways are biological dependent variables influenced by social stimuli?

I am convinced that three ideas about play pursued only briefly in this text deserve further consideration; they are reviewed here.

(1) Interspecies interactions as a selective force on behavioral development and on play: Interspecific competition and cooperation are significant evolutionary phenomena, and many phenotypic properties of individuals cannot be interpreted without reference to the ecological community to which the individual's species belongs (Cody & Diamond 1975). The evolutionary study of behavior still draws more heavily on population ecology than on community ecology, probably because behavioral ecology and sociobiology branched off from ecology in the late 1960's when community ecology was still considered somewhat disreputable by most evolutionary ecologists. I cannot accept continued isolation of evolutionary ethology from evolutionary community ecology. In my one attempt, to date, to synthesize these fields, I noted that many of the most apparently playful species have life-history patterns that ecologists used to label "K-selected," and I wondered whether vigorous play might simply be a mechanism of overexploitation. Animals may be selected to exploit resources more efficiently even if such efficiency results in a decline in the size of the population to which the animal belongs, provided that competing species are damaged to a greater extent by the resulting decrease in re-

source abundance (Diamond 1975, Levins 1975, M. L. Rosenzweig 1973a,b). I was intrigued by the image of young animals competitively burning up resources in play [a community analog of Schiller's (Beach 1945) surplus energy hypothesis], but such exploitation cannot evolve unless it allows the exploiter's population to grow faster (for the same reasons that damaging play is evolutionarily unstable, see above p. 474). Further study is needed, so the synthesis of community ecology and developmental behavioral ecology must remain a challenge to future generations of evolutionary biologists.

(2) Play and cohesion: As discussed above (Chapters 5, 7) there must be something to the idea that play facilitates individual recognition, prediction, cooperation, trust, and friendship, but this hypothesis lacks adequate mechanistic and evolutionary bases. Social grooming itself is still insufficiently understood from this point of view. Presumably, high-frequency grooming is an evolutionarily stable strategy of social behavior that works because it monopolizes the groomer and penalizes an individual deviant, but two separate behavioral mechanisms for cohesion, one for adults and one for immatures, seem a bit extravagant, like so much else in nature. Perhaps the assessment hypothesis (see above, p. 485) resolves this objection, since immatures must also learn to predict their friends' future behavior.

(3) Play motivation and drug use: Is there a possible relationship? Evidence for existence of play pheromones has been reviewed (Chapter 7). Elephants appear to enjoy the effects of ingesting fermented plant matter (Leyhausen 1973a, Sikes 1971), as do dairy cattle (J. Fagen pers. comm.). Catnip and other plants contain substances that may mimic felid reproductive pheromones (Leyhausen 1973a, Todd 1963). The felid catnip response includes rolling and rubbing, behaviors used in courtship and in soliciting play; catnip also heightens responsiveness to moving or movable objects, "making cats playful" (as advertised). However, evidence to date suggests that humans are the champion drug users of the animal kingdom. Is this evolutionary progress?

Does continuance of human motivation for play into the competitive environment of adulthood help explain the tendency of certain humans in certain environments to use mood-changing drugs (including alcohol)? Do such drugs make humans playful? Perhaps someone will pursue this idea to interesting and socially relevant ends.

Although it may not please sociobiological readers to see it discussed in print, one additional limitation of this work should be identified. Sociobiology is a revolutionary and powerful approach to behavior, including both old and new ideas, but I doubt that it could have enjoyed anything like its present success in any culture other than the current American culture of narcissism (Lasch 1978). The biology of selfishness is an intellectual counterpart of the romantic self-love expressed most arrogantly by such popular self-fulfillment books as Gail Sheehy's *Passages*. Moreover, the inevitable stages of human development posited by such books differ little in a philosophical sense from sociobiology's infamous insistence that fundamental biological constraints limit human behavior. When sociobiology ceases to reap the insights into behavior gained from the rediscovery of neo-Darwinist biology by the culture of narcissism and becomes culture bound, it will be time for some different ideas and for a radically new approach to social behavior. Whoever can successfully predict the shape of that approach and the time of its advent is welcome to do so.

Even status-conscious academics occasionally need liberation from the selfish crowd and from their own selfish egoes. What better antidote for professional competitiveness and academic jealousy than the twin enchantments of art and play? W. D. Hamilton (1977), for example, cites nonscientific perspectives on the problems of consciousness, purpose, identity, and continuity that sociobiology is now preparing to address. Because of its importance to aesthetics, play behavior seems especially suited to this dual perspective.

NONSCIENTIFIC PERSPECTIVES ON PLAY

The first observation to be discussed in this section is relatively simple, but has far-reaching implications. Symons (1978a) criticizes flexibility hypotheses of play, arguing that play serves to practice species-typical fighting skills, not to produce behavioral novelty or individual versatility. To Symons, play implies developmental resilience and variation-reduction, not developmental plasticity or variation-enhancement. This critique reflects the sociobiological concept of biological constraints: human potential is not unbounded, flexibility is not infinite, and even such protean behavior as play must be interpreted as an essentially conservative force. Flexibility theorists are almost all psychologists. Basic to their intellectual training is the idea that human potential is unlimited and that human behavior is infinitely malleable. I do not wish to obscure important distinctions within psychology; it would be exceedingly naive to lump Bruner, who views play as a source of behavioral flexibility, with Piaget, for whom play is epiphenomenal, and with Skinner, who seems not to recognize play as a behavioral entity. However, the division between Symons's approach to play as practice of species-typical skill and the flexibility theorists' speculations on play and individual versatility is deep if philosophically thorny, mirroring fundamental differences between psychological and biological approaches to behavior.

My second observation results from personal experience. The occasion was one of several opportunities I have had to date to observe play in an undomesticated species living in an undisturbed environment, but this experience, my first, proved particularly memorable, as the following account indicates.

One's initial reaction to wild animals at play need not be euphoric. When I first saw white-tailed deer (*Odocoileus virginianus*) fawns running repeatedly back and forth through shallow water, making rapid turns and performing frequent rear-kicks, body-twists, and headshakes, my immediate response was that they had gone mad or that I was seeing things. It was actually annoying and baffling to watch them. The sight made me uneasy and insecure, as if a crack had appeared in the fortifications of my well-ordered world to admit chaos and misrule. Only later did I realize that the fawns had been playing.

In the past, scientists who studied play sometimes encountered deep hostility from their colleagues and from the public. I can only explain this phenomenon by analogy to my own initial reaction to the deer. Play is protean behavior (Chance & Russell 1959, Driver & Humphries 1970, Humphries & Driver 1967, 1970). Apparently purposeless and patternless, its flurries of spirited activity strike no responsive chord. Starting and stopping capriciously and without apparent cause, they are protean and impossible to parse or to anticipate. "The sense behind the scene was not our sense" (Vladimir Nabokov, "Pale Fire," lines 709–710). The most irritating feature of play, however, is not the abyss, not perceptual incoherence as such, but rather that play taunts us with its inaccessibility. We feel that something is behind it all, but we do not know, or we have forgotten, how to see it.

Human responses to play characteristically juxtapose recollections of childhood with forebodings of mortality. The freedom of play often suggests greater autonomy. For example, the theme of play as liberation and the contrast of play with mortality are evident in the following verses ("Becoming," Sidor 1962):

> I am not now what I shall become
> at least, what I shall come to be
> and it is that one consolation
> that lately has set me free
>
> free from wisdom, free from care
> a bumblepuppy in the air
> a tumbleweed that rolls nowhere
> a careless step on a grassy stair
>
> I shall become what I am to be
> in the earth's own time, and see
> the green trunks bend in wind
> who have grown, but have not sinned
>
> growing as the spring fruit grows
> heedless of war, in scattered rows
> and kept by time in her windy breath
> waiting for ripeness, and for death

Perplexity and sporadic hostility continue to greet research on play, but, its perceptual insults notwithstanding, animal play itself has long been viewed with unashamed delight. Human fascination with the antics of puppies or kittens, or amusement at the chases and wrestling matches of young monkeys, may be as widespread as human hostility toward play. Playing animals can entertain a sympathetic onlooker, although this enjoyment is difficult or even socially unacceptable outside a few specified contexts like that of a zoo visit. Indeed, naturalists occasionally let the mask of scientific objectivity slip to hint at the pleasure they gain watching animals play. Even the most rigorous and thorough observers have stated, contrary to scientific convention and, as it were, despite themselves, that watching animal play is aesthetically pleasing (e.g., Leuthold 1977 p. 174 on black rhinoceros). Often such statements are appropriately made in nontechnical works (e.g., Lorenz 1955 p. 154 on domestic cats, Murie 1954 p. 74 on river otter).

In the play of animals we find a pure aesthetic that frankly defies science. Why kittens or puppies chase and vigorously paw at each other in reciprocal fashion without inflicting injury, repeating this behavior almost to the point of physical exhaustion, is not known. Yet this behavior fascinates, indeed enchants. Ernest Hemingway's Santiago, an old fisherman who had fought human and marlin and had known adversity, remembered, above all else, the play of wild lions. With Santiago, we may well ask "Why are the lions the main thing that is left?"

Santiago, Nabokov's Humbert Humbert, and Padraic Pearse himself are all sociobiological stock characters at first glance. There was no shortage of sex and violence in Santiago's long, adventurous life. Pearse, revolutionary and man of action, heroically sacrificed himself when young to win freedom for his people. To a sociobiologist, Humbert's amorous adventures with nymphets would seem as natural and normal as the male hamadryas baboon's attraction to prepubescent females. But when Humbert (*Lolita*, Chapter 36), Santiago, and Pearse looked back over their lives, remembered play—dreamed, regretted, cherished, and hopelessly poignant—was the sole thing that mattered any more.

Mere nostalgia for childhood is not the issue. The basis of these responses is that suggested by human delight in animal play. Recollections of childhood experience can be powerful, lucid, compelling, and paradoxically immediate. Questions about behaviors called play and processes called development are inseparably linked to questions about phenomena called time. Perhaps time itself might profitably be viewed from a sociobiological perspective, in which it (at least subjective time) represents a biological dependent variable, subject in each individual to effects of competition, manipulation by others, misinformation, and self-deception. I am not speaking of time merely as a limiting resource (as in time-limited foraging), but of the quality of perceived time. From a literary point of view, time appears considerably more malleable and qualitatively heterogeneous than we scientists have led ourselves to believe. In this context, play acquires unique status: it represents timeless experience (Fink 1968). Heraclitus (Diels, fragment 52), Rilke (Fourth Duino Elegy, lines 37–39), Yeats ("Among School Children"), Nabokov ("Pale Fire," lines 806–829), and especially Wordsworth ("Ode: Intimations of Immortality from Recollections of Early Childhood") offer this insight. Here, as in Hemingway, Nabokov, and Pearse, play, remembered and observed, holds the key to understanding human development in its broadest sense. To capture current experimental, psychological, and literary insights into play in an evolutionary framework consistent with current epigenetic theory now represents a manageable problem for biologists. If our efforts are successful, their implications for the study of behavioral development, especially human development, will be virtually unlimited.

List of
Special Notation

Symbol	Explanation	Chapter/Appendix
a	Cost-benefit ratio for innovativeness	8
a	Conditional probability that a nonperformer will become a performer, given that one or more performances are observed	8/II
B	Birth rate per individual per unit time	6
$B_i(x)$	Benefit to subject of play of quality x with individual i	7
C	Maximum possible viability at age two	4
C	Cost of playing at partner's optimal quality	7
$C(\gamma)$	Function measuring cost of play in units of juvenile survivorship	6
$C_i(x)$	Cost to subject of play of quality x with individual i	7
c	Probability of surviving to age of first reproduction	6
c	Probability of horizontal transmission of new behavior	8
D	Hypothetical allele for noninnovativeness	8/II
DD	Nonperforming animal having DD genotype	8/II
$\overline{DD}$	Performing animal having DD genotype	8/II
Dd	Nonperforming animal having Dd genotype	8/II
$\overline{Dd}$	Performing animal having Dd genotype	8/II
d	Hypothetical allele for innovativeness	8/II
dd	Nonperforming animal having dd genotype	8/II
$\overline{dd}$	Performing animal having dd genotype	8/II
E	Total resource at age one	4

496

E	Length of life	8/III
f	Probability of discovery	8/II
$f(x)$	Function measuring effectiveness of resources devoted to increased survivorship	4
g	Number of generations separating first instances of successive types of discoveries	8
$H(k)$	Random probability that a nonperformer will become a performer given k contacts with performers	8/II
h	Probability of vertical transmission of new behavior	8
h	Expected value of random probability $H(k)$, taking expectation over all possible k values	8/II
J	Length of immature period	8/III
j	Age at first reproduction	6
j	Random number of animals becoming performers in a given generation	8/II
K	Carrying capacity of an environment for a population	4
k	Cost of compromise in sib self-handicapping model	7
k	Random number of performing animals contacted by a nonperformer in a given generation	8/II
L	Total number of contacts with other individuals per individual per generation	8/II
l_x	Probability of survival from birth to age x	6
m_x	Fecundity at age x	6
N	Population size	8/II
n	Generation number	8/II
n	Number of individual play groups observed	8/III
P	Maximum possible immediate viability if all resource units used	4
$P_a(n)$	Probability of observing a given rare event in adult animals	8/III
$P_j(n)$	Probability of observing a given rare event in immature (juvenile) animals	8/III
$P_{ja}(n)$	Probability of observing separate rare events in immature (juvenile) period and in adulthood of same individual	8/III
P_1	Optimal amount of play at age one	6
P_1	Fraction of immature period spent in play	8/III

$P_1(n)$	Frequency of skilled DD individuals in generation n before selection	8/II
P_2	Optimal amount of play at age two	6
P_2	Fraction of adulthood spent in behavior made possible by innovative play as juvenile	8/III
$P_2(n)$	Frequency of unskilled DD individuals in generation n before selection	8/II
P_3	Optimal amount of play at age three	6
p	Fraction of population consisting of Tough animals	7
p	Fraction of Dd individuals exhibiting playful phenotype	8/I
p_i	Subject's probability of achieving play with individual i	7
p_1	Chance of survival from age one to age two	4
p_2	Chance of survival from age two to age three	4
$Q_1(n)$	One-half the frequency of skilled Dd individuals in generation n before selection	8/II
$Q_2(n)$	One-half the frequency of unskilled Dd individuals in generation n before selection	8/II
$R_1(n)$	Frequency of skilled dd individuals in generation n before selection	8/II
$R_2(n)$	Frequency of unskilled dd individuals in generation n before selection	8/II
r	Intrinsic rate of population growth	4, 6
r	Conditional probability of skill acquisition by a nonperforming individual given a single contact with a performer	8/II
T	Total amount of observation time	8/III
V_A	Value to a Tough animal of play at its personal optimal play quality	7
V_B	Value to a Gentle animal of play at its personal optimal play quality	7
x	Fraction of total resource used to improve immediate viability	4
x	Quality of play interaction	7
x	Frequency of DD animals	8/I
x	Probability that an individual having two performing parents will learn skills from them	8/II
x^*	Optimal amount of total resource used to improve immediate viability	4
$\bar{x}$	Frequency of $\overline{DD}$ animals	8/I

Y_j	Time spent watching play	8/III
γ	Absolute amount of resources devoted to play during the juvenile period	6
γ	Frequency of Dd animals	8/I
γ	Probability that an individual having exactly one performing parent will learn skills from it	8/II
$\overline{\gamma}$	Frequency of $\overline{Dd}$ animals	8/I
z	Frequency of dd animals	8/I
$\overline{z}$	Frequency of $\overline{dd}$ animals	8/I
α	Fraction of maximal return from play obtained by a Tough animal when self-handicapping	7
α'	Fraction of maximal return from play obtained by a Gentle animal when self-handicapping	7
α	Frequency of D allele	8/I
$\alpha(n)$	Frequency of D allele in generation n before selection	8/II
β	Fraction of maximal possible return from play obtained by a Tough animal when playing at a Gentle animal's preferred quality	7
β'	Fraction of maximal possible return from play obtained by a Gentle animal when playing at a Tough animal's preferred quality	7
β	Frequency of d allele	8/I
$\beta(n)$	Frequency of d allele in generation n before selection	8/II
$\gamma(n)$	Frequency of performers in generation n before selection	8/II
$\delta(n)$	Frequency of nonperformers in generation n before selection	8/II
ρ	Force of adult mortality	6
Σ	Normalization factor used in converting numbers of individuals to class frequencies	8/II
σ	Value of skill	8/II
τ	Cost of playfulness	8/II

Appendix I

Representative Definitions of Play

This list includes both nominal and real definitions. It demonstrates that approaches to play vary significantly even at a basic definitional level. Many real definitions have functional theories or unstated assumptions in tow. These assumptions concern definition itself, behavior, and behavioral research. Gilmore (1966) lists additional definitions of play.

Aldis (1975):

Almost everyone would agree that chasing and playfighting in young animals is play. The serious counterparts of these behaviors may be broadly classified as agonistic behaviors—predation, aggression, and flight. In play, these behaviors are usually accompanied by play signals and are modified in certain ways (lower intensity, relaxed muscle tone) from their serious counterparts. In addition, some serious behaviors may be omitted, new behaviors may be added, and the order may be changed. The causation of play may also differ from that of serious behavior.

Allin (in Gilmore 1966):

Play refers to those activities which are accompanied by a state of comparative pleasure, exhilaration, power, and the feeling of self-initiative.

G. Bateson (1955, 1956):

Play is a phenomenon in which the actions of "play" are related to, or denote, other actions of "not play."

The play of two individuals on a certain occasion would then be defined as the set of all messages exchanged by them within a limited period of time and modified by the paradoxical premise system which we have described.

Play is an onionskin system—a set of rules for communication—with an extra degree of freedom.

Bekoff (1972):

Social play is that behavior which is performed during social interactions in which there is a decrease in social distance between the interactants, and no evidence of social investigation or of agonistic (offensive or defensive) or passive-submissive behaviors on the part of the members of a dyad (triad, etc.), although these actions may occur as derived acts during play. In addition, there is a lability of the temporal sequences of action patterns, actions from various motivational contexts (e.g., sexual and agonistic) being combined.

Bolwig (1963):

[Play is] any action which is performed as an outlet for surplus energy which is not required by the animal for its immediate vital activities such as collecting food, eating, mating, nursing, and other activities which further its own survival and that of its species.

K. Buhler (1924):

[Play is] functional pleasure.

Buytendijk (1933):

Play expresses the essence of the youthful.

Claparede (1934):

[Play is] an expressive exercising of the ego and the rest of the personality, an exercising that strengthens developing cognitive skills and aids the emergence of additional cognitive skills.

Curti (in Gilmore 1966):

[Play is] highly motivated activity which, as free from conflicts, is usually though not always pleasurable.

Dewey (in Gilmore 1966):

[Play is] activities not consciously performed for the sake of any result beyond themselves.

Eibl-Eibesfeldt (1970):

We may consider play as an experimental dialogue with the environment.

Fink (1968):

Play is finite creativity in the magic dimension of illusion.

Hall (1968):

Play . . . is a very broad term which includes almost any activity which, to the observer, seems to have no immediate objective. It therefore includes the manipulation of non-food objects, and the whole variety of sensorimotor performances that are "exploratory." It also includes the complex social interactions that take place among young animals and sometimes between young animals and adults, these being thought to be highly important in the process of socialization of the young and possible in establishing relative ranks amongst the young which might carry over into the adult hierarchy.

Helanko (1958):

A system is play if the individual subject can freely choose the object of the system and no other system interferes with the resulting system.

Hinde (1970):

[Play is] a general term for activities which seem to the observer to make no immediate contribution to survival.

Huizinga (1950):

[Play is] a voluntary activity or occupation executed within fixed limits of time and place, according to rules freely accepted but absolutely binding, having its aim in itself and accompanied by a feeling of tension, joy, and the consciousness that it is "different" from "ordinary life."

Kaufman & Rosenblum (1966):

[Play is] those behaviors which are characterized by lightness, freedom of movement, a lack of tenseness, and an absence of stereotyped sequences. During such activities components are varied, and the roles of participants may be rapidly changing. Inanimate-object play . . . vigorous chewing, biting, manipulation, or carriage of an environmental object or toy.

Lazarus (in Gilmore 1966):

Play is an activity which is in itself free, aimless, amusing or diverting.

Loizos (1967):

[Play is] a positive approach towards and non-rigidified interaction with any feature of the animal's environment, including conspecifics, involving stimulation through most sensory modalities.

Marler & Hamilton (1966):

Play consists of elements drawn from other types of behavior and rearranged in new patterns of timing and sequence, along with certain (primarily communicative) behavior patterns evinced only in play.

Monod (1971):

[Play is] the outward expression of subjective simulation.

Müller-Schwarze (1978a):

"Ludic behavior" comprises "true play," simplest forms of play, and other "play-like" behaviors such as vacuum activities, or responses transferred to other objects. . . . The hallmark of "true play" is the presence of the three features: (a) true social interaction (as opposed to mere social facilitation), often including a code of rules such as inhibited use of teeth or claws; (b) play signals; and (c) a typical sequential organization of a play bout. These three characteristics signify the highest level of play behavior in nonhuman mammals. Simple forms of play, i.e., behaviors that do not meet these criteria (again, it is difficult to draw clear boundaries), could be called "protopedic" behavior.

Patrick (in Gilmore 1966):

[Play is] those human activities which are free and spontaneous and which are pursued for their own sake alone; interest in them is self-sustaining, and they are not continued under any internal or external compulsion.

Rasa (1971):

[Play is] behavior with no immediate reward other than its performance.

Schaller (1963):

[Play is] any relatively unstereotyped behavior in which an animal was involved in vigorous actions seemingly without definite purpose.

Schlosberg (1947):

Behavior is called playful only if it seems useless in the eyes of an observer.

Schneirla, Rosenblatt, & Tobach (in Rheingold 1963):

[Play is] casual joint activities.

Seashore (in Gilmore 1966):

[Play is] free self-expression for the pleasure of expression.

Spencer (in Gilmore 1966):

[Play is] activity performed for the immediate gratification derived, without regard for ulterior benefits.

Stern (in Gilmore 1966):

Play is voluntary, self-sufficient activity.

Sutton-Smith (1971):

[Play is] an exercise of voluntary control systems with disequilibrial outcomes.

Symons (1978a):

"Play" patterns differ from similar patterns in the species repertoire, and either lack immediate function or, at the least, lack the function of the patterns they resemble. . . . "Play," then, exists in contrasts: playfighting contrasts with fighting; immature sexual patterns contrast with mature sexual patterns; the infant caretaking activities of juvenile females contrast with the infant caretaking activities of mothers. . . . Individual animals exhibit some behavior patterns in two distinct modes; the term "play" seems to be a reasonable designation for one of these modes.

Valentine (in Berlyne 1969):

Play is any activity which is carried out entirely for its own sake.

Vygotsky (1967):

The essential attribute of play is a rule which has become an affect.

Welker (1961):

[Play is] a wide variety of vigorous and spirited activities: those that move the organism or its parts through space such as running, jumping, rolling and sommersaulting, pouncing upon and chasing objects or other animals, wrestling, and vigorous manipulation of body parts or objects in a variety of ways. The goals and incentives of vigorous play consist of certain patterns of variable or changing stimulation of the sensory surfaces.

E. P. Wheeler 2nd, from Buhler (in Gilmore 1966):

Play is an exercise of functions simply for the pleasure the exercise gives.

E. O. Wilson (1971b):

In mammals, play is comprised largely of rehearsals performed in a nonfunctional context of the serious activities of searching, fighting, courtship, hunting, and copulation.

Appendix II

Representative Lists of Characteristics of Play

Bekoff (1974c):

1. Actions from various contexts are incorporated into unpredictable (labile) temporal sequences.

2. The "play bout" is typically preceded by a meta-communicative signal which indicates "what follows is play"; these signals are also observed during the bout.

3. Certain actions may be repeated and performed in an exaggerated manner.

4. The activity appears "pleasurable" to the participants.

Henry & Herrero (1974):

1. Motor patterns during play often have an incomplete appearance when compared to their appearance during non-playful behavior. For example, during play-fighting, in dogs, snarling may be dissociated from the piloerection which inevitably accompanies it during true fighting.

2. Social play may be identified by certain motor patterns which are observed only during social play.

3. Social play is characterized by the exaggerated and uneconomical quality of the motor patterns involved.

4. A characteristic of play is that certain motor patterns are repeated more frequently in the play sequences than during the non-play sequences.

5. In play, normal temporal groupings of functionally related motor patterns break down, so that different kinds of behaviors intermingle in the same sequence; for example, prey capture behavior and sexual behavior may be combined in the same play sequence.

Loizos (1966):

1. The sequence may be reordered.

2. The individual movements making up the sequence may become exaggerated.

3. Certain movements within the sequence may be repeated more often than they would usually be.

4. The sequence may be broken off altogether by the introduction of irrelevant activities, and resumed later. This could be called fragmentation.

5. Movements may be both exaggerated and repeated.

6. Individual movements within the sequence may never be completed, and this incomplete element may be repeated many times. This applies equally to both the beginning of a movement (the intention element) and to its ending (the completion element).

Marler & Hamilton (1966):

1. The normal temporal groupings of functionally related actions break down. Elements of a number of different types intermingle in the same behavioral sequence.

2. The sequences of a given type are disrupted.

3. The pattern of behavior in play is less dependent on the normal S-R relationship.

Meyer-Holzapfel (1956b):

1. Play runs along in the tracks of instinctive movements as well as of learned actions without taking over their functions. This means that as a rule play lacks the immediate, biologically adaptive consequences that normally are associated with serious behavior. The consummatory act that usually terminates an instinctive behavior is lacking in play.

2. During play the orderly execution of a pattern sequence and the order of the instinctive movement typical for the serious context are dissolved. The partial activities can be freely interchanged. Play partners can change their roles frequently, for instance during fighting play.

3. During play, instinctive movements of different functional systems that exclude each other can occur together (mixed types of fighting and sexual play behavior, or hunting and fleeing behavior, etc.). The usual inhibitory mechanisms that as a rule block the simultaneous occurrence of instinctive movements are largely inactivated.

4. Play behavior is practically repeatable ad libitum and often assumes the role of a rhythmical sequence of the same behavioral elements. It lacks the reaction-specific fatigue that is characteristic of the instinctive behavior.

5. During play behavior facial expression and gestures—as far as we can interpret them—express a state of pleasure.

6. The search for a play partner, be it a conspecific, a person, or an object, is typical for many forms of play behavior.

7. Only in fighting play does the statement make sense that all species-specific social inhibitions appear.

8. There are transitions or rather mixed forms between play in the sense of an allochthonous activity and the gratification of autochthonous response transferences.

9. The animal that is ready to play is actively looking for an opportunity to play; there is a play appetence.

Rensch (1973):

1. the occurrence of components of normal instinctive behavior in incomplete and variable sequences lacking consummative actions
2. repetition of single components
3. tendency to exaggerate certain movements
4. social inhibitions, particularly avoidance of injuring the partner
5. possible use of inanimate objects or individuals of other species as substitute playmates
6. possible interruption in every stage by stronger stimuli (loud noise, appearance of enemies, need of defecation, and so on)
7. transmission of playing mood to other individuals, particularly to playmates
8. a certain degree of freedom in inventing new individual or experimental play, sometimes leading to new nervous and muscular coordinations
9. correspondence to simple human play

Symons (1978a):

1. The motor patterns observed in play are similar to those observed in other functional contexts.
2. Play patterns may be incomplete.
3. Play patterns may be inhibited.
4. Certain behavior patterns are unique to the play context.
5. Play patterns may be exaggerated or uneconomical.
6. Play occurs in different situations or as a result of different stimuli.
7. Patterns may be repeated more during play than during the nonplay activity it resembles.
8. Play patterns may be relatively unordered.
9. Patterns from different functional spheres may alternate during play.
10. Play is characteristic of immature animals.

Wasser (1978):

1. Short sequences and repetitious motor patterns
2. Lack of agonistic signals, especially vocalizations
3. Inhibited bites (judged by responses of recipients)
4. Elicitation of movements by stimuli inadequate to elicit such movements in their normal functional context
5. Animals repeatedly returning to the stimulus source (distinguishing the behavior from a withdrawal from noxious or dangerous stimuli)

XIII International Ethological Conference (1973) (cited in Weiss-Burger 1975; translated from the German):

1. Exaggerated motor patterns
2. Fragmented sequences of behavior

3. Rapid alternation of behavior

4. Mixture of behavioral acts from different social contexts (e.g., role change between flight and pursuit)

5. The same behavior directed at different stimuli (e.g., an animal bites and claws at part of its surroundings, then immediately treats a social partner likewise)

6. Occurs most often in immature animals

7. Occurs in a relaxed motivational field

8. Lacks consummatory act

Appendix III

An Incomplete List of English-language Terms Used in Reference to Play

Specialized terms are used for many movements, postures, or gaits characteristic of play. In addition, particular words exist for playful people, for places used for play, and for objects used in play. Languages other than English also offer terms unique to play. For example, in French a play-kick directed to the rear is a *ruade* and a skipping gambol is a *gambade*. German words used to denote playfighting include *balgen*, *toben*, and *tollen*. Definitions below are my own unless another source is cited. The source for most of these definitions is the *Shorter Oxford English Dictionary* (SOED). The *Oxford English Dictionary* (OED) was also consulted.

Antic A grotesque gesture, posture, or trick. (SOED)
Badinage Playful banter. (SOED)
Banter Good-humored raillery. (SOED)
Bear-play Rough tumultuous behavior. (SOED)
Blithe Jocund, gay, sprightly, merry. (SOED)
Boisterous Abounding in rough but good-natured activity bordering upon excess. (SOED)
Bumble To move clumsily.
Capade A tricky or unconventional movement or sequence of movements (from Ice Capades, a professional ice-skating troupe).
Caper A frolicsome leap, as of a kid; a frisky movement. (OED) (To cut capers. To perform capers.)
Caprice A sudden turn of the mind without apparent motive; a freak, whim, mere fancy. (The word "caprice" derives originally from an Italian word for horror: capricchio, capo = head + riccio = hedgehog, literally "head with the hair standing on end.") (SOED)

Later associated with capra, goat (see Caper)—a nice example of the ambivalent human attitude toward play that enables us to find this behavior both horrifying and delightful.

Capricious Subject to caprice; whimsical. (SOED)

Capriole A leap or caper, as in dancing. In horsemanship, a high leap made by a horse without advancing, the hind legs being jerked out together at the height of the leap. (SOED)

Caracole A half-turn to the right or left made by a horseman. (SOED)

Cavort To curvet, caper about, frisk. (SOED)

Curvet A leap of a horse in which the forelegs are raised together and equally advanced, and the hindlegs raised with a spring before the forelegs reach the ground. (SOED)

Fair play Equitable play (SOED); play without cheating.

Frisk To make brisk and lively movements. (SOED)

Fritter To waste time on trifles. (SOED)

Frivolous Characterized by lack of seriousness, sense, or reverence; given to trifling, silly. (SOED)

Frolic To make merry; later, to play pranks, gambol, caper about. (SOED)

Galumph To march exultingly with irregular bounding movements. (SOED)

Gambade A leap or bound of a horse. (SOED)

Gambol A leap or spring; a caper, frisk. (SOED) "To turn in the air against herself" (Th. Gautier 1839, of the dancer Augusta Maywood).

Gamesome Full of play; frolicsome, sportive. (SOED)

Horse-play Rough, coarse, or boisterous play. (SOED) (Horsing around. Indulging in horse-play.)

Jackanapes A pert, impertinent fellow. (SOED)

Jink A quick turn. (High jinks, high pranks. Lively or boisterous sport.) (SOED)

Jump for joy Said (literally) of children, etc.; also (figuratively) to be joyfully excited. (SOED) Also, leap for joy.

Kick up one's heels To caper, capriole, or curvet; but to "kick one's heels" is to wait idly or impatiently. (SOED)

Lark To play tricks, frolic; to tease sportively. (SOED)

Lightsome Light-hearted, cheerful. Flighty, frivolous. (SOED)

Lollop To go with a lounging gait. (SOED)

Make merry To be festive, to indulge in jollity. (SOED)

Mischief Playful tricks. (This word originally had far more serious connotations.)

Monkey business Mischief.
Monkey shines Monkey-like tricks or antics. (SOED)
Playboy A pleasure-loving man. (SOED)
Playfellow A playmate.
Playful Full of play, frolicsome, sportive. (SOED)
Playground A piece of ground used for playing on. (SOED)
Play-group A group of children who play together under adult supervision, usually at a regular time and place. (SOED) More generally, a group of animals that forms for the purpose of play.
Playmate A companion in play. (SOED)
Playpen An enclosure in which a young child may play in safety. (SOED)
Playsome Playful. (SOED)
Playstead A playground. (OED)
Playstow A playground. (OED)
Playsuit Informal, durable, one-piece children's apparel.
Plaything A toy to play with. (SOED) More generally, an item used in play.
Playtime A time for play or recreation. (SOED)
Playward Playful. (OED)
Rag An extensive display of noisy disorderly conduct. (SOED)
Rage To romp, to play wantonly (obs.).
Ragrowther To romp. (Also ragrowster: Williamson, *Tarka the Otter*.)
Raillery Good-humored ridicule, banter. (SOED)
Ramp To bound, rush, or range about in a wild or excited manner. (SOED)
Rascal A rogue, knave, or scamp (used playfully). (SOED)
Revel Riotous or noisy mirth or merry-making. (SOED)
Riot Unrestrained revelry, mirth, or noise. (SOED)
Rollick A sportive frolic or escapade. (SOED)
Rollicking Boisterously sportive. (SOED)
Romp. To play, sport, or frolic in a very lively, merry, or boisterous manner. (SOED)
Rough-and-tumble Having the character of a scuffle or scramble. (SOED)
Roughhouse To play roughly and boisterously, usually with body contact.
Scamp A rascal. (SOED)
Scamper To run or caper about nimbly. (SOED)
Silly Foolish, empty-headed; evincing or associated with foolishness. (SOED)

Skylark To frolic or play; to play tricks; to indulge in rough sport or horse-play. (SOED)

Skylarking A term originally used by seamen, to denote wanton play about the rigging, and tops, or in any part of the ship. (SOED) Now also used by aviators to denote wantonly playful aerobatics.

Sportful Having an inclination or a tendency to engage in sport or play. (SOED)

Sportive Disposed to be playful or frolicsome. (SOED) [The sportive lemur (*Lepilemur mustelinus*) is so named not because it is especially playful, but because of its characteristic raised-fist threat posture resembling a human boxer's stance.]

Spree A lively or boisterous frolic. (SOED)

Sprightly Lively, sportive. (SOED)

Tickle To touch or poke (a person) lightly in a sensitive part so as to excite spasmodic laughter. (SOED)

Toy A plaything. (As a verb: to play, sport; to frisk about.) (SOED)

Trifle To dally, fiddle, or toy with.

Tussle To push or pull about roughly. (SOED)

Wanton (Of young animals): Frisky, frolicsome. (SOED)

Word play Punning or, more generally, whimsical use of language.

Appendix IV

Game Theory Models of Conflict and Compromise in Social Play

In the first model (Model I), any individual occupies each role (Tough, Gentle) one-half of the time. Expected changes in fitness for each strategy in encounters with each of the four strategies considered are given in Table 7-2. If the fitness of one strategy played against itself is greater than the fitnesses of other strategies played against it, then the first strategy is said to be evolutionarily stable against the alternatives considered. This stability means that rare mutant or rare immigrant genes for the other strategies cannot invade a population of animals whose genes determine the first strategy. Recalling that $\beta < \alpha < 1$ and $\beta' < \alpha' < 1$, we see by inspection of Table 7-2 that GS and SG are the only evolutionarily stable strategies for this game.

In the second model (Model Ia), individual roles are fixed. The population consists of fraction p Tough, fraction $(1-p)$ Gentle. The alternative strategies considered are stinginess and generosity to animals of opposite type. In a population made up of the four possible combinations of type and strategy, expected fitness changes from the sixteen possible types of encounters are given by Table IV-1.

This model differs from the usual model of evolutionarily stable strategies because the frequencies of Tough and Gentle remain fixed. The model seeks to ascertain what combinations of strategies played by Tough and by Gentle will be evolutionarily stable in this environment. Therefore, we write a time recursion for the frequencies of the four possible phenotypes, solve for the fixed (equilibrium) points of this recursion, and analyze stability of these fixed points. Let (r, s, u, v) denote the population frequencies of Tough-Stingy, Tough-Generous, Gentle-Stingy, and Gentle-Generous phenotypes, respec-

Table IV-1 PAYOFF MATRIX FOR MODEL Ia

		Against:			
		Tough, Stingy	Tough, Generous	Gentle, Stingy	Gentle, Generous
To:	Tough, Stingy	V_A	V_A	0	V_A
	Tough, Generous	V_A	V_A	βV_A	αV_A
	Gentle, Stingy	0	V_B	V_B	V_B
	Gentle, Generous	$\beta' V_B$	$\alpha' V_B$	V_B	V_B

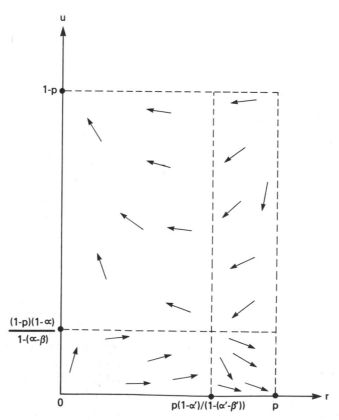

Fig. IV-1 Dynamic properties of Model Ia.

tively. Because the frequencies of Tough and Gentle types are fixed, $r+s=p$, $u+v=1-p$. Fixed points of the recursion are

$$(p, \ 0, \ 1-p, \ 0), \ (p, \ 0, \ 0, \ 1-p), \ (0, \ p, \ 1-p, \ 0), \ (0, \ p, \ 0, \ 1-p),$$

and $$\left[\frac{1-\alpha'}{1-(\alpha'-\beta')} \, p, \ \frac{\beta'}{1-(\alpha'-\beta')} \, p, \ \frac{1-\alpha}{1-(\alpha-\beta)} \, (1-p), \right.$$

$$\left. \frac{\beta}{1-(\alpha-\beta)} (1-p) \right]$$

Only the second and third equilibrium points are stable. Figure IV-1 illustrates the dynamic behavior of the model. For $\alpha-\beta \approx \alpha'-\beta'$ and $\alpha' \ll \alpha$, an equilibrium of Gentle-Stingy and Tough-Generous has the largest zone of attraction. This result means that when the biologically important quality of play is physical roughness, Tough animals should self-handicap in play with Gentle animals and should be tolerant of their partners, whereas Gentle animals should play as they would with other Gentle animals and should protest any roughness by their partner.

Appendix V

Innovation Models Without Cultural Transmission

These models of the evolution of innovative play in the absence of cultural transmission of the discovered skill (no observation learning and no intragenerational or intergenerational transmission) assume that the tendency to perform innovative play is determined by two alleles at the autosomal (D,d) locus. Dominant, recessive, and intermediate play alleles will be considered. Other modeling assumptions are infinite population size, random mating, natural selection following the discovery phase in each generation, a discovery probability f for all playful individuals, fitness increase σ for those playful animals who discover the skill, and cost τ for all playful animals.

Each generation is assumed to consist of three phases: (1) random mating and offspring production, (2) discovery, and (3) natural selection.

The following notation will be used: $\overline{DD}$, $\overline{Dd}$, and $\overline{dd}$ are *performers* (animals who have discovered the skill); DD, Dd, and dd denote the corresponding genotypes in *non-performing* phenotypes. The frequencies of the six classes of individuals $\overline{DD}$, DD, $\overline{Dd}$, Dd, $\overline{dd}$, and dd, computed after mating, but before discovery are $\overline{x}$, x, $\overline{y}$, y, $\overline{z}$, and z, respectively; α is the frequency of the D allele and β the frequency of d, computed at this same stage of a generation. Doubly primed frequencies are computed after selection and before mating; singly primed frequencies after discovery and before selection; and Σ will denote a normalization factor used in converting numbers of individuals to class frequencies.

The first model assumes varying penetrance: a fraction p of the Dd individuals $(0 \leq p \leq 1)$ is identical to the playful homozygote. If $p = 0$ then the play allele is recessive; if $p = 1$ then the play allele is domi-

516

Table V-1 FITNESS TABLEAU FOR VARYING PENETRANCE MODEL

Type of Animal	Fitness
$\overline{DD}$	$1+\sigma-\tau$
DD	$1-\tau$
$\overline{Dd}$	$1+\sigma-\tau$
Dd	$1-p\tau$
$\overline{dd}$	1
dd	1

nant. This is a model of intermediate penetrance for $0<p<1$. Table V-1 gives the appropriate fitness tableau. This model is identical to that resulting from the assumptions that all heterozygotes discover the skill with probability fp and that the fitness of a heterozygous animal that fails to discover the skill is $1-p\tau$.

Calculations for the model are as follows.

Mating and Offspring

$$\overline{x}=\overline{y}=\overline{z}=0$$
$$x = [\overline{x}''+x''+\tfrac{1}{2}\,(\overline{y}''+y'')]^2 = \alpha''^2$$
$$y = 2\alpha''\beta'',\ z = \beta''^2$$

and

$$\alpha=\alpha'',\ \beta=\beta''$$

where unprimed quantities are valid in generation n and doubly primed in $(n-1)$. Random mating leaves gene frequencies unchanged in the population.

Discovery

$$\overline{x}'=fx$$
$$x'=(1-f)x$$
$$\overline{y}'=fpy$$
$$y'=(1-fp)y$$
$$\overline{z}'=0$$
$$z'=z$$

Selection

$$\bar{x}'' + x'' = \alpha^2 \, (1 + f\sigma - \tau)/\Sigma$$
$$\bar{y}'' + y'' = 2\alpha\beta \, [1 + p \, (f\sigma - \tau)]/\Sigma$$

where $\Sigma = 1 + \alpha(f\sigma - \tau) \, (\alpha + 2\beta p)$.

A recursion specifying α in generation $(n + 1)$ as a function of α in generation n is therefore

$$\alpha(n + 1) = \{\alpha(n)^2 \, (1 + f\sigma - \tau) + \alpha(n)\beta(n) \, [1 + p(f\sigma - \tau)]\}/\Sigma(n)$$

or

$$\alpha(n + 1) = [\alpha(n)/\Sigma \, (n)] \, \{1 + [\alpha(n) + p(1 - \alpha(n))] \, (f\sigma - \tau)\}$$

A necessary and sufficient condition for increase of the play (D) allele is that $\alpha(n + 1) > \alpha \, (n)$ for all n, or, by analysis of the above recursion,

$$(f\sigma - \tau)(1 - \alpha(n))[\alpha(n) \, (1 - 2p) + p] > 0$$

Because the second and third terms are positive, play will be selected provided $f\sigma > \tau$ (equivalently, $f > a = \tau/\sigma$), no matter what the value of p. The only meaningful equilibrium points for the model are $\alpha = 0$, $\alpha = 1$. When $f\sigma > \tau$, $\alpha = 1$ is a stable equilibrium point, $\alpha = 0$ unstable; the converse holds when $f\sigma < \tau$. When $f\sigma = \tau$ any initial α will remain in effect; this final case is trivial.

OTHER MODELS

If all heterozygotes play at fitness cost $p\tau$ and all discover, with probability p, a skill having fitness value σ, the resulting model is identical to that above. More general intermediate cases can be derived using a three-parameter model in which a heterozygote's discovery probability is $\lambda_1 f$, its fitness cost is $\lambda_2 \tau$, and the fitness value of the skill discovered is $\lambda_3 \sigma$ $(0 < \lambda_1 < 1, \ 0 < \lambda_2 < 1, \ 0 < \lambda_3 < 1)$. There can be endless variations on the theme of any simple model and it seems pointless to pursue this particular line of analysis, since its other assumptions are more unrealistic than the assumption that play is determined by alleles obeying the dictates of the variable penetrance model.

Appendix VI

Innovation Models including Cultural Transmission

It is interesting to investigate the implications of cultural transmission of skill, using modifications of the simple framework outlined in Appendix V. Even this unrealistic formalism rapidly becomes complex and quite difficult to analyze without the aid of a computer. However, certain principles do seem to emerge. These more complex models offer stable polymorphic equilibria in place of monomorphism of models assuming no cultural transmission. One such model will now be outlined. See Fagen (1974b) for mathematical details.

MODEL DEFINITION

Assume a one-locus, two-allele (D, d) system in which individuals of any genotype may either be performers ["skilled" individuals in Feldman & Cavalli-Sforza's (1976) terminology] or nonperformers ("unskilled" individuals). The allele (d) determining inventiveness is assumed to be recessive. The events occurring in each generation are, in order: discovery of novel acts by a fraction f of the potentially inventive phenotypes; intragenerational transmission, with a parameter the conditional probability a that a nonperformer will acquire the ability to perform the novel skill, given observation of one or more performances; transgenerational transmission, with probability x that a given offspring will learn the skill from its parents if both parents are performers, with probability y that a given offspring will learn from its parents if one and only one parent is a performer, and with probability 0 that transgenerational transmission will occur if neither parent is a performer; and selection, with the relative survival rates of different phenotypes defined by $1+\sigma$, $1+\sigma$, $1+\sigma-\tau$, 1, 1, and $1-\tau$ for $\overline{DD}$, $\overline{Dd}$, $\overline{dd}$, DD, Dd, and dd, respectively, where horizontal superscript bars denote skilled phenotypes. Discoveries are assumed to be equally advantageous to the inventor and to other performers, conferring fitness increment $\sigma>0$ to all performers regardless of genotype; and each inventive individual, whether or not it happens to be a performer, loses fitness $\tau>0$. This fitness cost could result from expenditure of limiting time and energy resources on behaviors facilitating discovery and even from bodily harm incurred accidentally during investigation, experimentation, or play.

Initial discovery of novel skills or tools, subsequent dissemination of these discoveries, and their eventual augmentation or even replacement, in later generations, by further innovation complete one cycle of a periodic process driving cultural evolution. The frequency at which successive types of discoveries are made is a characteristic of inventive individuals in a given population. For the purposes of the current model, $g \geqslant 1$ generations are assumed to elapse between the first instance of discovery and the first instance of the next. This formulation does not recognize the possibility of multiple discoveries (Kawai 1965) per generation, nor does it seek to simulate those longer-term sociobiological cycles in which stagnation alternates with renaissance in a kind of punctuated cultural equilibrium (Hutchinson 1965 p. 92; see Eldredge & Gould 1972 for a possible evolutionary analog).

Having discussed the conceptual basis of the model, I now establish some further notation:

$P_1(n)$ = frequency of $\overline{DD}$ phenotype in generation n before selection

$P_2(n)$ = frequency of DD phenotype in generation n before selection

$2Q_1(n)$ = frequency of $\overline{Dd}$ phenotype in generation n before selection

$2Q_2(n)$ = frequency of Dd phenotype in generation n before selection

$R_1(n)$ = frequency of $\overline{dd}$ phenotype in generation n before selection

$R_2(n)$ = frequency of dd phenotype in generation n before selection

$\alpha(n)$ = frequency of D allele in generation n before selection
$= P_1(n) + P_2(n) + Q_1(n) + Q_2(n)$

$\beta(n)$ = frequency of d allele in generation n before selection
$= Q_1(n) + Q_2(n) + R_1(n) + R_2(n)$

$\gamma(n)$ = frequency of performers in generation n before selection
$= P_1(n) + 2Q_1(n) + R_1(n)$

$\delta(n)$ = frequency of nonperformers in generation n before selection
$= P_2(n) + 2Q_2(n) + R_2(n)$

The four components of the model (discovery, intragenerational transmission, mating and transgenerational transmission, and selection) will now be specified

Discovery

It is assumed that each *dd* animal will invent the current skill with probability f per generation, independently of all other animals. If the product of f and population size N is sufficiently large, the fraction of all animals in generation n that discovers the skill will be $fR_2(n)$, where singly primed state variables are current after selection, but before discovery. The process of discovery changes the state of the population according to the following equations:

$$P_1''(n) = P_1'(n)$$
$$P_2''(n) = P_2'(n)$$
$$2Q_1''(n) = 2Q_1'(n)$$
$$2Q_2''(n) = 2Q_2'(n)$$
$$R_1''(n) = R_1'(n) + fR_2'(n)$$
$$R_2''(n) = (1 - f)R_2'(n)$$

where the primed quantities denote pre-discovery and the doubly primed quantities post-discovery frequencies.

In order to model discovery simply, it was necessary to assume an identical probability f of discovery for all *dd* nonperformers and for all types of discoveries and to assume that discoveries occur independently and that population size is sufficiently large to make stochastic effects negligible. Since all innovative nonperformers are assumed to have an equal chance to make discoveries, the model cannot account for "geniuses" like Imo (Kawai 1965), who made two remarkable discoveries (sweet-potato washing and wheat-washing) within the short space of three years. In discussing the assumptions it is also important to point out how large "sufficiently large" is when probability of discovery and proportion of *dd* nonperformers are both small, as the following analysis indicates.

Suppose stochastic effects are considered negligible when the ratio of standard deviation to mean of the quantity j/N, the fraction of all animals making the discovery in a given generation, is less than 0.1. The mean of j/N is fR_2' and the standard deviation of j/N is

$$\left[\frac{R_2'f(1-f)}{N}\right]^{\frac{1}{2}}$$

since j is binomially distributed with parameters $R_2' N$ and f. The condition on the standard deviation to mean ratio can be written

$$\frac{1-f}{fR_2'N} < .01$$

which means we must require

$$N > \frac{100}{fR_2'}$$

For instance, if $f = .001$ and $R_2 = .001$, N must be greater than 10^8. This calculation demonstrates how unrealistic the assumption of negligible stochastic effects really is. Compare 10^8 with the size of most animal social groups [5 to 100, e.g., 30 in human hunter-gatherers (Lee & DeVore 1968, 1976)]. Clearly, when the d allele is rare (low R_2) and when discoveries are difficult or rare (low f), stochastic effects will be important even in rather large populations. In neglecting such effects, I may be painting an unrealistic picture of the process of discovery. In reality, many generations may elapse before a discovery is made.

Intragenerational Observational Learning

A nonperformer's chance of acquiring a new behavior from conspecifics other than parents by intragenerational observational learning will clearly depend on (1) the probability of observing performances and (2) the likelihood that the nonperformers will learn to perform the behavior as a result of exposure to such performances. Acquisition may also depend on such additional factors as the number of performances witnessed, the particular behavior being considered, and the social structure of the population. For simplicity, however, I assume a random mixing model (Boorman & Levitt 1973) in which the relevant probabilities depend at most on the overall frequency of performers in the population. In particular, time, the observer's genotype and previous experience, the nature of the new behavior, and the identity of the performer observed will all be neglected.

The simplest non-trivial relationship between observation probability and performer frequency is the identity: Prob {one or more performances observed by a given individual in generation n} = $\gamma''(n) = P_1''(n) + 2Q_1''(n) + R_1''(n)$. If c is the conditional probability (assumed constant) of acquisition of a new behavior by an individual given that observation of one or more performances has occurred, the probability of acquisition is given by $c\gamma''(n)$ (SMM). This simple multiplicative model (SMM) will be used here. I will further simplify the model by allowing $c = 1$. This additional assumption will hold provided that the act is easily mastered once observed. Novel behaviors spread rapidly in species as diverse as chimpanzees (Menzel 1973), northern blue jays (*Cyanocitta cristata*) (Jones & Kamil 1973), and common ravens (Gwinner 1966). In the blue jays, for example, tool-making and tool-using were acquired in one to two years by five of eight animals, and two others displayed some components of the behavior. A single individual is felt to have originated the pattern. It will be shown below that the model with $c = 1$ implies $\gamma(n + 1) = 2\gamma(n) - \gamma(n)^2$; if the study period is considered to span two years and a blue jay generation one year and if we assume the initial frequency of performers was $\frac{1}{8}$, after two generations λ will be 0.41 compared to the observed value of $\frac{5}{8} = 0.63$. If a generation is assumed to last two years rather than one, the SMM cannot account for the rapidity with which observation learning took place. The difficulty appears to be that in the blue jay colony all animals were housed in close proximity, so that the probability of observing a performance of tool-making and tool-use was not γ'', but 1, contrary to a basic assumption of the random mixing model. If this assumption, which does not appear to be obeyed by laboratory colonies of blue jays, is violated in other situations as well, and one still wishes to employ the random mixing model, the best policy is to set $c = 1$, realizing that the model may underestimate the rate of spread of behaviors in situations where animals are housed in close proximity for extended periods of time.

The analysis of a more detailed model of intragenerational observational learning also suggests that this process should be assumed to occur very rapidly. This detailed model again assumes a randomly mixing population. In the detailed model, as in Boorman & Levitt (1973), each individual is assumed to have $L \geq 1$ contacts per generation with other individuals, and population size is assumed to be large with respect to L. Let k be the number of contacts made by a nonperformer with individuals exhibiting the behavior. The probability that an individual contacted is a performer is precisely $\gamma''(n)$, since mixing is assumed to be random with respect to performance and nonperformance. Computed following the stage of discovery, $\gamma''(n)$ is given by $P_1''(n) + 2Q_1''(n) + R_1''(n)$. In a randomly mixing population, contacts are also assumed to occur independently of one another. Since the probability of contacting a performer on a given trial is $\gamma''(n)$, and since each contact is an independent event, the contacts can be modeled as independent Bernoulli trials with probability $\gamma''(n)$ of success (where success is defined as contact with a performer). As a result of these assumptions, k will be distributed binomially:

$$P_k(k_0) = \binom{N}{k_0} [\gamma''(n)]^{k_0} [1 - \gamma''(n)]^{N-k_0}$$

Let $H(k)$ be the probability that an individual will acquire the novel behavior given k contacts with performers. Function H should be monotonically increasing in k, with $H(0) = 0$, $H(\infty) = 1$. I will assume $H(k) = 1 - \exp(-sk)$. This function is concave downward, reflecting smaller increments in probability for large k, with the most rapid increase in the probability of learning occurring during the first few contacts. Parameter $s > 0$ determines the probability $H(1) = 1 - \exp(-s)$ of learning given a single contact with a performer. The expected probability of learning, taking the expectation over all possible k values, is

$$h \equiv E_k[H(k)] = E_k[1 - \exp(-sk)]$$

where E denotes the expected value operator.

$$h = 1 - E_k[\exp(-sk)] = 1 - M_k(-s)$$

where $M_k(t) = (1 - \delta''(n) + \gamma''(n)\exp t)^L$, the moment generating function of the binomial distribution. To obtain an expression for h in terms of previously defined quantities, substitute $t = -s$ in the expression above for $M_k(t)$:

$$h = 1 - [1 - \gamma'' + \gamma''\exp(-s)]^L = 1 - \{1 - \gamma''[1 - \exp(-s)]\}^L$$
$$= 1 - (1 - r\gamma'')^L$$

where $r = 1 - \exp(-s)$. The meaning of this result is that if the overall frequency of performers in the population is γ'', if mixing is random and a total of L contacts are made with other individuals, and if the probability of acquisition given a single contact with a performer is r, then the probability that a nonperforming individual will acquire a specified novel behavior by intragenerational observation learning is given by $h = 1 - (2 - r\gamma'')^L$. Under the standard large-sample approximation and with the random number of nonperformers acquiring the novel behavior per generation replaced by the expected value of this quantity, the following recursion defines the dynamics of intragenerational observation learning:

$$\gamma'''(n) = \gamma''(n) + \{1 - [1 - r\gamma''(n)]^L\}\delta''(n)$$

where $\delta''(n) = 1 - \gamma''(n)$.

An approximate γ''' recursion quadratic in γ'' is obtained using the first two terms of the binomial expansion of $(1 - r\gamma''(n))^L \approx 1 - rL\gamma''(n)$. Substituting this approximation in the above equation and substituting for $\delta''(n)$, we obtain

$$\gamma'''(n) \approx (1 + rL)\gamma''(n) - [\gamma''(n)]^2$$

These approximations resemble the simple multiplicative model, identifying c with rL. However, if $rL > 1$, c can no longer be interpreted as a probability, and if $rL\gamma > 1$ then the linear approximation is meaningless. A modified $\bar{h} \equiv \min(rL\gamma, 1)$ makes the approximation probabilistically meaningful.

The linear approximation to h is excellent for r and L small, but will significantly overestimate h as $r \to 1$, $L \to \infty$. Approximation $\bar{h}$ always overestimates the probability of learning by intragenerational observation, and since such an overestimate will tend to lower the computed equilibrium frequency of the d allele, the use of $\bar{h}$ in a place of h will always result in conservative predictions regarding the likelihood that play behavior will evolve (i.e., that the d allele will persist in a given population).

Novel behaviors have been observed to propagate through a small population in a single generation as a result of observational learning (both intragenerational and transgenerational processes may be involved, Kawai 1965). If a single animal discovers the behavior in a population of size N, then $\gamma''(n) = 1/N$, and the SMM predicts

$$\gamma'''(n) = \frac{2}{N} - \frac{1}{N^2} = \frac{1}{N}\left(2 - \frac{1}{N}\right)$$

Clearly for reasonable N, $\gamma''(n) \ll 1$, a contradiction of the observation that γ can go from $1/N$ to 1 in one generation.

As was the case in the model of discovery, the fraction of animals acquiring the new behavior is a random variable, but I will again use the expected value of this fraction, consistent with my previous assumption that population size is sufficiently large to make such an approximation valid. Under these assumptions, intragenerational observational learning will change the state of the population as follows:

$$P_1'''(n) = P_1''(n) + \gamma''(n)P_2''(n)$$
$$P_2'''(n) = [1 - \gamma''(n)]P_2''(n)$$
$$2Q_1'''(n) = 2Q_1''(n) + 2\gamma''(n)Q_2''(n)$$
$$2Q_2'''(n) = 2[1 - \gamma''(n)]Q_2''(n)$$
$$R_1'''(n) = R_1''(n) + \gamma''(n)R_2''(n)$$
$$R_2'''(n) = [1 - \gamma''(n)]R_2''(n)$$

Doubly primed frequencies apply after discovery and before observational learning, whereas triply primed frequencies are valid after observational learning takes place.

A recursion in $\gamma''(n)$ may be obtained as follows using the equations above:

$$\begin{aligned}
\gamma'''(n) &= P_1'''(n) + 2Q_1'''(n) + R_1'''(n) \\
&= P_1''(n) + 2Q_1''(n) + R_1''(n) + \gamma''(n)[P_2''(n) + 2Q_2''(n) + R_2''(n)] \\
&= \gamma''(n) + \gamma''(n)\delta''(n)
\end{aligned}$$

This expression is identical to the γ'' recursion in the full intragenerational learning model for $rL = 1$.

Although use of the SMM may tend to make the basic model's predictions regarding the evolution of play overoptimistic, the random mixing assumption has two other aspects both of which will tend to make the model more pessimistic. What if, contrary to the random mixing assumption, transmission of novel behaviors occurs primarily within litters or within playgroups? (I am assuming that playgroups are composed entirely of dd, "playful", animals.) The effect of this nonrandomness will be to make new behaviors accessible only to dd animals or to siblings of dd animals. In such populations, any new adaptive behavior will remain the exclusive property of the innovators, of their siblings, and of their offspring. I would conjecture that in closed societies of this type the evolution of potentially innovative play behavior would be highly likely.

Reproduction and Transgenerational (Parent-Offspring) Transmission of Behavior

If mating is random, both with respect to genotype and with respect to performance or nonperformance, the frequencies of the 21 possible types of matings are specified by the individual terms in the algebraic expansion of

$$[P_1'''(n) + P_2'''(n) + 2Q_1'''(n) + 2Q_2'''(n) + R_1'''(n) + R_2'''(n)]^2$$

The mechanism of transmission of D and d genes is specified by Mendel's laws, but transmission of behavior traits requires further assumptions. The behavioral state of young offspring will be assumed to depend on the presence or absence of the new behavior in their parents. Let x be the probability that a given offspring will learn the behavior from its parents if both parents are performers, and y be the probability that a given offspring will learn the behavior from its parent if one and only one parent is a performer. Obviously, an offspring will not learn the new behavior from its parents if neither parent is a performer, though it still may learn the behavior through observation of models other than its parents. Again replacing random variables by their expected values, and assuming that the transmission or nontransmission of the behavior to each individual offspring occurs independently, we obtain the following model for reproduction and transmission to offspring:

$$P_1(n+1) = [P_1'''(n) + Q_1'''(n)]\{x[P_1'''(n) + Q_1'''(n)] + 2y[P_2'''(n) + Q_2'''(n)]\}$$

$$P_2(n+1) = (1-x)[P_1'''(n) + Q_1'''(n)]^2 + 2(1-y)[P_1'''(n) + Q_1'''(n)] \\ \times [P_2'''(n) + Q_2'''(n)] + [P_2'''(n) + Q_2'''(n)]^2$$

$$2Q_1(n+1) = 2x[P_1'''(n) + Q_1'''(n)][Q_1'''(n) + R_1'''(n)] + 2y[P_1'''(n) + Q_1'''(n)] \\ \times [Q_2'''(n) + R_2'''(n)] + [Q_1'''(n) + R_1'''(n)][P_2'''(n) + Q_2'''(n)]$$

$$2Q_2(n+1) = 2(1-x)[P_1'''(n) + Q_1'''(n)][Q_1'''(n) + R_1'''(n)] + 2(1-y) \\ \times \{[P_1'''(n) + Q_1'''(n)][Q_2'''(n) + R_2'''(n)] + [Q_1'''(n) + R_1'''(n)] \\ \times [P_2'''(n) + Q_2'''(n)]\} + 2[P_2'''(n) + Q_2'''(n)][Q_2'''(n) + R_2'''(n)]$$

$$R_1(n+1) = x[Q_1'''(n) + R_1'''(n)]^2 + 2y[Q_1'''(n) + R_1'''(n)] \\ \times [Q_2'''(n) + R_2'''(n)]$$

$$R_2(n+1) = (1-x)[Q_1'''(n) + R_1'''(n)]^2 + 2(1-y)[Q_1'''(n) + R_1'''(n)] \\ \times [Q_2'''(n) + R_2'''(n)] + [Q_2'''(n) + R_2'''(n)]^2$$

Parameters x and y are assumed constant over individuals over time and over behavior type.

In domestic cats, as was mentioned earlier, observation learning when the model is one's own parent occurs even more rapidly than observation learning from a strange model (Chesler 1969). Mother-offspring relations in mammalian species as diverse as humans, elk, and monkeys afford frequent opportunities for close observation of the parent by the offspring (Rheingold 1963). Kawai (1965) claims that young macaques in the Koshima troop now learn sweet-potato washing from their mothers routinely and that only in one lineage did learning fail to occur in some individuals. The apprent efficiency of transgenerational observation learning when facilitated by the presence of a parent-offspring bond suggests that acquisition probabilities x and y should be very close to 1. I chose $x = y = 1$. This simplification may obscure intraspecific, interspecific, or interbehavior differences. Trivers (1974), for example, has argued that offspring may actually be selected in some cases to reject parental behavior as a model for their own.

It should be stressed that the transgenerational observation learning process of this model is specifically a parent-offspring type of transmission. In the discrete, non-overlapping generations framework of the basic model, parents are assumed to survive just long enough to act as models for their offspring's behavior. Learning from adults other than parents is not considered in this model.

The assumption $x = y = 1$ implies, for animals in which several young are born at a time, that if at least one parent is a performer all offspring will be performers. The actual mechanism could be parent → all littermates simultaneously or parent → offspring → other offspring.

Selection

The effects of natural selection are defined by a contingent fitness model (Boorman & Levitt 1973), in which individual fitness depends on the two components of the behavioral phenotype: performance or nonperformance, determined by individual experience, and potential innovativeness or noninnovativeness, determined by genotype. This dependence may be argued as follows. The fitness of performers would be increased if performance of a novel behavior allowed the animal to exploit a new habitat or food source or to use a tool or to adopt an improved form of social manipulation. Innovative ability, on the other hand, might be accompanied by specific risks. For instance, play behavior, which appears to be a component of the processes of discovery and invention, requires time, energy, and material, which could otherwise be used for growth and reproduction. Indeed, play may expose the animal to such environmental hazards as predation or dangerous physical conditions (Chapter 5).

In the contingent fitness model developed here, the fitness of a noninventive nonperformer is assumed to be 1. Performance is assumed to result in a fitness increment of $\sigma > 0$ for all performing animals, whereas inventiveness (dd genotype) is assumed to result in a fitness decrease τ, $0 < \tau < 1$, independent of performance status. These assumptions about fitness may be summarized in the following tableau:

Genotype	Performance Status	Fitness
DD	Performer	$1 + \sigma$
DD	Nonperformer	1
Dd	Performer	$1 + \sigma$
Dd	Nonperformer	1
dd	Performer	$1 + \sigma - \tau$
dd	Nonperformer	$1 - \tau$

The effect of selection according to the above tableau is specified by the following equations:

$$\tilde{P}'_1(n+1) = P_1(n+1)(1+\sigma); \ P'_1(n+1) = \tilde{P}'_1(n+1)/\Sigma$$
$$\tilde{P}'_2(n+1) = P_2(n+1); \ P'_2(n+1) = \tilde{P}'_2(n+1)/\Sigma$$
$$2\tilde{Q}'_1(n+1) = 2Q_1(n+1)(1+\sigma); \ 2Q'_1(n+1) = 2\tilde{Q}'_1(n+1)/\Sigma$$
$$2\tilde{Q}'_2(n+1) = 2Q_2(n+1); \ 2Q'_2(n+1) = 2\tilde{Q}'_2(n+1)/\Sigma$$
$$\tilde{R}'_1(n+1) = R_1(n+1)(1+\sigma-\tau); \ R'_1(n+1) = \tilde{R}'_1(n+1)/\Sigma$$
$$\tilde{R}'_2(n+1) = R_2(n+1)(1-\tau); \ R'_2(n+1) = \tilde{R}'_2(n+1)/\Sigma$$

where the normalizing factor

$$\Sigma = \tilde{P}'_1(n+1) + \tilde{P}'_2(n+1) + 2\tilde{Q}'_1(n+1) + 2\tilde{Q}'_2(n+1) + \tilde{R}'_1(n+1) + \tilde{R}'_2(n+1)$$

The selection process defined by the above assumptions is assumed to be the only agent affecting gene-frequency change in the population. In particular, population size is assumed large, making genetic drift effects negligible. Boorman & Levitt (1973) and Feldman & Cavalli-Sforza (1976) make this same assumption in their models of social evolution.

Although data on hunting success as a function of pack size in African lions tend to support the Boorman and Levitt fitness model, there are few data illustrating possible selective advantage resulting from performance of a newly discovered behavior. One possible example involves a chimpanzee whose dominance status increased after he discovered how to make loud noises using a metal can (van Lawick-Goodall 1971). Data bearing directly on possible selective disadvantage resulting from participation in play behavior are reviewed in Chapter 5.

Sequences of Discoveries

After a particular type of discovery (e.g., sweet-potato washing) has become fixed in a population, innovative animals, if present, will eventually make other types of discoveries (e.g., wheat-washing). The following very simple model of a sequence of discoveries will be used. (Although multiple discovery types may occur within a single generation, as was the case on Koshima Islet (Kawai 1965), I will not treat this case, which would require a population state vector with a dimension that doubled each time a new type of discovery was made. Note that accumulation of discoveries within a generation would always favor the d allele, since the initial advantage of discovery would be repeated. Again, the basic model is conservative about the evolution of play behavior.) Assume that distinguishable types of discoveries occur every g generations and that the other model parameters f, σ, and τ are independent of the particular type of discovery. The quantity g is assumed constant. When g generations have elapsed, I will redefine performance and nonperformance in terms of the new type of discovery that is about to spread through the population. Initially all animals will be nonperformers with respect to this new class of behavior. Gene frequencies, of course, will not be changed by this redefinition. In the basic model, γ is set equal to 0 every g generations after selection takes place, and the population state changes accordingly:

$$P_2'(n+1) = [P_1'(n+1) + P_2'(n+1)] \text{ after selection}$$
$$P_1'(n+1) = 0$$
$$2Q_2'(n+1) = [2Q_1'(n+1) + 2Q_2'(n+1)] \text{ after selection}$$
$$2Q_1'(n+1) = 0$$
$$R_2'(n+1) = [R_1'(n+1) + R_2'(n+1)] \text{ after selection}$$
$$R_1'(n+1) = 0$$

What if g is not sufficiently large to permit fixation of one discovery type before the next type occurs? Here, the dimension of the population state vector would have to double to account for two classes of performance and two classes of nonperformance. But if this situation occurs, the d allele will be a relative advantage, because d genes are preferentially concentrated in performers until $\gamma = 1$. By resetting γ to 0 even if it has not yet become equal to 1, I place the d alleles at an additional disadvantage whenever fixation of a new behavior type does not occur in g generations. The basic model may thus be seen to contain this additional conservative assumption. The value of a new discovery in this formulation will not depend on the number of cycles of discovery previously experienced by the population, but only on the constant g.

I have now defined two composite models. Both portray (1) discovery, (2) intragenerational learning, (3) reproduction and transgenerational learning, and (4) selection. The difference between the two models is as follows. The first complete model contains very general formulations of phases (2) and (3) along with eight parameters: f, g, L, r (or s), x, y, σ, and τ. The second model, the "basic model," assumes that intragenerational observation learning occurs according to the SMM with $a = 1$ and that transgenerational learning is perfect

$(x = y = 1)$. The basic model therefore has four parameters: f, g, σ, and τ.

Because analytic solution of derived equilibrium equations for even the basic model proved infeasible, properties of this model were investigated by computer simulation.

Appendix VII

Observing
Rare Events

Suppose (1) that novel behavior discovered in play is needed only rarely in adulthood, but that when the animal does make use of a novelty in meeting a rare contingency, this event is instantly recognizable. Assuming that we can appreciate the significance of a rare event if and when it does happen, what is our chance of observing in the juvenile stage the playful behavior that equipped the animal to meet the rare contingency successfully? (Ordinarily it would be very difficult for a human observer to recognize such an event, so this assumption will lead to overoptimistic predictions about the probability of actually observing playful innovation.)

Alternatively, suppose (2) that play results only rarely in some particular type of experience (a "programmed discovery"), but that once this experience has occurred, the adult animal uses the resulting information in a regular, recognizable part of its daily routine. Assuming that we can identify the relevance to adult life of rare discoveries in play, what is the chance that we will observe such a discovery taking place? Again, because this identification is difficult, the predictions of this analysis are expected to be overoptimistic.

Or, suppose (3) that rare discoveries made in play are used to meet rare contingencies in adulthood. Can we expect to observe both the discovery and the contingency?

Models of these three cases will be used to calculate our chance of observing the adaptive effects of play given a certain amount of observation time. The fundamental assumptions of the models should be stressed: that (case 1) a rare playful discovery can always be recognized and interpreted in terms of its use in everyday adult life; that (case 2) a rare event in adulthood can always be recognized and interpreted in

532

terms of known playful experience during the juvenile stage; that (case 3) rare discoveries and rare contingencies can both be recognized and related provided we are able to observe their occurrence.

These assumptions are grossly optimistic. This optimism is intentional because the purpose of this section is to derive lower bounds for the amount of time we must spend watching an animal before we can hope to observe some event that indicates the significance of play in its life. That is, if even in the most favorable case we are required to spend 100,000 hours watching an animal before our chance of observing the function of play becomes greater or equal to 5%, the expectation is clearly impractical.

Assume that an animal's life history consists of an immature and an adult stage. The dividing line could be identified with age at first reproduction, but the exact characterization may vary from species to species. This scheme [Nice (1943) and Horwich (1972) discuss more detailed ontogenetic schemes] is adequate for the present model.

Let

E = total length (years) of the life of an animal
J = total length (years) of the immature period
P_1 = percentage of the immature period spent in play
P_2 = percentage of the adult period spent in behavior made possible by innovative play in the juvenile stage
n = number of individual animals or playgrounds under observation

The *time* of occurrence of a rare event is assumed to be a random variable with a distribution uniform over the relevant portion of the life-history stage. That is, an immature who plays for a total of $P_1 J$ years will encounter the rare event once and only once during this period, but the event may happen at any time with equal probability. Thus, if an observer watches the play of one immature for $Y_j \leqslant P_1 J$ years, the probability that he will observe the rare event is

$$P_j \overset{\triangle}{=} Y_j / (P_1 J)$$

If an observer watches, simultaneously, n youngsters playing, and if the rare events are assumed to occur independently in the lives of the various immatures (an assumption that may be totally false for social animals, in whose case the unit observed might be identified with a play group rather than with an individual), the probability that the event will be observed [Case (1)] is 1 minus the probability that the event will not be observed in any one of the n independent "trials":

$$P_j(n) \overset{\triangle}{=} 1 - (1-P_j)^n = 1 - \left(1 - \frac{Y_j}{P_1 J}\right)^n$$

Similarly [Case (2)], if the rare event is assumed to occur during adulthood,

$$P_a(n) \overset{\triangle}{=} 1 - (1-P_a)^n = 1 - \left(1 - \frac{Y_a}{P_2(E-J)}\right)^n$$

Finally, suppose that we can hope to observe the function of a behavior if and only if we observe a rare event in the immature period and another, independent rare event in the adult period *of the same individual*. (An equally plausible model, not developed here, might assume that it would be sufficient to observe the rare juvenile event in one individual and the rare adult event in another individual.) Thus, for one individual, if the occurrence times of the two rare events, one taking place in the immature period and one in adulthood, are assumed to be independent, uniformly distributed random variables, the probability of observing both events given Y_j years of observation of the immature and Y_a years of observation of the adult is

$$P_{ja} \overset{\triangle}{=} P_a P_j = \frac{Y_a Y_j}{P_1 P_2 J(E-J)}$$

and it follows that for observations [Case (3)] on n independent units (individuals or stable groups of individuals)

$$P_{ja}(n) \overset{\triangle}{=} 1 - (1-P_a P_j)^n = 1 - \left[1 - \frac{Y_a Y_j}{P_1 P_2 J(E-J)}\right]^n$$

Optimal allocation of observation time in Case (3)

The formula in Case (3) depends both on Y_a and on Y_j. For a fixed total amount $T = Y_a + Y_j$ of observation time, the optimal allocation of observation time to juvenile-watching and to adult-watching in order to maximize P_{ja} is $Y_a = Y_j = T/2$, since $P_{ja}(n)$ is symmetric in Y_a and Y_j.

The three formulas derived above make it possible to calculate the chance that a rare event indicating the biological significance of innovative play will be observed. For example, this chance is 15% if we watch two juveniles' innovative play for 1000 hours each when the juvenile stage is three years long and if 5% of their time is spent in innovative play. This result is important because it is entirely feasible to observe play for 1000 hours, although no one has yet done so. If ethologists watched play for thousands of hours instead of the tens of hours currently in vogue, our concept of the biological function of play might change substantially.

Appendix VIII

Common/Scientific Names of Animals Cited in Text

Common Name	Scientific Name
Aardwolf	*Proteles cristatus* (Carnivora: Hyaenidae)
Accipiters	*Accipiter* spp. (Falconiformes: Accipitridae)
African buffalo	*Syncerus caffer* (Artiodactyla: Bovidae)
African civet	*Civettictis civetta* (Carnivora: Viverridae)
African dwarf mongoose	*Helogale parvula* (Carnivora: Viverridae)
African elephant	*Loxodonta africana* (Proboscidea: Elephantidae)
African giant rat	*Cricetomys gambianus* (Rodentia: Muridae)
African golden cat	*Profelis aurata* (Carnivora: Felidae)
African ground squirrel	*Xerus erythropus* (Rodentia: Sciuridae)
African hunting dog	*Lycaon pictus* (Carnivora: Canidae)
African porcupine	*Hystrix cristata* (Rodentia: Hystricidae)
African tree pangolin	*Manis tricuspis* (Pholidota: Manidae)
African wildcat	*Felis libyca* (Carnivora: Felidae)
Agouti	*Dasyprocta aguti* (Rodentia: Dasyproctidae)
Alaska fur seal	*Callorhinus ursinus* (Pinnipedia: Otariidae)
Allen's galago	*Galago alleni* (Primates: Lemuridae)
Allen's swamp monkey	*Allenopithecus nigroviridis* (Primates: Cercopithecidae)
Alpine marmot	*Marmota marmota* (Rodentia: Sciuridae)
American beaver	*Castor canadensis* (Rodentia: Castoridae)
American red squirrel	*Tamiasciurus hudsonicus* (Rodentia: Sciuridae)
Angola colobus	*Colobus angolensis* (Primates: Cercopithecidae)
Anna hummingbird	*Calypte anna* (Apodes: Trochilidae)
Aoudad	*Capra lervia* (Artiodactyla: Bovidae)
Arctic fox	*Alopex lagopus* (Carnivora: Canidae)
Arctic hare	*Lepus timidus* (Lagomorpha: Leporidae)
Ascension Island frigate-bird	*Fregata aquila* (Pelecaniformes: Fregatidae)

Ashy-grey mouse	*Pseudomys albocinereus* (Rodentia: Muridae)
Asian elephant	*Elephas maximus* (Proboscidea: Elephantidae)
Asian golden cat	*Profelis temmincki* (Carnivora: Felidae)
Australian magpie	*Gymnorhina tibicien* (Passeriformes: Cracticidae)
Australian sea lion	*Neophoca cinerea* (Pinnipedia: Otariidae)
Aye-aye	*Daubentonia madagascarensis* (Primates: Lemuridae)
Bald eagle	*Haliaeetus leucocephalus* (Falconiformes: Accipitridae)
Banded mongoose	*Mungos mungo* (Carnivora: Viverridae)
Barasingha	*Cervus duvauceli* (Artiodactyla: Cervidae)
Barbary ape	*Macaca sylvanus* (Primates: Cercopithecidae)
Barn owl	*Tyto alba* (Strigiformes: Tytonidae)
Bat-eared fox	*Otocyon megalotis* (Carnivora: Canidae)
Beech marten	*Martes foina* (Carnivora: Mustelidae)
Bharal	*Pseudois nayaur* (Artiodactyla: Bovidae)
Binturong	*Arctictis binturong* (Carnivora: Viverridae)
Bison	*Bison bison* (Artiodactyla: Bovidae)
Black & white colobus	*Colobus guereza* (Primates: Cercopithecidae)
Blackbacked jackal	*Canis mesomelas* (Carnivora: Canidae)
Black bear	*Ursus americanus* (Carnivora: Ursidae)
Blackbird	*Turdus merula* (Passeriformes: Muscicapidae)
Blackbuck	*Antilope cervicapra* (Artiodactyla: Bovidae)
Black colobus	*Colobus satanas* (Primates: Cercopithecidae)
Black-footed cat	*Felis nigripes* (Carnivora: Felidae)
Black-footed ferret	*Mustela nigripes* (Carnivora: Mustelidae)
Black-handed tamarin	*Saguinus midas* (Primates: Callitrichidae)
Black lemur	*Lemur macaco* (Primates: Lemuridae)
Black rat	*Rattus rattus* (Rodentia: Muridae)
Black rhinoceros	*Diceros bicornis* (Perissodactyla: Rhinocerotidae)
Black-tailed deer	*Odocoileus hemionus* (Artiodactyla: Cervidae)
Black-tailed prairie dog	*Cynomys ludovicianus* (Rodentia: Sciuridae)
Blesbok & bontebok	*Damaliscus dorcas* (Artiodactyla: Bovidae)
Blue whale	*Balaenoptera musculus* (Cetacea: Balaenopteridae)
Blunt-faced rat	*Pseudomys shortridgei* (Rodentia: Muridae)
Bobak marmot	*Marmota bobak* (Rodentia: Sciuridae)
Bokombouli	*Hapalemur griseus* (Primates: Lemuridae)
Bonnet macaque	*Macaca radiata* (Primates: Cercopithecidae)
Bottle-nosed dolphin	*Tursiops truncatus* (Cetacea: Delphinidae)
Brahminy kite	*Haliastur indus* (Falconiformes: Accipitridae)
Brazilian giant otter	*Pteronura brasiliensis* (Carnivora: Mustelidae)
Broad-toothed field mouse	*Apodemus mystacinus* (Rodentia: Muridae)
Brown antechinus	*Antechinus stuartii* (Marsupiala: Dasyuridae)
Brown bear, grizzly bear	*Ursus arctos* (Carnivora: Ursidae)

Brown capuchin	*Cebus apella* (Primates: Cebidae)
Brown desert mouse	*Pseudomys desertor* (Rodentia: Muridae)
Brown hyena	*Hyaena brunnea* (Carnivora: Hyaenidae)
Brown lemur	*Lemur fulvus* (Primates: Lemuridae)
Budgerigar	*Melopsittacus undulatus* (Psittaciformes: Psittacidae)
Bushbuck	*Tragelaphus scriptus* (Artiodactyla: Bovidae)
Bush dog	*Speothos venaticus* (Carnivora: Canidae)
Buzzard	*Buteo buteo* (Falconiformes: Accipitridae)
Buzzards	*Buteo* spp. (Falconiformes: Accipitridae)
Cacomistle	*Bassariscus sumichrasti* (Carnivora: Procyonidae)
Cactus finch	*Camarhynchus pallidus* (Passeriformes: Emberizidae)
California condor	*Gymnogyps californianus* (Falconiformes: Cathartidae)
California ground squirrel	*Otospermophilus beecheyi* (Rodentia: Sciuridae)
California sea lion	*Zalophus californianus* (Pinnipedia: Otariidae)
Cape fur seal	*Arctocephalus pusillus* (Pinnipedia: Otariidae)
Celebes black ape	*Macaca nigra* (Primates: Cercopithecidae)
Chacma baboon	*Papio ursinus* (Primates: Cercopithecidae)
Chaffinch	*Fringilla coelebs* (Passeriformes: Fringillidae)
Chamois	*Rupicapra rupicapra* (Artiodactyla: Bovidae)
Cheetah	*Acinonyx jubatus* (Carnivora: Felidae)
Chimney swift	*Chaetura pelagica* (Apodes: Apodidae)
Chimpanzee	*Pan troglodytes* (Primates: Pongidae)
Chipmunk	*Eutamias* spp. (Rodentia: Sciuridae)
Chital	*Axis axis* (Artiodactyla: Cervidae)
Choz-choz	*Octodontomys gliroides* (Rodentia: Octodontidae)
Clouded leopard	*Neofelis nebulosa* (Carnivora: Felidae)
Coati	*Nasua narica* (Carnivora: Procyonidae)
Collared dove	*Streptopelia decaocto* (Columbiformes: Columbidae)
Collared lemming	*Dicrostonyx groenlandicus* (Rodentia: Cricetidae)
Collared peccary	*Dictyles tajacu* (Artiodactyla: Tayassuidae)
Columbian ground squirrel	*Spermophilus columbianus* (Rodentia: Sciuridae)
Common crow	*Corvus brachyrhynchos* (Passeriformes: Corvidae)
Common dunnart	*Sminthopsis murina* (Marsupiala: Dasyuridae)
Common eider	*Somateria mollissima* (Anseriformes: Anatidae)
Common genet	*Genetta genetta* (Carnivora: Viverridae)
Common hamster	*Cricetus cricetus* (Rodentia: Cricetidae)

Common seal	*Phoca vitulina vitulina* (Pinnipedia: Phocidae)
Common tree shrew	*Tupaia glis* (Primates: Tupaiidae)
Common whitethroat	*Sylvia communis* (Passeriformes: Muscicapidae)
Coquerel's mouse lemur	*Microcebus coquereli* (Primates: Lemuridae)
Corsac fox	*Vulpes corsac* (Carnivora: Canidae)
Cotton-top marmoset	*Saguinus oedipus* (Primates: Callitrichidae)
Coyote	*Canis latrans* (Carnivora: Canidae)
Coypu or nutria	*Myocastor coypus* (Rodentia: Myocastoridae)
Crab-eating raccoon	*Procyon cancrivorus* (Carnivora: Procyonidae)
Crowned guenon	*Cercopithecus pogonias* (Primates: Cercopithecidae)
Cusimanse	*Crossarchus obscurus* (Carnivora: Viverridae)
Dall sheep	*Ovis dalli* (Artiodactyla: Bovidae)
DeBrazza monkey	*Cercopithecus neglectus* (Primates: Cercopithecidae)
Defassa waterbuck	*Kobus defassa* (Artiodactyla: Bovidae)
Degu	*Octodon degus* (Rodentia: Octodontidae)
Desert cavy	*Microcavia australis* (Rodentia: Caviidae)
Dhole	*Cuon alpinus* (Carnivora: Canidae)
Diana monkey	*Cercopithecus diana* (Primates: Cercopithecidae)
Dingo	*Canis dingo* (Carnivora: Canidae)
Domestic cat	*Felis catus* (Carnivora: Felidae)
Domestic cattle	*Bos taurus* (Artiodactyla: Bovidae)
Domestic dog	*Canis familiaris* (Carnivora: Canidae)
Domestic fowl	*Gallus gallus* (Galliformes: Phasianidae)
Domestic goat	*Capra hircus* (Artiodactyla: Bovidae)
Domestic guinea pig	*Cavia porcellus* (Rodentia: Caviidae)
Domestic horse	*Equus caballus* (Perissodactyla: Equidae)
Domestic sheep	*Ovis aries* (Artiodactyla: Bovidae)
Domestic swine	*Sus scrofa* (Artiodactyla: Suidae)
Dorcas gazelle	*Gazella dorcas* (Artiodactyla: Bovidae)
Douc langur	*Pygathrix nemaeus* (Primates: Cercopithecidae)
Drill	*Mandrillus leucophaeus* (Primates: Cercopithecidae)
Dromedary	*Camelus dromedarius* (Artiodactyla: Camelidae)
Dugong	*Dugong dugon* (Sirenia: Dugongidae)
Dusky titi	*Callicebus moloch* (Primates: Cebidae)
Eagle owl	*Bubo bubo* (Strigiformes: Strigidae)
East Caucasian tur	*Capra cylindricornis* (Artiodactyla: Bovidae)
Eastern chipmunk	*Tamias striatus* (Rodentia: Sciuridae)
Eastern wood rat	*Neotoma floridiana* (Rodentia: Cricetidae)
Elk, wapiti	*Cervus canadensis* (Artiodactyla: Cervidae)
Endrina	*Indri indri* (Primates: Lemuridae)

Ermine	*Mustela erminea* (Carnivora: Mustelidae)
Euro	*Macropus robustus* (Marsupiala: Macropodidae)
European badger	*Meles meles* (Carnivora: Mustelidae)
European beaver	*Castor fiber* (Rodentia: Castoridae)
European otter	*Lutra lutra* (Carnivora: Mustelidae)
European rabbit	*Oryctolagus cuniculus* (Lagomorpha: Leporidae)
European red squirrel	*Sciurus vulgaris* (Rodentia: Sciuridae)
European wildcat	*Felis silvestris* (Carnivora: Felidae)
Falanouc	*Eupleres goudotii* (Carnivora: Viverridae)
Falcon	*Falco* spp. (Falconiformes: Falconidae)
Fallow deer	*Cervus dama* (Artiodactyla: Cervidae)
Fanaloka	*Fossa fossana* (Carnivora: Viverridae)
Fat-tailed dunnart	*Sminthopsis crassicaudata* (Marsupiala: Dasyuridae)
Fennec	*Vulpes zerda* (Carnivora: Canidae)
Ferret	*Mustela furo* (Carnivora: Mustelidae)
Fisher	*Martes pennanti* (Carnivora: Mustelidae)
Fishing cat	*Prionailurus viverrinus* (Carnivora: Felidae)
Flat-headed cat	*Ictailurus planiceps* (Carnivora: Felidae)
Flying squirrel	*Glaucomys volans* (Rodentia: Sciuridae)
Fossa	*Cryptoprocta ferox* (Carnivora: Viverridae)
Fox squirrel	*Sciurus niger* (Rodentia: Sciuridae)
Galapagos sea lion	*Zalophus wollebaeki* (Pinnipedia: Otariidae)
Garden warbler	*Sylvia borin* (Passeriformes: Muscicapidae)
Gaur	*Bos gaurus* (Artiodactyla: Bovidae)
Gayal	*Bos frontalis* (Artiodactyla: Bovidae)
Gelada	*Theropithecus gelada* (Primates: Cercopithecidae)
Gerenuk	*Litocranius walleri* (Artiodactyla: Bovidae)
Giant anteater	*Myrmecophaga tridactyla* (Edentata: Myrmecophagidae)
Giant flying squirrel	*Petaurista petaurista* (Rodentia: Sciuridae)
Giant panda	*Ailuropoda melanoleuca* (Carnivora: Ursidae)
Giraffe	*Giraffa camelopardalis* (Artiodactyla: Giraffidae)
Golah	*Kakatoe roseicapilla* (Psittaciformes: Psittacidae)
Golden eagle	*Aquila chrysaetos* (Falconiformes: Accipitridae)
Golden hamster	*Mesocricetus auratus* (Rodentia: Cricetidae)
Golden jackal	*Canis aureus* (Carnivora: Canidae)
Golden langur	*Presbytis geei* (Primates: Cercopithecidae)
Gorilla	*Gorilla gorilla* (Primates: Pongidae)
Grant's gazelle	*Gazella granti* (Artiodactyla: Bovidae)
Great black-backed gull	*Larus marinus* (Charadriiformes: Laridae)
Greater bush-baby	*Galago crassicaudatus* (Primates: Lemuridae)

Greater dormouse — *Glis glis* (Rodentia: Gliridae)

Greater dwarf lemur — *Cheirogaleus major* (Primates: Lemuridae)

Great frigate-bird — *Fregata minor* (Pelecaniformes: Fregatidae)

Great grey kangaroo — *Macropus giganteus* (Marsupiala: Macropodidae)

Green acouchi — *Myoprocta pratti* (Rodentia: Caviidae)

Green monkey — *Cercopithecus sabaeus* (Primates: Cercopithecidae)

Grevy's zebra — *Equus grevyi* (Perissodactyla: Equidae)

Grey-cheeked mangabey — *Cercocebus albigena* (Primates: Cercopithecidae)

Grey fox — *Urocyon cinereoargenteus* (Carnivora: Canidae)

Grey mazama — *Mazama gouazoubira* (Artiodactyla: Cervidae)

Grey seal — *Halichoerus grypus* (Pinnipedia: Phocidae)

Grey squirrel — *Sciurus carolinensis* (Rodentia: Sciuridae)

Grison — *Galictis cuja* (Carnivora: Mustelidae)

Guadalupe fur seal — *Arctocephalus townsendi* (Pinnipedia: Otariidae)

Guanaco — *Lama guanacoe* (Artiodactyla: Camelidae)

Guinea pig — *Cavia aperea* (Rodentia: Caviidae)

Gunnison's prairie dog — *Cynomys gunnisoni* (Rodentia: Sciuridae)

Hairy-nosed wombat — *Lasiorhinus latifrons* (Marsupiala: Phascolomidae)

Hamadryas baboon — *Papio hamadryas* (Primates: Cercopithecidae)

Hanuman langur — *Presbytis entellus* (Primates: Cercopithecidae)

Harbor seal — *Phoca vitulina concolor* (Pinnipedia: Phocidae)

Harvest mouse — *Micromys minutus* (Rodentia: Muridae)

Hazel mouse — *Muscardinus avellanarius* (Rodentia: Gliridae)

Hedgehog — *Erinaceus europaeus* (Insectivora: Erinaceidae)

Herring gull — *Larus argentatus* (Charadriiformes: Laridae)

Hippopotamus — *Hippopotamus amphibius* (Artiodactyla: Hippopotamidae)

Hispaniolan solenodon — *Solenodon paradoxus* (Insectivora: Solenodontidae)

Hog badger — *Arctonyx collaris* (Carnivora: Mustelidae)

Hog deer — *Axis porcinus* (Artiodactyla: Cervidae)

Hooded crow — *Corvus corone* (Passeriformes: Corvidae)

Hoary marmot — *Marmota caligata* (Rodentia: Sciuridae)

Horned lark — *Eremophila alpestris* (Passeriformes: Alaudidae)

House crow — *Corvus splendens* (Passeriformes: Corvidae)

House mouse — *Mus musculus* (Rodentia: Muridae)

Humans — *Homo sapiens* (Primates: Hominidae)

Humboldt's spider monkey — *Ateles belzebuth* (Primates: Cebidae)

Hyacinthine macaw — *Anodorhynchus hyacinthinus* (Psittaciformes: Psittacidae)

Ibex — *Capra ibex* (Artiodactyla: Bovidae)

Ichneumon	*Herpestes ichneumon* (Carnivora: Viverridae)
Impala	*Aepyceros melampus* (Artiodactyla: Bovidae)
Inca tern	*Larosterna inca* (Charadriiformes: Laridae)
Indian flying fox	*Pteropus giganteus* (Chiroptera: Pteropidae)
Indian grey mongoose	*Herpestes edwardsi* (Carnivora: Viverridae)
Indian rhinoceros	*Rhinoceros unicornis* (Perissodactyla: Rhinocerotidae)
Jaguar	*Panthera onca* (Carnivora: Felidae)
Jaguarundi	*Herpailurus yagouaroundi* (Carnivora: Felidae)
Jamaican hutia	*Geocapromys brownii* (Rodentia: Capromyidae)
Japanese macaque	*Macaca fuscata* (Primates: Cercopithecidae)
Jay	*Garrulus glandarius* (Passeriformes: Corvidae)
Jungle babbler	*Turdoides striatus* (Passeriformes: Muscicapidae)
Jungle crow	*Corvus* sp. (Passeriformes: Corvidae)
Kaka	*Nestor meridionalis* (Psittaciformes: Psittacidae)
Kashmir markhor	*Capra falconeri* (Artiodactyla: Bovidae)
Kea	*Nestor notabilis* (Psittaciformes: Psittacidae)
Kerguelen fur seal	*Arctocephalus gazella* (Pinnipedia: Otariidae)
King colobus	*Colobus polykomos* (Primates: Cercopithecidae)
Kingfisher	*Alcedo atthis* (Coraciiformes: Alcedinidae)
Kinkajou	*Potos flavus* (Carnivora: Procyonidae)
Kirk's dik-dik	*Madoqua kirki* (Artiodactyla: Bovidae)
Kowari	*Dasyuroides byrnei* (Marsupiala: Dasyuridae)
Kra macaque	*Macaca fascicularis* (Primates: Cercopithecidae)
Kudu	*Tragelaphus* spp. (Artiodactyla: Bovidae)
Lammergeier	*Gypaetus barbatus* (Falconiformes: Accipitridae)
Large grey babbler	*Turdoides malcolmi* (Passeriformes: Muscicapidae)
Large-spotted genet	*Genetta tigrina* (Carnivora: Viverridae)
Lar gibbon	*Hylobates lar* (Primates: Pongidae)
Least weasel	*Mustela nivalis* (Carnivora: Mustelidae)
Lechwe	*Kobus leche* (Artiodactyla: Bovidae)
Leopard	*Panthera pardus* (Carnivora: Felidae)
Leopard cat	*Prionailurus bengalensis* (Carnivora: Felidae)
Leopard seal	*Hydrurga leptonyx* (Pinnipedia: Phocidae)
Lesser black-backed gull	*Larus fuscus* (Charadriiformes: Laridae)
Lesser bush-baby	*Galago senegalensis* (Primates: Lemuridae)
Lesser kudu	*Tragelaphus imberbis* (Artiodactyla: Bovidae)
Linsang	*Prionodon linsang* (Carnivora: Viverridae)
Lion	*Panthera leo* (Carnivora: Felidae)

Lion tamarin	*Leontopithecus rosalia* (Primates: Callitrichidae)
Little brown bat	*Myotis lucifugus* (Chiroptera: Vespertilionidae)
Little owl	*Athene noctua* (Strigiformes: Strigidae)
Llama	*Lama peruana* (Artiodactyla: Camelidae)
Loggerhead shrike	*Lanius ludovicianus* (Passeriformes: Laniidae)
Long-eared owl	*Asio otus* (Strigiformes: Strigidae)
Longtail weasel	*Mustela frenata* (Carnivora: Mustelidae)
Lowe's guenon	*Cercopithecus campbelli* (Primates: Cercopithecidae)
Lutong	*Presbytis cristatus* (Primates: Cercopithecidae)
Lynx	*Lynx lynx* (Carnivora: Felidae)
Macaw	*Ara ararauna* (Psittaciformes: Psittacidae)
Manatee	*Trichechus manatus* (Sirenia: Trichechidae)
Mandrill	*Mandrillus sphinx* (Primates: Cercopithecidae)
Maned wolf	*Chrysocyon brachyurus* (Carnivora: Canidae)
Mantled ground squirrel	*Callospermophilus lateralis* (Rodentia: Sciuridae)
Mantled howling monkey	*Alouatta villosa* (Primates: Cebidae)
Mara	*Dolichotis patagonum* (Rodentia: Caviidae)
Marbled cat	*Pardofelis marmorata* (Carnivora: Felidae)
Marbled polecat	*Vormela peregusna* (Carnivora: Mustelidae)
Marco Polo wild sheep	*Ovis polii* (Artiodactyla: Bovidae)
Margay	*Leopardus wiedi* (Carnivora: Felidae)
Marsh harrier	*Circus cyaneus* (Falconiformes: Accipitridae)
Masked shrew	*Sorex cinereus* (Insectivora: Soricidae)
Meerkat	*Suricata suricatta* (Carnivora: Viverridae)
Mexican prairie dog	*Cynomys mexicanus* (Rodentia: Sciuridae)
Milan, kite	*Milvus migrans* (Falconiformes: Accipitridae)
Mink	*Mustela vison* (Carnivora: Mustelidae)
Mongolian gazelle	*Procapra gutturosa* (Artiodactyla: Bovidae)
Mongolian gerbil	*Meriones unguiculatus* (Rodentia: Cricetidae)
Moose	*Alces alces* (Artiodactyla: Cervidae)
Mountain gazelle	*Gazella gazella* (Artiodactyla: Bovidae)
Mountain sheep	*Ovis canadensis* (Artiodactyla: Bovidae)
Mountain zebra	*Equus zebra* (Perissodactyla: Equidae)
Mouse lemur	*Microcebus murinus* (Primates: Lemuridae)
Muntjac	*Muntiacus muntjac* (Artiodactyla: Cervidae)
Muskox	*Ovibos moschatus* (Artiodactyla: Bovidae)
Muskrat	*Ondatra zibethicus* (Rodentia: Cricetidae)
Narrow-striped mongoose	*Mungotictis lineatus* (Carnivora: Viverridae)
New Zealand fur seal	*Arctocephalus forsteri* (Pinnipedia: Otariidae)
New Zealand sea lion	*Neophoca hookeri* (Pinnipedia: Otariidae)
Night monkey	*Aotus trivirgatus* (Primates: Cebidae)

Nilgiri langur — *Presbytis johnii* (Primates: Cercopithecidae)

North American porcupine — *Erethizon dorsatum* (Rodentia: Erethizontidae)

Northern blue jay — *Cyanocitta cristata* (Passeriformes: Corvidae)

Northern elephant seal — *Mirounga angustirostris* (Pinnipedia: Phocidae)

Northern hopping mouse — *Notomys alexis* (Rodentia: Muridae)

Norway rat — *Rattus norvegicus* (Rodentia: Muridae)

Okapi — *Okapia johnstoni* (Artiodactyla: Giraffidae)

Olingo — *Bassaricyon* spp. (Carnivora: Procyonidae)

Olive baboon — *Papio anubis* (Primates: Cercopithecidae)

Olympic marmot — *Marmota olympus* (Rodentia: Sciuridae)

Oncilla — *Leopardus tigrinus* (Carnivora: Felidae)

Orangutan — *Pongo pygmaeus* (Primates: Pongidae)

Orca — *Orcinus orca* (Cetacea: Delphinidae)

Oryx — *Oryx gazella* (Artiodactyla: Bovidae)

Oyster-catcher — *Haematopus ostralegus* (Charadriiformes: Haematopodidae)

Paca — *Cuniculus paca* (Rodentia: Dasyproctidae)

Pacarana — *Dinomys branickii* (Rodentia: Dinomyidae)

Pacific whitesided dolphin — *Lagenorhynchus obliquidens* (Cetecea: Delphinidae)

Pallas cat — *Felis manul* (Carnivora: Felidae)

Patas — *Erythrocebus patas* (Primates: Cercopithecidae)

Peregrine — *Falco peregrinus* (Falconiformes: Falconidae)

Persian jird — *Meriones persicus* (Rodentia: Cricetidae)

Peruvian desert fox — *Dusicyon sechurae* (Carnivora: Canidae)

Phayre's leaf monkey — *Presbytis phayrei* (Primates: Cercopithecidae)

Pigtail macaque — *Macaca nemestrina* (Primates: Cercopithecidae)

Pine marten — *Martes martes* (Carnivora: Mustelidae)

Plains zebra — *Equus burchelli* (Perissodactyla: Equidae)

Platypus — *Ornithorhynchus anatinus* (Monotremata: Ornithorhynchidae)

Polar bear — *Thalarctos maritimus* (Carnivora: Ursidae)

Polecat — *Mustela putorius* (Carnivora: Mustelidae)

Potto — *Perodicticus potto* (Primates: Lemuridae)

Prairie vole — *Microtus ochrogaster* (Rodentia: Cricetidae)

Proboscis monkey — *Nasalis larvatus* (Primates: Cercopithecidae)

Pronghorn — *Antilocapra americana* (Artiodactyla: Antilocapridae)

Pudu — *Pudu pudu* (Artiodactyla: Cervidae)

Puma — *Puma concolor* (Carnivora: Felidae)

Pygmy hippopotamus — *Choeropsis liberiensis* (Artiodactyla: Hippopotamidae)

Pygmy marmoset — *Cebuella pygmaea* (Primates: Callitrichidae)

Pygmy white-toothed shrew	*Suncus etruscus* (Insectivora: Soricidae)
Quaker parakeet	*Myiopsitta monachus* (Psittaciformes: Psittacidae)
Quenda	*Isoodon obesulus* (Marsupiala: Peramelidae)
Quoll	*Dasyurus viverrinus* (Marsupiala: Dasyuridae)
Raccoon	*Procyon lotor* (Carnivora: Procyonidae)
Raccoon-dog	*Nyctereutes procyonoides* (Carnivora: Canidae)
Ratel	*Mellivora capensis* (Carnivora: Mustelidae)
Raven	*Corvus corax* (Passeriformes: Corvidae)
Red colobus	*Colobus badius* (Primates: Cercopithecidae)
Red deer	*Cervus elaphus* (Artiodactyla: Cervidae)
Red fox	*Vulpes fulva* (Carnivora: Canidae)
Red fox	*Vulpes vulpes* (Carnivora: Canidae)
Red hartebeest	*Alcelaphus buselaphus* (Artiodactyla: Bovidae)
Red howling monkey	*Alouatta seniculus* (Primates: Cebidae)
Red kangaroo	*Macropus rufus* (Marsupiala: Macropodidae)
Red-legged partridge	*Alectoris rufa* (Galliformes: Phasianidae)
Red panda	*Ailurus fulgens* (Carnivora: Procyonidae)
Red spider monkey	*Ateles geoffroyi* (Primates: Cebidae)
Redtail monkey	*Cercopithecus ascanius* (Primates: Cercopithecidae)
Reed warbler	*Acrocephalus scirpaceus* (Passeriformes: Muscicapidae)
Reeves' muntjac	*Muntiacus reevesi* (Artiodactyla: Cervidae)
Reindeer, caribou	*Rangifer tarandus* (Artiodactyla: Cervidae)
Rhesus macaque	*Macaca mulatta* (Primates: Cercopithecidae)
Richardson's ground squirrel	*Spermophilus richardsoni* (Rodentia: Sciuridae)
Ringtail	*Bassariscus astutus* (Carnivora: Procyonidae)
Ringtailed lemur	*Lemur catta* (Primates: Lemuridae)
Ringtailed mongoose	*Galidia elegans* (Carnivora: Viverridae)
River otter	*Lutra canadensis* (Carnivora: Mustelidae)
Roan antelope	*Hippotragus equinus* (Artiodactyla: Bovidae)
Roe deer	*Capreolus capreolus* (Artiodactyla: Cervidae)
Royal antelope	*Neotragus pygmaeus* (Artiodactyla: Bovidae)
Ruffed lemur	*Varecia variegata* (Primates: Lemuridae)
Russet-eared guenon	*Cercopithecus erythrotis* (Primates: Cercopithecidae)
Rusty-spotted cat	*Prionailurus rubiginosus* (Carnivora: Felidae)
Rusty-spotted genet	*Genetta rubiginosa* (Carnivora: Viverridae)
Sable	*Martes zibellina* (Carnivora: Mustelidae)
Saddle-back tamarin	*Saguinus fuscicollis* (Primates: Callitrichidae)
Salt desert cavy	*Dolichotis salinicola* (Rodentia: Caviidae)
Sand cat	*Felis margarita* (Carnivora: Felidae)
Sand fox	*Vulpes rueppelli* (Carnivora: Canidae)
Sand rat	*Psammomys obesus* (Rodentia: Cricetidae)
Satanellus	*Dasyurus hallucatus* (Marsupiala: Dasyuridae)

Sea otter	*Enhydra lutris* (Carnivora: Mustelidae)
Sedge warbler	*Acrocephalus schoenobaenus* (Passeriformes: Muscicapidae)
Serval	*Leptailurus serval* (Carnivora: Felidae)
Short-tailed shrew	*Blarina brevicauda* (Insectivora: Soricidae)
Short-tailed vole	*Microtus agrestis* (Rodentia: Cricetidae)
Siamang	*Symphalangus syndactylus* (Primates: Pongidae)
Sifaka	*Propithecus verreauxi* (Primates: Lemuridae)
Sika deer	*Cervus nippon* (Artiodactyla: Cervidae)
Silvery marmoset	*Callithrix argentata* (Primates: Callitrichidae)
Sitatunga	*Tragelaphus spekei* (Artiodactyla: Bovidae)
Sloth bear	*Melursus ursinus* (Carnivora: Ursidae)
Slow loris	*Nycticebus coucang* (Primates: Lemuridae)
Smooth-coated otter	*Lutra perspicillata* (Carnivora: Mustelidae)
Snow leopard	*Uncia uncia* (Carnivora: Felidae)
Song sparrow	*Zonotrichia melodia* (Passeriformes: Fringillidae)
Sooty mangabey	*Cercocebus atys* (Primates: Cercopithecidae)
Southern elephant seal	*Mirounga leonina* (Pinnipedia: Phocidae)
Southern sea lion	*Otaria flavescens* (Pinnipedia: Otariidae)
Spectacled bear	*Tremarctos ornatus* (Carnivora: Ursidae)
Spectacled langur	*Presbytis obscurus* (Primates: Cercopithecidae)
Sperm whale	*Physeter catodon* (Cetacea: Physeteridae)
Sportive lemur	*Lepilemur mustelinus* (Primates: Lemuridae)
Spot-nosed guenon	*Cercopithecus nictitans* (Primates: Cercopithecidae)
Spotted hyena	*Crocuta crocuta* (Carnivora: Hyaenidae)
Spotted-necked otter	*Lutra maculicollis* (Carnivora: Mustelidae)
Spotted skunk	*Spilogale putorius* (Carnivora: Mustelidae)
Springbok	*Antidorcas marsupialis* (Artiodactyla: Bovidae)
Square-lipped rhinoceros	*Ceratotherium simum* (Perissodactyla: Rhinocerotidae)
Squirrel monkey	*Saimiri* spp. (Primates: Cebidae)
Steller sea lion	*Eumetopias jubatus* (Pinnipedia: Otariidae)
Striped hyena	*Hyaena hyaena* (Carnivora: Hyaenidae)
Striped skunk	*Mephitis mephitis* (Carnivora: Mustelidae)
Stumptail macaque	*Macaca arctoides* (Primates: Cercopithecidae)
Sykes' monkey	*Cercopithecus mitis* (Primates: Cercopithecidae)
Tahr	*Hemitragus jemlahicus* (Artiodactyla: Bovidae)
Tailless tenrec	*Tenrec ecaudatus* (Insectivora: Tenrecidae)
Takin	*Budorcas taxicolor* (Artiodactyla: Bovidae)
Talapoin	*Miopithecus talapoin* (Primates: Cercopithecidae)

Tamarisk gerbil — *Meriones tamariscinus* (Rodentia: Cricetidae)

Tapir — *Tapirus* spp. (Perissodactyla: Tapiridae)

Taruca — *Hippocamelus antisensis* (Artiodactyla: Cervidae)

Tasmanian barred bandicoot — *Perameles gunnii* (Marsupiala: Peramelidae)

Tasmanian devil — *Sarcophilus harrisii* (Marsupiala: Dasyuridae)

Tayra — *Eira barbara* (Carnivora: Mustelidae)

Thomson's gazelle — *Gazella thomsoni* (Artiodactyla: Bovidae)

Tiger — *Panthera tigris* (Carnivora: Felidae)

Tiger quoll — *Dasyurus maculatus* (Marsupiala: Dasyuridae)

Timber wolf — *Canis lupus* (Carnivora: Canidae)

Topi — *Damaliscus lunatus* (Artiodactyla: Bovidae)

Turaco — *Turaco fischeri* (Cuculiformes: Musophagidae)

Two-spotted palm civet — *Nandinia binotata* (Carnivora: Viverridae)

Two-toed sloth — *Choloepus didactylus* (Edentata: Bradypodidae)

Uganda kob — *Kobus kob* (Artiodactyla: Bovidae)

Uinta ground squirrel — *Spermophilus armatus* (Rodentia: Sciuridae)

Urial, mouflon — *Ovis ammon* (Artiodactyla: Bovidae)

Vampire bat — *Desmodus rotundus* (Chiroptera: Desmodontidae)

Vervet — *Cercopithecus aethiops* (Primates: Cercopithecidae)

Vicuna — *Vicugna vicugna* (Artiodactyla: Camelidae)

Walrus — *Odobenus rosmarus* (Pinnipedia: Odobenidae)

Warthog — *Phacochoerus aethiopicus* (Artiodactyla: Suidae)

Water chevrotain — *Hyemoschus aquaticus* (Artiodactyla: Tragulidae)

Weddell seal — *Leptonychotes weddelli* (Pinnipedia: Phocidae)

Wedge-tailed eagle — *Aquila audax* (Falconiformes: Accipitridae)

Weeper capuchin — *Cebus nigrivittatus* (Primates: Cebidae)

West African sun squirrel — *Heliosciurus gambianus* (Rodentia: Sciuridae)

Whiptail wallaby — *Macropus parryi* (Marsupiala: Macropodidae)

White-faced capuchin — *Cebus capucinus* (Primates: Cebidae)

White-fronted capuchin — *Cebus albifrons* (Primates: Cebidae)

White-necked raven — *Corvus albicollis* (Passeriformes: Corvidae)

White-nosed saki — *Chiropotes albinasus* (Primates: Cebidae)

White-rumped magpie — *Pica pica* (Passeriformes: Corvidae)

White-tailed deer — *Odocoileus virginianus* (Artiodactyla: Cervidae)

White-tailed mongoose — *Ichneumia albicauda* (Carnivora: Viverridae)

White-throat wood rat — *Neotoma albigula* (Rodentia: Cricetidae)

White tufted-ear marmoset — *Callithrix jacchus* (Primates: Callitrichidae)

White-winged chough — *Corcorax melanorhampos* (Passeriformes: Corvidae)

Wildebeest	*Connochaetes taurinus* (Artiodactyla: Bovidae)
Wisent	*Bison bonasus* (Artiodactyla: Bovidae)
Wolverine	*Gulo gulo* (Carnivora: Mustelidae)
Woodchuck	*Marmota monax* (Rodentia: Sciuridae)
Wood mouse	*Apodemus sylvaticus* (Rodentia: Muridae)
Woolly monkey	*Lagothrix lagothricha* (Primates: Cebidae)
Yellow baboon	*Papio cynocephalus* (Primates: Cercopithecidae)
Yellow-bellied marmot	*Marmota flaviventris* (Rodentia: Sciuridae)
Yellow-billed magpie	*Pica nuttalli* (Passeriformes: Corvidae)
Yellow-handed titi	*Callicebus torquatus* (Primates: Cebidae)
Yellow-necked field mouse	*Apodemus flavicollis* (Rodentia: Muridae)
Yellow-throated marten	*Martes flavigula* (Carnivora: Mustelidae)
Yellow-toothed cavy	*Galea musteloides* (Rodentia: Caviidae)

Appendix IX

Scientific/Common Names of Animals Cited in Text

Scientific Name	Common Name
Accipiter spp. (Falconiformes: Accipitridae)	accipiters
Acinonyx jubatus (Carnivora: Felidae)	cheetah
Acrocephalus schoenobaenus (Passeriformes: Muscicapidae)	sedge warbler
Acrocephalus scirpaceus (Passeriformes: Muscicapidae)	reed warbler
Aepyceros melampus (Artiodactyla: Bovidae)	impala
Ailuropoda melanoleuca (Carnivora: Ursidae)	giant panda
Ailurus fulgens (Carnivora: Procyonidae)	red panda
Alcedo atthis (Coraciiformes: Alcedinidae)	kingfisher
Alcelaphus buselaphus (Artiodactyla: Bovidae)	red hartebeest
Alces alces (Artiodactyla: Cervidae)	moose
Alectoris rufa (Galliformes: Phasianidae)	red-legged partridge
Allenopithecus nigroviridis (Primates: Cercopithecidae)	Allen's swamp monkey
Alopex lagopus (Carnivora: Canidae)	Arctic fox
Alouatta seniculus (Primates: Cebidae)	red howling monkey
Alouatta villosa (Primates: Cebidae)	mantled howling monkey
Anodorhynchus hyacinthinus (Psittaciformes: Psittacidae)	hyacinthine macaw
Antechinus stuartii (Marsupiala: Dasyuridae)	brown antechinus
Antidorcas marsupialis (Artiodactyla: Bovidae)	springbok
Antilocapra americana (Artiodactyla: Antilocapridae)	pronghorn
Antilope cervicapra (Artiodactyla: Bovidae)	blackbuck
Aotus trivirgatus (Primates: Cebidae)	night monkey
Apodemus flavicollis (Rodentia: Muridae)	yellow-necked field mouse
Apodemus mystacinus (Rodentia: Muridae)	broad-toothed field mouse

Apodemus sylvaticus (Rodentia: Muridae) — wood mouse

Aquila audax (Falconiformes: Accipitridae) — wedge-tailed eagle

Aquila chrysaetos (Falconiformes: Accipitridae) — golden eagle

Ara ararauna (Psittaciformes: Psittacidae) — macaw

Arctictis binturong (Carnivora: Viverridae) — binturong

Arctocephalus australis (Pinnipedia: Otariidae) — South American fur seal

Arctocephalus forsteri (Pinnipedia: Otariidae) — New Zealand fur seal

Arctocephalus gazella (Pinnipedia: Otariidae) — Kerguelen fur seal

Arctocephalus pusillus (Pinnipedia: Otariidae) — Cape fur seal

Arctocephalus townsendi (Pinnipedia: Otariidae) — Guadalupe fur seal

Arctonyx collaris (Carnivora: Mustelidae) — hog badger

Asio otus (Strigiformes: Strigidae) — long-eared owl

Ateles belzebuth (Primates: Cebidae) — Humboldt's spider monkey

Ateles geoffroyi (Primates: Cebidae) — red spider monkey

Athene noctua (Strigiformes: Strigidae) — little owl

Axis axis (Artiodactyla: Cervidae) — chital

Axis porcinus (Artiodactyla: Cervidae) — hog deer

Balaenoptera musculus (Cetacea: Balaenopteridae) — blue whale

Bassaricyon sp. (Carnivora: Procyonidae) — olingo

Bassariscus astutus (Carnivora: Procyonidae) — ringtail

Bassariscus sumichrasti (Carnivora: Procyonidae) — cacomistle

Bison bison (Artiodactyla: Bovidae) — bison

Bison bonasus (Artiodactyla: Bovidae) — wisent

Blarina brevicauda (Insectivora: Soricidae) — short-tailed shrew

Bos frontalis (Artiodactyla: Bovidae) — gayal

Bos gaurus (Artiodactyla: Bovidae) — gaur

Bos taurus (Artiodactyla: Bovidae) — domestic cattle

Bubo bubo (Strigiformes: Strigidae) — eagle owl

Budorcas taxicolor (Artiodactyla: Bovidae) — takin

Buteo buteo (Falconiformes: Accipitridae) — buzzard

Buteo spp. (Falconiformes: Accipitridae) — buzzards

Callicebus moloch (Primates: Cebidae) — dusky titi

Callicebus torquatus (Primates: Cebidae) — yellow-handed titi

Callithrix argentata (Primates: Callitrichidae) — silvery marmoset

Callithrix jacchus (Primates: Callitrichidae) — white tufted-ear marmoset

Callorhinus ursinus (Pinnipedia: Otariidae) — Alaska fur seal

Callospermophilus lateralis (Rodentia: Sciuridae) — mantled ground squirrel

Calypte anna (Apodes: Trochilidae) — Anna hummingbird

Camarhynchus pallidus (Passeriformes: Emberizidae) — cactus finch

Camelus dromedarius (Artiodactyla: Camelidae) — dromedary

Canis aureus (Carnivora: Canidae) — golden jackal
Canis dingo (Carnivora: Canidae) — dingo
Canis familiaris (Carnivora: Canidae) — domestic dog
Canis latrans (Carnivora: Canidae) — coyote
Canis lupus (Carnivora: Canidae) — timber wolf
Canis mesomelas (Carnivora: Canidae) — blackbacked jackal
Capra cylindricornis (Artiodactyla: Bovidae) — East Caucasian tur
Capra falconeri (Artiodactyla: Bovidae) — Kashmir markhor
Capra hircus (Artiodactyla: Bovidae) — domestic goat
Capra ibex (Artiodactyla: Bovidae) — ibex
Capra lervia (Artiodactyla: Bovidae) — aoudad
Capreolus capreolus (Artiodactyla: Cervidae) — roe deer
Castor canadensis (Rodentia: Castoridae) — American beaver
Castor fiber (Rodentia: Castoridae) — European beaver
Cavia aperea (Rodentia: Caviidae) — guinea pig
Cavia porcellus (Rodentia: Caviidae) — domestic guinea pig
Cebuella pygmaea (Primates: Callitrichidae) — pygmy marmoset
Cebus albifrons (Primates: Cebidae) — white-fronted capuchin
Cebus apella (Primates: Cebidae) — brown capuchin
Cebus capucinus (Primates: Cebidae) — white-faced capuchin
Cebus nigrivittatus (Primates: Cebidae) — weeper capuchin
Ceratotherium simum (Perissodactyla: Rhinocerotidae) — square-lipped rhinoceros
Cercocebus albigena (Primates: Cercopithecidae) — grey-cheeked mangabey
Cercocebus atys (Primates: Cercopithecidae) — sooty mangabey
Cercopithecus aethiops (Primates: Cercopithecidae) — vervet
Cercopithecus ascanius (Primates: Cercopithecidae) — redtail monkey
Cercopithecus campbelli (Primates: Cercopithecidae) — Lowe's guenon
Cercopithecus diana (Primates: Cercopithecidae) — diana monkey
Cercopithecus erythrotis (Primates: Cercopithecidae) — russet-eared guenon
Cercopithecus mitis (Primates: Cercopithecidae) — Sykes' monkey
Cercopithecus neglectus (Primates: Cercopithecidae) — De Brazza monkey
Cercopithecus nictitans (Primates: Cercopithecidae) — spot-nosed guenon
Cercopithecus pogonias (Primates: Cercopithecidae) — crowned guenon
Cercopithecus sabaeus (Primates: Cercopithecidae) — green monkey
Cervus canadensis (Artiodactyla: Cervidae) — elk, wapiti
Cervus dama (Artiodactyla: Cervidae) — fallow deer

Cervus duvauceli (Artiodactyla: Cervidae)	barasingha
Cervus elaphus (Artiodactyla: Cervidae)	red deer
Cervus nippon (Artiodactyla: Cervidae)	sika deer
Chaetura pelagica (Apodes: Apodidae)	chimney swift
Cheirogaleus major (Primates: Lemuridae)	greater dwarf lemur
Chiropotes albinasus (Primates: Cebidae)	white-nosed saki
Choeropsis liberiensis (Artiodactyla: Hippopotamidae)	pygmy hippomotamus
Choloepus didactylus (Edentata: Bradypodidae)	two-toed sloth
Chrysocyon brachyurus (Carnivora: Canidae)	maned wolf
Circus cyaneus (Falconiformes: Accipitridae)	marsh harrier
Civettictis civetta (Carnivora: Viverridae)	African civet
Colobus angolensis (Primates: Cercopithecidae)	Angola colobus
Colobus badius (Primates: Cercopithecidae)	red colobus
Colobus guereza (Primates: Cercopithecidae)	black & white colobus
Colobus polykomos (Primates: Cercopithecidae)	king colobus
Colobus satanas (Primates: Cercopithecidae)	black colobus
Connochaetes taurinus (Artiodactyla: Bovidae)	wildebeest
Corcorax melanorhampos (Passeriformes: Corvidae)	white-winged chough
Corvus albicollis (Passeriformes: Corvidae)	white-necked raven
Corvus brachyrhynchos (Passeriformes: Corvidae)	common crow
Corvus corax (Passeriformes: Corvidae)	raven
Corvus corone (Passeriformes: Corvidae)	hooded crow
Corvus sp. (Passeriformes: Corvidae)	jungle crow
Corvus splendens (Passeriformes: Corvidae)	house crow
Cricetomys gambianus (Rodentia: Muridae)	African giant rat
Cricetus cricetus (Rodentia: Cricetidae)	common hamster
Crocuta crocuta (Carnivora: Hyaenidae)	spotted hyena
Crossarchus obscurus (Carnivora: Viverridae)	cusimanse
Cryptoprocta ferox (Carnivora: Viverridae)	fossa
Cuniculus paca (Rodentia: Dasyproctidae)	paca
Cuon alpinus (Carnivora: Canidae)	dhole
Cyanocitta cristata (Passeriformes: Corvidae)	northern blue jay
Cynomys gunnisoni (Rodentia: Sciuridae)	Gunnison's prairie dog
Cynomys ludovicianus (Rodentia: Sciuridae)	black-tailed prairie dog
Cynomys mexicanus (Rodentia: Sciuridae)	Mexican prairie dog
Damaliscus dorcas (Artiodactyla: Bovidae)	blesbok & bontebok
Damaliscus lunatus (Artiodactyla: Bovidae)	topi
Dasyprocta aguti (Rodentia: Dasyproctidae)	agouti
Dasyuroides byrnei (Marsupiala: Dasyuridae)	kowari
Dasyurus hallucatus (Marsupiala: Dasyuridae)	satanellus
Dasyurus maculatus (Marsupiala: Dasyuridae)	tiger quoll

Dasyurus viverrinus (Marsupiala: Dasyuridae) quoll
Daubentonia madagascarensis (Primates: Lemuridae) aye-aye
Desmodus rotundus (Chiroptera: Desmodontidae) vampire bat
Diceros bicornis (Perissodactyla: Rhinocerotidae) black rhinoceros
Dicrostonyx groenlandicus (Rodentia: Gricetidae) collared lemming
Dictyles tajacu (Artiodactyla: Tayassuidae) collared peccary
Dinomys branickii (Rodentia: Dinomyidae) pacarana
Dolichotis patagonum (Rodentia: Caviidae) mara
Dolichotis salinicola (Rodentia: Caviidae) salt desert cavy
Dugong dugon (Sirenia: Dugongidae) dugong
Dusicyon sechurae (Carnivora: Canidae) Peruvian desert fox
Eira barbara (Carnivora: Mustelidae) tayra
Elephas maximus (Proboscidea: Elephantidae) Asian elephant
Enhydra lutris (Carnivora: Mustelidae) sea otter
Equus burchelli (Perissodactyla: Equidae) plains zebra
Equus caballus (Perissodactyla: Equidae) domestic horse
Equus grevyi (Perissodactyla: Equidae) Grevy's zebra
Equus zebra (Perissodactyla: Equidae) mountain zebra
Eremophila alpestris (Passeriformes: Alaudidae) horned lark
Erethizon dorsatum (Rodentia: Erethizontidae) North American porcupine
Erinaceus europaeus (Insectivora: Erinaceidae) hedgehog
Erythrocebus patas (Primates: Cercopithecidae) patas
Eumetopias jubatus (Pinnipedia: Otariidae) Steller sea lion
Eupleres goudotii (Carnivora: Viverridae) falanouc
Eutamias sp. (Rodentia: Sciuridae) chipmunk
Falco peregrinus (Falconiformes: Falconidae) peregrine
Falco spp. (Falconiformes: Falconidae) falcons
Felis catus (Carnivora: Felidae) domestic cat
Felis libyca (Carnivora: Felidae) African wildcat
Felis manul (Carnivora: Felidae) Pallas cat
Felis margarita (Carnivora: Felidae) sand cat
Felis nigripes (Carnivora: Felidae) black-footed cat
Felis silvestris (Carnivora: Felidae) European wildcat
Fossa fossana (Carnivora: Viverridae) fanaloka
Fregata aquila (Pelecaniformes: Fregatidae) Ascension Island frigate-bird
Fregata minor (Pelecaniformes: Fregatidae) great frigate-bird
Fringilla coelebs (Passeriformes: Fringillidae) chaffinch
Galago alleni (Primates: Lemuridae) Allen's galago
Galago crassicaudatus (Primates: Lemuridae) greater bush-baby
Galago senegalensis (Primates: Lemuridae) lesser bush-baby

Galea musteloides (Rodentia: Caviidae) — yellow-toothed cavy
Galictus cuja (Carnivora: Mustelidae) — grison
Galidia elegans (Carnivora: Viverridae) — ringtailed mongoose
Gallus gallus (Galliformes: Phasianidae) — domestic fowl
Garrulus glandarius (Passeriformes: Corvidae) — jay
Gazella dorcas (Artiodactyla: Bovidae) — dorcas gazelle
Gazella gazella (Artiodactyla: Bovidae) — mountain gazelle
Gazella granti (Artiodactyla: Bovidae) — Grant's gazelle
Gazella thomsoni (Artiodactyla: Bovidae) — Thomson's gazelle
Genetta genetta (Carnivora: Viverridae) — common genet
Genetta rubiginosa (Carnivora: Viverridea) — rusty-spotted genet
Genetta tigrina (Carnivora: Viverridae) — large-spotted genet
Geocapromys brownii (Rodentia: Capromyidae) — Jamaican hutia
Giraffa camelopardalis (Artiodactyla: Giraffidae) — giraffe
Glaucomys volans (Rodentia: Sciuridae) — flying squirrel
Glis glis (Rodentia: Gliridae) — greater dormouse
Gorilla gorilla (Primates: Pongidae) — gorilla
Gulo gulo (Carnivora: Mustelidae) — wolverine
Gymnogyps californianus (Falconiformes: Cathartidae) — California condor
Gymnorhina tibicien (Passeriformes: Cracticidae) — Australian magpie
Gypaetus barbatus (Falconiformes: Accipitridae) — lammergeier
Haematopus ostralegus (Charadriiformes: Haematopodidae) — oyster-catcher
Haliaeetus leucocephalus (Falconiformes: Accipitridae) — bald eagle
Haliastur indus (Falconiformes: Accipitridae) — Brahminy kite
Halichoerus grypus (Pinnipedia: Phocidae) — grey seal
Hapalemur griseus (Primates: Lemuridae) — bokombouli
Heliosciurus gambianus (Rodentia: Sciuridae) — West African sun squirrel
Helogale parvula (Carnivora: Viverridae) — African dwarf mongoose
Hemitragus jemlahicus (Artiodactyla: Bovidae) — thar
Herpailurus yagouaroundi (Carnivora: Felidae) — jaguarundi
Herpestes edwardsi (Carnivora: Viverridae) — Indian grey mongoose
Herpestes ichneumon (Carnivora: Viverridae) — ichneumon
Hippocamelus antisensis (Artiodactyla: Cervidae) — taruca
Hippopotamus amphibius (Artiodactyla: Hippopotamidae) — hippopotamus
Hippotragus equinus (Artiodactyla: Bovidae) — roan antelope
Homo sapiens (Primates: Hominidae) — humans
Hyaena brunnea (Carnivora: Hyaenidae) — brown hyena
Hyaena hyaena (Carnivora: Hyaenidae) — striped hyena

Hydrurga leptonyx (Pinnipedia: Phocidae) — leopard seal
Hyemoschus aquaticus (Artiodactyla: Tragulidae) — water chevrotain
Hylobates lar (Primates: Pongidae) — lar gibbon
Hystrix cristata (Rodentia: Hystricidae) — African porcupine
Ichneumia albicauda (Carnivora: Viverridae) — white-tailed mongoose
Ictailurus planiceps (Carnivora: Felidae) — flat-headed cat
Indri indri (Primates: Lemuridae) — endrina
Isoodon obesulus (Marsupiala: Peramelidae) — quenda
Kakatoe roseicapilla (Psittaciformes: Psittacidae) — golah
Kobus defassa (Artiodactyla: Bovidae) — Defassa waterbuck
Kobus kob (artiodactyla: Bovidae) — Uganda kob
Kobus leche (Artiodactyla: Bovidae) — lechwe
Lagenorhynchus obliquidens (Cetecea: Delphinidae) — Pacific whitesided dolphin
Lagothrix lagothricha (Primates: Cebidae) — woolly monkey
Lama guanacoe (Artiodactyla: Camelidae) — guanaco
Lama peruana (Artiodactyla: Camelidae) — llama
Lanius ludovicianus (Passeriformes: Laniidae) — loggerhead shrike
Larosterna inca (Charadriiformes: Laridae) — Inca tern
Larus argentatus (Charadriiformes: Laridae) — herring gull
Larus fuscus (Charadriiformes: Laridae) — lesser black-backed gull
Larus marinus (Charadriiformes: Laridae) — great black-backed gull
Lasiorhinus latifrons (Marsupiala: Phascolomidae) — hairy-nosed wombat
Lemur catta (Primates: Lemuridae) — ringtailed lemur
Lemur fulvus (Primates: Lemuridae) — brown lemur
Lemur macaco (Primates: Lemuridae) — black lemur
Leontopithecus rosalia (Primates: Callitrichidae) — lion tamarin
Leopardus tigrinus (Carnivora: Felidae) — oncilla
Leopardus wiedi (Carnivora: Felidae) — margay
Lepilemur mustelinus (Primates: Lemuridae) — sportive lemur
Leptailurus serval (Carnivora: Felidae) — serval
Leptonychotes weddelli (Pinnipedia: Phocidae) — Weddell seal
Lepus timidus (Lagomorpha: Leporidae) — Arctic hare
Litocranius walleri (Artiodactyla: Bovidae) — gerenuk
Loxodonta africana (Proboscidea: Elephantidae) — African elephant
Lutra canadensis (Carnivora: Mustelidae) — river otter
Lutra lutra (Carnivora: Mustelidae) — European otter
Lutra maculicollis (Carnivora: Mustelidae) — spotted-necked otter
Lutra perspicillata (Carnivora: Mustelidae) — smooth-coated otter
Lycaon pictus (Carnivora: Canidae) — African hunting dog
Lynx lynx (Carnivora: Felidae) — lynx
Macaca arctoides (Primates: Cercopithecidae) — stumptail macaque

Macaca fascicularis (Primates: Cercopith- Kra macaque
ecidae)
Macaca fuscata (Primates: Cercopithecidae) Japanese macaque
Macaca mulatta (Primates: Cercopithecidae) rhesus macaque
Macaca nemestrina (Primates: Cercopith- pigtail macaque
ecidae)
Macaca nigra (Primates: Cercopithecidae) Celebes black ape
Macaca radiata (Primates: Cercopithecidae) bonnet macaque
Macaca sylvanus (Primates: Cercopithecidae) Barbary ape
Macropus giganteus (Marsupiala: Macropodi- great grey kangaroo
dae)
Macropus parryi (Marsupiala: Macropodidae) whiptail wallaby
Macropus robustus (Marsupiala: Macropodi- euro
dae)
Macropus rufus (Marsupiala: Macropodidae) red kangaroo
Madoqua kirki (Artiodactyla: Bovidae) Kirk's dik-dik
Mandrillus leucophaeus (Primates: Cercopith- drill
ecidae)
Mandrillus sphinx (Primates: Cercopith- mandrill
ecidae)
Manis tricuspis (Pholidota: Manidae) African tree pangolin
Marmota bobak (Rodentia: Sciuridae) Bobak marmot
Marmota caligata (Rodentia: Sciuridae) hoary marmot
Marmota flaviventris (Rodentia: Sciuridae) yellow-bellied marmot
Marmota marmota (Rodentia: Sciuridae) Alpine marmot
Marmota monax (Rodentia: Sciuridae) woodchuck
Marmota olympus (Rodentia: Sciuridae) Olympic marmot
Martes flavigula (Carnivora: Mustelidae) yellow-throated marten
Martes foina (Carnivora: Mustelidae) beech marten
Martes martes (Carnivora: Mustelidae) pine marten
Martes pennanti (Carnivora: Mustelidae) fisher
Martes zibellina (Carnivora: Mustelidae) sable
Mazama gouazoubira (Artiodactyla: Cer- grey mazama
vidae)
Meles meles (Carnivora: Mustelidae) European badger
Mellivora capensis (Carnivora: Mustelidae) ratel
Melopsittacus undulatus (Psittaciformes: Psit- budgerigar
tacidae)
Melursus ursinus (Carnivora: Ursidae) sloth bear
Mephitis mephitis (Carnivora: Mustelidae) striped skunk
Meriones persicus (Rodentia: Cricetidae) Persian jird
Meriones tamariscinus (Rodentia: Cricetidae) tamarisk gerbil
Meriones unguiculatus (Rodentia: Cricetidae) Mongolian gerbil
Mesocricetus auratus (Rodentia: Cricetidae) golden hamster
Microcavia australis (Rodentia: Caviidae) desert cavy
Microcebus coquereli (Primates: Lemuridae) Coquerel's mouse lemur
Microcebus murinus (Primates: Lemuridae) mouse lemur

Micromys minutus (Rodentia: Muridae)	harvest mouse
Microtus agrestis (Rodentia: Cricetidae)	short-tailed vole
Microtus ochrogaster (Rodentia: Cricetidae)	prairie vole
Milvus migrans (Falconiformes: Accipitridae)	milan, kite
Miopithecus talapoin (Primates: Cercopithecidae)	talapoin
Mirounga angustirostris (Pinnipedia: Phocidae)	northern elephant seal
Mirounga leonina (Pinnipedia: Phocidae)	southern elephant seal
Mungos mungo (Carnivora: Viverridae)	banded mongoose
Mungotictis lineatus (Carnivora: Viverridae)	narrow-striped mongoose
Muntiacus muntjac (Artiodactyla: Cervidae)	muntjac
Muntiacus reevesi (Artiodactyla: Cervidae)	Reeves' muntjac
Mus musculus (Rodentia: Muridae)	house mouse
Muscardinus avellanarius (Rodentia: Gliridae)	hazel mouse
Mustela erminea (Carnivora: Mustelidae)	ermine
Mustela frenata (Carnivora: Mustelidae)	longtail weasel
Mustela furo (Carnivora: Mustelidae)	ferret
Mustela nigripes (Carnivora: Mustelidae)	black-footed ferret
Mustela nivalis (Carnivora: Mustelidae)	least weasel
Mustela putorius (Carnivora: Mustelidae)	polecat
Mustela vison (Carnivora: Mustelidae)	mink
Myiopsitta monachus (Psittaciformes: Psittacidae)	Quaker parakeet
Myocastor coypus (Rodentia: Capromyidae)	coypu or nutria
Myoprocta pratti (Rodentia: Caviidae)	green acouchi
Myotis lucifugus (Chiroptera: Vespertilionidae)	little brown bat
Myrmecophaga tridactyla (Edentata: Myrmecophagidae)	giant anteater
Nandinia binotata (Carnivora: Viverridae)	two-spotted palm civet
Nasalis larvatus (Primates: Cercopithecidae)	proboscis monkey
Nasua narica (Carnivora: Procyonidae)	coati
Neofelis nebulosa (Carnivora: Felidae)	clouded leopard
Neophoca cinerea (Pinnipedia: Otariidae)	Australian sea lion
Neophoca hookeri (Pinnipedia: Otariidae)	New Zealand sea lion
Neotoma albigula (Rodentia: Cricetidae)	white-throat wood rat
Neotoma floridiana (Rodentia: Cricetidae)	Eastern wood rat
Neotragus pygmaeus (Artiodactyla: Bovidae)	royal antelope
Nestor meridionalis (Psittaciformes: Psittacidae)	kaka
Nestor notabilis (Psittaciformes: Psittacidae)	kea
Notomys alexis (Rodentia: Muridae)	northern hopping mouse
Nyctereutes procyonoides (Carnivora: Canidae)	raccoon-dog
Nycticebus coucang (Primates: Lemuridae)	slow loris
Octodon degus (Rodentia: Octodontidae)	degu
Octodontomys gliroides (Rodentia: Octodontidae)	choz-choz

Odobenus rosmarus (Pinnipedia: Odobenidae)	walrus
Odocoileus hemionus (Artiodactyla: Cervidae)	black-tailed deer
Odocòileus virginianus (Artiodactyla: Cervidae)	white-tailed deer
Okapia johnstoni (Artiodactyla: Giraffidae)	okapi
Ondatra zibethicus (Rodentia: Cricetidae)	muskrat
Orcinus orca (Cetacea: Delphinidae)	orca
Ornithorhynchus anatinus (Monotremata: Ornithorhynchidae)	platypus
Oryctolagus cuniculus (Lagomorpha: Leporidae)	European rabbit
Oryx gazella (Artiodactyla: Bovidae)	gemsbok
Otaria flavescens (Pinnipedia: Otariidae)	southern sea lion
Otocyon megalotis (Carnivora: Canidae)	bat-eared fox
Otospermophilus beecheyi (Rodentia: Sciuridae)	California ground squirrel
Ovibos moschatus (Artiodactyla: Bovidae)	muskox
Ovis ammon (Artiodactyla: Bovidae)	urial & mouflon
Ovis aries (Artiodactyla: Bovidae)	domestic sheep
Ovis canadensis (Artiodactyla: Bovidae)	mountain sheep
Ovis dalli (Artiodactyla: Bovidae)	Dall sheep
Ovis polii (Artiodactyla: Bovidae)	Marco Polo wild sheep
Pan troglodytes (Primates: Pongidae)	chimpanzee
Panthera leo (Carnivora: Felidae)	lion
Panthera onca (Carnivora: Felidae)	jaguar
Panthera pardus (Carnivora: Felidae)	leopard
Panthera tigris (Carnivora: Felidae)	tiger
Papio anubis (Primates: Cercopithecidae)	olive baboon
Papio cynocephalus (Primates: Cercopithecidae)	yellow baboon
Papio hamadryas (Primates: Cercopithecidae)	hamadryas baboon
Papio ursinus (Primates: Cercopithecidae)	chacma baboon
Pardofelis marmorata (Carnivora: Felidae)	marbled cat
Perameles gunnii (Marsupialia: Peramelidae)	Tasmanian barred bandicoot
Perodicticus potto (Primates: Lemuridae)	potto
Petaurista petaurista (Rodentia: Sciuridae)	giant flying squirrel
Phacochoerus aethiopicus (Artiodactyla: Suidae)	warthog
Phoca vitulina (Pinnipedia: Phocidae)	common seal (*P.v. vitulina*) harbor seal (*P.v. concolor*)
Physeter catodon (Cetacea: Physeteridae)	sperm whale
Pica nuttalli (Passeriformes: Corvidae)	yellow-billed magpie
Pica pica (Passeriformes: Corvidae)	white-rumped magpie
Pongo pygmaeus (Primates: Pongidae)	orangutan
Potos flavus (Carnivora: Procyonidae)	kinkajou
Presbytis cristatus (Primates: Cercopithecidae)	lutong
Presbytis entellus (Primates: Cercopithecidae)	hanuman langur

Presbytis geei (Primates: Cercopithecidae) — golden langur
Presbytis johnii (Primates: Cercopithecidae) — Nilgiri langur
Presbytis obscurus (Primates: Cercopithecidae) — spectacled langur
Presbytis phayrei (Primates: Cercopithecidae) — Phayre's leaf monkey
Prionailurus bengalensis (Carnivora: Felidae) — leopard cat
Prionailurus rubiginosus (Carnivora: Felidae) — rusty-spotted cat
Prionailurus viverrinus (Carnivora: Felidae) — fishing cat
Prionodon linsang (Carnivora: Viverridae) — linsang
Procapra gutturosa (Artiodactyla; bovidae) — Mongolian gazelle
Procyon cancrivorus (Carnivora: Procyonidae) — crab-eating raccoon
Procyon lotor (Carnivora: Procyonidae) — raccoon
Profelis aurata (Carnivora: Felidae) — African golden cat
Profelis temmincki (Carnivora: Felidae) — Asian golden cat
Propithecus verreauxi (Primates: Lemuridae) — sifaka
Proteles cristatus (Carnivora: Hyaenidae) — aardwolf
Psammomys obesus (Rodentia: Cricetidae) — sand rat
Pseudois nayaur (Artiodactyla: Bovidae) — bharal
Pseudomys albocinereus (Rodentia: Muridae) — ashy-grey mouse
Pseudomys desertor (Rodentia: Muridae) — brown desert mouse
Pseudomys shortridgei (Rodentia: Muridae) — blunt-faced rat
Pteronura brasiliensis (Carnivora: Mustelidae) — Brazilian giant otter
Pteropus giganteus (Chiroptera: Pteropidae) — Indian flying fox
Pudu pudu (Artiodactyla: Cervidae) — pudu
Puma concolor (Carnivora: Felidae) — puma
Pygathrix nemaeus (Primates: Cercopithecidae) — Douc langur
Pyrrhocorax hybrid (Passeriformes: Corvidae) — chough hybrid
Rangifer tarandus (Artiodactyla: Cervidae) — reindeer, caribou
Rattus norvegicus (Rodentia: Muridae) — Norway rat
Rattus rattus (Rodentia: Muridae) — black rat
Rhinoceros unicornis (Perissodactyla: Rhinocerotidae) — Indian rhinoceros
Rupicapra rupicapra (Artiodactyla: Bovidae) — chamois
Saguinus fuscicollis (Primates: Callitrichidae) — saddle-back tamarin
Saguinus midas (Primates: Callitrichidae) — black-handed tamarin
Saguinus oedipus (Primates: Callitrichidae) — cotton-top marmoset
Saimiri spp. (Primates: Cebidae) — squirrel monkey
Sarcophilus harrisii (Marsupiala: Dasyuridae) — Tasmanian devil
Sciurus carolinensis (Rodentia: Sciuridae) — grey squirrel
Sciurus niger (Rodentia: Sciuridae) — fox squirrel
Sciurus vulgaris (Rodentia: Sciuridae) — European red squirrel
Sminthopsis crassicaudata (Marsupiala: Dasyuridae) — fat-tailed dunnart
Sminthopsis murina (Marsupiala: Dasyuridae) — common dunnart
Solenodon paradoxus (Insectivora: Solenodontidae) — Hispaniolan solenodon

Somateria mollissima (Anseriformes: Anatidae) — common eider

Sorex cinereus (Insectivora: Soricidae) — masked shrew

Speothos venaticus (Carnivora: Canidae) — bush dog

Spermophilus armatus (Rodentia: Sciuridae) — Uinta ground squirrel

Spermophilus columbianus (Rodentia: Sciuridae) — Columbian ground squirrel

Spermophilus richardsoni (Rodentia: Sciuridae) — Richardson's ground squirrel

Spilogale putorius (Carnivora: Mustelidae) — spotted skunk

Streptopelia decaocto (Columbiformes: Columbidae) — collared dove

Suncus etruscus (Insectivora: Soricidae) — pygmy white-toothed shrew

Suricata suricatta (Carnivora: Viverridae) — meerkat

Sus scrofa (Artiodactyla: Suidae) — domestic swine

Sylvia borin (Passeriformes: Muscicapidae) — garden warbler

Sylvia communis (Passeriformes: Muscicapidae) — common whitethroat

Symphalangus syndactylus (Primates: Pongidae) — Siamang

Syncerus caffer (Artiodactyla: Bovidae) — African buffalo

Tamias striatus (Rodentia: Sciuridae) — Eastern chipmunk

Tamiasciurus hudsonicus (Rodentia: Sciuridae) — American red squirrel

Tapirus spp. (Perissodactyla: Tapiridae) — tapir

Tenrec ecaudatus (Insectivora: Tenrecidae) — tailless tenrec

Thalarctos maritimus (Carnivora: Ursidae) — polar bear

Theropithecus gelada (Primates: Cercopithecidae) — gelada

Tockus fasciatus (Coraciiformes: Bucerotidae) — hornbill

Tragelaphus imberbis (Artiodactyla: Bovidae) — lesser kudu

Tragelaphus scriptus (Artiodactyla: Bovidae) — bushbuck

Tragelaphus spekei (Artiodactyla: Bovidae) — sitatunga

Tragelaphus spp. (Artiodactyla: Bovidae) — kudu

Tremarctos ornatus (Carnivora: Ursidae) — spectacled bear

Trichechus manatus (Sirenia: Trichechidae) — manatee

Tupaia glis (Primates: Tupaiidae) — common tree shrew

Turaco fischeri (Cuculiformes: Musophagidae) — turaco

Turdoides malcolmi (Passeriformes: Muscicapidae) — large grey babbler

Turdoides striatus (Passeriformes: Muscicapidae) — jungle babbler

Turdus merula (Passeriformes: Muscicapidae) — blackbird

Tursiops truncatus (Cetacea: Delphinidae) — bottle-nosed dolphin

Tyto alba (Strigiformes: Tytonidae) — barn owl

Uncia uncia (Carnivora: Felidae) — snow leopard

Urocyon cinereoargenteus (Carnivora: Canidae) — grey fox

Ursus americanus (Carnivora: Ursidae) — black bear

Ursus arctos (Carnivora: Ursidae) — brown bear, grizzly bear
Varecia variegata (Primates: Lemuridae) — ruffed lemur
Vicugna vicugna (Artiodactyla: Camelidae) — vicuna
Vormela peregusna (Carnivora: Mustelidae) — marbled polecat
Vulpes corsac (Carnivora: Canidae) — corsac fox
Vulpes fulva (Carnivora: Canidae) — red fox
Vulpes rueppelli (Carnivora: Canidae) — sand fox
Vulpes vulpes (Carnivora: Canidae) — red fox
Vulpes zerda (Carnivora: Canidae) — fennec
Xerus erythropus (Rodentia: Sciuridae) — African ground squirrel
Zalophus californianus (Pinnipedia: Otariidae) — California sea lion
Zalophus wollebaeki (Pinnipedia: Otariidae) — Galapagos sea lion
Zonotrichia melodia (Passeriformes: Fringillidae) — song sparrow

References

Abeelen, J. H. F. van & A. H. Schoones. 1977. Ontogeny of behavior in two inbred lines of selected mice. Devel. Psychobiol. 10:17–23.

Abelson, P. H. 1976. Cost-effective health care. (Editorial). Science 192:619.

Accordi, B. & R. Colacicchi. 1962. Excavations in the pygmy elephants cave of Spinagallo (Siracusa). Geol. Romana 1:217–229.

Adamson, J. 1960. Born free. Pantheon, N.Y., London.

Albignac, R. 1969a. Naissance et élèvage en captivité des jeunes *Cryptoprocta ferox,* viverrides malagaches. Mammalia 33:93–97.

Albignac, R. 1969b. Notes éthologiques sur quelques carnivores malagaches: le *Galidia elegans* I. Geoffroy. La Terre et la Vie 23:202–215.

Albignac, R. 1970a. Notes éthologiques sur quelques carnivores malagaches: le *Fossa fossa* (Schreber). La Terre et la Vie 24:383–394.

Albignac, R. 1970b. Notes éthologiques sur quelques carnivores malagaches: le *Cryptoprocta ferox* (Bennett). La Terre et la Vie 24:395–402.

Albignac, R. 1971. Notes éthologiques sur quelques carnivores malagaches: le *Mungotictis lineata* Pocock. La Terre et la Vie 25:328–343.

Albignac, R. 1973. Mammifères carnivores. Faune de Madagascar, 36. ORSTOM, CNRS, Paris.

Albignac, R. 1974. Observations eco-éthologiques sur le genre *Eupleres,* viverride de Madagascar. La Terre et la Vie 28:321–351.

Albignac, R. 1975. Breeding the fossa, *Cryptoprocta ferox,* at Montpellier Zoo. Int. Zoo Yb. 15:147–150.

Albignac, R. 1976. L'écologie de *Mungotictis decemlineata* dans les forêts decidues de l'ouest de Madagascar. La Terre et la Vie 30:347–376.

Aldis, O. 1975. Play fighting. Academic Press, New York.

Aldous, S. E. 1937. A hibernating black bear with cubs. J. Mammal. 18:466–468.

Aldous, S. E. 1940. Notes on a black-footed ferret raised in captivity. J. Mammal. 21:23–26.

562

Aldrich-Blake, F. P. G. & D. J. Chivers. 1973. On the genesis of a group of siamang. Am. J. Phys. Anthropol. 38:631–636.

Alexander, B. K. 1970. Parental behavior of adult male Japanese monkeys. Behaviour 36:270–285.

Alexander, R. D. 1974. The evolution of social behavior. Ann. Rev. Ecol. Systemat. 5:325–383.

Alexander, R. D. 1975. The search for a general theory of behavior. Behav. Sci. 20:77–100.

Ali, S. A. 1927. The Moghul emperors of India as naturalists and sportsmen. Part II. J. Bombay Nat. Hist. Soc. 32:34–63.

Altman, J., R. L. Brunner & S. L. Bayer. 1973. The hippocampus and behavioral maturation. Behav. Biol. 8:557–596.

Altmann, D. 1970. Ethologische Studie an Mufflons, Ovis ammon musimon (Pallas). Zool. Gart. 39:297–303.

Altmann, D. 1971. Verhaltensanalyse der Ontogenese von Steppenfuchsen, Vulpes corsac L. Zool. Gart. 41:1–6.

Altmann, D. 1972. Verhaltensstudien an Mahnenwölfen, Chrysocyon brachyurus. Zool. Gart. 41:278–298.

Altmann, J. 1974. Observational study of behavior: sampling methods. Behaviour 49:227–267.

Altmann, J. 1980. Baboon mothers and infants. Harvard University Press, Cambridge, Mass.

Altmann, J., S. A. Altmann, G. Hausfater & S. A. McCuskey. 1977. Life history of yellow baboons: physical development, reproductive parameters, and infant mortality. Primates 18:315–330.

Altmann, M. 1952. Social behavior of elk, Cervus canadensis nelsoni, in the Jackson Hole area of Wyoming. Behaviour 4:116–143.

Altmann, M. 1958. Social integration of the moose calf. Anim. Behav. 6:155–159.

Altmann, M. 1963. Naturalistic studies of maternal care in moose and elk. Pages 233–253 in H. L. Rheingold, ed., Maternal behavior in mammals. Wiley, N.Y.

Altmann, S. A. 1959. Field observations on a howling monkey society. J. Mammal. 40:317–330.

Altmann, S. A. 1962a. A field study of the sociobiology of rhesus monkeys, Macaca mulatta. Ann. N. Y. Acad. Sci. 102:338–435.

Altmann, S. A. 1962b. Social behavior of anthropoid primates: analysis of recent concepts. Pages 277–285 in E. L. Bliss, ed., Roots of behavior. Harper & Bros., N.Y.

Altmann, S. A. 1965. Sociobiology of rhesus monkeys. II. Stochastics of social communication. J. Theoret. Biol. 8:490–522.

Altmann, S. A. 1974. Baboons, space, time and energy. Amer. Zool. 14:221–248.

Altmann, S. A. & J. Altmann. 1970. Baboon ecology. University of Chicago Press, Chicago.

Anderson, C. O. & W. A. Mason. 1974. Early experience and complexity of social organization in groups of young rhesus monkeys (*Macaca mulatta*). J. comp. physiol. Psychol. 87:681–690.

Anderson, D. 1977. Gestation period of Geoffroy's cat, *Leopardus geoffroyi*, bred at Memphis Zoo. Int. Zoo Yb. 17:164–166.

Anderson, M. A. 1971. A watched potto never grows: a chronicle of the prenatal and first months of a *Perodicticus potto*. Discovery 6:89–98.

Andersson, A. B. 1969. Communication in the lesser bushbaby (*Galago senegalensis moholi*). M. Sc. thesis, Witwatersrand Univ.

Andrew, R. J. 1962. The situations that evoke vocalization in primates. Ann. N. Y. Acad. Sci. 102:296–315.

Andrew, R. J. 1963. The origin and evolution of the calls and facial expressions of the primates. Behaviour 20:1–109.

Andrew, R. J. 1976. Review of R. A. Hinde. Biological bases of human social behaviour. Anim. Behav. 24:958–959.

Andrews, R. C. 1932. The new conquest of central Asia: a narrative of the explorations of the Central Asiatic expeditions in Mongolia and China, 1921–1930. Natural History of Central Asia, Vol. I. The American Museum of Natural History, N.Y.

Angot, M. 1954. Observations sur les mammifères marins de l'archipel de Kerguelen, avec une étude détaillée de l'éléphant de mer, *Mirounga leonina* (L.). Mammalia 18:1–111.

Angst, W. & D. Thommen. 1977. New data and a discussion of infant killing in Old World monkeys and apes. Folia primatol. 27:198–229.

Angus, S. 1971. Water-contact behavior of chimpanzees. Folia primatol. 14:51–58.

Anon. 1976. Nova Scotia duck tolling retriever. Gaines Progress, Fall:4–5.

Ansell, W. F. H. 1947. Notes on some Burmese mammals. J. Bombay Nat. Hist. Soc. 47:379–383.

Antonius, O. 1939. Über Symbolhandlungen und Verwandtes bei Säugetieren. Z. Tierpsychol. 3:263–278.

Apfelbach, R. 1969a. *Crocuta crocuta* (Hyaenidae)—Spiel der Jungtiere. Encyclopedia Cinematographica Film E1486. Institut für den Wissenschaftlichen Film, Göttingen.

Apfelbach, R. 1969b. *Lycaon pictus* (Canidae)—Spiel der Jungtiere. Encyclopedia Cinemtographica Film E1487. Institut für den Wissenschaftlichen Film, Göttingen.

Armitage, K. B. 1962. Social behaviour of a colony of the yellow-bellied marmot (*Marmota flaviventris*). Anim. Behav. 10:319–331.

Armitage, K. B. 1974. Male behaviour and territoriality in the yellow-bellied marmot. J. Zool. 172:233–265.

Armitage, K. B. 1977. Social variety in the yellow-bellied marmot: a population-behavioural system. Anim. Behav. 25:585–593.

Armstrong, J. 1975. Hand-rearing Black-footed cats, *Felis nigripes*, at the National Zoological Park, Washington. Int. Zoo Yb. 15:245–249.

Arshavsky, I. A. 1972. Musculoskeletal activity and rate of entropy in mammals. Adv. Psychobiol. 1:1–52.

Ashmole, N. P. & H. Tovar S. 1968. Prolonged parental care in Royal Terns and other birds. Auk 85:90–100.

Aslin, H. 1974. The behaviour of *Dasyuroides byrnei* (Marsupiala) in captivity. Z. Tierpsychol. 35:187–208.

Åstrand, P.-O. & K. Rodahl. 1970. Textbook of work physiology. McGraw-Hill, N.Y.

Audubon, J. J. & J. Bachman. 1851. The viviparous quadrupeds of North America. V. G. Audubon, N.Y.

Austin, H. 1974. A computational view of the skill of juggling. Memo No. 330, MIT Artificial Intelligence Laboratory, Cambridge, Mass.

Autenrieth, R. E. & E. Fichter. 1975. On the behavior and socialization of pronghorn fawns. Wildl. Monogr. 42.

Autuori, M. P. & L. A. Deutsch. 1977. Contribution to the knowledge of the giant Brazilian otter, *Pteronura brasiliensis* (Gmelin 1788), Carnivora: Mustelidae. Zool. Gart. 47:1–8.

Avedon, E. M. & B. Sutton-Smith. 1971. The study of games. Wiley, N.Y.

Ayer, A. J. 1970. Metaphysics and common sense. Freeman, Cooper, San Francisco.

Babault, G. 1949. Notes éthologiques sur quelques mammifères africains. Mammalia 13:105–124.

Backhaus, D. 1959. Beobachtungen über das Freileben von Lewel Kuhantilopen (*Alcelaphus buselaphus lewel* Heuglin 1877) und Gelegenheitsbeobachtungen an Sennar Pferdeantilopen (*Hippotragus equinus backeri* Heuglin 1863). Z. Säugetierk. 24:1–34.

Baker, E. C. S. 1899. Indian ducks and their allies. J. Bombay Nat. Hist. Soc. 12:593–620.

Baker, R. P. & D. G. Preston. 1973. The effects of interspecies infant interaction upon social behavior of *Macaca irus* and *Erythrocebus patas*. Primates 14:383–392.

Baker, R. R. 1978. The evolutionary ecology of animal migration. Holmes & Meier, N.Y.

Baldwin, J. D. 1968. The social behavior of adult male squirrel monkeys (*Saimiri sciureus*) in a seminatural environment. Folia Primatol. 9:281–314.

Baldwin, J. D. 1969. The ontogeny of social behavior of squirrel monkeys (*Saimiri sciureus*) in a seminatural environment. Folia Primatol. 11:35–79.

Baldwin, J. D. 1971. The social organization of a semifree-ranging troop of squirrel monkeys (*Saimiri sciureus*). Folia Primatol. 14:23–50.

Baldwin, J. D. & J. I. Baldwin. 1971. Squirrel monkeys (*Saimiri*) in natural habitats in Panama, Colombia, Brazil and Peru. Primates 12:45–61.

Baldwin, J. D. & J. I. Baldwin. 1972. The ecology and behavior of squirrel monkeys (*Saimiri oerstedi*) in a natural forest in western Panama. Folia Primatol. 18:161–184.

Baldwin, J. D. & J. I. Baldwin. 1973a. Interactions between adult female and infant howling monkeys (*Alouatta palliata*). Folia Primatol. 20:27–71.

Baldwin, J. D. & J. I. Baldwin. 1973b. The role of play in social organization:

comparative observations on squirrel monkeys (*Saimiri*). Primates 14:369–381.

Baldwin, J. D. & J. I. Baldwin. 1974. Exploration and social play in squirrel monkeys (*Saimiri*). Am. Zool. 14:303–315.

Baldwin, J. D. & J. I. Baldwin. 1976a. Effects of food ecology on social play: a laboratory simulation. Z. Tierpsychol. 40:1–14.

Baldwin, J. D. & J. I. Baldwin. 1976b. Vocalizations of howler monkeys (*Alouatta palliata*) in southwestern Panama. Folia Primatol. 26:81–108.

Baldwin, J. D. & J. I. Baldwin. 1977. The role of learning phenomena in the ontogeny of exploration and play. Pages 343–406 in S. Chevalier-Skolnikoff & F. E. Poirier, eds., Primate bio-social development. Garland, N.Y.

Baldwin, J. D. & J. I. Baldwin. 1978. Reinforcement theories of exploration, play, creativity and psychosocial growth. Pages 231–258 in E. O. Smith, ed., Social play in primates. Academic Press, N.Y.

Baldwin, L. & G. Teleki. 1976. Patterns of gibbon behavior on Hall's Island, Bermuda. Gibbon and Siamang 4:21–105.

Balgooyen, T. G. 1976. Behavior and ecology of the American kestrel (*Falco sparverius* L.). Univ. Calif. Publ. Zool. 103:1–83.

Ball, J. H. 1955. Rearing the two-spotted palm civet (*Nandina binotata*). Nigerian Field 20:64–68.

Ball, S. C. 1943. Chimney swifts at play? Auk 60:269–270.

Balph, D. F. & A. W. Stokes. 1963. On the ethology of a population of Uinta ground squirrels. Am. Midl. Nat. 68:106–126.

Banks, E. M. 1964. Some aspects of sexual behavior in domestic sheep, *Ovis aries*. Behaviour 23:249–279.

Barash, D. P. 1973a. The social biology of the Olympic marmot. Anim. Behav. Monogr. 6:171–245.

Barash, D. P. 1973b. Social variety in the yellow-bellied marmot (*Marmota flaviventris*). Anim. Behav. 21:579–584.

Barash, D. P. 1974a. The evolution of marmot societies: a general theory. Science 185:415–420.

Barash, D. P. 1974b. Mother-infant relations in captive woodchucks (*Marmota monax*). Anim. Behav. 22:446–448.

Barash, D. P. 1974c. The social behavior of the hoary marmot (*Marmota caligata*). Anim. Behav. 22:256–261.

Barash, D. P. 1975. Behavior as evolutionary strategy. Review of J. L. Brown, The evolution of behavior and J. Alcock, Animal behavior. Science 190:1084–1085.

Barash, D. P. 1976. Social behaviour and individual differences in free-living Alpine marmots (*Marmota marmota*). Anim. Behav. 24:27–35.

Barash. D. P. 1977. Sociobiology and behavior. Elsevier North-Holland, N.Y.

Barger, A. C., V. Richards, J. Metcalfe & B. Gunther. 1956. Regulation of the circulation during exercise. Am. J. Physiol. 184:613–623.

Barker, G. H. 1924. Among the birds at Redcliffe. Queensl. Nat. 4:75–76.

Barlow, G. W. 1977. Modal action patterns. Pages 94–125 in T. A. Sebeok, ed., How animals communicate. Indiana University Press, Bloomington.

Barnard, R. J., R. Macalpin, A. Kattus & G. Buckberg. 1973. Ischemic response to sudden strenuous exercise in healthy men. Circulation 48:936–942.

Barnett, S. A. 1958. Exploratory behavior. Br. J. Psychol. 49:289–310.

Barnett, S. A. 1969. Grouping and dispersive behaviour among wild rats. Pages 3–14 in S. Garattini & E. B. Sigg, eds., Aggressive behaviour. Excerpta Medica, Amsterdam.

Barnett, S. A. 1975. The rat: a study in behaviour. University of Chicago Press, Chicago.

Barrett, P. & P. Bateson. 1978. The development of play in cats. Behaviour 66:106–120.

Barrette, C. 1977. The social behaviour of captive muntjacs *Muntiacus reevesi* (Ogilby 1839). Z. Tierpsychol. 43:188–213.

Bartholomew, C. 1975. The only way to fly. Pages 118–124 in A. S. Burack, ed., The writer's handbook. The Writer, Boston.

Bartholomew, G. G. 1959. Mother-young relations and the maturation of pup development in the Alaska fur seal. Anim. Behav. 7:163–171.

Bartikova, J. 1973. Rozdily v bojovem chovani ruzne starych samcu sitatung (*Tragelaphus spekei* Sclater, 1864). (Differences in the belligerent behavior of males of *Tragelaphus spekei* Sclater, 1864 at different ages). J. Lynx (Prague) 14:117–120.

Bartlett, F. 1958. Thinking: an experimental and social study. George Allen and Unwin, London.

Basmajian, J. V. 1972. Electromyography comes of age. Science 176:603–609.

Basmajian, J. V. 1974. Muscles alive: their functions revealed by electromyography, 3rd ed. Williams & Wilkins, Baltimore.

Basset, C. A. L. 1972. Biophysical principles affecting bone structure. Pages 1–76 in G. Bourne, ed., The biochemistry and physiology of bone, Vol. 3, 2nd ed. Academic Press, N.Y.

Bastock, M. 1967. Courtship. Heinemann, London.

Bate, W. J. 1977. Samuel Johnson. Harcourt Brace Jovanovich, N.Y. & London.

Bates, H. W. 1969. (1st ed. 1863.) The naturalist on the river Amazons. Everyman's Library, Dent, London, and Dutton, N.Y.

Bates, M. 1944. Notes on a captive icticyon. J. Mammal. 25:152–154.

Bates, M. 1960. The forest and the sea. Vintage, N.Y.

Bateson, G. 1955. A theory of play and fantasy. Psychiatr. Res. Rep. 2:39–51.

Bateson, G. 1956. The message "This is play." Pages 145–246 in B. Schaffner, ed., Group Processes (Trans. 2nd Conf.). Macy Foundation, N.Y.

Bateson, G. 1963. The role of somatic change in evolution. Evolution 17:529–539.

Bateson, G. 1978. Mind and nature. Dutton, N.Y.

Bateson, P. P. G. 1973. Preferences for familiarity and novelty: a model for the simultaneous development of both. J. Theor. Biol. 41:249–259.

Bateson, P. P. G. 1976a. Rules and reciprocity in behavioural development. Pages 401–421 in P. P. G. Bateson & R. A. Hinde, eds., Growing points in ethology. University Press, Cambridge.

Bateson, P. P. G. 1976b. Specificity and the origins of behavior. Pages 1–20 in J. Rosenblatt, R. Hinde, E. Shaw & C. Beer, eds., Advances in the study of behavior, Vol. 6. Academic Press, N.Y.

Bateson, P. P. G. 1978a. Early experience and sexual preferences. Pages 29–53 in J. B. Hutchinson, ed., Biological determinants of sexual behaviour. Wiley, London.

Bateson, P. P. G. 1978b. How does behavior develop? Pages 55–66 in P. P. G. Bateson & P. H. Klopfer, eds., Perspectives in ethology, Vol. 3. Plenum, N.Y.

Bateson, P. P. G. 1979. How do sensitive periods arise and what are they for? Anim. Behav. 27:470–486.

Bateson, P. & M. Young. 1979. The influence of male kittens on the object play of their female siblings. Behav. Neur. Biol. 27:374–378.

Battersby, E. 1944. Do young birds play? Ibis 86:225.

Battersby, W. S., H. L. Teuber & M. B. Bender. 1953. Problem-solving behavior in men with frontal or occipital brain injuries. J. Psychol. 35:329–351.

Batzli, G. O., L. L. Getz & S. S. Hurley. 1977. Suppression of growth and reproduction of microtine rodents by social factors. J. Mammal. 58:583–591.

Beach, F. A. 1945. Current concepts of play in animals. Am. Nat. 79:523–541.

Beach, N. 1978. The new science of fitness. Pages 22, 74 in New York Times Magazine, Part 2: Men's Fashions.

Beadle, M. 1977. The cat. Simon & Schuster, N.Y.

Bearder, S. K. 1969. Territorial and intergroup behaviour of the lesser bushbaby, Galago senegalensis moholi (A. Smith), in semi-natural conditions and in the field. M. Sc. thesis, Witwatersrand University.

Beck, A. M. 1973. The ecology of stray dogs. York Press, Baltimore.

Beck, B. 1975. Primate tool behavior. Pages 413–447 in R. H. Tuttle, ed., Socioecology and psychology of primates. Mouton, The Hague.

Beck, B. 1976. Predatory shell dropping by herring gulls. Paper presented at annual meeting of Animal Behavior Society, Boulder, Colorado.

Beck, B. 1978. Ontogeny of tool use by nonhuman animals. Pages 405–419 in G. M. Burghardt & M. Bekoff, eds., The development of behavior: Comparative and evolutionary aspects. Garland STPM, N.Y.

Beck, B. B. & R. Tuttle. 1972. The behavior of gray langurs at a Ceylonese waterhole. Pages 351–377 in R. Tuttle, ed., The functional and evolutionary biology of primates. Aldine-Atherton, Chicago & N.Y.

Beer, C. G. 1977. What is a display? Am. Zool. 17:155–165.

Behm, U. 1953. Aufzucht von Vielfraßen. Zool. Gart. 20:77–81.

Bekoff, M. 1972. The development of social interaction, play, and metacommunication in mammals: an ethological perspective. Q. Rev. Biol. 47:412–434.

Bekoff, M. 1974a. Social play in coyotes, wolves and dogs. Bioscience 24:225–230.

Bekoff, M., ed. 1974b. Social play in mammals. Am. Zool. 14:265–436.

Bekoff, M. 1974c. Social play and play-soliciting by infant canids. Am. Zool. 14:323–340.

Bekoff, M. 1975. The communication of play intention: are play signals functional? Semiotica 15:231–239.

Bekoff, M. 1976a. Animal play: problems and perspectives. Pages 165–188 in P. P. G. Bateson & P. H. Klopfer, eds., Perspectives in ethology, Vol. 2. Plenum Press, N.Y. & London.

Bekoff, M. 1976b. The social deprivation paradigm: who's being deprived of what? Dev. Psychobiol. 9:497–498.

Bekoff, M. 1977a. Mammalian dispersal and the ontogeny of individual behavioral phenotypes. Am. Nat. 111:715–732.

Bekoff, M. 1977b. Social communication in canids: evidence for the evolution of a stereotyped mammalian display. Science 197:1097–1099.

Bekoff, M. 1978a. Behavioral development in coyotes and eastern coyotes. Pages 97–126 in M. Bekoff, ed., Coyotes: biology, behavior and management. Academic Press, N.Y.

Bekoff, M. 1978b. Social play: structure, function and the evolution of a cooperative social behavior. Pages 367–383 in G. Burghardt & M. Bekoff, eds., The development of behavior. Garland, N.Y.

Bekoff, M. & J. A. Byers. 1979. A critical reanalysis of the ontogeny and phylogeny of mammalian social play: An ethological hornet's nest. In K. Immelmann, G. Barlow, M. Main & L. Petrinovich, eds., Behavioral development in animals and man: The Bielefeld conference. Cambridge Univ. Press, Cambridge & N.Y.

Bell, G. 1976. On breeding more than once. Am. Nat. 110:57–77.

Bell, T. 1830. Account of a pair of living acouchies (Dasyprocta acuschy). Proc. Zool. Soc. Lond. 1830–1832:6–7.

Bennett, E. L. 1976. Cerebral effects of differential experience and training. Pages 279–287 in M. R. Rosenzweig & E. L. Bennett, eds., Neural mechanisms of learning and memory. MIT Press, Cambridge, Mass.

Bennett, E. T. 1834. Observations on the genus Cryptoprocta. Proc. Zool. Soc. Lond. 1833–1836:13.

Bennett, G. 1834. On the natural history and habits of the Ornithorhynchus paradoxus. Proc. Zool. Soc. Lond. 1833–1836:141–146.

Bennett, G. 1835. Notes on the natural history and habits of the Ornithorhynchus paradoxus, Blum. Trans. Zool. Soc. Lond. 1:229–258.

Bennett, G. 1859. Notes on the duck-bill (Ornithorhynchus anatinus). Proc. Zool. Soc. Lond. 27:213–218.

Bennett, G. 1860a. Gatherings of a naturalist in Australasia. (Cited in Burrell 1974; original not seen.)

Bennett, G. 1860b. Notes on the habits of the brown coati (Nasua fusca, Desm.) Ann. Mag. Nat. Hist. 6:391–392.

Bentham, J. 1840. The theory of legislation. Weeks, Jordan, Boston.

Benzon, T. A. & R. F. Smith. 1975. A case of programmed cheetah Acinonyx jubatus breeding. Int. Zoo Yb. 15:154–157.

Berger, J. 1979. Social ontogeny and behavioural diversity: consequences for Bighorn sheep *Ovis canadensis* inhabiting desert and mountain environments. J. Zool. Lond. 188:251–266.

Berger, J. 1980. Ecology, structure and functions of social play in bighorn sheep. J. Zool. Lond., in press.

Berger, M. E. 1972. Population structure of olive baboons (*Papio anubis*) in the Laikipia district of Kenya. East Afr. Wildl. J. 10:159–164.

Berlyne, D. E. 1969. Laughter, humor, and play. Pages 795–852 in G. Lindzey & E. Aronson, eds., Handbook of social psychology, Vol. 3, 2nd ed. Addison-Wesley, Reading, Mass.

Bernstein, I. S. 1964. A field study of the activities of howler monkeys. Anim. Behav. 12:92–97.

Bernstein, I. S. 1965. Activity patterns in a cebus monkey group. Folia Primatol. 3:211–224.

Bernstein, I. S. 1967. A field study of the pigtail monkey (*Macaca nemestrina*). Primates 8:217–228.

Bernstein, I. S. 1968. The lutong of Kuala Selangor. Behaviour 32:1–16.

Bernstein, I. S. 1969. Introductory techniques in the formation of pigtail monkey troops. Folia Primatol. 10:1–19.

Bernstein, I. S. 1970. Activity patterns in pigtail monkey groups. Folia Primatol. 12:187–198.

Bernstein, I. S. 1971. Activity profiles of primate groups. Behav. Nonhuman Primates 3:69–106.

Bernstein, I. S. 1972a. Daily activity cycles and weather influences on a pigtail monkey group. Folia Primatol. 18:390–415.

Bernstein, I. S. 1972b. The organization of primate societies: longitudinal studies of captive groups. Pages 399–487 in R. Tuttle, ed., The functional and evolutionary biology of primates. Aldine-Atherton, Chicago & N.Y.

Bernstein, I. S. 1975. Activity patterns in a gelada monkey group. Folia Primatol. 23:50–71.

Bernstein, I. S. 1976. Activity patterns in a sooty mangabey group. Folia Primatol. 26:185–206.

Bernstein, I. S. & W. A. Draper. 1964. The behavior of juvenile rhesus monkeys in groups. Anim. Behav. 12:84–91.

Bernstein, I. S. & R. J. Schusterman. 1964. The activity of gibbons in a social group. Folia Primatol. 2:161–170.

Bertram, B. C. R. 1976. Kin selection in lions and in evolution. Pages 281–301 in P. P. G. Bateson & R. A. Hinde, eds., Growing points in ethology. Cambridge Univ. Press, Cambridge & London.

Bertrand, M. 1969. The behavioral repertoire of the stumptail macaque. Bibliotheca primatologica, no. 11. Karger, Basel.

Bertrand, M., P. Hunkeler & F. Bourlière. 1969. Ecologie et comportement de la mone de Lowe. 16 mm film. Collection F. Bourlière, Paris.

Betts, B. J. 1976. Behaviour in a population of Columbian ground squirrels, *Spermophilus columbianus columbianus*. Anim. Behav. 24:652–680.

Biben, M. 1979. Predation and predatory play behaviour of domestic cats. Anim. Behav. 27:81–94.

Biddulph, C. H. 1954. Strange behaviour of a house crow. J. Bombay Nat. Hist. Soc. 52:208–209.

Bielański, A. 1977. Foetal behaviour and the possibility of its detection in farm animals. Appl. Anim. Ethol. 3:379–390.

Bierens de Haan, J. A. 1952. Das Spiel eines jugen solitären Schimpansen. Behaviour 4:144–157.

Bilodeau, E. A. ed. 1966. Acquisition of skill. Academic Press, N.Y.

Bingham, H. C. 1927. Parental play of chimpanzees. J. Mammal. 8:77–89.

Bingham, H. C. 1929. Observations on growth and development of chimpanzees. Am. J. Phys. Anthropol. 13:433–368.

Birch, H. G. 1945a. The relation of previous experience to insightful problem-solving. J. Comp. Psychol. 38:367–383.

Birch, H. G. 1945b. The role of motivational factors in insightful problem-solving. J. Comp. Psychol. 38:295–317.

Birdseye, C. 1956. Observations on a domesticated Peruvian desert fox Dusicyon. J. Mammal. 37:284–287.

Birkenmeier, E. & E. Birkenmeier. 1971. Hand-rearing the leopard cat, Felis bengalensis borneoensis. Int. Zoo Yb. 11:118–121.

Bishop, A. 1962. Control of the hand in lower primates. Ann. N. Y. Acad. Sci. 102:316–337.

Bishop, A. 1963. Use of the hand in lower primates. Pages 133–225 in J. Buettner-Janusch, ed., Evolutionary and genetic biology of primates, Vol. 2. Academic Press, N.Y.

Bishop, J. M. & D. Symons. 1978. Aggressive play in rhesus monkeys. 16 mm color sound film. University of California Extension Media Center, Berkeley.

Blaauw, F. E. 1889. (A letter to the society on the development of the horns of the white-tailed gnu [Catodepas gnu]). Proc. Zool. Soc. Lond. 1889:2–5.

Blaffer Hrdy, S. 1976. Care and exploitation of nonhuman primate infants by conspecifics other than the mother. Adv. Stud. Behav. 6:101–158.

Blaffer Hrdy, S. 1977. The langurs of Abu. Harvard University Press, Cambridge, Mass.

Blauvelt, H. 1956. Neonate-mother relationship in goat and man. Pages 94–140 in B. Schaffner, ed., Group processes (Trans. 2nd Conf.). Macy Foundation, N.Y.

Blaxter, K. L. 1960. Energy utilization in the ruminant. Pages 183–197 in D. Lewis, ed., Digestive physiology and nutrition of the ruminant. Butterworth, London.

Blomfield, S. & D. Marr. 1970. How the cerebellum may be used. Nature 227:1224–1228.

Bloor, C. M. & A. S. Leon. 1970. Interaction of age and exercise on the heart and its blood supply. Lab. Invest. 22:160–165.

Blurton Jones, N. G. 1967. An ethological study of some aspects of social behavior of children in nursery school. Pages 347–368 in D. Morris, ed., Primate ethology. Aldine, Chicago.

Blurton Jones, N. G. 1972. Comparative aspects of mother-child contact. Pages 305–328 in N. G. Blurton Jones, ed., Ethological studies of child behavior. University Press, Cambridge.

Blurton Jones, N. G. 1975. Ethology, anthropology and childhood. Pages 69–92 in R. Fox, ed., Biosocial anthropology. John Wiley & Sons, N.Y.

Blurton Jones, N. G. & M. J. Konner. 1973. Sex differences in behaviour of London and Bushman children. Pages 689–750 in R. P. Michael & J. H. Crook, eds., Comparative ecology and behaviour of primates. Academic Press, London & N.Y.

Bolles, R. C. & P. J. Woods. 1964. The ontogeny of behavior in the albino rat. J. Anim. Behav. 12:427–441.

Bolwig, N. 1959. A study of the behavior of the Chacma baboon (Papio ursinus). Behaviour 14:136–163.

Bolwig, N. 1963. Bringing up a young monkey (Erythrocebus patas). Behaviour 21:300–330.

Bolwig, N. 1965. Observations on the early behaviour of a young African elephant, Loxodonta africana. Int. Zoo Yb. 5:149–152.

Bond, R. M. 1942. Development of young goshawks. Wilson Bull. 54:81–88.

Bondarchuk, L. S., S. K. Matisheva & R. N. Skibnevskii. 1976. Development of behavior in young bottlenose dolphins (Tursiops truncatus). Zool. Zh. 55:276–281.

Boorman, S. A. & P. R. Levitt. 1973. A frequency-dependent natural selection model for the evolution of social cooperation networks. Proc. Nat. Acad. Sci. 70:187–189.

Bopp, P. 1968. Biologische Studien an Humboldts Marimondas (Atele belzebuth E. Geoffr.). Zool. Gart. 36:160–172.

Borell, A. E. & R. Ellis. 1934. Mammals of the Ruby Mountains region of North-Eastern Nevada. J. Mammal. 15:12–44.

Borowsky, R. 1978. Social inhibition of maturation in natural populations of Xiphophorus variatus (Pisces: Poeciliidae). Science 201:933–935.

Bourlière, F., M. Bertrand & C. Hunkeler. 1969. L'ecologie de la Mone de Lowe en Côte d'Ivoire. La Terre et la Vie 2:135–163.

Bourlière, F., C. Hunkeler & M. Bertrand. 1970. Ecology and behavior of Lowe's guenon (Cercopithecus campbelli lowei) in the Ivory Coast. Pages 297–350 in J. R. Napier & P. H. Napier, eds., Old world monkeys: evolution, systematics, and behavior. Academic Press, N.Y. & London.

Box, H. O. 1975a. Quantitative studies of behaviour within captive groups of marmoset monkeys (Callithrix jacchus). Primates 16:155–174.

Box, H. O. 1975b. A social developmental study of young monkeys (Callithrix jacchus) within a captive family group. Primates 16:419–435.

Bradbury, J. W. & S. L. Vehrecamp. 1976a. Social organization and foraging in emballonurid bats. I. Field studies. Behav. Ecol. Sociobiol. 1:337–381.

Bradbury, J. W. & S. L. Vehrencamp. 1976b. Social organization and foraging in emballonurid bats. II. A model for the determination of group size. Behav. Ecol. Sociobiol. 1:383–404.

Bradbury, J. W. & S. L. Vehrencamp. 1977a. Social organization and foraging in emballonurid bats. III. Mating systems. Behav. Ecol. Sociobiol. 2:1–17.

Bradbury, J. W. & S. L. Vehrencamp. 1977b. Social organization and foraging in emballonurid bats. IV. Parental investment patterns. Behav. Ecol. Sociobiol. 2:19–29.

Bradshaw, A. D. 1965. Evolutionary significance of phenotypic plasticity in plants. Adv. Genet. 13:115–155.

Bramblett, C. A. 1978. Sex differences in the acquisition of play among juvenile vervet monkeys. Pages 33–48 in E. O. Smith, ed., Social play in primates. Academic Press, N.Y.

Braun, H. 1952. Über das Unterscheidungsvermögen unbenannter Anzahlen bei Papageien. Z. Tierpsychol. 9:40–91.

Brazelton, T. B., B. Koslowski & M. Main. 1974. The origins of reciprocity. Pages 49–76 in M. Lewis & L. A. Rosenblum, eds., The effect of the infant on its caregiver. Wiley, N.Y.

Brazelton, T. B., E. Tronick, L. Adamson, H. Als & S. Wise. 1975. Early mother-infant reciprocity. Pages 137–149 in R. Porter & M. O'Connor, eds., Parent-infant interaction. Ciba Foundation Symposium no. 33 (new series). Associated Scientific Publishers, Amsterdam and American Elsevier, N.Y.

Brereton, J. L. 1971. Inter-animal control of space. Pages 69–91 in A. H. Esser, ed., Behavior and environment. Plenum, N.Y.

Breuggeman, J. A. 1978. The function of adult play in free-ranging *Macaca mulatta*. Pages 169–192 in E. O. Smith, ed., Social play in primates. Academic Press, N.Y.

Breidermann, L. 1967. Zum Ablauf der sommerlichen Aktivitätsperiodik des Gamswildes (*Rupicapra r. rupicapra* L. 1758) in freier Wildbahn. Zool. Gart. 33:279–305.

Brockelman, W. Y. 1975. Competition, the fitness of offspring and optimal clutch size. Am. Nat. 109:677–699.

Broderip, W. J. 1835. Observations on the habits, etc. of a male chimpanzee *Troglodytes niger*, Geoff., living in the Society's menagerie. Proc. Zool. Soc. Lond. 1833–1836:160–165.

Brody, E. J. & A. E. Brody. 1974. Breeding Müller's Bornean gibbon, *Hylobates lar mulleri*. Int. Zoo Yb. 14:110–113.

Brooks, A. 1961. A study of the Thomson's gazelle (*Gazella thomsonii* Guenther) in Tanganyika. H. M. Stationery Office, London.

Brooks, P. H. & D. J. A. Cole. 1970. The effect of the presence of a boar on the attainment of puberty in gilts. J. Reprod. Fertil. 23:435–440.

Brosset, A. 1968. Observations sur l'éthologie du tayra *Eira barbara* (Carnivora). La Terre et la Vie 22:29–50.

Brosset, A. 1973. Etude comparative de l'ontogenèse des comportements chez les rapaces Accipitridés et Falconidés. Z. Tierpsychol. 32:386–417.

Brown, D. H. & K. S. Norris. 1956. Observations of captive and wild cetaceans. J. Mammal. 37:311–326.

Brownlee, A. 1954. Play in domestic cattle: an analysis of its nature. Br. Vet. J. 110:48–68.

Brukoff, J. M. 1972. On being mother to a tiger. Field Mus. Nat. Hist. Bull. 43(11):6–8.

Bruner, J. S. 1969a. Eye, hand, and mind. Pages 223–235 in D. Elkind & J. H. Flavell, eds., Studies in cognitive development: essays in honor of Jean Piaget. Oxford University Press, N.Y.

Bruner, J. S. 1969b. Processes of growth in infancy. Pages 205–228 in A. Ambrose, ed., Stimulation in early infancy. Academic Press, London.

Bruner, J. S. 1970. The growth and structure of skill. Pages 63–94 in K. J. Connolly, ed., Mechanisms of motor skill development. Academic Press, N.Y.

Bruner, J. S. 1972. Nature and uses of immaturity. Am. Psychol. 27:687–708.

Bruner, J. S. 1973a. Competence in infants. Pages 297–308 in J. M. Anglin, ed., Beyond the information given. Norton, N.Y.

Bruner, J. S. 1973b. Organization of early skilled action. Child Dev. 44:1–11.

Bruner, J. S. 1976. Introduction. Pages 14–24 in J. S. Bruner, A. Jolly & K. Sylva, eds., Play: its role in development and evolution. Basic Books, N.Y.

Bruner, J. S. & B. M. Bruner. 1968. On voluntary action and its hierarchical structure. Int. J. Psychol. 3:239–255.

Bruner, J. S., J. J. Goodnow & G. A. Austin. 1956. A study of thinking. Wiley, N.Y.

Bruner, J. S., A. Jolly & K. Sylva, eds. 1976. Play: its role in development and evolution. Basic Books, N.Y.

Bruner, J. S. & V. Sherwood. 1976. Peekaboo and the learning of rule structures. Pages 277–285 in J. S. Bruner, A. Jolly & K. Sylva, eds., Play: its role in development and evolution. Basic Books, N.Y.

Bruns, E. H. 1977. Winter behavior of pronghorns in relation to habitat. J. Wildl. Manag. 41:560–571.

Bubenik, A. B. 1965. Beitrag zur Geburtskunde und zu den Mutter-Kind-Beziehungen des Reh- (*Capreolus capreolus* L.) und Rotwildes (*Cervus elaphus* L.). Z. Saeugetierkd. 30:65–128.

Budd, A., L. G. Smith & F. W. Shelley. 1943. On the birth and upbringing of the female chimpanzee "Jaqueline" (born 28/11/37 in the Zoological Gardens, London). Proc. Zool. Soc. Lond. 113:1–20.

Budich, G. 1971. Bei Kolkraben beobachtet. Falke 18:101.

Budnitz, N. & K. Dainis. 1975. *Lemur catta:* Ecology and behavior. Pages 219–235 in I. Tattersall & R. W. Sussman, eds., Lemur biology. Plenum Press, N.Y.

Buechner, H. K. 1950. Life history, ecology, and range use of the pronghorn antelope in Trans-Pecos Texas. Am. Midl. Nat. 43:257–354.

Buechner, H. K., S. F. Mackler, H. R. Stroman & W. A. Xanten. 1975. Birth of an Indian rhinoceros, *Rhinoceros unicornis,* at the National Zoological Park, Washington. Int. Zoo Yb. 15:160–165.

Bühler, K. 1924. Die geistige Entwicklung des Kindes. Fischer, Jena.

Buirski, P., H. Kellerman, R. Plutchik, R. Weininger & N. Buirski. 1973. A field study of emotions, dominance, and social behavior in a group of baboons (*Papio anubis*). Primates 14:67–78.

Buller, A. J. & R. Pope. 1977. Plasticity in mammalian skeletal muscle. Philos. Trans. R. Soc. Lond. Ser. B Biol. Sci. 278:295–305.

Burckhardt, D. 1958. Observations sur la vie sociale du cerf (*Cervus elaphus*) au parc national Suisse. Mammalia 22:226–244.

Burgers, J. M. 1966. Curiosity and play: basic factors in the development of life. Science 154:1680–1681.

Burghardt, G. 1977. Of iguanas and dinosaurs: social behavior and communication in neonate reptiles. Am. Zool. 17:177–190.

Burghardt, G. 1978. Behavioral ontogeny in reptiles: whence, whither and why. Pages 149–174 in G. M. Burghardt & M. Bekoff, eds., The development of behavior. Garland, N.Y.

Burghardt, G. M. & L. S. Burghardt. 1972. Notes on the behavioral development of two (female) black bear cubs: the first eight months. IUCN Bull. 23:207–220.

Burleigh, I. G. 1974. On the cellular regulation of growth and development in skeletal muscle. Biol. Rev. 49:267–320.

Burn, J. 1967. Play and exploration. Dev. Med. 9:347–348.

Burrell, H. 1974. The platypus. Rigby Ltd., Adelaide.

Burton, F. D. 1972. The integration of biology and behavior in the socialization of *Macaca sylvana* of Gibraltar. Pages 29–62 in F. E. Poirier, ed., Primate socialization. Random House, N.Y.

Burton, J. A., ed. 1973. Owls of the world. Dutton, N.Y.

Buytendijk, F. J. J. 1933. Wesen und Sinn des Spiels: Das Spielen des Menschen und der Tiere als Erscheinungsform der Lebenstriebe. (Het spel van mensch en dier als openbaring van levensdriften.) Wolff, Berlin.

Byers, J. A. 1977. Terrain preferences in the play behavior of Siberian ibex kids (*Capra ibex sibirica*). Z. Tierpsychol. 45:199–209.

Cade, T. J. 1953. Behavior of a young gyrfalcon. Wilson Bull. 65:26–31.

Cahalane, V. H. 1947. Mammals of North America. Macmillan, N.Y.

Caillois, R. 1961. Man, play and games. Free Press, N.Y.

Caldwell, D. K., M. C. Caldwell & D. W. Rice. 1966. Behavior of the sperm whale, *Physeter catodon* L. Pages 677–717 in K. S. Norris, ed., Whales, dolphins and porpoises. University of California Press, Berkeley & Los Angeles.

Cameron, A. W. 1967. Breeding behavior in a colony of Western Atlantic gray seals. Can. J. Zool. 45:161–173.

Candland, D. K., J. A. French & C. N. Johnson. 1978. Object-play: test of a categorized model by the genesis of object-play in *Macaca fuscata*. Pages 259–296 in E. O. Smith, ed., Social play in primates. Academic Press, N.Y.

Cane, V. R. 1978. On fitting low-order Markov chains to behaviour sequences. Anim. Behav. 26:332–338.

Canfield, C. A. 1866. On the habits of the prongbuck (*Antilocapra americana*), and the periodical shedding of its horns. Proc. Zool. Soc. Lond. 1866:105–110.

Carpenter, C. R. 1934. A field study of the behavior and social relations of red howling monkeys. Comp. Psychol. Monogr. 10:1–168.

Carpenter, C. R. 1935. Behavior of red spider monkeys in Panama. J. Mammal. 16:171–180.

Carpenter, C. R. 1937. An observational study of two captive mountain gorillas (Gorilla beringei). Hum. Biol. 9:175–196.

Carpenter, C. R. 1940. A field study in Siam of the behavior and social relations of the gibbon (Hylobates lar). Comp. Psychol. Monogr, 16:1–212.

Carpenter, C. R. 1965. The howlers of Barro Colorado Island. Pages 250–291 in I. DeVore, ed., Primate behavior. Holt, Rinehart, & Winston, N.Y.

Carpenter, C. R. 1969. Approaches to studies of the naturalistic communicative behavior in nonhuman primates. Pages 40–70 in T. A. Sebeok & A. Ramsay, eds., Approaches to animal communication. Mouton, The Hague.

Carpenter, C. R. 1971. Macaca fuscata (Cercopithecidae)—Play of the young. Encyclopedia Cinematographica Film E1467. Institut für den Wissenschaftlichen Film, Göttingen.

Carpenter, C. R. 1974. Activity characteristics of gibbons (Hylobates lar). Part 3: Social behavior. Psychological Cinema Register film PCR-2253K, Pennsylvania State University, University Park.

Cartmill, M. 1972. Arboreal adaptations and the origin of the order Primates. Pages 97–122 in R. H. Tuttle, ed., The functional and evolutionary biology of primates. Aldine-Atherton, Chicago.

Cartmill. M. 1974. Rethinking primate origins. Science 184:436–443.

Cartmill, M. 1975. Primate origins. Burgess Pub. Co., Minneapolis.

Casey, D. E. & T. W. Clark. 1976. Some spacing relations among the central males of a transplanted troop of Japanese macaques (Arashiyama West). Primates 17:433–450.

Caswell, H. 1978. Predator-mediated coexistence: a nonequilibrium model. Am. Nat. 112:127–154.

Caton, J. D. 1877. The antelope and deer of America. Forest and Stream Pub. Co., N.Y. 2nd ed.

Caughley, G. 1966. Mortality patterns in mammals. Ecology 47:906–918.

Cavalli-Sforza, L. L. & M. W. Feldman. 1973a. Cultural versus biological inheritance: phenotypic transmission from parents to children (A theory of the effect of parental phenotypes on children's phenotypes). Am. J. Hum. Genet. 25:618–637.

Cavalli-Sforza, L. L. & M. W. Feldman. 1973b. Models for cultural inheritance: I. Group mean and group variation. Theor. Pop. Biol. 4:42–55.

Chalmers, N. R. 1978. A comparison of play and non-play activities in feral olive baboons. Page 131–134 in D. J. Chivers & J. Herbert, eds., Recent advances in primatology, Vol. 1. Behaviour. Academic Press, N.Y.

Chamberlin, T. C. 1965. The method of multiple working hypotheses. Science 148:754–759.

Chance, M. R. A. & W. M. S. Russell. 1959. Protean displays: a form of allaesthetic behaviour. Proc. Zool. Soc. Lond. 132:65–70.

Chandler, C. F. 1975. Development and function of marking and sexual behavior in the Malagasy prosimian primate, *Lemur fulvus*. Primates 16:35–48.

Charles-Dominique, P. 1971. Eco-éthologie des prosimiens du Gabon. Rev. Biol. Gabonica 7:121–228.

Charles-Dominique, P. 1977. Ecology and behaviour of nocturnal primates: prosimians of equatorial West Africa. Trans. R. D. Martin. Columbia University Press, N.Y.

Charnov, E. L. & W. M. Schaffer. 1973. Life-history consequences of natural selection: Cole's result revisited. Am. Nat. 107:791–792.

Chartin, J. & F. Petter. 1960. Reproduction et élevage en captivité du ouistiti. Mammalia 24:153–154.

Chateaubriand, F.A.R. de. 1951. Mémoires d'outre-tombe. Édition nouvelle établie d'après l'édition originale et les deux dernières copies du texte. V. 1. Gallimard, Paris.

Cheney, D. L. 1977. The acquisition of rank and the development of reciprocal alliances among free-ranging immature baboons. Behav. Ecol. Sociobiol. 2:303–318.

Cheney, D. L. 1978. The play partners of immature baboons. Anim. Behav. 26:1038–1050.

Chepko, B. D. 1971. A preliminary study of the effects of play deprivation on young goats. Z. Tierpsychol. 28:517–526.

Chesler, P. 1969. Maternal influence in learning by observation in kittens. Science 166:901–903.

Chevalier-Skolnikoff, S. 1973. Visual and tactile communication in *Macaca arctoides* and its ontogenic development. Am. J. Phys. Anthropol. 38:515–518.

Chevalier-Skolnikoff, S. 1974. The ontogeny of communication in the stump-tail macaque. Contributions to Primatology, Vol. 2. Karger, Basel.

Chisholm, A. H. 1958. Bird wonders of Australia. Michigan State University Press, Lansing.

Chisholm, A. H. 1971. Further notes on tool-using by birds. Victorian Nat. 88:342–343.

Chivers, D. J. 1973. An introduction to the socio-ecology of Malayan forest primates. Pages 101–146 in R. R. Michael & J. H. Crook, eds., Comparative ecology and behaviour of primates. Academic Press, London & N.Y.

Chivers, D. J. 1974. The Siamang in Malaya: a field study of a primate in tropical rain forest. Contributions to Primatology, Vol. 4. Karger, Basel.

Chivers, D. J. 1976. Communication within and between family groups of siamang (*Symphalangus syndactylus*). Behaviour 57:116–135.

Christian, A. 1974. Fortpflanzungsbiologie und Verhalten bei *Cebuella pygmaea* und *Tamarin tamarin*. Z. Tierpsychol. Supp. 14.

Cicala, G. A., I. B. Albert & F. A. Ulmer, Jr. 1970. Sleep and other behaviours of the red kangaroo (*Megaleia rufa*). Anim. Behav. 18:787–790.

Claparède, E. 1934. Sur la nature et la fonction du jeu. Arch. Psychol. 24:350–369.

Clark, C. B. 1977. A preliminary report on weaning among chimpanzees of the Gombe National Park, Tanzania. Pages 235–260 in S. Chevalier-Skolnikoff & F. E. Poirier, eds., Primate bio-social development. Garland, N.Y.

Clark, D. L., J. R. Kreutzberg & F. K. W. Chee. 1977. Vestibular stimulation influence on motor development in infants. Science 196:1228–1229.

Cline, D. R., D. B. Siniff & A. W. Erickson. 1971. Underwater copulation of the Weddell seal. J. Mammal. 52:216–218.

Clive, J. 1976. Gibbon's humor. Daedalus 105:27–35.

Close, R. I. 1972. Dynamic properties of mammalian skeletal muscles. Physiol. Rev. 52:129–197.

Clutton-Brock, T. H. 1972. Feeding and ranging behaviour in the red colobus monkey. Ph.D. diss., Cambridge University.

Clutton-Brock, T. H. 1974. Activity patterns of red colobus (Colobus badius tephrosceles). Folia Primatol. 21:161–187.

Cobb, E. 1977. The ecology of imagination in childhood. Columbia University Press, N.Y.

Cody, M. & J. Diamond, eds. 1975. The ecology and evolution of communities. Harvard University Press, Cambridge, Mass.

Coe, M. J. 1967. "Necking" behaviour in the giraffe. J. Zool. Lond. 151:313–321.

Coelho, A. M., Jr. 1974. Socio-bioenergetics and sexual dimorphism in primates. Primates 15:263–269.

Coelho, A. M., Jr., C. A. Bramblett, L. B. Quick & S. S. Bramblett. 1976. Resource availability and population density in primates: a sociobioenergetic analysis of the energy budgets of Guatemalan howler and spider monkeys. Primates 17:63–80.

Cohen, J. E. 1976. Irreproducible results and the breeding of pigs. Bioscience 26:391–394.

Colbert, E. H. 1955. Evolution of the vertebrates. Wiley, N.Y.

Cole, L. C. 1954. The population consequences of life history phenomena. Q. Rev. Biol. 29:103–137.

Cole, L. W. 1912. Observations of the senses and instincts of the raccoon. J. Anim. Behav. 2:299–309.

Collard, R. R. 1967. Fear of strangers and play behavior in kittens with varied social experience. Child Dev. 38:877–891.

Collier, G. 1970. Work: A weak reinforcer. Trans. N.Y. Acad. Sci. 32:557–576.

Collins, L. R. & J. F. Eisenberg. 1972. Notes on the behaviour and breeding of pacaranas, Dinomys branickii, in captivity. Int. Zoo Yb. 12:108–114.

Conley, J. M. 1975. Activity pattern of Lemur fulvus. J. Mammal. 56:712–715.

Connolly, K. J. 1970. Skill development: problems and plans. Pages 3–21 in K. J. Connolly, ed., Mechanisms of motor skill development. Academic Press, N.Y.

Connolly, K. J., ed. 1971. Motor skills in infancy. Academic Press, N.Y.

Connolly, K. J. 1973. Factors influencing the manual skills of young children.

Pages 337–365. In R. A. Hinde & J. Stevenson-Hinde, eds., Constraints on learning: limitations and predispositions. Academic Press, London & N.Y.

Connolly, K. J. 1977. The nature of motor skill development. J. Human Movement Studies 3:128–143.

Connolly, K. J. & J. S. Bruner, eds. 1974. The growth of competence. Academic Press, N.Y.

Connolly, K. & J. Elliott. 1972. The evolution and ontogeny of hand function. Pages 329–383 in N. G. Blurton Jones, ed., Ethological studies of child behaviour. University Press, Cambridge.

Cooper, J. B. 1942. An exploratory study of African lions. Comp. Psychol. Monogr. 17:1–48.

Corbet, G. B. 1978. The mammals of the Palaearctic region. Cornell University Press, N.Y.

Corkill, N. L. 1929. On the occurrence of the cheetah (*Acononyx jubatus*) in Iraq. J. Bombay Nat. Hist. Soc. 33:700–702.

Coss, R. G. & A. Globus. 1979. Social experience affects the development of dendritic spines and branches on tectal interneurons in the jewel fish. Dev. Psychobiol. 12:347–358.

Coulon, J. 1971. Influence de l'isolement social sur le comportement du cobaye. Behaviour 38:93–120.

Cratty, B. J. 1964. Movement behavior and motor learning. Lea & Febiger, Philadelphia.

Crisler, L. 1958. Arctic wild. Harper, N.Y.

Croft, D. B. 1980. Social behaviour of the euro, *Macropus robustus,* in the Australian arid zone. Aust. J. Wildl. Res., in press.

Crook, J. H. 1967. Bleeding hearts and bone breakers. Film 40116 available through Audio-Visual Services, Pennsylvania State University, University Park.

Cropley, A. J. & E. Feuring. 1971. Training creativity in young children. Dev. Psychol. 4:105.

Cummins, M. S. & S. J. Suomi. 1976. Long-term effects of social rehabilitation in rhesus monkeys. Primates 17:43–51.

Cummins, R. A., P. J. Livesey, J. G. M. Evans & R. N. Walsh. 1977. A developmental theory of environmental enrichment. Science 197:692–694.

Cutter, W. L. 1957. A young jaguarundi in captivity. J. Mammal. 38:515–516.

Dagg, A. J. & J. B. Foster. 1976. The giraffe: its biology, behavior and ecology. Van Nostrand Reinhold, N.Y.

Dalquest, W. W. & J. H. Roberts. 1951. Behavior of young grisons in captivity. Am. Midl. Nat. 46:359–366.

Dalton, M. 1961. The adventures of Rikki Tikki. Wild Life, Nairobi 3:18–20.

Daly, M. 1976. Behavioral development in three hamster species. Dev. Psychobiol. 9:315–323.

Daly, M. & S. Daly. 1975. Behavior of *Psammomys obesus* (Rodentia: Gerbillinae) in the Algerian Sahara. Z. Tierpsychol. 37:298–321.

Daly, M. & M. Wilson. 1978. Sex, evolution, and behavior. Duxbury Press, N. Scituate, Mass.

Dansky, J. L. & I. W. Silverman. 1973. Effects of play on associative fluency in preschool-aged children. Dev. Psychol. 9:38–43.

Dansky, J. L. & I. W. Silverman. 1975. Play: a general facilitator of associative fluency. Dev. Psychol. 11:104.

Darling, F. F. 1937. A herd of red deer. Oxford University Press, London.

Darwin, C. R. 1874. The descent of man; and selection in relation to sex, 2nd ed. (1st ed., 1871). A. L. Burt, N.Y.

Darwin, C. R. 1877. A biographical sketch of an infant. Mind 2:285–294.

Darwin, C. R. 1896. Journal of researches into the natural history and geology of the countries visited during the voyage of H.M.S. Beagle round the world, under the command of Capt. Fitz Roy, R. N. New ed. Appleton, N.Y.

Darwin, C. R. 1898. The expression of the emotions in man and animals. Appleton, N.Y.

Dasmann, R. F. & R. D. Taber. 1956. Behavior of Columbian black-tailed deer with reference to population ecology. J. Mammal. 37:143–164.

Dathe, H. 1968. Zum Vorkommen mähnenloser Zebras. Zool. Gart. 35:67–68.

Davenport, R. K. 1967. The orang-utan in Sabah. Folia Primatol. 5:247–263.

David, J. H. M. 1973. The behaviour of the bontebok, *Damaliscus dorcas dorcas,* (Pallas 1766), with special reference to territorial behaviour. Z. Tierpsychol. 33:38–107.

David, J. H. M. 1975. Observations on mating behaviour, parturition, suckling and the mother-young bond in the bontebok (*Damaliscus dorcas dorcas*). J. Zool. Lond. 177:203–223.

David, K. 1940. Intelligenzversuche am Eichhörnchen. Z. Tierpsychol. 4:162–164.

Davies, N. B. & R. E. Green. 1976. The development and ecological significance of feeding techniques in the reed warbler (*Acrocephalus scirpaceus*). Anim. Behav. 24:213–229.

Davies, N. B. & T. R. Halliday. 1977. Optimal mate selection in the toad *Bufo bufo*. Nature 269:56–58.

Davis, J. A. 1979. Samaki. The story of an otter in Africa. E. P. Dutton, N.Y.

Davis, R. B., C. F. Herreid & H. L. Short. 1962. Mexican freetailed bats in Texas. Ecol. Monogr. 32:311–316.

Dawkins, M. & R. Dawkins. 1974. Some descriptive and explanatory stochastic models of decision-making. Pages 119–168 in D. J. McFarland, ed., Motivational control systems analysis. Academic Press, London.

Dawkins, R. 1976a. Hierarchical organisation: a candidate principle for ethology. Pages 7–54 in P. P. G. Bateson & R. A. Hinde, eds., Growing points in ethology. Cambridge University Press, Cambridge & N.Y.

Dawkins, R. 1976b. The selfish gene. Oxford University Press, N.Y.

Dawkins, R. & M. Dawkins. 1973. Decisions and the uncertainty of behavior. Behaviour 45:83–103.

Dawson, T. J. 1972. Primitive mammals and patterns in the evolution of thermoregulation. Pages 1–18 in J. Bligh & R. E. Moore, eds., Essays on temperature regulation. North-Holland, Amsterdam & London and American Elsevier, N.Y.

Dawson, W. L. & J. H. Bowles. 1909. The birds of Washington. Occidental Publishing Co., Seattle.

Deag, J. M. 1973. Intergroup encounters in the wild Barbary macaque *Macaca sylvanus* L. Pages 315–373 in R. P. Michael & J. H. Crook, eds., Comparative ecology and behaviour of primates. Academic Press, London & N.Y.

DeGhett, V. J. 1970. Ontogeny of play fighting in the Mongolian gerbil (*Meriones unguiculatus*). Am. Zool. 10:293.

Delacour, J. 1933. On the Indochinese gibbons (*Hylobates concolor*). J. Mammal. 14:71–73.

Delius, J. D. 1973. Agonistic behavior of juvenile gulls, a neuroethological study. Anim. Behav. 21:236–246.

Demetrius, L. 1969. The sensitivity of population growth rate to perturbations in the life cycle components. Math. Biosci. 4:129–136.

dePoncins, E., Baron. 1895. Shooting *Ovis polii* on the Pamirs. J. Bombay Nat. Hist. Soc. 10:53–62.

Derscheid, J. M. 1947. Strange parrots. Avic. Mag. 53:44–49.

DeVore, I. 1963. Mother-infant relations in free-ranging baboons. Pages 305–335 in H. L. Rheingold, ed., Maternal behavior in mammals. Wiley, N.Y.

DeVore, I. & M. J. Konner. 1974. Infancy in hunter-gatherer life: an ethological perspective. Pages 113–141 in N. F. White, ed., Ethology and psychiatry. University of Toronto Press, Toronto & Buffalo, N.Y.

DeVos, A. 1958. Summer observations on moose behavior in Ontario. J. Mammal. 39:128–139.

DeVos, A. 1960. Behavior of barren ground caribou on their calving grounds. J. Wildl. Manage. 24:250–258.

DeVos, A., P. Brokx & V. Geist. 1967. A review of social behavior of the North American cervids during the reproductive period. Am. Midl. Nat. 77:390–417.

DeVos, A. & R. J. Dowsett. 1966. The behaviour and population structure of three species of the genus *Kobus*. Mammalia 30:30–55.

Diamond, J. M. 1975. Assembly of species communities. Pages 342–444 in M. L. Cody & J. M. Diamond, eds., Ecology and evolution of communities. Belknap Press of Harvard University Press, Cambridge, Mass.

Dieterlen, F. 1959. Das Verhalten des syrischen Goldhamsters (*Mesocricetus auratus* Waterhouse). Z. Tierpsychol. 16:47–103.

Dieterlen, F. 1965. Von der Lebensweise und dem Verhalten der Felsenmaus, *Apodemus mystacinus* (Danford & Alston, 1877), nebst Beiträgen zur vergleichenden Ethologie der Gattung *Apodemus*. Säugetierk. Mitt. 13:153–161.

Dillard, A. 1973. Pilgrim at Tinker Creek. Bantam Books, N.Y.

Dimelow, E. J. 1963. The behavior of the hedgehog (*Erinaceus europaeus* L.)

in the routine of life in captivity. Proc. Zool. Soc. Lond. 141:281–289.

Dittrich, L. 1962. Versuchte künstliche Aufzucht eines Flußpferdes (*Hippopotamus amphibius* L.). Zool. Gart. 26:175–190.

Dittrich, L. 1967. Breeding the black rhinoceros, *Diceros bicornis,* at Hanover Zoo. Int. Zoo Yb. 7:161–162.

Dixson, A. F. 1977. Observations on the displays, menstrual cycles and sexual behaviour of the "Black ape" of Celebes (*Macaca nigra*). J. Zool. Lond. 182:63–84.

Dixson, A. F., D. M. Scruton & J. Herbert. 1975. Behaviour of the talapoin monkey (*Miopithecus talapoin*) studied in groups, in the laboratory. J. Zool. Lond. 176:177–210.

Dobroruka, L. J. & R. Horbowyjova. 1972. Poznamky k etologii pekari paskovaneho, *Dicotyles tajacu* (Linnaeus, 1766) v zoologicke zahrade v Praze. (Notes on the ethology of collared peccary, *Dicotyles tajacu* in the Prague Zoological Garden). Lynx (Prague) 13:85–94.

Dobzhansky, T. 1962. Mankind evolving. Yale University Press, New Haven.

Dodsworth, P. T. L. 1913. Notes on some mammals found in the Simla District, the Simla Hill States, and Kalka and adjacent country. J. Bombay Nat. Hist. Soc. 22:726–748.

Döhl, J. & D. Podolczak. 1973. Versuche zur Manipulierfreudigkeit von zwei jungen Orang-Utans (*Pongo pygmaeus*) im Frankfurter Zoo. Zool. Gart. 43:81–94.

Dolan, K. J. 1976. Metacommunication in the play of a captive group of Sykes monkeys. Paper presented at the 45th Annual Meeting, American Association of Physical Anthropologists, St. Louis.

Dolgin, K. G. 1978. The sequencing and activity patterns found in chimpanzee play behavior. M.A. thesis, University of Pennsylvania.

Dolhinow, P. J. & N. Bishop. 1970. The development of motor skills and social relationships among primates through play. Minn. Symp. Child Psychol. 4:141–198.

Dominis, J. & M. Edey. 1968. The cats of Africa. Time-Life Books, N.Y.

Donald, C. H. 1948. Jackals. J. Bombay Nat. Hist. Soc. 47:721–726.

Dorst, J. 1970. A field guide to the larger mammals of Africa. Houghton Mifflin, Boston.

Doughty, C. M. 1923. Travels in Arabia Deserta, 2 vols. (1st ed. 1888.) Boni & Liveright, N.Y.

Douglas-Hamilton, I. 1972. On the ecology and behaviour of the African elephant: the elephants of Lake Manyara. Ph.D. diss., Oriel College, Oxford.

Douglas-Hamilton, I. & O. Douglas-Hamilton. 1975. Among the elephants. Viking, N.Y.

Doyle, G. A. 1974a. The behaviour of the lesser bushbaby. Pages 213–231 in R. D. Martin, G. A. Doyle & A. C. Walker, eds., Prosimian biology. University of Pittsburgh Press, Pittsburgh.

Doyle, G. A. 1974b. Behavior of prosimians. Pages 155–353 in A. M. Schrier

& F. Stollnitz, eds., Behav. Nonhuman Prim., Vol. 5. Academic Press, N.Y.

Draper, P. 1976. Social and economic constraints on child life among the !Kung. Pages 199–217 in R. B. Lee & I. DeVore, eds., Kalahari hunter-gatherers: studies of the !Kung San and their neighbors. Harvard University Press, Cambridge, Mass. & London.

Draper, W. A. 1967. A behavioural study of the home-cage activity of the white rat. Behaviour 28:280–306.

Driver, P. M. & D. A. Humphries. 1970. Protean displays as inducers of conflict. Nature 226:968–969.

Drüwa, P. 1977. Beobachtungen zur Geburt und natürlichen Aufzucht von Waldhunden (Speothos venaticus) in der Gefangenschaft. Zool. Gart. 47:109–137.

Dubkin, L. 1952. The white lady. Putnam, N.Y.

Dubost, G. 1971. Observations éthologiques sur le Muntjak (Muntiacus muntjak Zimmermann 1780 et M. reevesi Ogilby 1839) en captivité et semiliberté. Z. Tierpsychol. 28:387–427.

Dubost, G. 1975. Le comportement du chevrotain africain, Hyemoschus aquaticus Ogilby (Artiodactyla, Ruminantia). Z. Tierpsychol. 37:403–501.

Dubost, G. & H. Genest. 1974. Le comportement social d'une colonie de Maras Dolichotis patagonum Z. dans le Parc de Branféré. Z. Tierpsychol. 35:225–302.

Dücker, G. 1960. Beobachtungen über das Paarungsverhalten des Ichneumons (Herpestes ichneumon L.) Z. Säugetierk. 25:47–51.

Dücker, G. 1962. Brutpflegeverhalten und Ontogenese des Verhaltens bei Surikaten (Suricata suricatta Schreb., Viverridae). Behaviour 19:305–340.

Dücker, G. 1968. Beobachtungen am kleinen Grison, Galictis (Grisonella) cuja (Molina). Z. Säugetierk. 33:288–297.

Dücker, G. 1971. Gefangenschaftsbeobachtungen an Pardelrollern Nandina binotata (Reinwardt). Z. Tierpsychol. 28:77–89.

Du Mond, F. V. 1968. The squirrel monkey in a seminatural environment. Pages 87–145 in L. A. Rosenblum & R. W. Cooper, eds., The squirrel monkey. Academic Press, N.Y.

Du Mond, F. V. 1970. Notes on primates in the Asiatic Primate Grotto at Miami Monkey Jungle. Int. Zoo Yb. 10:131–133.

Dunbar, R. I. M. & E. P. Dunbar. 1974. Ecological relations and niche separation between sympatric terrestrial primates in Ethiopia. Folia Primatol. 21:36–60.

Dunbar, R. & P. Dunbar. 1975. Social dynamics of gelada baboons. Contributions to Primatology 6. Karger, Basel.

Dunn, J. 1976. How far do early differences in mother-child relations affect later development? Pages 481–496 in P. P. G. Bateson & R. A. Hinde, eds., Growing points in ethology. University Press, Cambridge.

Dunn, J. & C. Wooding. 1977. Play in the home and its implications for learning. Pages 45–58 in B. Tizard & D. Harvey, eds., Biology of play. Heinemann, London & Lippincott, Philadelphia.

Durrell, G. 1966. Two in the bush. Viking Press, N.Y.

East, K. & J. D. Lockie. 1964. (Notes on British mammals No. 9). Observations on a family of weasels (Mustela nivalis) bred in captivity. Proc. Zool. Soc. Lond. 143:359–363.

East, K. & J. D. Lockie. 1965. Further observations on weasels (Mustela nivalis) and stoats (Mustela erminea) born in captivity. J. Zool. Lond. 147:234–238.

Eaton, R. L. 1969. Notes on breathing rates in wild cheetahs. Mammalia 33:543–544.

Eaton, R. L. 1974. The cheetah: the biology, ecology, and behavior of an endangered species. Van Nostrand Reinhold, N.Y.

Ebhardt, H. 1954. Verhaltensweisen von Islandpferden in einem norddeutschen Freigelände. Säugetierk. Mitt. 2:145–154.

Eck, S. 1969. Über das Verhalten eines im Dresdener Zoologischen Garten aufgezogenen Brillenbären (Tremarctos ornatus [Cuv.]) Zool. Gart. 37:81–92.

Eckerman, C. O., J. L. Whatley & S. L. Kutz. 1975. Growth of social play with peers during the second year of life. Dev. Psychol. 11:42–49.

Edgerton, V. R. 1978. Mammalian muscle fiber types and their adaptability. Am. Zool. 18:113–126.

Editors. 1935. The wild animals of the Indian Empire and the problem of their preservation. Part III. J. Bombay Nat. Hist. Soc. 37, Supp.: 112–188.

Editors. 1959. Notes on a tame Takin. J. Bombay Nat. Hist. Soc. 56:128–129.

Egan, J. 1976. Object-play in cats. Pages 161–165 in J. S. Bruner, A. Jolly & K. Sylva, eds., Play: its role in development and evolution. Basic Books, N.Y.

Egbert, A. L. & A. W. Stokes. 1976. The social behaviour of brown bears on an Alaskan salmon stream. Pages 41–56 in M. R. Pelton, J. W. Lentfer & G. E. Falk, eds., Bears—Their biology and management. Papers of the Third International Conference on Bear Research and Management. IUCN Publications (N.S.) No. 40. Morges, Switzerland.

Ehlers, K. 1964. Sorgen mit Eisbärennachwuchs. Zool. Gart. 29:231–240.

Ehrat, H., H. Wissdorf & E. Isenbügel. 1974. Postnatale Entwicklung und Verhalten von Meriones unguiculatus (Milne Edwards, 1867) vom Zeitpunkt der Geburt bis zum Absetzen der Jungtiere im Alter von 30 Tagen. Z. Säugetierk. 39:41–50.

Ehrlich, A. & A. Musicant. 1977. Social and individual behaviors in captive slow lorises. Behaviour 60:195–220.

Eibl-Eibesfeldt, I. 1950a. Beiträge zur Biologie der Haus- und der Ährenmaus nebst einigen Beobachtungen an anderen Nagern. Z. Tierpsychol. 7:558–587.

Eibl-Eibesfeldt, I. 1950b. Über die Jugendentwicklung des Verhaltens eines männlichen Dachses (Meles meles L.) unter besonderer Berücksichtigung des Spieles. Z. Tierpsychol. 7:327–355.

Eibl-Eibesfeldt, I. 1951a. Beobachtungen zur Fortpflanzungsbiologie und Jugendentwicklung des Eichhörnchens (Sciurus vulgaris L.) Z. Tierpsychol. 8:370–400.

Eibl-Eibesfeldt, I. 1951b. Gefangenschaftsbeobachtungen an der persischen Wüstenmaus (*Meriones persicus persicus* Blanford): ein Beitrag zur vergleichenden Ethologie der Nager. Z. Tierpsychol. 8:400–423.

Eibl-Eibesfeldt, I. 1952. Einige Beobachtungen an einer in Freiheit gehaltenen weiblichen Biberratte (*Myocastor coypus*). Zool. Gart. 19:277–283.

Eibl-Eibesfeldt, I. 1953. Zur Ethologie des Hamsters (*Cricetus cricetus* L.) Z. Tierpsychol. 10:204–254.

Eibl-Eibesfeldt, I. 1955. Ethologische Studien am Galápagos-Seelöwen, *Zalophus wollebaeki* Sivertsen. Z. Tierpsychol. 12:286–303.

Eibl-Eibesfeldt, I. 1967. Concepts of ethology and their significance in the study of human behavior. Pages 127–146 in H. W. Stevenson, E. H. Hess & H. L. Rheingold, eds., Early behavior: Comparative and developmental approaches. Wiley, N.Y.

Eibl-Eibesfeldt, I. 1970. Ethology: the biology of behavior. Holt, Rinehart & Winston, N.Y.

Eibl-Eibesfeldt, I. 1973. Taubblind geborenes Maedchen (Deutschland): Explorierverhalten und Spiel. Homo 24:48–49.

Eibl-Eibesfeldt, I. & H. Sielmann. 1962. Beobachtungen an Spechtfinken *Cactospiza pallida* (Sclater und Salvin) J. Ornithol. 103:92–101.

Einarsen, A. S. 1948. The pronghorn antelope and its management. Wildlife Management Institute, Washington, D.C.

Einon, D. & M. Morgan. 1976. Habituation of object contact in socially-reared and isolated rats (*Rattus norvegicus*). Anim. Behav. 24:415–420.

Einon, D. & M. Morgan. 1977. A critical period for social isolation in the rat. Dev. Psychobiol. 10:123–132.

Einon, D. & M. Morgan. 1978a. Early isolation produces enduring hyperactivity in the rat, but no effect upon spontaneous alternation. Q. J. Exp. Psychol. 30:151–156.

Einon, D. & M. Morgan. 1978b. Habituation under different levels of stimulation in socially reared and isolated rats: a test of the arousal hypothesis. Behav. Biol. 22:553–558.

Einon, D., M. Morgan & C. C. Kibbler. 1978. Brief periods of socialization and later behavior in the rat. Dev. Psychobiol. 11:213–225.

Einstein, A. 1945. Letter. In J. Hadamard, The psychology of invention in the mathematical field. Princeton University Press, Princeton, N.J.

Eisenberg, J. F. 1975. Phylogeny, behavior, and ecology in the Mammalia. Pages 47–68 in W. P. Luckett & F. S. Szalay, eds., Phylogeny of the primates: a multidisciplinary approach. Plenum, N.Y.

Eisenberg, J. F. 1976. Communication mechanisms and social integration in the black spider monkey, *Ateles fusciceps robustus,* and related species. Smithson. Contr. Zool. 213.

Eisenberg, J. F., L. R. Collins & C. Wemmer. 1975. Communication in the Tasmanian devil (*Sarcophilus harrisii*) and a survey of auditory communication in the Marsupiala. Z. Tierpsychol. 37:379–399.

Eisenberg, J. F. & E. Gould. 1970. The tenrecs: a study in mammalian behavior and evolution. Smithsonian Institution Press, Washington, D.C.

Eisenberg, J. F. & R. E. Kuehn. 1966. The behavior of *Ateles geoffroyi*. Smithson. Misc. Collect. 151:1–63.

Eisenberg, J. F. & P. Leyhausen. 1972. The phylogenesis of predatory behavior in mammals. Z. Tierpsychol. 30:59–93.

Eisenberg, J. F., G. M. McKay & M. R. Jainudeen. 1971. Reproductive behavior of the Asiatic elephant (*Elephas maximus maximus* L.). Behaviour 38:193–225.

Eisenberg, J. F., N. A. Muckenhirn & R. Rudran. 1972. The relation between ecology and social structure in primates. Science 176:863–874.

Ekblom, B. 1969a. Effect of physical training in adolescent boys. J. Appl. Physiol. 27:350–355.

Ekblom, B. 1969b. Effect of physical training on oxygen transport in man. Acta Physiol. Scand. Supp. 328.

Elbel, E. R. & W. J. Mikols. 1972. The effects of passive or active warmup upon certain physiological measures. Int. Z. Angew. Physiol. 31:41–52.

Eldredge, N. & S. J. Gould. 1972. Punctuated equilibria: an alternative to phyletic gradualism. Pages 82–115 in T. J. M. Schopf, ed., Models in paleobiology. Freeman, Cooper & Company, San Francisco.

Ellefson, J. H. 1966. A natural history of primates in the Malay peninsula. Ph.D. diss., University of California, Berkeley.

Ellefson, J. O. 1968. Territorial behavior in the common white-handed gibbon *Hylobates lar* Linn. Pages 180–199 in P. Jay, ed., Primates: studies in adaptation and variability. Holt, Rinehart & Winston, N.Y.

Elliot, D. G. 1871. A monograph of the Felidae. Privately printed by the author.

Elliott, J. M. & K. J. Connolly. 1974. Hierarchical structure in skill development. Pages 135–168 in K. J. Connolly & J. S. Bruner, eds., The growth of competence. Academic Press, London & N.Y.

Elliott, R. C. 1976. Observations on a small group of mountain gorillas (*Gorilla gorilla beringei*). Folia Primatol. 25:12–24.

Emlen, J. M. 1970. Age specificity and ecological theory. Ecology 51:588–601.

Emlen, J. M. 1975. Niches and genes: some further thoughts. Amer. Nat. 109:472–476.

Emory, G. R. 1975. The patterns of interaction between the young males and group members in captive groups of *Mandrillus sphinx* and *Theropithecus gelada*. Primates 16:317—334.

Emory, G. R. 1976. Aspects of attention, orientation, and status hierarchy in mandrills (*Mandrillus sphinx*) and gelada baboons (*Theropithecus gelada*). Behaviour 59:70–87.

Encke, W., R. Gandras & H.-J. Bienick. 1970. Beobachtungen am Mähnenwolf (*Chrysocyon brachyurus*). Zool. Gart. 38:47–67.

Engesser, U. 1977. Socialisation junger Wellensittiche (*Melopsittacus undulatus* Shaw). Z. Tierpsychol. 43:68–105.

English, W. L. 1934. Notes on the breeding of a Douroucouli (*Aotus trivirgatus*). Proc. Zool. Soc. Lond. 1934:143–144.

Epple, G. 1968. Comparative studies on vocalization in marmoset monkeys (Hapalidae). Folia Primatol. 8:1–40.

Erikson, E. H. 1977. Toys and reasons: stages in the ritualization of experience. Norton, N.Y.

Eriksson, B. 1972. Physical training, oxygen supply and muscle metabolism in 11–13 year old boys. Acta Physiol. Scand. Supp. 384.

Espmark, Y. 1969. Mother-young relations and development of behaviour in roe deer (*Capreolus capreolus* L.). Viltrevy 6:462–540.

Espmark, Y. 1971. Mother-young relationship and ontogeny of behaviour in reindeer (*Rangifer tarandus* L.). Z. Tierpsychol. 29:42–81.

Estes, R. D. 1967. The comparative behavior of Grant's and Thomson's gazelles. J. Mammal. 48:189–209.

Estes, R. D. 1969. Territorial behavior of the wildebeest (*Connochaetes taurinus* Burchell, 1823). Z. Tierpsychol. 26:284–370.

Estes, R. D. 1974. Social organization of the African Bovidae. Pages 166–205 in V. Geist & F. Walther, eds., The behavior of ungulates and its relation to management. IUCN Pubs., Morges, Switzerland.

Estes, R. D. 1976. The significance of breeding synchrony in the wildebeest. E. Afr. Wildl. J. 14:135–152.

Estes, R. D. & J. Goddard. 1967. Prey selection and hunting behavior of the African wild dog. J. Wildl. Manage. 31:52–70.

Ewer, R. F. 1963. The behavior of the meerkat, *Suricata suricatta* (Schreber) Z. Tierpsychol. 20:570–607.

Ewer, R. F. 1966. Juvenile behaviour in the African ground squirrel, *Xerus erythropus* (E. Geoff.). Z. Tierpsychol. 23:190–216.

Ewer, R. F. 1967. The behaviour of the African giant rat (*Cricetomys gambianus* Waterhouse). Z. Tierpsychol. 24:6–79.

Ewer, R. F. 1968a. Ethology of mammals. Plenum, N.Y.

Ewer, R. F. 1968b. A preliminary survey of the behaviour in captivity of the dasyurid marsupial, *Sminthopsis crassicaudata* (Gould). Z. Tierpsychol. 25:319–365.

Ewer, R. F. 1971. The biology and behaviour of a free-living population of black rats. Anim. Behav. Monogr. 4:125–174.

Ewer, R. F. 1973. The carnivores. Cornell University Press, Ithaca, N.Y.

Ewer, R. F. 1975. Why study small mammals? Int. Zoo Yb. 15:1–4.

Ewer, R. F. & C. Wemmer. 1974. The behaviour in captivity of the African civet, *Civettictis civetta* (Schreber). Z. Tierpsychol. 34:359–394.

Fady, J. C. 1969. Les jeux sociaux: le compagnon de jeux chez les jeunes. Observations chez *Macaca irus*. Folia Primatol. 11:134–143.

Faegri, K. & L. van der Pijl. 1972. The principles of pollination ecology, 2nd ed. Pergamon, Elmsford.

Fagen, R. 1972a. An optimal life-history strategy in which reproductive effort decreases with age. Am. Nat. 106:258–261.

Fagen, R. 1972b. The paradox of play. Unpublished manuscript prepared for Biology 244, Sociobiology, Harvard University.

Fagen, R. 1974a. Selective and evolutionary aspects of animal play. Am. Nat. 108:850–858.

Fagen, R. 1974b. Theoretical bases for the evolution of play in animals. Ph.D. diss., Division of Engineering and Applied Physics, Harvard University, Cambridge, Mass.

Fagen, R. 1975. Exercise, play, and physical training in animals. Paper presented at annual meeting of Animal Behavior Society, Wilmington, North Carolina.

Fagen, R. 1976a. Exercise, play, and physical training in animals. Pages 189–219 in P. P. G. Bateson & P. H. Klopfer, eds., Perspectives in ethology, Vol. 2. Plenum, N.Y.

Fagen, R. 1976b. Modelling how and why play works. Pages 96–115 in J. S. Bruner, A. Jolly & K. Sylva, eds., Play: its role in development and evolution. Basic Books, N.Y.

Fagen, R. 1977. Selection for optimal age-dependent schedules of play behavior. Am. Nat. 111:395–414.

Fagen, R. 1978a. Evolutionary biological models of animal play behavior. Pages 385–404 in G. M. Burghardt & M. Bekoff, eds., The development of behavior: Comparative and evolutionary aspects. Garland STPM, N.Y.

Fagen, R. M. 1978b. Population structure and social behavior in the domestic cat (Felis catus). Carnivore Genetics Newsletter 3:276–281.

Fagen, R. M. 1978c. Social behavior: A quantitative study. Review of D. Symons, Play and aggression. Science 200:658–659.

Fagen, R. M. & T. K. George. 1977. Play behavior and exercise in young ponies (Equus caballus L.). Behav. Ecol. Sociobiol. 2:267–269.

Fagen, R. M. & K. S. Wiley. 1978. Felid paedomorphosis, with special reference to Leopardus. Carnivore 1:72–81.

Fagen, R. M. & D. Y. Young. 1978. Temporal patterns of behavior: durations, intervals, latencies, and sequences. Pages 79–114 in P. W. Colgan, ed., Quantitative ethology. Wiley-Interscience, N.Y.

Farentinos, R. C. 1971. Some observations on the play behavior of the Steller sea lion (Eumetopias jubata). Z. Tierpsychol. 28:428–438.

Faugier, C. & B. Condé. 1973. Observations au cours de l'élevage au biberon de Genetta genetta L. Mammalia 37:515–516.

Faust, R. & I. Faust. 1959. Bericht über Aufzucht und Entwicklung einses isolierten Eisbären, Thalarctos maritimus (Phipps). Zool. Gart. 25:143–165.

Fedigan, L. M. 1972a. Roles and activities of male geladas (Theropithecus gelada). Behaviour 41:82–90.

Fedigan, L. M. 1972b. Social and solitary play in a colony of vervet monkeys (Cercopithecus aethiops). Primates 13:347–364.

Fedigan, L. M. 1976. A study of roles in the Arashiyama West troop of Japanese monkeys (Macaca fuscata). Contributions to Primatology 9. Karger, Basel.

Fedigan, L. M. & L. Fedigan. 1977. The social development of a handicapped infant in a free-living troop of Japanese monkeys. Pages 205–222 in S. Chevalier-Skolnikoff & F. E. Poirier, eds., Primate bio-social development: Biological, social, and ecological determinants. Garland, N.Y.

Feist, J. D. & D. R. McCullough. 1976. Behavior patterns and communication in feral horses. Z. Tierpsychol. 41:337–371.

Feitelson, D. & G. S. Ross. 1973. The neglected factor—play. Hum. Dev. 16:202–223.

Feldman, M. W. & L. L. Cavalli-Sforza. 1976. Cultural and biological evolutionary processes, selection for a trait under complex transmission. Theor. Pop. Biol. 9:238–259.

Feldman, M. W. & L. L. Cavalli-Sforza. 1977. Quantitative inheritance, stabilizing selection and cultural evolution. Pages 761–777 in E. Pollak, O. Kempthorne & T. B. Bailey, Jr., eds., Proceedings of the international conference on quantitative genetics. Iowa State University Press, Ames.

Fellner, K. 1965. Natural rearing of clouded leopards, Neofelis nebulosa, at Frankfurt Zoo. Int. Zoo Yb. 5:111–112.

Fellner, K. 1968. Erste natürliche Aufzucht von Nebelpardern (Neofelis nebulosa) in einem Zoo. Zool Gart. 35:105–137.

Fenson, L., J. Kagan, R. B. Kearsley & P. R. Zelazo. 1976. Developmental progression of manipulative play in the first 2 years. Child Dev. 47:232–236.

Fenson, L., V. Sapper & D. G. Minner. 1974. Attention and manipulative play in the one-year-old child. Child Dev. 45:757–764.

Fentress, J. C. 1967. Observations on the behavioral development of a hand-reared male timber wolf. Am. Zool. 7:339–351.

Fentress, J. C. 1973. Specific and nonspecific factors in the causation of behavior. Pages 155–224 in P. P. G. Bateson & P. H. Klopfer, eds., Perspectives in Ethology, Vol. 1. Plenum, N.Y.

Fentress, J. C. 1978. Mus musicus: the developmental orchestration of selected movement patterns in mice. Pages 321–342 in G. M. Burghardt & M. Bekoff, eds., The development of behavior: comparative and evolutionary aspects. Garland, N.Y.

Ferchmin, P. A., E. L. Bennett & M. R. Rosenzweig. 1975. Direct contact with enriched environment is required to alter cerebral weights in rats. J. Comp. Physiol. Psychol. 88:360–367.

Ferchmin, P. A. & V. E. Eterović. 1977. Brain plasticity and environmental complexity: role of motor skills. Physiol. Behav. 18:455–461.

Ferchmin, P. A. & V. A. Eterović. 1979. Mechanism of brain growth by environmental stimulation. Science 205:522.

Ferchmin, P. A., V. A. Eterović & L. E. Levin. 1980. Genetic learning deficiency does not hinder environment-dependent brain growth. Physiol. Behav. 24:45–50.

Ferron, J. 1975. Solitary play of the red squirrel (Tamiasciurus hudsonicus) Can. J. Zool. 53:1495–1499.

Fichter, E. 1950. Watching coyotes. J. Mammal. 31:66–73.

Ficken, M. S. 1977. Avian play. Auk 94:573–582.

Fink, E. 1968. The oasis of happiness: toward an ontology of play. Yale French Studies 41:19–30.

Fischer, C. E. C. 1921. The habits of the grey mongoose. J. Bombay Nat. Hist. Soc. 28:274.

Fisher, E. M. 1940. Early life of a sea otter pup. J. Mammal. 21:132–137.

Fisher, J. 1954. Evolution and bird sociality. Pages 71–83 in J. Huxley, A. C.

Hardy & E. B. Ford, eds., Evolution as a process. George Allen & Unwin, London.

Fisher, R. A. 1958. The genetical theory of natural selection, 2nd rev. ed. Dover, N.Y.

Fitts, P. M. & M. I. Posner. 1967. Human performance. Brooks/Cole, Belmont, Calif.

Fitzgerald, A. 1935. Rearing marmosets in captivity. J. Mammal. 16:181–188.

Fitzgerald, J. P. & R. R. Lechleitner. 1974. Observations on the biology of Gunnison's prairie dog in central Colorado. Am. Midl. Nat. 92:145–163.

Fitzsimons, F. W. 1920. The natural history of South Africa: Mammals. V. 3. Longmans, Green & Co., London.

Fleay, D. H. 1935. Breeding of *Dasyurus viverrinus* and general observations on the species. J. Mammal. 16:10–16.

Fleming, T. H., E. T. Hooper & D. E. Wilson. 1972. Three Central American bat communities: structure, reproductive cycles, and movement patterns. Ecology 53:555–569.

Florio, P. L. & L. Spinelli. 1967. Successful breeding of a cheetah *Acinonyx jubatus* in a private zoo. Int. Zoo. Yb. 7:150–152.

Florio, P. L. & L. Spinelli. 1968. Second successful breeding of cheetahs, *Acinonyx jubatus,* in a private zoo. Int. Zoo Yb. 8:76–78.

Flower, S. S. 1900. On the Mammalia of Siam and the Malay Peninsula. Proc. Zool. Soc. Lond. 1900:306–377.

Flux, J. E. C. 1970. Life history of the mountain hare (*Lepus timidus scoticus*) in north-east Scotland. J. Zool. Lond. 161:75–123.

Forbes, R. B. 1963. Care and early behavioral development of a lion cub. J. Mammal. 44:110–111.

Forrester, D. J. & R. S. Hoffmann. 1963. Growth and behavior of a captive bighorn lamb. J. Mammal. 44:116–118.

Fossey, D. 1972. Vocalizations of the mountain gorilla (*Gorilla gorilla beringei*). Anim. Behav. 20:36–53.

Fossey, D. 1979. Development of the mountain gorilla (*Gorilla gorilla beringei*): the first thirty-six months. Pages 139–184 in D. A. Hamburg & E. R. McCown, eds., The great apes (Perspectives on human evolution, Vol. 5). Benjamin/Cummings, Menlo Park, Calif.

Fouts, R. S., R. Mellgren & W. Lemmon. 1973. American sign language in the chimpanzee: chimpanzee-to-chimpanzee communication. Paper presented at the Midwestern Psychological Association meeting, Chicago.

Fouts, R. S. & R. L. Rigby. 1977. Man-chimpanzee communication. Pages 1034–1054 in T. A. Sebeok, ed., How animals communicate. Indiana University Press, Bloomington & London.

Fox, G. J. 1972. Some comparisons between siamang and gibbon behavior. Folia Primatol. 18:122–139.

Fox, M. W. 1969a. The anatomy of aggression and its ritualization in Canidae: a developmental and comparative study. Behaviour 35:242–257.

Fox, M. W. 1969b. Behavioral effects of rearing dogs with cats during the "critical period of socialization'." Behaviour 35:273–280.

Fox, M. W. 1970. A comparative study of the development of facial expressions in canids: wolf, coyote and foxes. Behaviour 36:49–73.

Fox, M. W. 1971. Socio-infantile and socio-sexual signals in canids: a comparative and ontogenetic study. Z. Tierpsychol. 28:185–210.

Fox, M. W. 1972a. Behavior of wolves, dogs, and related canids. Harper & Row, N.Y.

Fox, M. W. 1972b. Socio-ecological implications of individual differences in wolf litters: A developmental and evolutionary perspective. Behavior 41:298–313.

Fox, M. W. 1973. Social dynamics of three captive wolf packs. Behaviour 47:290–301.

Fox, M. W. & A. L. Clark. 1971. The development and temporal sequencing of agonistic behavior in the coyote (Canis latrans). Z. Tierpsychol. 28:262–278.

Fox, M. W. & J. A. Cohen. 1977. Canid communication. Pages 728–748 in T. A. Sebeok, ed., How animals communicate. Indiana University Press, Bloomington & London.

Fox, M. W., S. Halperin, A. Wise & E. Kohn. 1976. Species and hybrid differences in frequencies of play and agonistic actions in canids. Z. Tierpsychol. 40:194–209.

Fox, M. W. & D. Stelzner. 1966. Behavioral effects of differential early experience in the dog. Anim. Behav. 14:273–281.

Frädrich, H. 1965. Zur Biologie und Ethologie des Warzenschweines (Phacochoerus aethiopicus Pallas), unter Berücksichtigung des Verhalten anderer Suiden. Z. Tierpsychol. 22:328–393.

Frädrich, H. 1974. Notizen uber seltener gehaltene Cerviden. I. Zool. Gart. 44:189–200.

Frädrich, H. & H.-G. Klös. 1976. Zur Haltung und Zucht des Gaurs, Bos gaurus. Zool. Gart. 46:417–425.

Frädrich, H. & E. Thenius. 1972. Tapirs. Pages 17–33 in B. Grzimek's Animal Life Encyclopedia: Vol. 13 (Mammals IV). Van Nostrand Reinhold, N.Y.

Frank, F. 1957. Zucht und Gefangenschafts-Biologie der Zwergmaus. Z. Säugetierk. 22:1–44.

Frank, H. 1952. Uber die Jugendentwicklung des Eichhörnchens. Z. Tierpsychol. 9:12–22.

Frantz, J. 1963. Beobachtungen bei einer Löwenäffchen-Aufzucht. Zool. Gart. 28:115–120.

Fraser, A. F. 1976. Some features of an ultrasonic study of bovine fetal kinesis. Appl. Anim. Ethol. 2:379–383.

Fraser, A. F., H. Hastie, R. B. Callicott & S. Brownlie. 1975. An exploratory ultrasonic study on quantitative foetal kinesis in the horse. Appl. Anim. Ethol. 1:395–404.

Freeland, W. J. 1976. Pathogens and evolution of primate societies. Biotropica 8:12.

Freeman, H. E. 1975. A preliminary study of the behaviour of captive snow leopards, Panthera uncia. Int. Zoo Yb. 15:217–222.

Freeman, H. E. & J. Alcock. 1973. Play behaviour of a mixed group of juvenile gorillas and orangutans, *Gorilla g. gorilla* and *Pongo p. pygmaeus*. Int. Zoo Yb. 13:189–194.

Frere, A. G. 1929. Breeding habits of the common mongoose (*Herpestes edwardsi*). J. Bombay Nat. Hist. Soc. 33:426–428.

Freuchen, P. 1915. General observations as to natural conditions in the country visited by the expedition. In Report of the First Thule Expedition (Knud Rasmussen). Medd. om Grønland 51:390–400.

Fulk, G. W. 1976. Notes on the activity, reproduction and social behavior of *Octodon degus*. J. Mammal. 57:495–505.

Fuller, W. A. 1960. Behaviour and social organization of the wild bison in Wood Buffalo National Park, Canada. Arctic 13:2–19.

Gadgil, M. D. & W. H. Bossert. 1970. Life historical consequences of natural selection. Am. Nat. 104:1–24.

Galat-Luong, A. 1975. Notes préliminaires sur l'écologie de *Cercopithecus ascanius schmidti* dans les environs de Bangui (R.C.A.). La Terre et la Vie 29:288–297.

Galdikas, B. 1978. Orangutans and hominid evolution. Pages 287–309 in S. Udin, ed., Spectrum: Essays presented to Sultan Takdir Alisjahbana on his seventieth birthday. Dian Rakyat, Jakarta.

Gangloff, B. 1975. Beitrag zur Ethologie der Schleichkatzen (Bänderlinsang, *Prionodon linsang* (Hardw.), und Bänderpalmenroller, *Hemigalus derbyanus* (Gray)). Zool. Gart. 45:329–376.

Gangloff, B. & P. Ropartz. 1972. Le répertoire comportemental de la genette, *Genetta genetta* (Linné). La Terre et la Vie 26:489–560.

Gardner, R. A. & B. T. Gardner. 1969. Teaching sign language to a chimpanzee. Science 165:664–672.

Garner, R. L. 1919. Some observations on diseases and mental characteristics of apes. Am. J. Phys. Anthropol. 2:75–77.

Gartlan, J. S. 1968. Structure and function in primate society. Folia Primatol. 8:89–120.

Gartlan, J. S. 1969. Sexual and maternal behavior of the vervet monkey. *Cercopithecus aethiops*. J. Reprod. Fertil. Supp. 6:137–150.

Gartlan, J. S. 1970. Preliminary notes on the ecology and behavior of the drill, *Mandrillus leucophaeus* Ritgen 1824. Pages 445–480 in J. R. Napier & P. H. Napier, eds., Old world monkeys: evolution, systematics, and behavior. Academic Press, N.Y. & London.

Gartlan, J. S. 1975. Ecology and behavior of the patas monkey, *Erythrocebus patas*. Rockefeller University Film Service, N.Y.

Garvey, C. 1977. Play. Harvard University Press, Cambridge, Mass.

Gaston, A. J. 1977. Social behavior within groups of jungle babblers (*Turdoides striatus*). Anim. Behav. 25:828–848.

Gates, W. H. 1937. Spotted skunks and bobcat. J. Mammal. 18:240.

Gauthier-Pilters, H. 1959. Einige Beobachtungen zum Droh-, Angriffs-und Kampfverhalten des Dromedarhengstes, sowie über Geburt und Verhaltensentwicklung des Jungtieres, in der nordwestlichen Sahara. Z. Tierpsychol. 16:593–604.

Gauthier-Pilters, H. 1962. Beobachtungen an Feneks (*Fennecus zerda* Zimm.). Z. Tierpsychol. 19:440–464.

Gauthier-Pilters, H. 1966. Einige Beobachtungen über das Spielverhalten des Fenek (*Fennecus zerda* Zimm.). Z. Säugetierk. 31:337–350.

Gautier-Hion, A. 1970. L'organisation sociale d'une bande de talapoins (*Miopithecus talapoin*) dans le nord-est du Gabon. Folia Primatol. 12:116–141.

Gautier-Hion, A. 1971. Répertoire comportemental du talapoin (*Miopithecus talapoin*). Biol. Gabonica 7:295–391.

Gautier-Hion, A. & J. P. Gautier. 1971. La nage chez les Cercopithèques arboricales du Gabon. La Terre et la Vie 25:67–75.

Gautier-Hion, A. & J. P. Gautier. 1974. Les associations polyspécifiques de Cercopithèques du Plateau de M'Passa (Gabon). Folia Primatol. 22:134–177.

Gee, E. P. 1961. Some notes on the golden cat Felis temmincki. J. Bombay Nat. Hist. Soc. 58:1–12.

Geertz, C. 1972. Deep play: notes on the Balinese cockfight. Daedalus 101:1–38.

Gehring, C. B. 1976. Grußverhalten und Erkennen vertrauter Personen sowie weitere Verhaltensweisen einer Kapuziner-Gruppe (*Cebus apella*) im Zoo. Zool. Gart. 46:353–366.

Geidel, B. & W. Gensch. 1976. The rearing of clouded leopards, *Neofelis nebulosa,* in the presence of the male. Int. Zoo Yb. 16:124–126.

Geist, V. 1963. The North American moose (*Alces alces andersoni*). Behaviour 20:377–416.

Geist, V. 1966. The evolution of horn-like organs. Behaviour 27:175–214.

Geist, V. 1971. Mountain sheep: a study in behavior and evolution. University of Chicago Press, Chicago & London.

Geist, V. 1972. An ecological and behavioural explanation of mammalian characteristics, and their implication to therapsid evolution. Z. Säugetierk. 37:1–15.

Geist, V. 1974. On the relationship of social evolution and ecology in ungulates. Am. Zool. 14:205–220.

Geist, V. 1977. A comparison of social adaptations in relation to ecology in gallinaceous bird and ungulate societies. Ann. Rev. Ecol. Syst. 8:193–207.

Geist, V. 1978a. Life strategies, human evolution, environmental design: toward a biological theory of health. Springer Verlag, N.Y.

Geist, V. 1978b. On weapons, combat, and ecology. Pages 1–30 in L. Krames, P. Pliner & T. Alloway, eds., Aggression, dominance, and individual spacing. Plenum, N.Y.

Gentry, R. L. 1974. The development of social behavior through play in the Steller sea lion. Am. Zool. 14:391–403.

Gerall, A. A. 1963. An exploratory study of the effect of social isolation variables on the sexual behaviour of male guinea pigs. Anim. Behav. 11:274–282.

Gerall, H. D. 1965. Effects of social isolation and physical confinement on motor and sexual behavior of guinea pigs. J. Pers. Soc. Psychol. 2:460–464.

Ghobrial, L. I. & J. F. Cloudsley-Thompson. 1976. Daily cycle of activity of the dorcas gazelle in the Sudan. J. Interdiscipl. Cycle Res. 7:47–50.

Gianini, C. A. 1923. Caribou and fox. J. Mammal. 4:253–254.

Gibson-Hill, C. A. 1947. Notes on the birds of Christmas Island. Raffles Mus., Singapore, Bull. 18:87–165.

Gilmore, J. 1966. Play: a special behavior. Pages 343–355 in R. N. Haber, ed., Current research in motivation. Holt, Rinehart & Winston, N.Y.

Glickman, S. E. & R. W. Sroges. 1966. Curiosity in zoo animals. Behaviour 26:151–188.

Goethe, F. 1940. Beiträge zur Biologie des Iltis. Z. Säugetierk. 15:180–223.

Goethe, F. 1955. Beobachtungen bei der Aufzucht junger Silbermöwen. Z. Tierpsychol. 12:402–433.

Golani, I. 1976. Homeostatic motor processes in mammalian interactions: a choreography of display. Pages 69–134 in P. P. G. Bateson & P. H. Klopfer, eds., Perspectives in Ethology, Vol. 2. Plenum, N.Y.

Goldberg, S. 1978. Early experiences and behavior. Review of J. Kagan, R. B. Kearsley & P. R. Zelazo, eds., Infancy. Its place in human development. Science 202:1177–1178.

Goldberg, S. & M. Lewis. 1969. Play behavior in the year-old infant: early sex differences. Child Dev. 40:21–31.

Golding, R. R. 1972. A gorilla and chimpanzee exhibit at the University of Ibadan Zoo. Int. Zoo Yb. 12:71–76.

Goldman, L. & H. H. Swanson. 1975. Developmental changes in pre-adult behavior in confined colonies of golden hamsters. Dev. Psychobiol. 8:137–150.

Goldspink, G. 1970. Morphological adaptation due to growth and activity. Pages 521–536 in E. J. Briskey, R. G. Cassens & B. B. Marsh, eds., The physiology and biochemistry of muscle as a food, 2. University of Wisconsin Press, Madison.

Goldstein, I. P. & M. L. Miller. 1976. Structured planning and debugging: a linguistic theory of design. Memo 387, MIT Artificial Intelligence Laboratory, Cambridge, Mass.

Goldstein, I. & S. Papert. 1976. Artificial intelligence, language and the study of knowledge. Memo 337, MIT Artificial Intelligence Laboratory, Cambridge, Mass.

Golomb, C. & C. B. Cornelius. 1977. Symbolic play and its cognitive significance. Dev. Psychol. 13:246–252.

Goodall, J. 1965. Chimpanzees of the Gombe Stream Reserve. Pages 425–473 in I. DeVore, ed., Primate behavior. Holt, Rinehart & Winston, N.Y.

Goodwin, D. 1951. Some aspects of the behaviour of the jay Garrulus glandarius (Part II). Ibis 93:602–625.

Goodwin, D. 1953. Observations on voice and behavior of the red-legged partridge Alectoris rufa. Ibis 95:581–614.

Goodwin, M. K. 1979. Notes on caravan and play behavior in young Sorex cinereus. J. Mammal. 60:411–413.

Gordon, K. 1943. The natural history and behavior of the western chipmunk and the mantled ground squirrel. Oregon State Monographs, Studies in Zoology, no. 5. Oregon State College Press, Corvallis.

Gorgas, M. 1972. Zur Fortpflanzungsbiologie des Eisbären im natürlichen Verbreitungsgebiet und im Zoo. Zeitschrift des Kölner Zoo 15:3–12.

Gossow, H. & G. Schürholz. 1974. Social aspects of wallowing behaviour in red deer herds. Z. Tierpsychol. 34:329–336.

Goswell, M. J. & J. S. Gartlan. 1965. Pregnancy, birth and early infant behaviour in the captive patas monkey *Erythrocebus patas*. Folia Primatol. 3:189–200.

Gottlieb, G. 1976. The roles of experience in the development of behavior and the nervous system. Pages 25–54 in G. Gottlieb, ed., Neural and behavioral specificity. (Studies on the development of behavior and the nervous system, Vol. 3.) Academic Press, N.Y.

Gould, S. J. 1966. Allometry and size in ontogeny and phylogeny. Biol. Rev. 41:587–640.

Gould, S. J. 1977. Ontogeny and phylogeny. Harvard University Press, Cambridge, Mass.

Goy, R. W. 1968. Organising effects of androgen on the behavior of rhesus monkeys. Pages 12–31 in R. P. Michael, ed., Endocrinology and human behaviour. Oxford University Press, London.

Goy, R. W. 1970. Experimental control of psychosexuality. Philos. Trans. R. Soc. Lond. Ser. B Biol. Sci. 259:149–162.

Goy, R. W. & J. A. Resko. 1972. Gonadal hormones and behavior of normal and pseudohermaphroditic nonhuman female primates. Recent Prog. Horm. Res. 28:707–733.

Goy, R. W. & W. Wallen. 1979. Experiential variables influencing play, foot-clasp mounting and adult sexual competence in male rhesus monkeys. Psychoneuroendocrinology 4:1–12.

Gradl-Grams, M. 1977. Verhaltensstudien an Damwild in Gefangenschaft. Zool. Gart. 47:81–108.

Graf, W. 1966. The Axis deer in Hawaii. J. Bombay Nat. Hist. Soc. 63:629–733.

Grantham-McGregor, S., M. Stewart, C. Powell & W. N. Schofield. 1979. Effect of stimulation on mental development of malnourished child. Lancet 2:200–201.

Grau, G. A. & F. R. Walther. 1976. Mountain gazelle agonistic behaviour. Anim. Behav. 24:626–636.

Greenough, W. T. 1976. Enduring brain effects of differential experience and training. Pages 255–278 in M. R. Rosenzweig & E. L. Bennett, eds., Neural mechanisms of learning and memory. MIT Press, Cambridge, Mass.

Greenough, W. T. 1978. Development and memory: The synaptic connection. Pages 127–145 in T. Teyler, ed., Brain and learning. Greylock Publishers, Stamford, Conn.

Greer, J. K. 1965. Mammals of Malleco Province, Chile. Mich. State Univ. Mus. Ser. 3:49–152.

Gregory, R. L. 1969. On how so little information controls so much behaviour. Pages 236–247 in C. H. Waddington, ed., Towards a theoretical biology. 2: Sketches. Edinburgh University Press, Edinburgh.

Griffin, D. R. 1976. The question of animal awareness. Rockefeller University Press, N.Y.

Griffiths, M. 1978. The biology of the monotremes. Academic Press, N.Y.

Grime, J. P. 1977. Evidence for the existence of three primary strategies in plants and its relevance to ecological and evolutionary theory. Am. Nat. 111:1169–1194.

Groos, K. 1898. The play of animals. Trans. E. L. Baldwin. D. Appleton & Co., N.Y.

Groos, K. 1899. Die Spiele der Menschen. Fischer, Jena. (The play of man. Appleton, N.Y. 1901.)

Groves, C. P. 1973. Notes on the ecology and behaviour of the Angola colobus (*Colobus angolensis* P. L. Sclater 1860) in N. E. Tanzania. Folia Primatol. 20:12–26.

Grzimek, B. 1949. Rangordnungsversuche mit Pferden. Z. Tierpsychol. 6:455–464.

Grzimek, B. 1967. Four-legged Australians. Collins, London.

Grzimek, B. & M. Grzimek. 1960. Serengeti shall not die. Hamish Hamilton, London.

Gucwinska, H. & A. Gucwinski. 1968. Breeding the Zanzibar galago, *Galago senegalensis zanzibaricus,* at Wroclaw Zoo. Int. Zoo Yb. 8:111–114.

Guggisberg, C. A. W. 1975. Wild cats of the world. Taplinger, N.Y.

Guhl, A. M. 1958. The development of social organisation in the domestic chick. Anim. Behav. 6:92–111.

Gundlach, H. 1965. *Sus scrofa* (Suidae)—Spiele der Jungtiere. Encyclopedia Cinematographica Film E949. Institut für den Wissenschaftlichen Film, Göttingen.

Gundlach, H. 1968. Brutfürsorge, Brutpflege, Verhaltensontogenese und Tagesperiodik beim Europäischen Wildschwein (*Sus scrofa* L.) Z. Tierpsychol. 25:955–995.

Gunston, D. 1971. The golden eagle in England. Birds Ctry. 24:7–8.

Gunter, G. 1953. Observations on fish turning flips over a line. Copeia(3):188–190.

Gurney, G. H. 1909. Notes on a collection of birds made in British East Africa. Ibis 3:484–532.

Gwinner, E. 1964. Untersuchungen über das Ausdrucks- und Sozialverhalten des Kolkraben (*Corvus corax corax* L.) Z. Tierpsychol. 21:657–748.

Gwinner, E. 1966. Über einige Bewegungsspiele des Kolkraben (*Corvus corax* L.). Z. Tierpsychol. 23:28–36.

Haas, E. 1967. Pride's progress. Avon, N.Y.

Haas, G. 1959. Untersuchungen über angeborene Verhaltensweisen bei Mähnenspringern (*Ammotragus lervia* Pallas). Z. Tierpsychol. 16:218–242.

Haas, G. 1963. Beitrag zum Verhalten des Bambusbären (*Ailuropus melanoleucus*). Zool. Gart. 27:225–233.

Hadamard, J. 1945. Psychology of invention in the mathematical field. Princeton University Press, Princeton, N.J.

Hafez, E. S. E., ed. 1962. The behaviour of domestic animals. Williams & Wilkins, Baltimore.

Hafez, E. S. E. & J. P. Signoret. 1969. The behaviour of swine. Pages 349–390 in E. S. E. Hafez, ed., The behaviour of domestic animals, 2nd ed. Ballière, Tindall & Cassell, London.

Hall, G. S. 1906. Youth, its education, regimen and hygiene. Appleton, N.Y.

Hall, K. R. L. 1962. The sexual, agonistic, and derived social behaviour patterns of the wild chacma baboon, Papio ursinus. Proc. Zool. Soc. Lond. 139:283–327.

Hall, K. R. L. 1963. Tool-using performances as indicators of behavioral adaptability. Curr. Anthropol. 4:479–494.

Hall, K. R. L. 1965. Behaviour and ecology of the wild patas monkey, Erythrocebus patas, in Uganda. J. Zool. 148:15–87.

Hall, K. R. L. 1968. Behavior and ecology of the wild patas monkey, Erythrocebus patas, in Uganda. Pages 32–120 in P. Jay, ed., Primates: studies in adaptation and variability. Holt, Rinehart & Winston, N.Y.

Hall, K. R. L., R. C. Boelkins & M. J. Goswell. 1965. Behaviour of patas monkeys in captivity. Folia Primatol. 3:22–49.

Hall, K. R. L. & C. R. Carpenter. 1967. Chacma baboons (Papio ursinus): Ecology and behavior. Psychological Cinema Register Film PCR–2167. Pennsylvania State University, University Park.

Hall, K. R. L. & B. Mayer. 1967. Social interactions in a group of captive patas monkeys (Erythrocebus patas). Folia Primatol. 5:213–236.

Hall, K. R. L. & G. B. Schaller. 1964. Tool-using behavior of the California sea otter. J. Mammal. 45:287–298.

Halsman, P. 1959. The jump book. Simon & Schuster, N.Y.

Hamburger, V. 1973. Anatomical and physiological basis of embryonic motility in birds and mammals. Pages 51–76 in G. Gottlieb, ed., Studies on the development of behavior and the nervous system, Vol. 1: Behavioral embryology. Academic Press, N.Y.

Hamilton, C. 1967. Play in the context of social relationships in a herd of ponies. M.S. thesis, University of Pennsylvania.

Hamilton, J. E. 1934. The southern sea-lion. Discovery Rep. VIII:269–318.

Hamilton, W. D. 1964. The genetical evolution of social behaviour. Parts I & II. J. Theor. Biol. 7:1–16, 17–52.

Hamilton, W. D. 1966. The moulding of senescence by natural selection. J. Theor. Biol 12:12–45.

Hamilton, W. D. 1971. Selection of selfish and altruistic behaviour in some extreme models. Pages 57–91 in J. F. Eisenberg & W. S. Dillon, eds., Man and beast: comparative social behavior. Smithsonian Press, Washington, D.C.

Hamilton, W. D. 1975. Innate social aptitudes of man: an approach from evolutionary genetics. Pages 133–155 in R. Fox, ed., Biosocial anthropology. Wiley, N.Y.

Hamilton, W. D. 1977. The play by nature. Review of R. Dawkins, The selfish gene. Science 196:757–759.

Hamilton, W. J. 1933. The weasels of New York. Their natural history and economic status. Am. Midl. Nat. 14:289–344.

Hamilton, W. J., III. 1972. Reproductive adaptations of the red tree mouse. J. Mammal. 43:486–502.

Hamilton, W. J., III, R. E. Buskirk & W. H. Buskirk. 1975. Chacma baboon tactics during intertroop encounters. J. Mammal. 56:857–870.

Hämmerling, F. & W. Lippert. 1975. Beobachtung des Geburts- und Mutter-Kind-Verhaltens beim Mähnenwolf (*Chrysocyon brachyurus*) über ein Nachtsichgerat (Infrarot-Beobachtungsanlage). Zool. Gart. 45:393–415.

Hampton, J. K., Jr., S. H. Hampton & B. T. Landwehr. 1966. Observations on a successful breeding colony of the marmoset, *Oedipomidas oedipus*. Folia Primatol. 4:265–287.

Hanby, J. P. 1974. Male-male mounting in Japanese monkeys (*Macaca fuscata*). Anim. Behav. 22:836–849.

Hanby, J. P. 1976. Sociosexual development in primates. Pages 1–67 in P. P. G. Bateson & P. H. Klopfer, eds., Perspectives in ethology, Vol. 2. Plenum, N.Y.

Hanby, J. P. & C. E. Brown. 1974. The development of sociosexual behaviours in Japanese macaques *Macaca fuscata*. Behaviour 49:152–196.

Hanks, J., M. S. Price & R. W. Wrangham. 1969. Some aspects of the ecology and behaviour of the Defassa waterbuck (*Kobus defassa*) in Zambia. Mammalia 33:471–494.

Hansen, E. W. 1966. The development of maternal and infant behavior in the rhesus monkey. Behaviour 27:107–149.

Hansen, E. W. 1974. Some aspects of behavioral development in evolutionary perspective. Pages 182–186 in N. F. White, ed., Ethology and psychiatry. University of Toronto Press, Toronto, Ont. & Buffalo, N.Y.

Hansen, J. W. 1967. Effect of dynamic training on the isometric endurance of the elbow flexors. Int. Z. Angew. Physiol. 23:367–370.

Happold, M. 1976a. The ontogeny of social behavior in four conilurine rodents (Muridae) of Australia. Z. Tierpsychol. 40:265–278.

Happold, M. 1976b. Social behavior of the conilurine rodents (Muridae) of Australia. Z. Tierpsychol. 40:113–182.

Hardy, P. 1975. A lifetime of badgers. David & Charles, N. Pomfret, Vt.

Harlow, H. F. 1974. Induction and alleviation of depressive states in monkeys. Pages 197–208 in N. F. White, ed., Ethology and psychiatry. University of Toronto Press, Toronto, Ont. & Buffalo, N.Y.

Harlow, H. F. & M. K. Harlow. 1965. The affectional systems. Behav. Nonhuman Primates 2:287–334.

Harper, L. V. 1970. Ontogenetic and phylogenetic functions of the parent-offspring relationship in mammals. Adv. Stud. Behav. 3:75–117.

Harrington, J. E. 1975. Field observations of social behavior of *Lemur fulvus fulvus* E. Geoffroy 1812. Pages 259–279 in I. Tattersall & R. W. Sussman, eds., Lemur biology. Plenum Press, N.Y.

Harris, C. J. 1968. Otters. Weidenfeld & Nicolson, London.

Harrisson, B. 1963. A study of orang-utan behavior in the semi-wild state, 1959–60. Int. Zoo Yb. 3:57–68.

Hartley, L. H., G. Grimby, Å. Kilbom, N. J. Nilsson, I. Åstrand, J. Bjure,

B. Ekblom & B. Saltin. 1969. Physical training in sedentary middle-aged and older men. III. Cardiac output and gas exchange at submaximal and maximal exercise. Scand. J. Clin. Lab. Invest. 24:335–344.

Hartman, D. S. 1979. Ecology and behavior of the manatee (*Trichechus manatus*) in Florida. Spec. Pub. No. 5, Am. Soc. of Mammalogists.

Hartman, L. 1964. The behaviour and breeding of captive weasels (*Mustela nivalis* L.). N. Z. J. Sci. 7:147–156.

Hasler, J. F. & M. W. Sorenson. 1974. Behavior of the tree shrew, *Tupaia chinensis* in captivity. Am. Midl. Nat. 91:294–314.

Hassenberg, L. 1977. Zum Fortpflanzungsverhalten des mesopotamischen Damhirsches, *Cervus dama mesopotamica* Brooke, 1875, in Gefangenschaft. Säugetierk. Mitt. 25:161–194.

Hassenberg, L. & H.-G. Klös. 1975. Zur Fortpflanzung und Jungenaufzucht des Barasingha, *Cervus duvauceli* G. Cuvier, 1823, in Gefangenschaft. Säugetierk. Mitt. 23:64–73.

Hausfater, G. 1976. Predatory behavior of yellow baboons. Behaviour 56:44–68.

Hayes, C. 1952. The ape in our house. Gollancz, London.

Hayes, K. J. & C. Hayes, 1952. Imitation in a home-raised chimpanzee. J. Comp. Physiol. Psychol. 45:450–459.

Heath, R. H. 1908. Shooting notes from the Garwhal Himalayas. J. Bombay Nat. Hist. Soc. 18:931–934.

Hedhammar, Å., F.-M. Wu, L. Krook, H. F. Schryver, A. de Lahunta, J. P. Whalen, F. A. Kallfelz, E. A. Nunez, H. F. Hintz, B. E. Sheffy & G. D. Ryan. 1974. Overnutrition and skeletal disease: an experimental study in growing Great Dane dogs. Cornell Vet. 64, Supp. 5.

Hediger, H. 1958. Verhalten der Beuteltiere (Marsupiala). Handb. d. Zool. 8, 10(9), 1–28.

Heglund, N. C., C. R. Taylor & T. A. McMahon. 1974. Scaling stride frequency and gait to animal size: mice to horses. Science 186:1112–1113.

Heidt, G. A., M. K. Petersen & G. L. Kirkland. 1968. Mating behavior and development of least weasels (*Mustela nivalis*) in captivity. J. Mammal. 49:413–419.

Heimburger, N. 1961. Beobachtungen an handaufgezogenen Wildcaniden (Wölfin und Schakalin) und Versuche über ihre Gedächtnisleistungen. Z. Tierpsychol. 18:265–284.

Heinrich, B. 1975. Thermoregulation in bumblebees. II. Energetics of warm-up and free flight. J. Comp. Physiol. 96:155–166.

Heinroth, O. & M. Heinroth. 1928. Die Vögel Mitteleuropas, Vol. 2. Bermühler Verlag, Berlin-Lichterfelde.

Heinsohn, G. E. 1966. Ecology and reproduction of the Tasmanian bandicoots. Univ. Cal. Pub. Zool. 80.

Helanko, R. 1958. Theoretical aspects of play. Akateeminen Kirjakauppa, Helsinki.

Hemmer, H. 1968. Untersuchungen zur Stammesgeschichte der Pantherkatzen (Pantherinae). Teil II: Studien zur Ethologie des Nebelparders *Neofelis nebulosa* (Griffith 1821) und des Irbis *Uncia uncia* (Schreber 1775). Veröff. Zool. Staatssamml. München 12:155–247.

Hemmer, H. 1976. Gestation period and postnatal development in felids. World's Cats 3(2):143–165.

Henisch, B. A. & H. K. Henisch. 1970. Chipmunk portrait. Carnation Press, State College, Pa.

Henry, J. D. & S. M. Herrero. 1974. Social play in the American black bear, Am. Zool. 14:371–389.

Herrero, S. & D. Hamer. 1977. Courtship and copulation of a pair of grizzly bears, with comments on reproductive plasticity and strategy. J. Mammal. 58:441–444.

Herrick, F. H. 1924a. The daily life of the American eagle: late phase. Auk 41:389–422.

Herrick, F. H. 1924b. The daily life of the American eagle: late phase. Auk 41:517–541.

Herrick, F. H. 1924c. An eagle observatory. Auk. 41:89–105.

Herrick, F. H. 1934. The American eagle. A study in natural and civil history. Appleton Century, N.Y.

Herrmann, D. 1971. Beobachtungen des Gruppenlebens ostaustralischer Graugroßkänguruhs, *Macropus giganteus* (Zimmermann, 1777) und Bennetkänguruhs, *Protemnodon rufogrisea* (Desmarest, 1817). Säugetierk. Mitt. 19:352–362.

Hershkovitz, P. 1977. Living New World monkeys (Platyrrhini), Vol. 1. University of Chicago Press, Chicago.

Herter, K. 1958. Die säugetierkundlichen Arbeiten aus dem Zoologischen Institut der Freien Universität Berlin. Z. Säugetierk. 23:1–32.

Herter, K. 1965. Hedgehogs. Phoenix House, London.

Herter, K. & M. Herter. 1955. Über eine scheinträchtige Iltisfähe mit untergeschobenem Katzenjungen. Zool. Gart. 22:33–46.

Herter, K. & I.-D. Ohm-Kettner. 1954. Über die Aufzucht und das Verhalten zweier Baummarder (*Martes martes* L.) Z. Tierpsychol. 11:113–137.

Hess, J. P. 1973. Some observations on the sexual behaviour of captive lowland gorillas, *Gorilla g. gorilla* (Savage and Wyman). Pages 507–581 in R. P. Michael & J. H. Crook, eds., Comparative ecology and behaviour of primates. Academic Press, London & N.Y.

Hewlett, K. G. & M. A. Newman. 1968. "Skana," the killer whale, *Orcinus orca,* at Vancouver Public Aquarium. Int. Zoo Yb. 8:209–211.

Hick, U. 1962. Beobachtungen über das Spielverhalten unseres Hyazinth Ara (*Anodorhynchus hyacinthus*). Freunde d. Kölner Zoo 5:8–9.

Hick, U. 1968. The collection of Saki monkeys at Cologne Zoo. Int. Zoo Yb. 8:192–194.

Hick, U. 1969. Successful raising of a pudu *Pudu pudu* at Cologne Zoo. Int. Zoo Yb. 9:110–112.

Hick, U. 1972. Breeding and maintenance of Douc langurs, *Pygathrix nemaeus nemaeus,* at Cologne Zoo. Int. Zoo Yb. 12:98–103.

Hickman, J. C. 1975. Environmental unpredictability and plastic energy allocation strategies in the annual *Polygonum cascadense* (Polygonaceae). J. Ecol. 63:689–701.

Hill, C. 1946. Playtime at the zoo. Zoo–Life 1:24–26.

Hill, H. L. & M. Bekoff. 1977. The variability of some motor components of social play and agonistic behavior in infant Eastern coyotes. *Canis latrans* var. Anim. Behav. 25:907–909.

Hill, W. C. O. 1966. Laboratory breeding, behavioural development and relations of the talapoin (*Miopithecus talapoin*). Mammalia 30:353–370.

Hinde, R. A. 1966. Animal behaviour. McGraw-Hill, N.Y.

Hinde, R. A. 1970. Animal behaviour, 2nd ed. McGraw-Hill, N.Y.

Hinde, R. A. 1971. Development of social behavior. Behav. Nonhuman Primates 3:1–68.

Hinde, R. A. 1974. Biological bases of human social behaviour. McGraw-Hill, N.Y.

Hinde, R. A. 1975. The concept of function. Pages 3–15 in G. Baerends, C. Beer & A. Manning, eds., Function and evolution in behaviour. Clarendon Press, Oxford.

Hinde, R. A. 1976. Multiple review of Wilson's *Sociobiology*. Anim. Behav. 24:706–707.

Hinde, R. A. 1978. Dominance and role—two concepts with dual meanings. J. Social Biol. Struct. 1:27–38.

Hinde, R. A. & L. M. Davies. 1972. Changes in mother-infant relationship after separation in Rhesus monkeys. Nature 239:41–42.

Hinde, R. A., T. E. Rowell & Y. Spencer-Booth. 1964. Behavior of socially living rhesus monkeys in their first six months. Proc. Zool. Soc. Lond. 143:609–649.

Hinde, R. A. & M. J. A. Simpson. 1975. Qualities of mother-infant relationships in monkeys. Pages 39–57 in R. Porter & M. O'Connor, eds., Parent-infant interaction. Ciba Foundation Symposium No. 33 (new series). Associated Scientific Publishers, Amsterdam & American Elsevier, N.Y.

Hinde, R. A. & J. Stevenson-Hinde, eds. 1973. Constraints on learning: limitations and predispositions. Academic Press, London & N.Y.

Hines, M. 1942. The development and regression of reflexes, postures, and progression in the young macaque. Carnegie Institution of Washington. Contributions to Embryology 30:153–210.

Hingston, R. W. G. 1913. A study of emotional expression in *Felis pardus*. J. Bombay Nat. Hist. Soc. 22:230–236.

Hinton, H. E. & A. M. S. Dunn. 1967. Mongooses: their natural history and behaviour. University of California Press, Berkeley & Los Angeles.

Hirsch, J. 1976. Multiple review of Wilson's *Sociobiology*. Anim. Behav. 24:707–709.

Hirth, D. H. 1977. Social behavior of white-tailed deer in relation to habitat. Wildl. Monogr. 53.

Hladik, C. M. 1973. Alimentation et activité d'un groupe de Chimpanzés réintroduits en forêt gabonaise. La Terre et la Vie 27:343–413.

Ho, Y.-C. 1970. Differential games, dynamic optimization, and generalized control theory. J. Optimization Theory and Appl. 6:179–209.

Hodl-Rohn, I. 1974. Verhaltensstudien an drei zahmen Glattottern, *Lutra* (*Lutrogale*) *perspicillata* (I. Geoffroy 1826). Säugetierk. Mitt. 22:17–28.

Hoesch, W. 1964. Beobachtungen an einem zahmen Honigdachs (*Mellivora capensis*). Zool. Gart. 28:182–188.

Højgaard, A. M. 1954. Nogle Iagttagelser over en tam Grønlansk Ravn. (*Corvus corax principalis*). Dansk orn. Foren. Tiddskr. 48:38–47. (Cited in Thorpe 1966.)

Holloszy, J. O. 1973. Biochemical adaptations to exercise: aerobic metabolism. Exercise and Sport Sciences Reviews 1:45–71.

Holloszy, J. O. 1975. Adaptation of skeletal muscle to endurance exercise. Med. Sci. Sports 7:155–164.

Holloszy, J. O. & G. W. Booth. 1976. Biochemical adaptations to endurance exercise in muscle. Annu. Rev. Physiol. 38:273–291.

Holt, J. 1967. How children learn. Pitman, N.Y.

Holt, K. S. 1975. Movement and child development, Heinemann, London.

Honigmann, H. 1935. Beobachtungen am Grossen Ameisenbären (*Myrmecophaga tridactyla* L.). Z. Säugetierk. 10:78–104.

Hooijer, D. A. 1967. Indo-Australian insular elephants. Genetica 38:143–162.

Hopf, S. 1971. New findings on the ontogeny of social behaviour in the squirrel monkey. Psychiatr. Neurol. Neurochir. 74:21–34.

Hopf, S., E. Hartmann-Wiesner, B. Kühlmorgen & S. Mayer. 1974. The behavioral repertoire of the squirrel monkey (*Saimiri*). Folia Primatol. 21:225–249.

Hornaday, W. T. 1934. The minds and manners of wild animals. C. Scribner's Sons, N.Y.

Horr, D. A. 1977. Orang-utan maturation: growing up in a female world. Pages 289–321 in S. Chevalier-Skolnikoff & F. E. Poirier, eds., Primate biosocial development. Garland, N.Y.

Horwich, R. 1972. The ontogeny of social behavior in the gray squirrel (*Sciurus carolinensis*). Z. Tierpsychol. Supp. 8.

Horwich, R. 1974. Development of behaviors in a male spectacled langur (*Presbytis obscurus*). Primates 15:151–178.

Horwich, R. & L. LaFrance. 1972. The mountain guereza. Field Mus. Nat. Hist. Bull. 43(11):2–5.

Howard, H. E. 1907. The British warblers: a history with problems of their lives. R. H. Porter, London.

Howard, W. J. 1935. Notes on the hibernation of a captive black bear. J. Mammal. 16:321.

Hubbard, W. D. 1963. Ibamba. Gollancz, London.

Huber, P. 1810. Recherches sur les moeurs des fourmis indigènes. J. J. Paschoud, Paris.

Hubl, H. 1952. Beiträge zur Kenntnis der Verhaltensweisen junger Eulenvögel in Gefangenschaft: (Schleiereule *Tyto alba*, Steinkauz *Athene noctua* und Waldkauz *Strix aluco aluco*). Z. Tierpsychol. 9:102–119.

Hughes, F. 1977. Hand-rearing a Sumatran tiger, *Panthera tigris sumatrae*, at Whipsnade Park. Int. Zoo Yb. 17:219–221.

Huizinga, J. 1950. Homo Ludens. Beacon, Boston.

Humphries, D. A. & P. M. Driver. 1967. Erratic display as a device against predators. Science 156:1767–1768.

Humphries, D. A. & P. M. Driver. 1970. Protean defence by prey animals. Oecologia 5:285–302.

Hunkeler, C., F. Bourlière & M. Bertrand. 1972. Le comportement social de la Mone de Lowe (*Cercopithecus campbelli lowei*). Folia Primatol. 17:218–236.

Hurlock, E. B. 1972. Child development, 5th ed. McGraw-Hill, N.Y.

Hurrell, H. G. 1962. Foxes. Sunday Times Publications, London.

Hutchinson, G. E. 1965. The ecological theater and the evolutionary play. Yale University Press, New Haven.

Hutchinson, G. E. 1976. Man talking or thinking. Am. Sci. 64:22–27.

Hutson, H. P. W. 1945. Roosting procedure of *Corvus corax laurencei* Hume. Ibis 87:456–459.

Hutt, C. 1966. Exploration and play in children. Symp. Zool. Soc. Lond. 18:23–44.

Hutt, C. 1970. Specific and diversive exploration. Adv. Child Dev. Behav. 5:119–180.

Hutt, C. & R. Bhavnani. 1972. Predictions from play. Nature 237:171–172.

Huxley, J. & E. M. Nicholson. 1963. Lammergeir, *Gypaetus barbatus*, breaking bones. Ibis 105:106–107.

Ibscher, L. 1967. Geburt und frühe Entwicklung zweier Gibbons (*Hylobates lar* L.). Folia Primatol. 5:43–69.

Inhelder, E. 1955. Zur Psychologie einiger Verhaltensweisen—besonders des Spiels—von Zootieren. Z. Tierpsychol. 12:88–144.

Innis, A. C. 1958. The behaviour of the giraffe, *Giraffa camelopardalis* in the Eastern Transvaal. Proc. Zool. Soc. Lond. 131:245–278.

Itani, J. 1959. Paternal care in the wild Japanese monkey, *Macaca fuscata fuscata*. Primates 2:61–93.

Itoigawa, N. 1975. Variables in male leaving a group of Japanese macaques. Proc. Symp. Congr. Int. Primatol. Soc. 5th., 233–245.

Izard, C. E. 1975. Patterns of emotions and emotion communication in "hostility" and aggression. Adv. Stud. Commun. and Affect 2:77–101.

Jackson, J. R. 1963a. Nesting of Keas. Notornis 10:319–326.

Jackson, J. R. 1963b. Studies at a Kaka's nest. Notornis 10:168–176.

Jackson, V. A. 1918. Other notes on the hog-badger. J. Bombay Nat. Hist. Soc. 26:281.

Jaeger, E. C. 1929. Denizens of the mountains. Charles C. Thomas, Springfield, Ill.

Jantschke, F. 1972. Orang-utans in Zoologischen Gärten. Piper, München.

Jantschke, F. 1973. On the breeding and rearing of bush dogs, *Speothos venaticus*, at Frankfurt Zoo. Int. Zoo Yb. 13:141–143.

Jarman, M. V. & P. J. Jarman. 1973. Daily activity of impala. E. Afr. Wildl. J. 11:75–92.

Jarman, P. J. 1972. The development of a dermal shield in impala. J. Zool. Lond. 166:349–356.

Jarman, P. J. 1974. The social organisation of antelope in relation to their ecology. Behaviour 58:215–267.

Jaworowska, M. 1976. Verhaltensbeobachtungen an primitiven polnischen Pferden, die in einem polnischen Wald-Schutzgebiet—in Freiheit leben— erhalten werden. Säugetierk. Mitt. 24:241–268.

Jay, P. 1963. Mother-infant relations in langurs. Pages 282–304 in H. L. Rheingold, ed., Maternal behavior in mammals. Wiley, N.Y.

Jay, P. 1965. The common langur of North India. Pages 197–249 in I. De-Vore, ed., Primate behavior. Holt, Rinehart & Winston, N.Y.

Jenkins, P. F. 1978. Cultural transmissinn of song patterns and dialect development in a free-living bird population. Anim. Behav. 26:50–78.

Jensen, S. 1904. Mammals observed on Amdrup's journeys to East Greenland, 1898–1900. Medd. om Grønland 29:27–61.

Jerdon, T. C. 1874. Mammals of India. John Wheldon, London.

Jerison, H. J. 1973. Evolution of the brain and intelligence. Academic Press, N.Y.

Jewell, P. A. & C. Loizos, eds. 1966. Play, exploration, and territory in mammals. Symp. Zool. Soc. Lond. 18.

Johnson, M. J. & R. Gayden. 1975. Breeding the bald eagle, *Haliaeetus leucocephalus* at the National Zoological Park, Washington. Int. Zoo Yb. 15:98–100.

Johnson, R. C. & G. R. Medinnus. 1965. Child psychology: behavior and development. Wiley, N.Y.

Jolly, A. 1966a. Lemur behavior. University of Chicago Press, Chicago & London.

Jolly, A. 1966b. Lemur social behavior and primate intelligence. Science 153:501–506.

Jolly, A. 1972. Troop continuity and troop spacing in *Propithecus verreauxi* and *Lemur catta* at Berenty (Madagascar). Folia Primatol. 17:335–362.

Jolly, H. 1968. Play and the sick child: a comparative study of its role in a teaching hospital in London and one in Ghana. Lancet 2:1286–1287.

Jones, T. B. & A. C. Kamil. 1973. Tool-making and tool-using in the Northern blue jay. Science 180:1076–1077.

Joubert, E. 1972a. Activity patterns shown by Hartmann zebra *Equus zebra hartmannae* in South West Africa with reference to climatic factors. Madoqua Ser. I (5):33–52.

Joubert, E. 1972b. The social organization and associated behaviour in the Hartmann zebra *Equus zebra hartmannae*. Madoqua Ser. I(6):17–56.

Jouventin, P. 1975. Observations sur la socio-écologie du Mandrill. La Terre et la Vie 29:493–532.

Junge, C. 1966. Pudu, *Pudu pudu,* at Chillan Viejo Zoo. Int. Zoo Yb. 6:263–264.

Kagan, J. 1978. The growth of the child: reflections on human development. Norton, N.Y.

Kagan, J., R. B. Kearsley & P. R. Zelazo. 1978. Infancy. Its place in human development. Harvard University Press, Cambridge, Mass.

Kalacheva, E. L. 1965. The effect of muscular loads on the play activity in golden hamsters. Pages 88–92 in Slozhnye Formy Povedeniya, Nauka, Moscow & Leningrad.

Kalas, K. 1976. Beobachtungen bei der Handaufzucht eines kanadischen Bibers, *Castor canadensis* Kuhl, 1820. Säugetierk. Mitt. 24:304–315.

Kammer, A. 1968. Motor patterns during flight and warm-up in Lepidoptera. J. Exp. Biol. 48:89–109.

Kaplan, J. 1972. Differences in the mother-infant relations of squirrel monkeys housed in social and restricted environments. Dev. Psychobiol. 5:43–52.

Kaufman, G. W., D. B. Siniff & R. Reichle. 1975. Colony behavior of Weddell seals, *Leptonychotes weddelli*, at Hutton Cliffs, Antarctica. Pages 228–246 in K. Ronald & A. W. Mansfield, eds., Symposium on the biology of the seal. Rapp. P.-V. Cons. Int. Explor. Mer 169.

Kaufman, I. C. & L. A. Rosenblum. 1965. The waning of the mother-infant bond in two species of macaque. Pages 41–59 in B. M. Foss, ed., Determinants of infant behaviour, Vol. 4. Methuen, London.

Kaufman, I. C. & L. A. Rosenblum. 1966. A behavioral taxonomy for *Macaca nemestrina* and *Macaca radiata*, based on longitudinal observation of family groups in the laboratory. Primates 7:205–258.

Kaufman, I. C. & L. A. Rosenblum. 1967a. Depression in infant monkeys separated from their mothers. Science 155:1030–1031.

Kaufman, I. C. & L. A. Rosenblum. 1967b. The reaction to separation in infant monkeys: anaclitic depression and conservation-withdrawal. Psychosom. Med. 29:648–675.

Kaufmann, H. 1965. Definitions and methodology in the study of aggression. Psychol. Bull. 64:351–364.

Kaufmann, J. H. 1962. The ecology and social behavior of the coati, *Nasua narica*, on Barro Colorado Island, Panama. Univ. Calif. Publ. Zool. 60:95–222.

Kaufmann, J. H. 1974. Social ethology of the whiptail wallaby, *Macropus parryi*, in northeastern New South Wales. Anim. Behav. 22:281–369.

Kaufmann, J. H. 1975. Field observations of the social behavior of the eastern grey kangaroo, *Macropus giganteus*. Anim. Behav. 23:214–221.

Kavanagh, M. 1978. The social behaviour of doucs (*Pygathrix nemaeus nemaeus*) at San Diego Zoo. Primates 19:101–114.

Kavanagh, M. & L. Dresdale. 1975. Observations on the woolly monkey (*Lagothrix lagothrica*) in northern Colombia. Primates 16:285–294.

Kawai, M. 1965. Newly acquired pre-cultural behavior of the natural troop of Japanese monkeys on Koshima Island. Primates 6:1–30.

Kay, D. E. 1978. A preliminary investigation into play in young Greenland collared lemmings (*Dicrostonyx groenlandicus*). Unpublished manuscript.

Kay, H. 1970. Analyzing motor skill performance. Pages 139–159 in K. Connolly, ed., Mechanisms of motor skill development. Academic Press, N.Y.

Keller, R. 1975. Das Spielverhalten der Keas (*Nestor notabilis* Gould) des Zürcher Zoos. Z. Tierpsychol. 38:393–408.

Keller, R. 1976. Beitrag zur Biologie und Ethologie der Keas (*Nestor notabilis*) des Zürcher Zoos. Zool. Beitr. 22:111–156.

Keller, R. 1977. Beitrag zur Ethologie des kleinen Pandas (*Ailurus fulgens*, Cuvier, 1825). Inaugural-Dissertation, Universität Zürich.

Kelly, R. M. 1975. Playing the weight away. Woman's Day, Nov. 1975:2.

Kennion, T. A. 1921. A baby hog deer in captivity. J. Bombay Nat. Hist. Soc. 28:271–273.

Keul, J., E. Doll & D. Keppler. 1972. Energy metabolism of human muscle (Med. & Sport 7). University Park Press, Baltimore, London & Tokyo.

Kiley-Worthington, M. 1976. The tail movements of ungulates, canids and felids with particular reference to their causation and function as displays. Behaviour 56:69–115.

Kilham, L. 1974. Play in hairy, downy, and other woodpeckers. Wilson Bull. 86:35–42.

King, J. A. 1955. Social behavior, social organization, and population dynamics in a black-tailed prairie dog town in the Black Hills of South Dakota. Univ. Michigan, Contrib. Lab. Vert. Biol., No. 67.

Kinloch, A. P. 1926. The Nilgiri Tahr (*Hemitragus hylocrius*). J. Bombay Nat. Hist. Soc. 31:520–521.

Kinsey, K. P. 1977. Agonistic behavior and social organization in a reproductive population of Allegheny woodrats, *Neotoma floridana magister*. J. Mammal. 58:417–419.

Kinzey, W. G., A. L. Rosenberger, P. S. Heisler, D. L. Prowse & J. S. Trilling. 1977. A preliminary field investigation of the yellow handed titi monkey, *Callicebus torquatus torquatus*, in northern Peru. Primates 18:159–181.

Kirchshofer, R. 1960. Einige Verhaltensbeobachtungen an einem Guereza-Jungen *Colobus polykomos kikuyuensis* unter besonderer Berücksichtigung des Spiels. Z. Tierpsychol. 17:506–514.

Kirchshofer, R., H. Frädrich, D. Podolczak & G. Podolczak. 1967. An account of the physical and behavioural development of the hand-reared gorilla infant *Gorilla g. gorilla* born at Frankfurt Zoo. Int. Zoo Yb. 7:108–113.

Kirchshofer, R., K. Weisse, K. Berenz, H. Klose & I. Klose. 1968. A preliminary account of the physical and behavioural development during the first 10 weeks of the hand-reared gorilla twins, *Gorilla g. gorilla*, born at Frankfurt Zoo. Int. Zoo Yb. 8:121–128.

Kirsch, J. A. W. 1977. Biological aspects of the marsupial-placental dichotomy: a reply to Lillegraven. Evolution 31:898–900.

Kitchen, D. W. 1974. Social behavior and ecology of the pronghorn. Wildl. Monogr. 38.

Kleiman, D. G. 1971. The courtship and copulatory behaviour of the green acouchi, *Myoprocta pratti*. Z. Tierpsychol. 29:259–278.

Kleiman, D. G. 1972. Social behavior in the bush dog (*Speothos venaticus*) and maned wolf (*Chrysocyon brachyurus*): a study in contrast. J. Mammal. 53:791–806.

Kleiman, D. G. 1974. Patterns of behaviour in hystricomorph rodents. Symp. Zool. Soc. Lond. 34:171–209.

Kleiman, D. G. 1977. Progress and problems in lion tamarin, *Leontopithecus rosalia rosalia,* reproduction. Int. Zoo Yb. 17:92–97.

Kleiman, D. G. 1979. Parent-offspring conflict and sibling competition in a monogamous primate. Am. Nat. 114:753–760.

Kleiman, D. G. & C. A. Brady. 1978. Coyote behavior in the context of recent canid research: problems and perspectives. Pages 163–188 in M. Bekoff, ed., Coyotes: biology, behavior, and management. Academic Press, N.Y.

Kleiman, D. G. & L. R. Collins. 1972. Preliminary observations on scent-marking, social behavior, and play in the juvenile giant panda, *Ailuropoda melanoleuca.* Am. Zool. 12:644.

Kleiman, D. G. & J. F. Eisenberg. 1973. Comparisons of canid and felid social systems from an evolutionary perspective. Anim. Behav. 21:637–659.

Klein, L. L. 1971. Observations on copulation and seasonal reproduction of two species of spider monkeys, *Ateles belzebuth* and *A. geoffroyi.* Folia Primatol. 15:233–248.

Klingel, H. 1967. Soziale Organisation und Verhalten freilebender Steppenzebras. Z. Tierpsychol. 24:580–624.

Klingel, H. 1968. Soziale Organisation und Verhaltensweisen von Hartmann- und Bergzebras (*Equus zebra hartmannae* und *E. z. zebra*). Z. Tierpsychol. 25:76–88.

Klingel, H. 1969. The social organisation and population ecology of the Plains zebra (*Equus quagga*). Zoologica Afr. 4:249–263.

Klingel H. 1974a. A comparison of the social behaviour of the Equidae. Pages 124–132 in V. Geist & F. Walther, eds., The behavior of ungulates and its relation to management. IUCN Publications, N. S. 24. Morges, Switzerland.

Klingel, H. 1974b. Soziale Organisation und Verhalten des Grevy-Zebras (*Equus grevyi*). Z. Tierpsychol. 36:37–70.

Klinger, E. 1971. Structure and functions of fantasy. Wiley, N.Y.

Klopfer, P. H. 1970. Sensory physiology and esthetics. Am. Sci. 58:399–403.

Klopfer, P. H. 1972. Patterns of maternal care in lemurs: II. Effects of group size and early separation. Z. Tierpsychol. 30:277–296.

Klopfer, P. H. & M. S. Klopfer. 1970. Patterns of maternal care in lemurs. I. Normative description. Z. Tierpsychol. 27:984–996.

Klugh, A. B. 1927. Ecology of the red squirrel. J. Mammal. 8:1–32.

Knappen, P. 1930. Play instinct in gulls. Auk. 47:551–552.

Knight, S. W. 1976. Dominance hierarchies of coyote litters. Paper presented at annual meeting of Animal Behavior Society, Boulder, Colo., June 20–25, 1976.

Koehler, O. 1966. Vom Spiel bei Tieren. Freiburger Dies Universitatis 13:1–32.

Koenig, L. 1957. Beobachtungen über Reviermarkung sowie Droh-, Kampf- und Abwehrverhalten des Murmeltieres (*Marmota marmota* L.). Z. Tierpsychol. 14:510–521.

Koenig, L. 1960. Das Aktionssystem des Siebenschläfers (*Glis glis* L.). Z. Tierpsychol. 17:427–505.

Koenig, L. 1970. Zur Forpflanzung und Jugendentwicklung des Wüstenfuchses (*Fennecus zerda* Zimm. 1780). Z. Tierpsychol. 27:205–246.

Koenig, O. 1959. *Phacochoerus aethiopicus africanus*—Spiel der Jungtiere. Encyclopedia Cinematographica Film E202. Institut für den Wissenschaftlichen Film, Göttingen.

Koenig, O. 1972. *Thalarctos maritimus* (Ursidae)-Spielen. Encyclopedia Cinematographica Film E1308. Institut für den Wissenschaftlichen Film, Göttingen.

Koestler, A. 1964. The act of creation. Macmillan, N. Y.

Koestler, A. 1976. Association and bisociation. Pages 643–649 in J. S. Bruner, A. Jolly & K. Sylva, eds., Play: Its role in development and evolution. Basic Books, N.Y.

Koestler, A. & J. R. Smythies, eds. 1970. Beyond reductionism: new perspectives in the life sciences. Macmillan, N.Y.

Koford, C. B. 1957. The vicuña and the puna. Ecol. Monogr. 27:153–219.

Koford, C. B. 1963. Group relations in an island colony of rhesus monkeys. Pages 136–152 in C. H. Southwick, ed., Primate social behavior. Van Nostrand-Reinhold, Princeton, N.J.

Köhler, W. 1926. The mentality of apes. Harcourt, Brace & Co., N.Y.

Konner, M. 1972. Aspects of the developmental ethology of a foraging people. Pages 285–304 in N. Blurton Jones, ed., Ethological studies of child behavior. University Press, Cambridge.

Konner, M. 1975. Relations among infants and juveniles in comparative perspective. Pages 99–129 in M. Lewis & L. A. Rosenblum, eds., Friendship and peer relations. Wiley, N.Y.

Konner, M. 1976a. Relations among infants and juveniles in comparative perspective. Soc. Sci. Inform. 15:371–402.

Konner, M. 1976b. Maternal care, infant behavior and development among the !Kung. Pages 218–245 in R. B. Lee & I. DeVore, eds., Kalahari huntergatherers: Studies of the !Kung San and their neighbors. Harvard University Press, Cambridge, Mass. & London.

Konner, M. 1977a. Evolution of human behavior development. Pages 69–109 in P. H. Leiderman & S. Tulkin, eds., Culture and infancy: variations in the human experience. Academic Press, N.Y.

Konner, M. 1977b. Infancy among the Kalahari Desert San. Pages 287–328 in P. H. Leiderman & S. Tulkin, eds., Culture and infancy: variations in the human experience. Academic Press, N.Y.

Krämer, A. 1969. Soziale Organisation und Sozialverhalten einer Gemspopulation (*Rupicapra rupicapra* L.) der Alpen. Z. Tierpsychol. 26:889–964.

Krämer, G. 1961. Beobachtungen an einem von uns aufgezogenem Wolf. Z. Tierpsychol. 18:91–109.

Krebs, J. 1978. Beau Geste and song repetition: a reply to Slater. Anim. Behav. 26:304–305.

Krebs, J. R. & N. B. Davies, eds. 1978. Behavioural ecology: an evolutionary approach. Sinauer, Sunderland, Mass.

Krishnan, M. 1972. An ecological survey of the larger mammals of peninsular India. J. Bombay Nat. Hist. Soc. 69:26–54.

Kritzler, H. 1952. Observations on the pilot whale in captivity. J. Mammal. 33:321–334.

Krott, P. 1953. Das Wiederfinden eines auf einem Transport entsprungenem Vielfraßes. Z. Tierpsychol. 10:254–268.

Krott, P. 1960. Der Vielfrass. Ziemsen Verlag. Wittenberg.

Krott, P. 1961. Der gefahrliche Braunbär (Ursus arctos L. 1758). Z. Tierpsychol. 18:245–256.

Krott, P. & G. Krott. 1963. Zum Verhalten des Braunbären (Ursus arctos L. 1758) in den Alpen. Z. Tierpsychol. 20:160–206.

Kruijt, J. P. 1964. Ontogeny of social behaviour in Burmese red jungle fowl (Gallus gallus spadiceus). Behaviour, Suppl. 12:1–201.

Kruuk, H. 1972. The spotted hyena. Univ. of Chicago Press, Chicago.

Kruuk, H. 1975. Hyaena. Oxford University Press, London & N.Y.

Kuczura, A. 1973. Piecewise Markov processes. SIAM J. Appl. Math. 24:169–181.

Kühme, W. 1961. Beobachtungen an afrikanischen Elefanten (Loxodonta africana Blumenbach 1797) in Gefangenschaft. Z. Tierpsychol. 18:285–296.

Kühme, W. 1963. Ergänzende Beobachtungen an afrikanischen Elefanten (Loxodonta africana Blumenbach 1797) in Freigehege. Z. Tierpsychol. 20:66–79.

Kühme, W. 1965. Freilandstudien zur Soziologie des Hyänenhundes. Z. Tierpsychol. 22:495–541.

Kuhn, T. S. 1962. The structure of scientific revolutions. University of Chicago Press, Chicago.

Kummer, H. 1967. Tripartite relations in hamadryas baboons. Pages 63–71 in S. A. Altmann, ed., Social communication among primates. University of Chicago Press, Chicago.

Kummer, H. 1968a. Social organization of hamadryas baboons: a field study. Bibliotheca primatologica, no. 6. Karger, Basel.

Kummer, H. 1968b. Two variations in the social organization of baboons. Pages 293–312 in P. C. Jay, ed., Primates: studies in adaptation and variability. Holt, Rinehart & Winston, N.Y.

Kummer, H. 1971. Primate societies. Aldine, Chicago.

Kummer, H. & F. Kurt. 1965. A comparison of social behavior in captive and wild hamadryas baboons. Pages 65–80 in H. Vagtborg, ed., The baboon in medical research. University of Texas Press, Austin.

Kunkel, P. & I. Kunkel. 1964. Beiträge zur Ethologie des Hausmeerschweinchens Cavia aperea f. porcellus (L.). Z. Tierpsychol. 21: 602–641.

Kuo, Z. Y. 1930. The genesis of the cat's responses to the rat. J. Comp. Psychol. 11:1–35.

Kurland, J. A. 1973. A natural history of kra macaques (Macaca fascicularis Raffles, 1821) at the Kutai Reserve, Kalimantan Timur, Indonesia. Primates 14:245–262.

Kurland, J. A. 1977. Kin selection in the Japanese monkey. Contributions to Primatology 12. Karger, Basel & N.Y.

Kuschinski, L. 1974. Breeding binturongs, *Arctictis binturong,* at Glasgow Zoo. Int. Zoo Yb. 14:124–126.

Lack, D. 1966. Population studies of birds. Oxford University Press, Oxford.

Lahiri, R. K. & C. H. Southwick. 1966. Parental care in *Macaca sylvana.* Folia Primatol. 4:257–264.

Lamond, H. G. 1953. Kangaroo. John Day, N.Y.

Lancaster, J. B. 1971. Play-mothering: the relation between juvenile females and young infants among free-ranging vervet monkeys (*Cercopithecus aethiops*). Folia Primatol. 15:161–182.

Landowski, J. 1972. Erste gelungene künstliche Aufzucht von Binturongs (*Arctictis binturong* Raffl.) im Warschauer Zoo. Zool. Gart. 42:38–50.

Langfeldt, T. 1974. Diazepam-induced play behavior in cats during prey killing. Psychopharm. 36:181–184.

Langman, V. A. 1977. Cow-calf relationships in giraffe (*Giraffa camelopardalis giraffa*). Z. Tierpsychol. 43:264–286.

Lasch, C. 1978. Culture of narcissism: American life in an age of diminishing expectations. Norton, N.Y.

Latta, J., S. Hopf & D. Ploog. 1967. Observation on mating behaviour and sexual play in the squirrel monkey (*Saimiri sciureus*). Primates 8:229–245.

Laundré, J. 1977. The daytime behaviour of domestic cats in a free-roaming population. Anim. Behav. 25:990–998.

Laws, R. M. 1959. Accelerated growth in seals, with special reference to the Phocidae. Norsk Hvalfangsttidende 9:425–451.

Laws, R. M., I. S. C. Parker & R. C. B. Johnstone. 1975. Elephants and their habitats: the ecology of elephants in North Bunyoro, Uganda. Clarendon Press, Oxford.

Lazar, J. W. & G. D. Beckhorn. 1974. Social play or the development of social behavior in ferrets (*Mustela putorius*)? Am. Zool. 14:405–414.

Lazell, J. D. jr. & N. C. Spitzer. 1977. Apparent play behavior in the American alligator. Copeia, 188.

Lebret, T. 1948. The diving play of surface-feeding duck. Br. Birds 41:247.

Lee, R. B. & I. DeVore, eds. 1968. Man the hunter. Aldine, Chicago.

Lee, R. B. & I. DeVore, eds. 1976. Kalahari hunter-gatherers: studies of the !Kung San and their neighbors. Harvard University Press, Cambridge, Mass.

Leen, N. & A. Novick. 1969. The world of bats. Holt, Rinehart & Winston, N.Y.

Lehner, P. N. 1978. Coyote communication. Pages 127–162 in M. Bekoff, ed., Coyotes: biology, behavior, and management. Academic Press, N.Y.

Leighton, A. H. 1933. Notes on the relations of beavers to one another and to the muskrat. J. Mammal. 14:27–35.

Lent, P. C. 1966. Calving and related social behavior in the barren-ground caribou. Z. Tierpsychol. 23:701–756.

Lent, P. C. 1969. A preliminary study of the Okavango lechwe (*Kobus leche leche* Gray). E. Afr. Wildl. J. 7:147–157.

Lent, P. C. 1974. Mother-infant relationships in ungulates. Pages 14–55 in V.

Geist & F. Walther, eds., The behavior of ungulates and its relation to management, Vol. 1. International Union for the Conservation of Nature. Morges, Switzerland.

Leresche, L. A. 1976. Dyadic play in hamadryas baboons. Behaviour 57:190–205.

Lethmate, J. 1976. Gebrauch und Herstellung von Trinkwerkzeugen bei Orangutans. Zool. Anz. 197:251–263.

Lethmate, J. 1977. Werkzeugherstellung eines jungen Orang-Utans. Behaviour 62:174–189.

Leuthold, W. 1970. Observations on the social organization of Impala (*Aepyceros melampus*). Z. Tierpsychol. 27:693–721.

Leuthold, W. 1977. African ungulates: A comprehensive review of their ethology and behavioral ecology. Springer-Verlag, Berlin, Heidelberg & New York.

Leuthold, W. & B. M. Leuthold. 1973. Notes on the behaviour of two young antelopes reared in captivity. Z. Tierpsychol. 32:418–424.

Levine, S. 1962. Psychophysiological effects of infantile stimulation. Pages 246–253 in E. L. Bliss, ed., Roots of behavior. Harper & Row, N.Y.

Levins, R. 1968. Evolution in changing environments. Princeton University Press, Princeton, N.J.

Levins, R. 1969. Thermal acclimation and heat resistance in *Drosophila* species. Am. Nat. 103:493–500.

Levins, R. 1975. Evolution in communities near equilibrium. Pages 16–50 in M. L. Cody & J. M. Diamond, eds., Ecology and evolution of communities. Belknap Press of Harvard University Press, Cambridge.

Levitsky, D. A. & R. H. Barnes. 1972. Nutritional and environmental interactions in the behavioral development of the rat: Long-term effects. Science 176:68–72.

Levy, J. 1979. Play behavior and its decline during development in rhesus monkeys (*Macaca mulatta*). Ph.D. diss., University of Chicago.

Lewontin, R. C. 1978. Adaptation. Sci. Am. 239 (3):212–230.

Leyhausen, P. 1949. Beobachtungen an einem jungen Schwarzbären (*Ursus americanus* Pall.) Z. Tierpsychol. 6:433–444.

Leyhausen, P. 1953. Beobachtungen an einer brasilianischen Tigerkatze. Z. Tierpsychol. 10:77–91.

Leyhausen, P. 1963. Über südamerikanische Pardelkatzen. Z. Tierpsychol. 20:627–640.

Leyhausen, P. 1965. The communal organization of solitary mammals. Symp. Zool. Soc. Lond. 14:249–263.

Leyhausen, P. 1973a. Addictive behavior in free-ranging animals. Bayer Symposium IV, Psychic Dependence: 58–64.

Leyhausen, P. 1973b. On the function of the relative hierarchy of moods (as exemplified by the phylogenetic and ontogenetic development of prey-catching in carnivores). Pages 144–247 in K. Lorenz & P. Leyhausen, Motivation of human and animal behavior. Van Nostrand Reinhold N.Y.

Leyhausen, P. 1973c. Verhaltensstudien an Katzen. Z. Tierpsychol., Supp. 2, 3rd ed. Paul Parey, Berlin & Hamburg.

Leyhausen, P. 1979. Cat behavior. (Trans. B. A. Tonkin.) Garland STPM Press, N.Y.

Lichstein, L. 1973. Play in rhesus monkeys: I. Definition. II. Diagnostic significance. Ph. D. diss., Univ. Wisconsin, Madison.

Lieberman, J. N. 1977. Playfulness. Its relationship to imagination and creativity. Academic Press, N.Y.

Liers, E. E. 1951. Notes on the river otter (*Lutra canadensis*). J. Mammal. 32:1–9.

Limbaugh, C. 1961. Observations on the California sea otter. J. Mammal. 42:271–273.

Lindemann, W. 1955. Über die Jugendentwicklung beim Luchs (*Lynx l. lynx* Kerr) und bei der Wildkatze (*Felis s. sylvestris* Schreb.). Behaviour 8:1–45.

Lindemann, W. & W. Rieck. 1953. Beobachtungen bei der Aufzucht von Wildkatzen. Z. Tierpsychol. 10:92–119.

Linsdale, J. M. 1946. The California ground squirrel. A record of observations made on the Hastings Natural History Reservation. University of California Press, Berkeley.

Linsdale, J. M. & P. Q. Tomich. 1953. A herd of mule deer. University of California Press, Berkeley.

Lockley, R. M. 1974. The private life of the rabbit. Macmillan. N.Y.

Lockwood, R. 1976. An ethological analysis of social structure and affiliation in captive wolves (*Canis lupus*). Ph.D. diss., Washington Univ. St. Louis, Mo.

Löhmer, R. 1976. Zur Verhaltensontogenese bei *Procyon cancrivorus cancrivorus* (Procyonidae). Z. Säugetierk. 41:42–58.

Loizos, C. 1966. Play in mammals. Symp. Zool. Soc. Lond. 18:1–9.

Loizos, C. 1967. Play behaviour in higher primates: a review. Pages 176–218 in D. Morris, ed., Primate ethology. Aldine, Chicago.

Loizos, C. 1969. An ethological study of chimpanzee play. Pages 87–93 in C. R. Carpenter, ed., Proceedings of the Second International Congress on Primatology, Atlanta, Georgia, Vol. 1. Karger, N.Y.

Lombardi, J. R. & J. G. Vandenbergh. 1977. Pheromonally induced sexual maturation in females: regulation by the social environment of the male. Science 196:545–546.

Łomnicki, A. & L. B. Slobodkin. 1966. Floating in *Hydra littoralis*. Ecology 47:881–889.

Lorenz, K. Z. 1952. King Solomon's ring. Thomas Y. Crowell, N.Y.

Lorenz, K. Z. 1955. Man meets dog. Trans. M. K. Wilson. Methuen, London.

Lorenz, K. Z. 1956. Plays and vacuum activities. Pages 633–645 in M. Autuori et al., eds., L'instinct dans le comportement des animaux et de l'homme. Masson et Cie., Paris.

Lorenz, K. Z. & P. Leyhausen. 1973. Motivation of human and animal behavior. Van Nostrand Reinhold, N.Y.

Lorenz, R. 1969. Zur Ethologie des Kalifornischen Seelöwen (*Zalophus califor-*

nianus Lesson 1828) und anderer Robben in Zoologischen Gärten. Zool. Gart. 37:181–192.

Losey, G. S., Jr. 1978. Information theory and communication. Pages 43–78 in P. W. Colgan, ed., Quantitative ethology. Wiley-Interscience, N.Y.

Loveridge, A. 1923. Notes on East African mammals, collected 1920–1923. Proc. Zool. Soc. Lond. 1923:685–739.

Low, B. S. 1978. Environmental uncertainty and the parental strategies of marsupials and placentals. Am. Nat. 112:197–213.

Low, R. 1970. The Massena's parrot. Avic. Mag. 76:153–154.

Lowther, F. de L. 1940. A study of the activities of a pair of *Galago senegalensis moholi* in captivity, including the birth and postnatal development of twins. Zoologica 25:433–462.

Loy, J. 1970. Behavioral responses of free-ranging rhesus monkeys to food shortage. Am. J. Phys. Anthropol. 33:263–271.

Lucas, N. S., E. M. Hume & H. Henderson. 1927. On the breeding of the common marmoset (*Hapale jacchus*) in captivity when irradiated with ultraviolet rays. Proc. Zool. Soc. Lond. 1927:447–451.

Ludwig, J. 1965. Beobachtungen über das Spiel bei Boxern. Z. Tierpsychol. 22:813–838.

Lumia, A. R. 1972. The relationship between dominance and play behavior in the American buffalo, *Bison bison*. Z. Tierpsychol. 30:416–419.

Lumsden, C. J. & E. O. Wilson. 1981. Genes, mind, and culture: The coevolutionary process. Harvard University Press, Cambridge, Mass.

Lunt, D. C. 1968. Taylors Gut, in the Delaware state. Knopf, N.Y.

Lyon, M. W. 1936. Mammals of Indiana. Am. Midl. Nat. 17:1–373.

MacArthur, R. H. 1972. Geographical ecology. Harper & Row, N.Y.

MacArthur, R. H. & E. R. Pianka. 1966. On optimal use of a patchy environment. Am. Nat. 100:603–609.

MacArthur, R. H. & E. O. Wilson. 1967. The theory of island biogeography. Princeton University Press, Princeton, N.J.

McBride, A. F. & H. Kritzler. 1951. Observations on pregnancy, parturition, and postnatal behavior in the bottlenose dolphin. J. Mammal. 32:251–266.

McBride, G. 1963. The "teat order" and communication in young pigs. Anim. Behav. 11:53–56.

McBride, G. & J. W. James. 1964. Social behavior of domestic animals. IV. Growing pigs. Anim. Prod. 6:129–139.

McBride, G., I. P. Parer & F. Foenander. 1969. The social organization and behaviour of the feral domestic fowl. Anim. Behav. Monogr. 2:127–181.

McCabe, R. A. & A. S. Hawkins. 1946. The Hungarian partridge in Wisconsin. Am. Midl. Nat. 36:1–75.

McCall, R. B. 1974. Exploratory manipulation and play in the human infant. Mon. Soc. Res. Child Dev. 39:1.

McCann, C. 1933. Observation on some of the Indian langurs. J. Bombay Nat. Hist. Soc. 36:618–628.

McClintock, M. 1974. Sociobiology of reproduction in the Norway rat (*Rattus norvegicus*): Estrous synchrony and the role of the female rat in copulatory behavior. Ph.D. diss., University of Pennsylvania.

McClure, H. E. 1964. Some observation of primates in climax diptocarp forest near Kuala Lumpur, Malaya. Primates 5:39–58.

McCrane, M. P. 1966. Birth, behavior, and development of a hand-reared two-toed sloth, *Choleopus didactylus.* Int. Zoo Yb. 6:156–163.

McCulloch, W. S. 1945. A heterarchy of values determined by the topology of nervous nets. Bull. Math. Biophys. 7:89–93.

McCullough, D. F. 1969. The tule elk: its history, behavior and ecology. Univ. Calif. Publ. Zool. 88:1–209.

McDonald, D. 1977. Play and exercise in the California ground squirrel (*Spermophilus beecheyi*). Anim. Behav. 25:782–784.

McDougall, P. 1975. The feral goats of Kielderhead Moor. J. Zool. Lond. 176:215–246.

McFarland, D. 1973. Discussion of paper by K. V. Connolly. Pages 363–365 in Hinde, R. A. & J. G. Stevenson-Hinde, eds., Constraints on learning: limitations and predispositions. Academic Press, London & N.Y.

McFarland, D., ed. 1974. Motivational control systems analysis. Academic Press, N.Y.

McFarland, D. 1976. Form and function in the temporal organisation of behaviour. Pages 55–93 in P. P. G. Bateson & R. A. Hinde, eds., Growing points in ethology. Cambridge University Press, Cambridge & N.Y.

McFarland, D. & R. M. Sibly. 1975. The behavioural final common path. Philos. Trans. R. Soc. Lond. Ser. B. Biol. Sci. 270:265–293.

McGrew, W. C. 1972a. Aspects of social development in nursery school children with emphasis on introduction to the group. Pages 129–156 in N. Blurton Jones, ed., Ethological studies of child behavior. Cambridge University Press, Cambridge.

McGrew, W. C. 1972b. An ethological study of children's behavior. Academic Press, New York & London.

McGrew, W. C. 1977. Socialization and object manipulation of wild chimpanzees. Pages 261–288 in S. Chevalier-Skolnikoff & F. E. Poirier, eds., Primate bio-social development. Garland, N.Y.

McGuire, M. T. & members of the Behavioral Sciences Foundation. 1974. The St. Kitts vervet. Contributions to Primatology, Vol. 1. Karger, Basel.

McHugh, T. 1958. Social behaviour of the American buffalo (*Bison bison bison*). Zoologica 43:1–40.

McKay, G. M. 1973. Behavior and ecology of the Asiatic elephant in southeastern Ceylon. Smithson. Contrib. Zool. 125.

MacKinnon, J. 1974a. The behaviour and ecology of wild orang-utangs (*Pongo pygmaeus*). Anim. Behav. 22:3–74.

MacKinnon, J. 1974b. In search of the red ape. Holt, Rinehart and Winston, N.Y.

McNab, A. G. & M. C. Crawley. 1975. Mother and pup behavior of the New Zealand fur seal, *Arctocephalus forsteri* (Lesson). Mauri Ora 3:77–88.

MacNair, M. R. & G. A. Parker. 1978. Models of parent-offspring conflict. I. Monogamy. Anim. Behav. 26:97–110.

MacRoberts, M. H. 1970. The social organization of Barbary apes (*Macaca sylvana*) on Gibraltar. Am. J. Phys. Anthropol. 33:83–99.

Manns, T. 1978. Query on raven play. Harvard Magazine 80(3):71.

Maple, T. & E. L. Zucker. 1978. Ethological studies of play behavior in captive great apes. Pages 113–142 in E. O. Smith, ed., Social play in primates. Academic Press, N.Y.

Marler, P. 1956. Behavior of the chaffinch, Fringilla coelebs. Behaviour Supp. 5.

Marler, P. 1970. A comparative approach to vocal development: song learning in the white-crowned sparrow. J. Comp. Physiol. Psychol. 71:1–25.

Marler, P. 1975. On strategies of behavioural development. Pages 254–275 in G. Baerends, C. Beer & A. Manning, eds., Function and evolution in behaviour. Clarendon Press, Oxford.

Marler, P. 1977. The evolution of communication. Pages 45–70 in T. A. Sebeok, ed., How animals communicate. Indiana University Press, Bloomington & London.

Marler, P. & W. J. Hamilton, III. 1966. Mechanisms of animal behavior. Wiley, N.Y.

Marler, P. & R. Tenaza. 1977. Signaling behavior of apes with special reference to vocalization. Pages 965–1033 in T. A. Sebeok, ed., How animals communicate. Indiana University Press, Bloomington & London.

Marler, P. & J. van Lawick-Goodall. 1971. Vocalizations of wild chimpanzees. Rockefeller University Film Service, N.Y.

Marlow, B. J. 1975. The comparative behaviour of the Australasian sea lions Neophoca cinerea and Phocarctos hookeri (Pinnipedia: Otariidae). Mammalia 39:159–230.

Marr, D. 1969. A theory of cerebellar cortex. J. Physiol. 202:437–470.

Marr, D. 1970. A theory for cerebral neocortex. Proc. R. Soc. Lond. Ser. B. Biol. Sci. 176:161–234.

Marr, D. 1971. Simple memory: a theory for archicortex. Philos. Trans. R. Soc. Lond. Ser. B. Biol. Sci. 262:23–81.

Marsden, H. M. 1972. The effect of food deprivation on intergroup relations in rhesus monkeys. Behav. Biol. 7:369–374.

Martin, R. D. 1968. Reproduction and ontogeny in tree shrews (Tupaia belangeri) with reference to their general behavior and taxonomic relationships. Z. Tierpsychol. 25:409–495, 505–532.

Martin, R. D. 1972. Adaptive radiation and behaviour of the Malagasy lemurs. Philos. Trans. R. Soc. Lond. Ser. B. Biol. Sci. 264:295–352.

Martin, R. D., G. A. Doyle & A. C. Walker, eds. 1974. Prosimian biology. University of Pittsburgh Press, Pittsburgh.

Marx, D., H.-J. Schrenk & C. Schmidtborn. 1977. Spiel- und Eliminationsverhalten von Säugferkeln und frühabgesetzten Ferkeln in Käfiggruppenhaltung (Flatdecks). Dtsch. Tierärztl. Wochenschr. 84(4):141–149.

Mason, W. A. 1965a. Determinants of social behavior in young chimpanzees. Pages 335–364 in A. M. Schrier, H. F. Harlow & F. Stollnitz, eds., Behavior of nonhuman primates, Vol. 2. Academic Press, N.Y.

Mason, W. A. 1965b. The social development of monkeys and apes. Pages 514–543 in I. DeVore, ed., Primate behavior. Holt, Rinehart & Winston, N.Y.

Mason, W. A. 1967. Motivational aspects of social responsiveness in young chimpanzees. Pages 103–126 in H. W. Stevenson, ed., Early behavior: comparative and developmental approaches. Wiley, N.Y.

Mason, W. A. 1971. Motivational factors in psychosocial development. Pages 35–67 in W. J. Arnold & M. M. Page, eds., Nebraska symposium on motivation. University of Nebraska Press, Lincoln.

Mason, W. A. 1974. Comparative studies of social behavior in *Callicebus* and *Saimiri:* Behavior of male-female pairs. Folia Primatol. 22:1–8.

Mason, W. A. 1978. Social experience and primate cognitive development. Pages 223–251 in G. M. Burghardt & M. Bekoff, eds., The development of behavior: comparative and evolutionary aspects. Garland STPM, N.Y.

Mason, W. A. & G. Berkson. 1975. Effects of maternal mobility on the development of rocking and other behaviors in rhesus monkeys: a study with artificial mothers. Dev. Psychobiol. 8:197–211.

Mason, W. A., J. H. Hollis & L. G. Sharpe. 1962. Differential responses of chimpanzees to social stimulation. J. Comp. Physiol. Psychol. 55:1105–1110.

Maxwell, G. 1961. Ring of bright water. Dutton, N.Y.

May, R. M. 1974. Biological populations with nonoverlapping generations: stable points, stable cycles, and chaos. Science 186:645–647.

May, R. M. 1977. Population genetics and cultural inheritance. Nature 268:11–13.

Maynard Smith, J. 1977. Parental investment: a prospective analysis. Anim. Behav. 25:1–9.

Maynard Smith, J. 1978. The evolution of sex. Cambridge University Press, Cambridge & N.Y.

Maynard Smith, J. & G. A. Parker. 1976. The logic of asymmetric contests. Anim. Behav. 24:159–175.

Maynard Smith, J. & G. R. Price. 1973. The logic of animal conflict. Nature 246:15–18.

Mayr, E. 1963. Animal species and evolution. Belknap Press of Harvard University Press, Cambridge.

Mead, M. 1976. Towards a human science. Science 191:903–910.

Mears, C. E. & H. F. Harlow. 1975. Play: early and eternal. Proc. Nat. Acad. Sci. (U.S.A.) 72:1878–1882.

Mech, L. D. 1966. The wolves of Isle Royale. Fauna of the National Parks of the US—Fauna Series 7. US Government Printing Office, Washington.

Mech, L. D. 1970. The wolf: the ecology and behavior of an endangered species. Natural History Press, Garden City, N.Y.

Meder, E. 1958. *Gnathonemus petersii* (Günther). Z. Vivaristik (Mannheim) 4:161–171.

Meester, J. & H. W. Setzer, eds. 1971. The mammals of Africa: an identification manual. Smithsonian Institution Press, Washington, D.C.

Meier, E. 1973. Beiträge zur Geburt des Damwildes (*Cervus dama* L.). Z. Säugetierk. 38:348–373.

Meischner, I. 1959. Verhaltensstudien an Pelikanen. Zool. Gart. 25:104–126.

Melchior, F. 1976. Künstliche Aufzucht von Wüstenfüchsen (*Fennecus zerda*). Zool. Gart. 46:431–440.

Mellers, W. 1973. Twilight of the gods: the music of the Beatles. Viking, N.Y.

Menzel, E. W., Jr. 1963. The effects of cumulative experience on responses to novel objects in young isolation-reared chimpanzees. Behaviour 21:1–12.

Menzel, E. W., Jr. 1969. Chimpanzee utilization of space and responsiveness to objects: age differences and comparison with macaques. Pages 72–80 in C. R. Carpenter, ed., Proceedings of the Second International Congress of Primatology, Vol. 1. Karger, Basel.

Menzel, E. W., Jr. 1971. Group behavior in young chimpanzees: responsiveness to cumulative novel changes in a large outdoor enclosure. J. Comp. Physiol. Psychol. 74:46–51.

Menzel, E. W., Jr. 1972. Spontaneous invention of ladders in a group of young chimpanzees. Folia Primatol. 17:87–106.

Menzel, E. W., Jr. 1973. Further observations on the use of ladders in a group of young chimpanzees. Folia Primatol. 19:450–457.

Menzel, E. W., Jr. 1974. A group of young chimpanzees in a one-acre field. Behav. Nonhuman Primates 5:83–153.

Menzel, E. W., Jr. 1975. Communication and aggression in a group of young chimpanzees. Adv. Stud. Commun. and Affect 2:103–133.

Menzel, E. W., Jr., R. K. Davenport, Jr. & C. M. Rogers. 1961. Some aspects of behavior toward novelty in young chimpanzees. J. Comp. Physiol. Psychol. 54:16–19.

Menzel, E. W., Jr., R. K. Davenport, Jr. & C. M. Rogers. 1970. The development of tool using in wild-born and restriction-reared chimpanzees. Folia Primatol. 12:273–283.

Menzel, E. W., Jr., R. K. Davenport, Jr. & C. M. Rogers. 1972. Protocultural aspects of chimpanzee responsiveness to novel objects. Folia Primatol. 17:161–170.

Merkt, H. & A.-R. Günzel. 1979. A survey of early pregnancy losses in West German thoroughbred mares. Equine Vet. J. 11:256–258.

Merrick, N. J. 1977. Social grooming and play behavior of a captive group of chimpanzees. Primates 18:215–224.

Mesarovic, M. D., D. Macko & Y. Takahara. 1971. Theory of hierarchical, multilevel systems. Academic Press, N.Y.

Meshkova, N. N. 1970. Ob ontogeneze igrovogo povedeniya u nektotorykh khishchnykh i parnokopytnykh mlekopitayushchikh. (Ontogenesis of play behavior in some predatory and artiodactyl mammals.) Zool. Zh. 49:907–915.

Messmer, E. & I. Messmer. 1956. Die Entwicklung der Lautäußerungen und einiger Verhaltensweisen der Amsel (*Turdus merula merula* L.) unter natürlichen Bedingungen und nach Einzelaufzucht in schalldichten Raumen. Z. Tierpsychol. 13:341–441.

Metz, H. 1974. Stochastic models for the temporal fine structure of behaviour sequences. Pages 5–86 in D. J. McFarland, ed., Motivational control systems analysis. Academic Press, London.

Meyer, P. 1972. Zur Biologie und Okologie des Atlashirsches *Cervus elaphus barbarus*, 1833. Z. Säugetierk. 37:101–116.

Meyer-Holzapfel, M. 1956a. Über die Bereitschaft zu Spiel- und Instinkthandlungen. Z. Tierpsychol. 13:442–462.

Meyer-Holzapfel, M. 1956b. Das Spiel bei Säugetieren. Handb. Zool. Berl. 8(10):1–36.

Meyer-Holzapfel, M. 1958. Boquetins en captivité. Mammalia 22:90–103.

Meyer-Holzapfel, M. 1960. Über das Spiel bei Fischen, insbesondere beim Tapirrüsselfisch (*Mormyrus kannume* Forskål) Zool. Gart. 25:189–202.

Meyer-Holzapfel, M. 1968. Zur Bedeutung verschiedener Holz- und Laubarten für den Braunbären. Zool. Gart. 36:12–33.

Meyer-Holzapfel, M. & H. Räber. 1976. Zur Ontogenese des Beutefangs beim Waldkauz (*Strix a. aluco* L.). Beobachtungen und Experimente. Behaviour 57:1–50.

Michael, E. D. 1968. Playing by white-tailed deer in south Texas. Am. Midl. Nat. 80:535–537.

Michod, R. E. 1979. Evolution of life histories in response to age-specific mortality factors. Am. Nat. 113:531–550.

Milhaud, C., M. Klein & G. Chapouthier. 1973. Le comportement social de jeu des chimpanzes en tant que test psychopharmacologique. Psychopharm. 32:293–300.

Millar, J. S. 1977. Adaptive features of mammalian reproduction. Evolution 31:370–386.

Miller, E. H. 1975a. Annual cycle of fur seals, *Arctocephalus forsteri* (Lesson), on the Open Bay Islands, New Zealand. Pac. Sci. 29:139–152.

Miller, E. H. 1975b. A comparative study of facial expressions of two species of pinnipeds. Behaviour 53:268–284.

Miller, F. L. 1975. Play activities of black-tailed deer in northwestern Oregon. Can. Field Nat. 89:149–156.

Miller, S. 1973. Ends, means and galumphing: some leitmotifs of play. Am. Anthropol. 75:87–98.

Miller, W. J. & L. S. Miller. 1958. Synopsis of behaviour traits of the ring neck dove. Anim. Behav. 6:3–8.

Millikan, G. C. & R. I. Bowman. 1967. Observations on Galapagos tool-using finches in captivity. Living Bird 6:23–42.

Milne, L. 1924. The home of an Eastern clan: A study of the Palaungs of the Shan States. Clarendon Press, Oxford.

Minett, F. C. 1947. Notes on a flying squirrel (*Petaurista* sp.). J. Bombay Nat. Hist. Soc. 47:52–56.

Minsky, M. 1975. A framework for representing knowledge. Pages 211–277 in P. H. Winston, ed., The psychology of computer vision. McGraw-Hill, N.Y.

Minsky, M. 1977. Plain talk about neurodevelopmental epistemology. MIT AI Laboratory Artificial Intelligence Memo No. 430. Massachusetts Institute of Technology, Cambridge, Mass.

Mitchell, A. W. 1977. Preliminary observations on the daytime activity patterns of lesser kudu in Tsavo National Park, Kenya. E. Afr. Wildl. J. 15:199–206.

Mitchell, G. D. 1968. Persistent behavior pathology in rhesus monkeys following early social isolation. Folia Primatol. 8:132–147.

Mizuhara, H. 1964. Social changes of Japanese monkey troops in the Takasakiyama. Primates 5:27–52.

Moehlman, P. D. 1979. Jackal helpers and pup survival. Nature 277:382–383.

Mohr, E. 1928. Epimys rattus in captivity. J. Mammal. 9:113–117.

Mohr, E. 1936a. Biologische Beobachtungen an Solenodon paradoxus Brandt in Gefangenschaft. I. Zool. Anz. 113:177–188.

Mohr, E. 1936b. Biologische Beobachtungen an Solenodon paradoxus Brandt in Gefangenschaft. II. Zool. Anz. 116:65–76.

Mohr, E. 1965. Altweltliche Stachelschweine. Neue Brehm Buch. No. 350.

Mohr, E. 1968. Spielbereitschaft bei Wisent. Z. Säugetierk. 33:116–121.

Mohr, H. 1960. Über die Entwicklung einiger Verhaltensweisen bei handaufgezogenen Sperbern (Accipiter n. nisus L.) und Baumfalken (Falco s. subbuteo L.) Z. Tierpsychol. 17:700–727.

Monfort-Braham, N. 1975. Variations dans la structure sociale du topi, Damaliscus korrigum Ogilby, au Parc National de l'Akagera, Rwanda. Z. Tierpsychol. 39:332–364.

Monod, J. 1971. Chance and necessity. Knopf, N.Y.

Moody, M. I. & E. W. Menzel, Jr. 1976. Vocalizations and their behavioral contexts in the Tamarin Saguinus fuscicollis. Folia Primatol. 25:73–94.

Moore, J. C. 1956. Observations of manatees in aggregations. Am. Mus. Novit. 1811:1–24.

Moreau, R. E. 1938. A contribution to the biology of the Musophagiformes, the so-called plantain-eaters. Ibis 2:639–671.

Moreau, R. E. & W. M. Moreau. 1941. Breeding biology of silvery-cheeked hornbill. Auk. 58:13–27.

Moreau, R. E. & W. M. Moreau. 1944. Do young birds play? Ibis 86:93–94.

Morgan, M. J. 1973. Effects of post-weaning environment on learning in the rat. Anim. Behav. 21:429–442.

Morgan, M. J., D. F. Einon & R. G. M. Morris. 1977. Inhibition and isolation rearing in the rat: extinction and satiation. Physiol. Behav. 18:1–5.

Morgan, M. J., D. F. Einon & D. Nicholas. 1975. Effects of isolation rearing on behavioural inhibition in the rat. Q. Jl. Exp. Psychol. 27:615–634.

Morgan, P. D. & G. W. Arnold. 1974. Behavioural relationships between merino ewes and lambs during the four weeks after birth. Anim. Prod. 19:169–176.

Mori, Umeyo. 1974. The inter-individual relationships involved in social play of the young Japanese monkeys of the natural troop in Koshima Islet. J. Anthropol. Soc. Nippon 82:303–318.

Mörike, D. 1973. Verhalten einer Gruppe von Dianameerkatzen im Frankfurter Zoo. Primates 14:263–300.

Mörike, D. 1976. Verhalten einer Gruppe von Brazzameerkatzen (Cercopithecus neglectus) im Heidelberger Zoo. Primates 17:475–512.

Morris, D. 1962a. The behaviour of the green acouchi (*Myoprocta pratti*) with special reference to scatter hoarding. Proc. Zool. Soc. Lond. 139:701–732.

Morris, D. 1962b. The biology of art. Knopf, N.Y.

Morris, D. 1964. The response of animals to a restricted environment. Symp. Zool. Soc. Lond. 13:99–120.

Morris, D. 1968. Play and the sick child. Lancet 2:1391.

Morris, I. C. 1977. Young blue wren playing? Victorian Nat. 94:131.

Morris, K. & J. Goodall. 1977. Competition for meat between chimpanzees and baboons of the Gombe National Park. Folia Primatol. 28:109–121.

Morris, R. & D. Morris. 1966. Men and pandas. McGraw-Hill, N.Y.

Mowat, F. 1963. Never cry wolf. Atlantic Little Brown, Boston.

Moynihan, M. 1964. Some behaviour patterns of platyrrhine monkeys. I. The night monkey (*Aotus trivirgatus*). Smithson. Misc. Collect. 146:1–84.

Moynihan, M. 1966. Communication in the titi monkey, *Callicebus*. J. Zool. Lond. 150:77–127.

Moynihan, M. 1970. Some behavior patterns of platyrrhine monkeys. II. *Saguinus geoffroyi* and other tamarins. Smithson. Contr. Zool. 28.

Mueller, H. C. 1974. The development of prey recognition and predatory behaviour in the American kestrel *Falco sparverius*. Behaviour 49:313–324.

Mukherjee, R. P. 1969. A field study on the behaviour of two roadside groups of rhesus macaque (*Macaca mulatta* Zimmerman) in northern Uttar Pradesh. J. Bombay Nat. Hist. Soc. 66:47–56.

Mukherjee, R. P. & S. S. Saha. 1974. The golden langurs (*Presbytis geei* Kajuria, 1956) of Assam. Primates 15:327–340.

Müller, H. 1970. Beiträge zur Biologie des Hermelins, *Mustela erminea* Linné, 1758. Säugetierk. Mitt. 18:293–380.

Müller-Schwarze, D. 1966. Experimente zur Triebspezifität des Säugetierspiels. Naturwissenschaften 53:137–138.

Müller-Schwarze, D. 1968. Play deprivation in deer. Behaviour 31:144–162.

Müller-Schwarze, D. 1971. Ludic behavior in young mammals. Pages 229–249 in M. B. Sterman, D. J. McGinty & A. M. Adinolfi, eds., Brain development and behavior. Academic Press, N.Y.

Müller-Schwarze, D. 1978a. Evolution of play behavior. Dowden, Hutchinson & Ross, Stroudsburg, Pa.

Müller-Schwarze, D. 1978b. Play behavior in Adélie penguin fledglings. Pages 375–377 in D. Müller-Schwarze, ed., Evolution of play behavior. Dowden, Hutchinson & Ross, Stroudsburg, Pa.

Müller-Schwarze, D. & C. Müller-Schwarze. 1969. Spielverhalten und allgemeine Aktivität bei Schwarzwedelhirschen. Bonner Zool. Beitr. 10:282–289.

Müller-Schwarze, D. & C. Müller-Schwarze. 1973. Behavioural development of hand-reared pronghorn *Antilocapra americana*. Int. Zoo Yb. 13:217–220.

Müller-Using, D. 1956. Zum Verhalten des Murmeltieres (*Marmota marmota*(L.)). Z. Tierpsychol. 13:135–142.

Müller-Using, D. 1972. Die afrikanische Pelz- oder Bärenrobbe (*Arctocephalus pusillus*). Newsletter S.W.A. Scient. Soc. 11(9–10) 1971:3–7.

Münch, H. 1958. Zur Ökologie und Psychologie von *Marmota m. marmota*. Z. Säugetierk. 23:129–138.

Munro, D. A. 1954. Prairie falcon "playing?" Auk. 71:333–334.

Murie, A. 1944. The wolves of Mt. McKinley. U. S. Government Printing Office, Washington, D.C.

Murie, J. 1869. Report on the eared seals collected by the Society's keeper François Lecomte in the Falkland Islands. Proc. Zool. Soc. Lond. 1869:100–109.

Murie, O. 1954. A field guide to animal tracks. Houghton Mifflin, Boston.

Murray, L. T. 1932. Notes on personal experiences with pronghorn antelope in Texas. J. Mammal. 13:41–45.

Muul, I. & Lim Boo-Liat. 1970. Ecological and morphological observations on *Felis planiceps*. J. Mammal. 51:806–808.

Nabokov, V. 1962. Pale fire; a novel. Putnam, N.Y.

Nabokov, V. 1970. The annotated Lolita. A. Appel, Jr., ed. McGraw-Hill, N.Y.

Nabokov, V. 1974. Look at the Harlequins! McGraw Hill, N.Y.

Nadel, L., J. O'Keefe & A. Black. 1975. Slam on the brakes: a critique of Altman, Brunner and Bayer's response-inhibition model of hippocampal function. Behav. Biol. 14:151–162.

Namier, Sir L. 1961. England in the age of the American revolution, 2nd ed. St. Martin's Press, London.

Nance, J. 1975. The gentle Tasaday: A Stone Age people in the Philippine rain forest. Harcourt Brace Jovanovich, N.Y.

Narayanan, C. H., M. W. Fox & V. Hamburger. 1971. Prenatal development of spontaneous and evoked activity in the rat (*Rattus norvegicus albinus*). Behaviour 40:100–134.

Nash, L. T. 1978. The development of the mother-infant relationship in wild baboons (*Papio anubis*). Anim. Behav. 26:746–759.

Naundorff, E. 1929. Der Dachs als Hausgenosse. Z. Säugetierk. 4:122–124.

Neal, E. 1948. The badger. Collins, London.

Neal, E. 1962. Badgers. The Sunday Times, London.

Neal, E. 1970. The banded mongoose, *Mungos mungo* Gmelin. E. Afr. Wildl. J. 8:63–71.

Neelakantan, K. K. 1952. Juvenile Brahminy kites (*Haliastur indus*) learning things the modern way. J. Bombay Nat. Hist. Soc. 51:739.

Neelakantan, K. K. 1969. Black jackals (*Canis aureus*) in Kerala. J. Bombay Nat. Hist. Soc. 66:612–614.

Nelson, J. E. & G. Smith. 1971. Notes on growth rates in native cats of the family Dasyuridae. Int. Zoo Yb. 11:38–41.

Neuweiler, G. 1969. Verhaltensbeobachtungen an einer indischen Flughundkolonie (*Pteropus g. giganteus* Brünn). Z. Tierpsychol. 26:166–199.

Neville, M. K. 1972a. The population structure of red howler monkeys (*Alouatta seniculus*) in Trinidad and Venezuela. Folia Primatol. 17:56–86.

Neville, M. K. 1972b. Social relations within troops of red howler monkeys (*Alouatta seniculus*). Folia Primatol. 18:47–76.

Newell, T. G. 1971. Social encounters in two prosimian species: *Galago crassicaudatus* and *Nycticebus coucang*. Psychonomic Sci. 24:128–130.

Newton, I. 1973. Finches. Taplinger, N.Y.

Nice, M. M. 1943. Studies in the life history of the song sparrow. II. Trans. Linn. Soc. N.Y., 6:1–238.

Nichols, A. L. 1977. Coalitions and learning. Applications to a simple game in the triad. Behav. Sci. 22:391–402.

Nichols, J. D., W. Conley, B. Batt & A. R. Tipton. 1976. Temporally dynamic reproductive strategies and the concept of r- and K-selection. Am. Nat. 110:995–1005.

Nicolson, N. A. 1977. A comparison of early behavioral development in wild and captive chimpanzees. Pages 529–560 in S. Chevalier-Skolnikoff & F. E. Poirier, eds., Primate bio-social development. Garland, N.Y.

Niemitz, C. 1974. A contribution to the postnatal behavioral development of *Tarsius bancanus,* Horsfield, 1821, studied in two cases. Folia Primatol. 21:250–276.

Nissen, H. W. 1931. A field study of the chimpanzee. Comp. Psychol. Monogr. 8:1–22.

Noble, R. C. 1945. The nature of the beast. A popular account of animal psychology from the point of view of a naturalist. Doubleday, Doran, Garden City, N.Y.

Nolte, A. 1955a. Field observations on the daily routine and social behaviour of common Indian monkeys, with special reference to the bonnet monkey (*Macaca radiata*). J. Bombay Nat. Hist. Soc. 53:177–184.

Nolte, A. 1955b. Freilandbeobachtungen über das Verhalten von *Macaca radiata* in Südindien. Z. Tierpsychol. 12:77–87.

Nolte, A. 1958. Beobachtungen über das Instinktverhalten von Kapuzineraffen (*Cebus apella* L.) in der Gefangenschaft. Behaviour 12:183–207.

Norikoshi, K. 1974. The development of peer-mate relationships in Japanese macaque infants. Primates 15:39–46.

Norton-Griffiths, M. 1966. Oystercatchers and mussels. Ibis 108:455–456.

Norton-Griffiths, M. 1967. Some ecological aspects of the feeding behaviour of the oyster catcher, *Haematopus ostralegus,* on the edible mussel, *Mytilus edulis*. Ibis 109:412–424.

Norton-Griffiths, M. 1969. The organisation, control and development of parental feeding in the oystercatcher (*Haematopus ostralegus*). Behaviour 34:55–114.

Oakley, F. B. & P. C. Reynolds. 1976. Differing responses to social play deprivation in two species of macaque. Pages 179–188 in D. F. Lancy & B. A. Tindall, eds., The anthropological study of play: problems and prospects. Leisure Press, Cornwall, N.Y.

Oates, J. F. 1974. The ecology and behaviour of the black-and-white colobus monkey (*Colobus guereza* Rüppell) in East Africa. Ph.D. diss., University of London.

Oberholtzer, E. C. 1911. Some observations on moose. Proc. Zool. Soc. London. 1911:362–364.

O'Connor, R. J. 1978. Broad reduction in birds: selection for fratricide, infanticide and suicide? Anim. Behav. 26:79–96.

O'Gara, B. W. 1969. Unique aspects of reproduction in the female pronghorn *Antilocapra americana* Ord. Am. J. Anat. 125:217–232.

Ognev, S. I. 1962. Mammals of U.S.S.R. and adjacent countries. Vol. III: Carnivora, Fissipedia and Pinnipedia. Israel Program for Scientific Translations, Jerusalem.

Oke, V. R. 1967. A brief note on the dugong, *Dugong dugon,* at Cairns Oceanarium. Int. Zoo Yb. 7:220–221.

Olioff, M. & J. Stewart. 1978. Sex differences in the play behavior of prepubescent rats. Physiol. Behav. 20:113–115.

Oliver, W. L. R. 1975. The Jamaican hutia (*Geocapromys brownii brownii*). Jersey Wildlife Preservation Trust 12th Annual Report:10–17.

Olivier, G. 1958. Notes sur la biologie du cerf (*Cervus elaphus*). Mammalia 22:245–250.

Olomon, C. M., M. D. Breed & W. J. Bell. 1976. Ontogenetic and temporal aspects of agonistic behavior in a cockroach, *Periplaneta americana.* Behav. Biol. 17:243–248.

Ooshima, A., G. Fuller, G. Cardinale, S. Spector & S. Udenfriend. 1975. Collagen biosynthesis in blood vessels of brain and other tissues of the hypertensive rat. Science 190:898–900.

Opie, I. & P. Opie. 1967. The lore and language of schoolchildren. Clarendon Press, Oxford.

Opie, I. & P. Opie. 1969. Children's games in street and playground. Clarendon Press, Oxford.

Oppenheimer, J. R. 1968. Behavior and ecology of the white-faced monkey, *Cebus capucinus,* on Barro Colorado Island, C. Z. Ph.D. diss., University of Illinois.

Oppenheimer, J. R. 1969. Behavior and ecology of the white-faced monkey, *Cebus capucinus,* on Barro Colorado Island, C. Z. Diss. Abstr. Intl. B 30:442–443.

Oppenheimer, J. R. 1974. Cebus monkeys of Barro Colorado Island: ecology and behavior. Psychological Cinema Register film PCR–2258K, Pennsylvania State University, University Park.

Oppenheimer, J. R. & E. C. Oppenheimer. 1973. Preliminary observations of *Cebus nigrivittatus* (Primates: Cebidae) on the Venezuelan llanos. Folia Primatol. 19:409–436.

Orr, R. T. 1967. The Galapagos sea lion. J. Mammal. 48:62–69.

Oster, G. F. & E. O. Wilson. 1978. Caste and ecology in the social insects. Princeton University Press, Princeton, N.J.

Otte, D. 1975. On the role of intraspecific deception. Am. Nat. 109:239–242.

Owen, J. 1973. Behaviour and diet of a captive royal antelope, *Neotragus pygmaeus* L. Mammalia 37:56–65.

Owen, Prof. 1848. Remarks on the "Observations sur l'Ornithorhynque" par M. Jules Verraux. Ann. Mag. Nat. Hist. 2:317–322.

Owens, N. 1974. The development of behaviour in free-living baboons (*Papio anubis*). Ph.D. diss., Selwyn College, Cambridge.

Owens, N. W. 1975a. A comparison of aggressive play and aggression in free-living baboons, *Papio anubis*. Anim. Behav. 23:757–765.

Owens, N. W. 1975b. Social play behaviour in free-living baboons, *Papio anubis*. Anim. Behav. 23:387–408.

Owens, N. W. 1976. The development of sociosexual behaviour in free-living baboons, *Papio anubis*. Behaviour 57:241–259.

Owen-Smith, N. 1973. The behavioural ecology of the white rhinoceros. Ph.D. diss. University of Wisconsin. Diss. Abstr. Intl. 34:74–3542.

Owen-Smith, N. 1975. The social ethology of the white rhinoceros *Ceratotherium simum*. Z. Tierpsychol. 38:337–384.

Packer, C. 1979a. Inter-troop transfer and inbreeding avoidance in *Papio anubis*. Anim. Behav. 27:1–36.

Packer, C. 1979b. Male dominance and reproductive activity in *Papio anubis*. Anim. Behav. 27:37–45.

Page, F. J. T. 1962. Roe deer. Sunday Times Publications, London.

Pagès, E. 1972. Comportement maternel et développement du jeune chez un pangolin arboricole (*M. tricuspis*). Biol. Gabon. 8:63–120.

Pagès, E. 1975. Étude éco-éthologique de *Manis tricuspis* par radiotracking. Mammalia 39:613–641.

Paillard, J. 1960. The patterning of skilled movements. Pages 1679–1708 in Handbook of physiology. I. Neurophysiology. Vol. III. American Physiological Society, Washington, D.C.

Paillard, J. 1976. Le codage nerveux des commandes motrices. Revue E.E.G. Neurophysiologie 6:453–472.

Paillard, J. 1977. Introduction. J. Human Movement Studies 3:127.

Pakenham, R. H. W. 1936. Field notes on the birds of Zanzibar and Pemba. Ibis 6:249–272.

Paluck, R. J., J. D. Lieff & A. H. Esser. 1970. Formation and development of a group of juvenile *Hylobates lar*. Primates 11:185–194.

Panday, D. J. 1952. Strange behaviour of a house crow (*Corvus splendens*). J. Bombay Nat. Hist. Soc. 50:939.

Papoušek, H. & M. Papoušek. 1975. Cognitive aspects of preverbal social interaction between human infants and adults. Pages 241–260 in R. Porter & M. O'Connor, eds., Parent-infant interaction. Ciba Foundation Symposium No. 33 (new series). Associated Scientific Publishers, Amsterdam and American Elsevier, N.Y.

Parker, A. 1975. Young male peregrines passing vegetation fragments to each other. Br. Birds 68:242–243.

Parker, G. A. 1974. Assessment strategy and the evolution of animal conflicts. J. Theor. Biol. 47:223–243.

Parker, G. A. & M. R. MacNair. 1978. Models of parent-offspring conflict. II. Promiscuity. Anim. Behav. 26:111–122.

Parker, G. A. & R. A. Stuart. 1976. Animal behavior as a strategy optimizer: evolution of resource assessment strategies and optimal emigration thresholds. Am. Nat. 110:1055–1076.

Parker, P. 1977. An ecological comparison of marsupial and placental patterns of reproduction. Pages 273–286 in B. Stonehouse & G. Gilmore, eds., The biology of marsupials. Macmillan, N.Y. & London.

Parker, S. T. 1977. Piaget's sensorimotor period series in an infant macaque: a model for comparing unstereotyped behavior and intelligence in human and nonhuman primates. Pages 43–112 in S. Chevalier-Skolnikoff & F. E. Poirier, eds., Primate bio-social development. Garland, N.Y.

Paulian, P. 1964. Contribution a l'étude de l'otarie de l'ile Amsterdam. Mammalia 28, Suppl. 1.

Pawlby, S. J. 1977. Imitative interaction. Pages 203–224 in H. R. Schaffer, ed., Studies in mother-infant interaction. Academic Press, N.Y.

Pearce, J. 1937. A captive New York weasel. J. Mammal. 18:483–488.

Pelosse, J. L. 1977. Une observation sur le comportement de la mère et du veau chez le renne sauvage de forêt finlandais, Rangifer tarandus fennicus Lonnberg, 1909. Säugetierk. Mitt. 25:108–114.

Perry, R. 1966. The world of the polar bear. University of Washington Press, Seattle.

Persson, C. 1942. Two cases of playful behavior observed in hooded crows (Corvus c. cornix L.). Vår Fågelvärld Stockholm 1942:27–28.

Peters, J. L. 1931. Check-list of birds of the world. 15 vols. Harvard University Press, Cambridge, Mass.

Petersen, M. K. 1979. Behavior of the margay. Carnivore 2:69–76.

Peterson, R. S. 1968. Social behavior in pinnipeds. Pages 3–53 in R. J. Harrison, ed., The behavior and physiology of pinnipeds. Appleton-Century-Crofts, N.Y.

Peterson, R. S. & G. A. Bartholomew. 1967. The natural history and behavior of the California sea lion. Am. Soc. Mammalogists Spec. Pub. 1.

Peterson, R. S., C. L. Hubbs, R. L. Gentry & R. L. DeLong. 1968. The Guadalupe fur seal: habitat, behavior, population size and field identification. J. Mammal. 49:665–675.

Petter, F. 1952. Le renard famélique. La Terre et la Vie 1952:191–193.

Petter, F. & J.-J. Petter. 1963. Cheirogaleus major (Lemuridae)—Jeu de jeunes animaux. Encyclopedia Cinematographica Film E551. Institut für den Wissenschaftlichen Film, Göttingen.

Petter, J.-J. 1965. The lemurs of Madagascar. Pages 292–319 in I. DeVore, ed., Primate behavior. Holt, Rinehart & Winston, N.Y.

Petter, J.-J., R. Albignac & Y. Rumpler. 1977. Mammifères lémuriens (primates prosimiens). Faune de Madagascar, V. 44. ORSTOM, CNRS, Paris.

Petter, J.-J. & A. Peyrieras. 1970a. Nouvelle contribution a l'étude d'un lémurien malagache, le Aye-aye (Daubentonia madagascarensis E. Geoffroy). Mammalia 34:167–193.

Petter, J.-J. & A. Peyrieras. 1970b. Observations éco-éthologiques sur les lémuriens malagaches du genre Hapalemur. La Terre et la Vie 24:356–382.

Petter-Rousseaux, A. 1964. Reproductive physiology and behaviour of the Lemuroidea. Pages 92–131 in J. Buettner-Janusch, ed., Evolutionary and genetic biology of primates, Vol. 2. Academic Press, N.Y.

Pfeffer, P. 1967. Le mouflon de Corse (Ovis ammon musimon Schreber, 1782);

position systématique, écologie et éthologie comparées. Mammalia 31, Supp.

Pfeffer, P. 1972. Observations sur le comportement social et predateur du lycaon (*Lycaon pictus*) en republique Centrafricaine. Mammalia 36:1–7.

Pfeiffer, J. E. 1978. The emergence of man, 3rd ed. Harper & Row, N.Y.

Phoenix, C. H. 1974. The role of androgens in the sexual behavior of adult male rhesus monkeys. Pages 249–258 in W. Montagna & W. A. Sadler, eds., Reproductive behavior. Plenum, N.Y.

Piaget, J. 1962. Play, dreams and imitation in childhood. Norton, N.Y.

Pillai, N. G. 1963. The Nilgiri tahr (*Hemitragus hylocrius*) in captivity. J. Bombay Nat. Hist. Soc. 60:451–454.

Pilleri, G. 1960. Zum Verhalten der Paka (*Cuniculus paca* Linnaeus). Z. Säugetierk. 25:107–111.

Pilters, H. 1954. Untersuchungen über angeborene Verhaltensweisen bei Tylopoden, unter besonderer Berücksichtigung der neuweltlichen Formen. Z. Tierpsychol. 11:213–303.

Pimlott, D. H., J. A. Shannon & G. B. Kolenosky. 1969. Res. Branch Res. Rept. (Wildlife) No. 87.

Pinto, D. 1972. Patterns of activity in three nocturnal prosimians: *Galago senegalensis moholi, Galago crassicaudatus umbrosus,* and *Microcebus murinus murinus*. M.Sc. thesis, Witwatersrand Univ.

Pitcairn, T. K. 1976. Attention and social structure in *Macaca fascicularis*. Pages 51–81 in M. R. A. Chance & R. R. Larsen, eds., The structure of social attention. Wiley, N.Y.

Pizzimenti, J. J. & L. R. McClenaghan. 1974. Reproduction, growth and development, and behavior in the Mexican prairie dog, *Cynomys mexicanus* (Merriam). Am. Midl. Nat. 92:130–145.

Plato. 1926. Laws. R. G. Bury tr. 2 vols. (Loeb Classical Library, 9.) Harvard University Press, Cambridge, Mass. & Heinemann, London.

Ploog, D., S. Hopf & P. Winter. 1967. Ontogenese des Verhaltens von Totenkopf-Affen (*Saimiri sciureus*). Psychol. Forsch. 31:1–41.

Ploog, D., K. Hupfer, V. Jürgens & J. D. Newman. 1975. Neuroethologic studies of vocalization in squirrel monkeys with special references to genetic differences of calling in two subspecies. Pages 231–254 in M. A. B. Brazier, ed., Growth and development of the brain. Raven Press, N.Y.

Ploog, D. & M. Maurus. 1973. Social communication among squirrel monkeys: analysis by sociometry, bioacoustics and cerebral radio-stimulation. Pages 211–233 in R. P. Michael & J. H. Crook, eds., Comparative ecology and behaviour of primates. Academic Press, London & N.Y.

Pluta, G. & B. B. Beck. 1979. Object play by captive polar bears. Paper presented at annual meeting of Animal Behavior Society, New Orleans.

Poduschka, W. 1969. Ergänzungen zum Wissen über *Erinaceus e. roumanicus* und kritische Überlegungen zur bisheriger Literatur über europäische Igel. Z. Tierpsychol. 26:761–804.

Poglayen-Neuwall, I. 1962. Beiträge zu einem Ethogramm des Wickelbären (*Potos flavus* Schreber). Z. Säugetierk. 27:1–44.

Poglayen-Neuwall, I. 1973. Preliminary notes on maintenance and behaviour of the Central American cacomistle *Bassariscus sumichrasti* at Louisville Zoo. Int. Zoo Yb. 13:207–211.

Poglayen-Neuwall, I. & I. Poglayen-Neuwall. 1965. Gefangenschaftsbeobachtungen an Makibären (*Bassaricyon* Allen, 1876). Z. Säugetierk. 30:321–366.

Pohle, C. 1973. Zur Zucht von Bengalkatzen (*Felis bengalensis*) im Tierpark Berlin. Zool. Gart. 43:110–126.

Poirier, F. E. 1968. The Nilgiri langur (*Presbytis johnii*) mother-infant dyad. Primates 9:45–68.

Poirier, F. E. 1969a. Behavioral flexibility and intertroop variation among Nilgiri langurs (*Presbytis johnii*) of South India. Folia Primatol. 11:119–133.

Poirier, F. E. 1969b. The Nilgiri langur (*Presbytis johnii*) troop: its composition, structure, formation and change. Folia Primatol. 10:20–47.

Poirier, F. E. 1970a. Nilgiri langur ecology and social behavior. Primate Behav. 1:251–383.

Poirier, F. E. 1970b. The communication matrix of the Nilgiri langur (*Presbytis johnii*) of South India. Folia Primatol. 13:92–136.

Poirier, F. E. 1972a. The St. Kitts green monkey (*Cercopithecus aethiops sabaeus*): ecology, population dynamics, and selected behavioral traits. Folia Primatol. 17:20–55.

Poirier, F. E., ed. 1972b. Primate socialization. Random House, N.Y.

Poirier, F. E. & E. O. Smith. 1974a. The crab-eating macaques (*Macaca fascicularis*) of Angaur Island, Palau, Micronesia. Folia Primatol. 22:258–306.

Poirier, F. E. & E. O. Smith. 1974b. Socializing functions of primate play. Am. Zool. 14:275–287.

Pollock, J. I. 1975. Field observations on *Indri indri*. A preliminary report. Pages 287–311 in I. Tattersall & R. Sussman, eds., Lemur biology. Plenum Press, N.Y.

Pond, C. M. 1977. The significance of lactation in the evolution of mammals. Evolution 31:177–199.

Pond, C. M. 1978. Morphological aspects and the ecological and mechanical consequences of fat deposition in wild vertebrates. Ann. Rev. Ecol. Syst. 9:519–570.

Ponugaeva, A. G. 1961. The physiologic characteristics of play activity in the golden hamster. Page 158 in III. Soveshchanie po evolyutsionnoi fiziologii, Leningrad.

Poole, T. B. 1966. Aggressive play in polecats. Symp. Zool. Soc. Lond. 18:23–44.

Poole, T. B. 1967. Aspects of aggressive behavior in polecats. Z. Tierpsychol. 24:351–369.

Poole, T. B. 1972. Diadic interactions between pairs of male polecats (*Mustela furo* and *Mustela furo* x *putorius* hybrids) under standardized environmental conditions during the breeding season. Z. Tierpsychol. 30:45–58.

Poole, T. B. 1973. The aggressive behavior of individual male polecats (*Mustela putorius, M. furo* and hybrids) towards familiar and unfamiliar opponents. J. Zool. Lond. 170:395–414.

Poole, T. B. 1974. The effects of oestrous condition and familiarity on the sexual behaviour of polecats (*Mustela putorius* and *M. furo* x *M. putorius* hybrids). J. Zool. Lond. 172:357–362.

Poole, T. B. 1978. An analysis of social play in polecats (Mustelidae) with comments on the form and evolutionary history of the open mouth play face. Anim. Behav. 26:36–49.

Poole, T. B. & J. Fish. 1975. An investigation of playful behavior in *Rattus norvegicus* and *Mus musculus*. J. Zool. Lond. 175:61–71.

Poole, T. B. & J. Fish. 1976. An investigation of individual, age and sexual differences in the play of *Rattus norvegicus* (Mammalia: Rodentia). J. Zool. Lond. 179:249–260.

Popp, J. L. & I. DeVore. 1974. Aggressive competition and social dominance theory. Ms. for Wenner-Gren Conference "The Behavior of the Great Apes."

Popp, J. L. & I. DeVore. 1979. Aggressive competition and social dominance theory: synopsis. Pages 317–338 in D. A. Hamburg & E. R. McCown, eds., The great apes. (Perspectives on human evolution, Vol. 5) Benjamin/Cummings, Menlo Park, Calif.

Porter, S. 1947. The breeding of the Kea. Avic. Mag. 53:50–55.

Portmann, A. & W. Stingelin. 1961. The central nervous system. Pages 1–36 in A. J. Marshall, ed., Biology and comparative physiology of birds, Vol. 2. Academic Press, N.Y.

Poupa, O., K. Rakušan & B. Ošťádal. 1970. The effect of physical activity upon the heart of vertebrates. Med. Sport 4:202–233.

Pournelle, G. H. 1962. Observations on the birth and early development of Allen's monkey. J. Mammal. 43:265–266.

Pournelle, G. H. 1967. Observations on reproductive behaviour and early postnatal development of the proboscis monkey, *Nasalis larvatus orientalis*, at San Diego Zoo. Int. Zoo Yb. 7:90–93.

Prater, S. H. 1935. The wild animals of the Indian Empire. J. Bombay Nat. Hist. Soc. 37:112–166.

Pratt, D. M. & V. H. Anderson. 1979. Giraffe cow-calf relationships and social development of the calf in the Serengeti. Z. Tierpsychol. 51:233–251.

Prior, R. 1968. The roe deer of Cranbourne Chase. Oxford University Press, London & N.Y.

Procter, J. 1963. A contribution to the natural history of the spotted-necked otter (*Lutra maculicollis* Lichtenstein) in Tanganyika. E. Afr. Wildl. J. 1:93–102.

Pruitt, C. H. 1976. Play and agonistic behavior in young captive black bears. Pages 79–86 in M. R. Pelton, J. W. Lentfer & G. E. Folk, eds., Bears— Their biology and management. Papers of the Third International Conference on Bear Research and Management. Morges, Switzerland. IUCN Publications (N.S.) No. 40.

Purton, A. C. 1978. Ethological categories of behavior and some consequences of their conflation. Anim. Behav. 26:653–670.

Quine, W. 1960. Word and object. Technology Press, Cambridge, Mass. & Wiley, N.Y.

Quiring, D. P. 1941. The scale of being according to the power formula. Growth 5:301–327.

Radcliffe, H. D. 1909. Intelligence in birds. J. Bombay Nat. Hist. Soc. 19:526–527.

Rahaman, H. 1973. The langurs of the Gir Santuary (Gujarat)—A preliminary survey. J. Bombay Nat. Hist. Soc. 70:295–314.

Rahaman, H. & M. D. Parthasarathy. 1969. Studies on the social behavior of bonnet monkeys. Primates 10:149–162.

Raibert, M. H. 1977. Motor control and learning by the state space model. Technical report 439, MIT Artificial Intelligence Laboratory, Cambridge, Mass.

Ralls, K. 1977. Sexual dimorphism in mammals: avian models and unanswered questions. Am. Nat. 111:917–938.

Rand, R. W. 1967. The cape fur-seal (Arctocephalus pusillus). 3. General behaviour on land and at sea. South Afr. Div. Sea Fish. Invest. Rep., No. 60.

Randolph, M. C. & B. A. Brooks. 1967. Conditioning of a vocal response in a chimpanzee through social reinforcement. Folia Primatol. 5:70–79.

Ranger, G. 1950. Life of the crowned hornbill. Part III. Ostrich 21:1–14.

Ransom, T. W. 1971. Ecology and behaviour of the baboon, Papio anubis, at the Gombe Stream National Park, Tanzania. Ph.D. diss., Dept. of Psychology, University of California at Berkeley.

Ransom, T. W. & B. S. Ransom. 1971. Adult male-infant relations among baboons (Papio anubis). Folia Primatol. 16:179–195.

Ransom, T. W. & T. E. Rowell. 1972. Early social development of feral baboons. Pages 105–144 in F. E. Poirier, ed., Primate socialization. Random House, N.Y.

Rarick, G. L. & G. L. Larsen. 1959. The effect of variations in the intensity and frequency of isometric muscular effort on the development of static muscular strength in pre-pubescent males. Int. Z. Angew. Physiol. 18:13–21.

Rasa, O. A. E. 1971. Social interaction and object manipulation in weaned pups of the Northern elephant seal Mirounga angustirostris. Z. Tierpsychol. 29:82–102.

Rasa, O. A. E. 1973. Prey capture, feeding techniques, and their ontogeny in the African dwarf mongoose, Helogale undulata rufula. Z. Tierpsychol. 32:449–488.

Rasa, O. A. E. 1977. The ethology and sociology of the dwarf mongoose (Helogale undulata rufula). Z. Tierpsychol. 43:337–406.

Rauch, H.-G. 1957. Zum Verhalten von Meriones tamariscinus Pall. (1778). Z. Säugetierk. 22:218–240.

Read, M. S. 1975. Behavioral correlates of malnutrition. Pages 335–354 in M. A. B. Brazier, ed., Growth and development of the brain. Raven Press, N.Y.

Redican, W. K. & G. Mitchell. 1974. Play between adult male and infant rhesus monkeys. Am. Zool. 14:295–302.

Redshaw, M. & K. Locke. 1976. The development of play and social behaviour in two lowland gorilla infants. Jersey Wildlife Preservation Trust 13th Annual Report: 71–85.

Remington, J. D. 1952. Food habits, growth and behavior of two captive pine martens. J. Mammal. 33:66–70.

Rensch, B. 1973. Play and art in apes and monkeys. Pages 102–123 in E. W. Menzel, ed., Precultural primate behavior. Karger, Basel.

Rensch, B. & G. Dücker. 1959. Die Spiele von *Mungo* und *Ichneumon*. Behaviour 14:185–213.

Repenning, C. A., R. S. Peterson & C. L. Hubbs. 1971. Contributions to the systematics of the Southern fur seals, with particular reference to the Juan Fernández and Guadalupe species. Pages 1–34 in W. H. Burt, ed., Antarctic pinnipedia. Antarctic Research Series, no. 18. American Geophysical Union, N.Y.

Reynolds, P. C. 1976. Play, language and human evolution. Pages 621–635 in J. S. Bruner, A. Jolly & K. Sylva, eds., Play: its role in development and evolution. Basic Books, N.Y.

Reynolds, V. 1961. The social life of a colony of rhesus monkeys (*Macaca mulatta*). Ph.D. diss., Univ. of London. (Cited by Loizos 1967; original not seen.)

Reynolds, V. & G. Luscombe. 1976. Greeting behaviour, displays and rank order in a group of free-ranging chimpanzees. Pages 105–115 in M. R. A. Chance & R. R. Larsen, eds., The structure of social attention. Wiley, N.Y. & London.

Reynolds, V. & F. Reynolds. 1965. Chimpanzees of the Budongo forest. Pages 368–424 in I. DeVore, ed., Primate behavior. Holt, Rinehart & Winston, N.Y.

Rheingold, H. L., ed. 1963. Maternal behavior in mammals. Wiley, N.Y.

Rhine, R. J. 1972. Changes in the social structure of two groups of stumptail macaques (*Macaca arctoides*). Primates 13:181–194.

Rhine, R. J. 1973. Variation and consistency in the social behavior of two groups of stumptail macaques (*Macaca arctoides*). Primates 14:21–35.

Rhine, R. J. & C. Kronenwetter. 1972. Interaction patterns of two newly formed groups of stumptail macaques (*Macaca arctoides*). Primates 13:19–33.

Richard, A. 1970. A comparative study of the activity patterns and behavior of *Alouatta villosa* and *Ateles geoffroyi*. Folia Primatol. 12:241–263.

Richard, A. 1974. Intra-specific variation in the social organization and ecology of *Propithecus verreauxi*. Folia Primatol. 22:178–207.

Richard, A. F. 1976. Preliminary observations on the birth and development of *Propithecus verreauxi* to the age of six months. Primates 17:357–366.

Richard, A. F. & R. Heimbuch. 1975. An analysis of the social behavior of three groups of *Propithecus verreauxi*. Pages 313–333 in I. Tattersall & R. W. Sussman, eds., Lemur biology. Plenum Press, N.Y.

Richardson, W. B. 1943. Wood rats (*Neotoma albigula*): their growth and development. J. Mammal. 24:130–143.

Richter, W. von. 1966. Untersuchungen über angeborene Verhaltensweisen des Schabrackentapirs (*Tapirus indicus*) und des Flachlandtapirs (*Tapirus terrestrus*). Zool. Beitr. 12:67–159.

Ricklefs, R. E. 1977. On the evolution of reproductive strategies in birds: reproductive effort. Am. Nat. 111:453–478.

Ride, W. D. L. 1970. A guide to the native mammals of Australia. Oxford University Press, Melbourne.

Rijksen, H. D. 1978. A field study on Sumatran orangutans (*Pongo pygmaeus Abelii* Lesson 1827). H. Veenman & Zonen B. V., Wageningen.

Ring, T. P. A. 1923. The elephant-seals of Kerguelen Land. Proc. Zool. Soc. Lond. 1923:431–443.

Ripley, S. 1967. The leaping of langurs. Am. J. Phys. Anthropol. 26:149–170.

Riss, D. & J. Goodall. 1976. Sleeping behavior and associations in a group of captive chimpanzees. Folia Primatol. 25:1–11.

Ritchie, A. T. A. 1963. The black rhinoceros (*Diceros bicornis* L.). E. Afr. Wildl. J. 1:54–62.

Roberts, B. 1934. Notes on the birds of central and southeast Iceland, with special reference to food habits. Ibis 4:239–264.

Roberts, M. G. 1915. The keeping and breeding of Tasmanian devils. Proc. Zool. Soc. Lond. 1915:575–581.

Roberts, M. S. 1975. Growth and development of mother-reared red pandas, *Ailurus fulgens*. Int. Zoo Yb. 15:57–63.

Roberts, P. 1971. Social interactions of *Galago crassicaudatus*. Folia Primatol. 14:171–181.

Roberts, T. J. 1970. A note on the yellow-throated marten *Martes flavigula* (Boddaert) in West Pakistan. J. Bombay Nat. Hist. Soc. 67:321–323.

Roberts, T. J. 1977. The mammals of Pakistan. Ernest Benn, London.

Robinson, R. 1950. Definition. Clarendon Press, Oxford.

Rodon, G. S. 1894. A bison calf. J. Bombay Nat. Hist. Soc. 9:226–227.

Rodon, G. S. 1898. Notes on a jackal cub. J. Bombay Nat. Hist. Soc. 12:220.

Roe, N. A. & W. E. Rees. 1976. Preliminary observations of the taruca (*Hippocamelus antisensis*: Cervidae) in southern Peru. J. Mammal. 57:722–730.

Romanes, G. J. 1884. Mental evolution in animals. With a posthumous essay on instinct by Charles Darwin. Appleton, N.Y.

Romer, A. S. 1966. Vertebrate paleontology, 3rd ed. University of Chicago Press, Chicago.

Rondinelli, R. & L. L. Klein. 1976. An analysis of adult social spacing tendencies and related social interactions in a colony of spider-monkeys (*Ateles geoffroyi*) at the San Francisco Zoo. Folia primatol. 25:122–142.

Rood, J. P. 1958. Habits of the short-tailed shrew in captivity. J. Mammal. 39:499–507.

Rood, J. P. 1970. Ecology and social behaviour of the desert cavy (*Microcavia australis*). Am. Midl. Nat. 83:415–454.

Rood, J. P. 1972. Ecological and behavioral comparisons of three genera of Argentine cavies. Anim. Behav. Monogr. 5:1–83.

Rood, J. P. 1975. Population dynamics and food habits of the banded mongoose. E. Afr. Wild. J. 13:89–111.

Rose, M. D. 1977a. Interspecific play between free ranging guerezas (*Colobus guereza*) and vervet monkeys (*Cercopithecus aethiops*). Primates 18:957–964.

Rose, M. D. 1977b. Positional behavior of olive baboons (*Papio anubis*) and its relationship to maintenance and social activities. Primates 18:59–116.

Rose, S. P. R. 1969. Neurochemical correlates of learning and environmental change. FEBS Letters 5:305–312.

Rosenblatt, J. S. 1971. Suckling and home orientation in the kitten: a comparative developmental study. Pages 345–410 in E. Tobach, L. Aronson & E. Shaw, eds., The biopsychology of development. Academic Press, N.Y.

Rosenblatt, J. S. 1972. Learning in newborn kittens. Sci. Am. 227(6):18–25.

Rosenblatt, J. S. 1974. Some features of early behavioural development in kittens. Pages 94–110 in N. F. White, ed., Ethology and psychiatry. University of Toronto Press, Toronto, Ont. & Buffalo, N.Y.

Rosenblatt, J. S., G. Turkewitz & T. C. Schneirla. 1962. Development of suckling and related behavior in neonate kittens. Pages 198–210 in E. L. Bliss, ed., Roots of behavior. Harper & Bros., N.Y.

Rosenblatt, L. 1977. Developmental trends in infant play. Pages 33–44 in B. Tizard & O. Harvey, eds., Biology of play. Heinemann, London & Lippincott, Philadelphia.

Rosenblum, L. A. 1968. Mother-infant relations and early behavioral development in the squirrel monkey. Pages 207–233 in L. A. Rosenblum & R. W. Cooper, eds., The squirrel monkey. Academic Press, N.Y.

Rosenblum, L. A. 1971a. Infant attachment in monkeys. Pages 85–109 in H. R. Schaffer, ed., The origins of human social relations. Academic Press, N.Y. & London.

Rosenblum, L. A. 1971b. Kinship interaction patterns in pigtail and bonnet macaques. Proc. 3rd Int. Congr. Primatol., Zurich, 1971 3:79–84.

Rosenblum, L. A., I. C. Kaufman & A. J. Stynes. 1969. Interspecific variations in the effects of hunger on diurnally varying behavior elements in macaques. Brain, Behav. Evol. 2:119–131.

Rosenblum, L. A. & A. Lowe. 1971. The influence of familiarity during rearing on subsequent partner preferences in squirrel monkeys. Psychonom. Sci. 23:35–37.

Rosenson, L. M. 1972. Observations of the maternal behaviour of two captive greater bushbabies (*Galago crassicaudatus argentatus*). Anim. Behav. 20:677–688.

Rosenzweig, M. L. 1968. The strategy of body size in mammalian carnivores. Am. Midl. Nat. 80:299–315.

Rosenzweig, M. L. 1973a. Evolution of the predator isocline. Evolution 27:84–94.

Rosenzweig, M. L. 1973b. Exploitation in three trophic levels. Am. Nat. 107:275–294.

Rosenzweig, M. R. 1966. Environmental complexity, cerebral change, and behavior. Am. Psychol. 21:321–332.

Rosenzweig, M. R. 1971. Role of experience in development of neurophysiological regulatory mechanisms and in organization of the brain. Pages 15–33 in D. N. Walcher & D. L. Peters, eds., The development of self-regulatory mechanisms. Academic Press, N.Y. & London.

Rosenzweig, M. R. & E. L. Bennett. 1976. Enriched environments: facts, factors, and fantasies. Pages 179–213 in L. Petrinovich & J. L. McGaugh, eds., Knowing, thinking and believing. Plenum Press, N.Y.

Rosenzweig, M. R. & E. L. Bennett. 1977a. Effects of environmental enrichment or impoverishment on learning and on brain values in rodents. Pages 163–196 in A. Oliverio, ed., Genetics, environment and intelligence. Elsevier North-Holland, N.Y.

Rosenzweig, M. R. & E. L. Bennett. 1977b. Experiential influences on brain anatomy and brain chemistry in rodents. Pages 289–327 in G. Gottlieb, ed., Studies on the development of behavior and the nervous system, Vol. 4, Early influences. Academic Press, N.Y.

Rosenzweig, M. R., E. L. Bennett, M. Hebert & H. Morimoto. 1978. Social grouping cannot account for cerebral effects of enriched environments. Brain Res. 153:563–576.

Rosenzweig, M. R., W. Love & E. L. Bennett. 1968. Effects of a few hours a day of enriched experience on brain chemistry and brain weights. Physiol. Behav. 3:819–825.

Rosevear, D. R. 1974. The carnivores of West Africa. The British Museum (Natural History), London.

Roth, H. H. 1970. Über die Jugendentwicklung von Waschbären, *Procyon lotor*. Säugetierk. Mitt. 18:7–8.

Roth, J. P. 1967. Great ape house at Albuquerque Zoo. Int. Zoo Yb. 7:50–54.

Roth-Kolar, H. 1957. Beiträge zu einem Aktionssystem des Aguti (*Dasyprocta aguti aguti* L.). Z. Tierpsychol. 14:362–375.

Rowe, E. G. 1947. The breeding biology of *Aquila verreauxi* Lesson. Ibis 89:576–606.

Rowell, L. B. 1974. Human cardiovascular adjustments to exercise and thermal stress. Physiol. Rev. 54:75–159.

Rowell, T. E. 1959. Maternal behaviour in the golden hamster. Ph.D. diss., University of Cambridge.

Rowell, T. E. 1961. The family group in golden hamsters: its formation and break-up. Behaviour 17:81–94.

Rowell, T. E. 1967. A quantitative comparison of the behavior of a wild and a caged baboon group. Anim. Behav. 15:499–509.

Rowell, T. E. 1972. The social behavior of monkeys. Penguin, Baltimore.

Rowell, T. E. 1975. Growing up in a monkey group. Ethos 3:113–128.

Rowell, T. E., N. A. Din & A. Omar. 1968. The social development of baboons in their first three months, J. Zool. Lond. 155:461–483.

Rowell, T. E., R. A. Hinde & Y. Spencer-Booth. 1964. "Aunt"-infant interaction in captive rhesus monkeys. Anim. Behav. 12:219–226.

Rowe-Rowe, D. T. 1971. The development and behaviour of a rusty-spotted genet, *Genetta rubiginosa* Pucheran. Lammergeyer No. 13:29–44.

Rowley, J. 1929. Life history of the sea-lions on the California coast. J. Mammal. 10:1–36.

Rudge, M. R. 1970. Mother and kid behaviour in feral goats. Z. Tierpsychol. 27:687–692.

Rudnai, J. A. 1973. The social life of the lion; a study of the behaviour of wild lions (*Panthera leo massaica* [Newmann]) in the Nairobi National Park, Kenya. Washington Square East, Wallingford, Pa.

Rumbaugh, D. M. 1974. Comparative primate learning and its contributions to understanding development, play, intelligence, and language. Pages 253–281 in A. B. Chiarelli, ed., Perspectives in primate biology. Plenum Press, N.Y. & London.

Ruppenthal, G. C., M. K. Harlow, C. D. Eisele, H. F. Harlow & S. J. Suomi. 1974. Development of peer interactions of monkeys reared in a nuclear-family environment. Child Dev. 45:670–682.

Rutherford, G. W. 1979. HIA hazard analysis: injuries associated with public playground equipment. U.S. Consumer Product Safety Commission, Directorate for Hazard Identification and Analysis—Epidemiology. U.S. Government Printing Office, Washington, D.C.

Rutishauer, I. H. E. & R. G. Whitehead. 1972. Energy intake and expenditure in 1–3 year old Ugandan children living in a rural environment. Br. J. Nutr. 28:145–152.

Saayman, G. S. 1970. Baboon's brother jackal. Animals 12:442–443.

Sabater Pi, J. 1972. Contribution to the ecology of *Mandrillus sphinx* Linnaeus 1758 of Rio Muni (Republic of Equatorial Guinea). Folia Primatol. 17:304–319.

Sacerdoti, E. D. 1977. A structure for plans and behavior. Elsevier, N.Y.

Sachs, B. D. & V. S. Harris. 1978. Sex differences and developmental changes in selected juvenile activities (play) of domestic lambs. Anim. Behav. 26:678–684.

Sade, D. S. 1967. Determinants of dominance in a group of free-ranging rhesus monkeys. Pages 99–115 in S. A. Altmann, ed., Social communication among primates. University of Chicago Press, Chicago.

Sade, D. S. 1973. An ethogram for rhesus monkeys. I. Antithetical contrasts in posture and movement. Am. J. Phys. Anthropol. 38:537–542.

Saltin, B., B. Blomqvist, J. H. Mitchell, R. L. Johnson Jr., K. Wildenthal & C. B. Chapman. 1968. Response to submaximal and maximal exercise after bed rest and training. Circulation 38, Suppl. 7.

Sauer, E. G. F. & E. M. Sauer. 1963. The South-West African bushbaby of the *Galago senegalensis* group. J. South West Afr. Sci. Soc. 16:5–35.

Sauer, F. 1954. Die Entwicklung der Lautäußerungen von Ei ab schalldicht gehaltener Dorngrasmücken (*Sylvia c. communis,* Latham) im Vergleich mit später isolierten und mit wildlebenden Artgenossen. Z. Tierpsychol. 11:10–93.

Sauer, F. 1956. Über das Verhalten junger Gartengrasmücken *S. borin.* J. f. Ornith. 97:156–189.

Savage, E. S. & C. Malick. 1977. Play and sociosexual behavior in a captive chimpanzee (*Pan troglodytes*) group. Behaviour 60:179–194.

Savage, E. S., J. W. Temerlin & W. B. Lemmon. 1973. Group formation among captive mother-infant chimpanzees (*Pan troglodytes*). Folia Primatol. 20:453–473.

Schaffer, W. M. 1974. Selection for optimal life histories: the effects of age structure. Ecology 55:291–303.

Schaffer, W. M. & M. D. Gadgil. 1975. Selection for optimal life histories in plants. Pages 142–157 in M. L. Cody and J. M. Diamond, eds., Ecology and evolution of communities. The Belknap Press of Harvard University Press, Cambridge, Mass.

Schaffer, W. M. & M. L. Rosenzweig. 1977. Selection for optimal life histories. II: Multiple equilibria and the evolution of alternative reproductive strategies. Ecology. 58:60–72.

Schaller, G. B. 1963. The mountain gorilla. University of Chicago Press, Chicago.

Schaller, G. 1967. The deer and the tiger. University of Chicago Press, Chicago.

Schaller, G. B. 1968. Hunting behavior of the cheetah in the Serengeti National Park, Tanzania. E. Afr. Wildl. J. 6:95–100.

Schaller, G. B. 1972. The Serengeti lion. University of Chicago Press, Chicago.

Schaller, G. B. 1975. Review of Laws, R. M., I. S. C. Parker & R. C. B. Johnstone, Elephants and their habitats. Science 190:263–264.

Schaller, G. B. 1977. Mountain monarchs: wild sheep and goats of the Himalaya. University of Chicago Press, Chicago.

Schein, M. W. 1954. Group behavior patterns in dairy cattle and their effect on production. Sc. D. Diss., Johns Hopkins University, Baltimore.

Schenkel, R. 1966a. Play, exploration and territoriality in the wild lion. Symp. Zool. Soc. Lond. 18:11–22.

Schenkel, R. 1966b. On sociology and behaviour in impala (Aepyceros melampus suara Matschie). Z. Säugetierk. 31:177–205.

Schenkel, R. 1966c. On sociology and behaviour in impala (Aepyceros melampus Lichtenstein). E. Afr. Wildl. J. 4:99–114.

Schenkel, R. & L. Schenkel-Hulliger. 1969. Ecology and behaviour of the black rhinoceros (Diceros bicornis L.). Paul Parey, Berlin & Hamburg.

Scheurmann, E. 1975. Beobachtung zur Fortpflanzung des Gayal, Bibos frontalis Lambert, 1837. Z. Säugetierk. 40:113–127.

Schifter, H. 1968. Zucht und Markierungsverhalten von Wohlaffen im Zoologischen Garten Zürich. Zool. Gart. 36:107–132.

Schiller, P. H., 1952. Innate constituents of complex responses in primates. Psychol. Rev. 59:177–191.

Schiller, P. H. 1957. Innate motor action as a basis of learning: manipulative patterns in the chimpanzee. Pages 264–287 in C. M. Schiller, ed., Instinctive behaviour. International Universities Press, N.Y.

Schloeth, R. 1958. Über die Mutter-Kind-Beziehungen beim halbwilden Camargue-Rind. Säugetierk. Mitt. 6:145–150.

Schloeth, R. 1961. Das Sozialleben des Camargue-Rindes. Qualitative und quantitative Untersuchungen über die sozialen Beziehungen—insbesondere die soziale Rangordnung—des halbwilden französischen Kampfrindes. Z. Tierpsychol. 18:574–627.

Schlosberg, H. 1947. The concept of play. Psychol. Rev. 54:229–231.

Schlottman, R. S. & B. Seay. 1972. Mother-infant separation in the Java monkey (*Macaca irus*). J. Comp. Physiol. Psychol. 79:334–340.

Schmalhausen, I. I. 1949. Factors of evolution. Blakiston, Philadelphia.

Schmid, B. 1919. Das Tier in seinen Spielen. Verlag Th. Thomas, Leipzig.

Schmid, B. 1932. Biologische und psychologische Beobachtungen an einem in Gefangenschaft gehaltenen weiblichen Dachs (*Meles meles* L.). Z. Säugetierk. 7:156–165.

Schmid, B. 1934. Vergleichend biologische und psychologische Beobachtungen und Versuche an drei Meerkatzenartigen (*Cercopithecidae*) und einem schwarzen Brüllaffen (*Alouatta caraya* Humboldt). Z. Säugetierk. 9:164–187.

Schmid, B. 1939. Psychologische Beobachtungen und Versuche an einem jungen, männlichen Ameisenbären (*Myrmecophaga tridactyla* L.). Z. Tierpsychol. 2:117–126.

Schmidt, F. 1934. Über die Fortpflanzungsbiologie von sibirischem Zobel (*Martes zibellina* L.) und europäischem Baummarder (*Martes martes* L.). Z. Säugetierk. 9:392–403.

Schmidt, F. 1943. Naturgeschichte des Baum- und des Steinmarders. Verlag Schöps, Leipzig.

Schmidt, R. A. 1975. A schema theory of discrete motor skill learning. Psychol. Rev. 82:225–260.

Schmidt, R. A. 1976. Control processes in motor skills. Exercise and Sports Sciences Reviews 4.

Schmidt, U. &. U. Manske. 1973. Die Jugendentwicklung der Vampirfledermäuse (*Desmodus rotundus*). Z. Säugetierk. 38:14–33.

Schmidt-Nielsen, K. 1975. Animal physiology: adaptation and environment. Cambridge University Press, London & N.Y.

Schmied, A. 1973. Beiträge zu einem Aktionssystem der Hirschziegenantilope (*Antilope cervicapra* Linné 1758). Z. Tierpsychol. 32:153–198.

Schneider, D. G., L. D. Mech & J. R. Tester. 1971. Movements of female raccoons and their young as determined by radio-tracking. Anim. Behav. Monogr. 4:1–43.

Schneirla, T. C., J. S. Rosenblatt & E. Tobach. 1963. Maternal behavior in the cat. Pages 122–168 in H. L. Rheingold, ed., Maternal behavior in mammals. Wiley, N.Y.

Schoen, A. M. S., E. M. Banks & S. E. Curtis. 1976. Behavior of young Shetland and Welsh ponies (*Equus caballus*). Biol. Behav. 1:192–216.

Schoener, T. W. 1971. Theory of feeding strategies. Annu. Rev. Ecol. Syst. 2:369–404.

Schoener, T. W. 1974a. Competition and the form of habitat shift. Theor. Pop. Biol. 6:265–307.

Schoener, T. W. 1974b. The compression hypothesis and temporal resource partitioning. Proc. Nat. Acad. Sci. (U.S.A.) 71:4169–4172.

Schoener, T. W. 1974c. Resource partitioning in ecological communities. Science 185:27–39.

Schomber, H. W. 1963. Beiträge zur Kenntnis der Giraffengazelle (*Litocranius walleri* Brooke, 1878). Säugetierk. Mitt. 11, Supp. 1.

Schönberner, D. 1965. Beobachtungen zur Fortpflanzungsbiologie des Wolfes, *Canis lupus*. Z. Säugetierk. 30:171–178.

Schramm, D. L. 1968. A field study of beaver behavior in E. Barnard, Vt. M.S. thesis, Dartmouth College, Hanover, N.H.

Schreitmüller, W. 1952. Einiges über zahme Fischottern (*Lutra lutra* L.). Z. Säugetierk. 17:172–173.

Schuller, L. 1961. Kampfspiel zwischen einem Grevy-Zebra-♂ und einem Watussirind-♂. Säugetierk. Mitt. 9:10–12.

Schulman, S. R. 1978. Kin selection, reciprocal altruism, and the principle of maximization: a reply to Sahlins. Quart. Rev. Biol. 53:283–286.

Sclater, P. L. 1863. On some new and interesting animals recently acquired for the society's menagerie. Proc. Zoo. Soc. Lond. 1863:374–378.

Sclater, W. L. & R. E. Moreau. 1933. Taxonomic and field notes on some birds of North-Eastern Tanganyika Territory. Part IV. Ibis 3:187–219.

Seay, B. 1966. Maternal behavior in primiparous and multiparous rhesus monkeys. Folia Primatol. 4:146–168.

Sedlag, U. 1973. Mauswiesel (*Mustela nivalis*) als Hausgenosse. Zool. Gart. 43:188–198.

Seidensticker, J. 1976. On the ecological separation between tigers and leopards. Biotropica 8:225–234.

Seitz, A. 1950a. Untersuchungen über angeborene Verhaltensweisen bei Caniden. I. Beobachtungen an Füchsen (*Vulpes* Briss.). Z. Tierpsychol. 7:1–32.

Seitz, A. 1950b. Verhaltensstudien an Caniden. II. Fortgesetzte Beobachtungen an Füchsen. Z. Tierpsychol. 7:32–46.

Seitz, A. 1955. Untersuchungen über angeborene Verhaltensweisen bei Caniden. III. Beobachtungen an Marderhunden (*Nyctereutes procyonides* Gray). Z. Tierpsychol. 12:463–489.

Seitz, A. 1959. Beobachtungen an handaufgezogenen Goldschakalen (*Canis aureus algirensis* Wagner 1843). Z. Tierpsychol. 16:747–771.

Seligman, M. E. P. & J. L. Hager, eds. 1972. Biological boundaries of learning. Appleton-Century-Crofts, N.Y.

Settle, G. A. 1977. The quiddity of tiger quolls. Aust. Nat. Hist. 19(5):166–169.

Severinghaus, C. & E. Cheatum. 1956. Life and times of the white-tailed deer. Pages 57–186 in W. P. Taylor, ed., The deer of North America. Stackpole Co., Harrisburg, Pa. & Wildlife Mgt. Inst., Washington, D.C.

Seyfarth, R. M. 1977. A model of social grooming among adult female monkeys. J. Theor. Biol. 65:671–698.

Shadle, A. R. 1944. The play of American porcupines (*Erethizon d. dorsatum* and *E. epizanthum*). J. Comp. Psychol. 37:145–150.

Shadle, A. R. & W. R. Ploss. 1943. An unusual porcupine parturition and development of the young. J. Mammal. 24:492–496.

Shank, C. C. 1972. Some aspects of social behaviour in a population of feral goats (*Capra hircus* L.). Z. Tierpsychol. 30:488–528.

Sheak, W. H. 1924. Some further observations on the chimpanzee. J. Mammal. 5:122–129.

Sheldon, C. 1921. A fox associating with mountain sheep on the Kenai Peninsula, Alaska. J. Mammal. 2:234.

Shelley, O. 1935. Notes on the growth, behavior and taming of young marsh hawks. Auk 52:287–299.

Shepherd, P. 1968. Some notes on breeding the Quaker parakeet (*Myiopsitta monarchus*). Avic. Mag. 74:210–211.

Shettleworth, S. J. 1972. Constraints on learning. Adv. Study Behav. 4:1–68.

Shillito, E. E. 1963. Exploratory behavior in the short-tailed vole (*Microtus agrestis*). Behaviour 21:145–154.

Sibly, R. 1975. How incentive and deficit determine feeding tendency. Anim. Behav. 23:437–446.

Sidor, E. S. 1962. Becoming. In a collection of poems awarded the Irene Glascock Poetry Prize for 1962. Mount Holyoke College, S. Hadley, Mass.

Sikes, S. K. 1971. Natural history of the African elephant. American Elsevier, N.Y.

Simmons, K. E. L. 1966. Anting and the problem of self-stimulation. J. Zool. Lond. 149:145–162.

Simonds, P. E. 1965. The bonnet macaque in South India. Pages 175–197 in I. DeVore, ed., Primate behavior: field studies of monkeys and apes. Holt, Rinehart & Winston, N.Y.

Simonds, P. E. 1977. Peers, parents, and primates: the developing network of attachments. Adv. Study Commun. Affect 3:145–176.

Simpson, C. 1964. Notes on *Mungos mungo*. Arnoldia (Bulawayo) 1:1–8.

Simpson, M. J. A. 1973. Social displays and the recognition of individuals. Pages 225–279 in P. P. G. Bateson & P. H. Klopfer, eds., Perspectives in ethology. Plenum Press, N.Y.

Simpson, M. J. A. 1976. The study of animal play. Pages 385–400 in P. P. G. Bateson & R. A. Hinde, eds., Growing points in ethology. Cambridge University Press, Cambridge.

Simpson, M. J. A. 1978. Tactile experience and sexual behavior: aspects of development with special reference to primates. Pages 785–807 in J. B. Hutchinson, ed., Biological determinants of sexual behaviour. Wiley, London.

Simpson, M. J. A. & A. E. Simpson. 1977. One-zero and scan methods for sampling behaviour. Anim. Behav. 25:726–731.

Sinclair, A. R. E. 1977. The African buffalo. University of Chicago Press, Chicago.

Singer, J. L. 1973. The child's world of make-believe; experimental studies of imaginative play. Academic Press, N.Y.

Singer, J. L. & D. G. Singer. 1976. Imaginative play and pretending in early childhood: some experimental approaches. Pages 69–112 in A. Davids, ed., Child personality and psychopathology: current topics, Vol. 3. Wiley, N.Y.

Skeldon, P. C. 1963. Exhibiting aquatic mammals: Brazilian giant otters (*Pteronura brasiliensis*) at Toledo Zoo. Int. Zoo Yb. 3:30–21.

Skinner, M. P. 1929. White-tailed deer formerly in the Yellowstone Park. J. Mammal. 10:101–115.

Skutch, A. F. 1951. Life history of the boat-billed flycatcher. Auk. 68:30–49.

Sladen, W. J. L. 1956. Social structure in penguins. Pages 28–93 in B. Schaffner, ed., Group processes (Trans. 2nd Conf.). Macy Foundation, N.Y.

Slater, P. J. B. 1974. A reassessment of ethology. Pages 89–113 in Broughton, W. B., ed., Biology of brains. Academic Press, N.Y. & London.

Slatkin, M. & R. Lande. 1976. Niche width in a fluctuating environment—density-independent model. Am. Nat. 110:31–55.

Slonim, A. D. 1972. Ecophysiologic studies of innate behavior in mammals. Int. J. Psychobiol. 2:37–64.

Smilansky, S. 1968. The effects of sociodramatic play on disadvantaged pre-school children. Wiley, NY.

Smith, E. A. 1968. Adoptive suckling in the grey seal. Nature 217:762–763.

Smith, E. O. 1978. A historical view of the study of play: statement of the problem. Pages 1–32 in E. O. Smith, ed., Social play in primates. Academic Press, N.Y.

Smith, E. O. & M. D. Fraser. 1978. Social play in rhesus macaques (Macaca mulatta): A cluster analysis. Pages 79–112 in E. O. Smith, ed., Social play in primates. Academic Press, N.Y.

Smith, G. A. 1971. The use of the foot in feeding, with especial reference to parrots. Avic. Mag. 77;93–100.

Smith, P. K. 1973. Temporal clusters and individual differences in the behaviour of preschool children. Pages 751–798 in R. P. Michael & J. H. Crook, eds., Comparative ecology and behaviour of primates. Academic Press, London & N.Y.

Smith, P. K. 1977. Social and fantasy play in young children. Pages 123–145 in B. Tizard & D. Harvey, eds., Biology of play. Heinemann, London & Lippincott, Philadelphia.

Smith, P. K. & K. Connolly. 1972. Patterns of play and social interaction in preschool children. Pages 65–95 in N. G. Blurton Jones, ed., Ethological studies of child behaviour. Cambridge University Press, Cambridge.

Smith, S. M. 1972. The ontogeny of impaling behaviour in the loggerhead shrike, Lanius ludovicianus L. Behaviour 42:232–247.

Smith, S. M. 1973. A study of prey attack behaviour in young loggerhead shrikes, Lanius ludovicianus L. Behaviour 44:113–141.

Smith, W. J. 1977. The behavior of communicating. Harvard University Press, Cambridge, Mass.

Smith, W. J., S. L. Smith, E. C. Oppenheimer, J. G. DeVilla & F. A. Ulmer. 1973. Behavior of a captive population of black-tailed prairie dogs. Annual cycle of social behavior. Behaviour 46:189–220.

Snow, C. J. 1967. Some observations on the behavioral and morphological development of coyote pups. Am. Zool. 7:353–355.

Snyder, P. A. 1972. Behavior of Leontopithecus rosalia (the golden lion marmoset) and related species: a review. Pages 23–49 in D. D. Bridgewater, ed., Saving the lion marmoset. The Wild Animal Propagation Trust, Wheeling, W. Va.

Soczka, L. 1974. Éthologie sociale et sociométrie: Analyse de la structure d'un groupe de singes crabiers (*Macaca fascicularis* = *irus*) en captivité. Behaviour 50:254–269.

Sohn, J. J. 1977. Socially induced inhibition of genetically determined maturation in the platyfish, *Xiphophorus maculatus*. Science 195:199–201.

Sorenson, M. W. 1970. Behavior of tree shrews. Primate Behav. 1:141–193.

Sorenson, M. W. & C. H. Conaway. 1966. Observations on the social behavior of tree shrews in captivity. Folia Primatol. 4:124–145.

Sorenson, M. W. & C. H. Conaway. 1968. The social and reproductive behavior of *Tupaia montana* in captivity. J. Mammal. 49:502–512.

Sosnovskii, J. P. 1967. Breeding the red dog or dhole, *Cuon alpinus*, at Moscow Zoo. Int. Zoo Yb. 7:120–122.

Southwick, C. H. 1967. An experimental study of intergroup agonistic behavior in rhesus monkeys (*Macaca mulatta*). Behaviour 28:182–209.

Southwick, C. H., M. A. Beg & M. R. Siddiqi. 1965. Rhesus monkeys in North India. Pages 111–159 in I. DeVore, ed., Primate behavior: field studies of monkeys and apes. Holt, Rinehart & Winston, N.Y.

Sparks, J. 1967. Allogrooming in primates: a review. Pages 148–175 in D. Morris, ed., Primate ethology. Aldine, Chicago.

Spencer, C. C. 1943. Notes on the life history of Rocky Mountain bighorn sheep in the Tanyall Mountains of Colorado. J. Mammal. 24:1–11.

Spencer, D. A. 1930. An interesting caesarean operation. J. Mammal. 11:84–86.

Spencer-Booth, Y. 1968. The behaviour of twin rhesus monkeys and comparisons with the behaviour of single infants. Primates 9:75–84.

Spencer-Booth, Y. 1970. The relationships between mammalian young and conspecifics other than mothers and peers: a review. Adv. Study Behav. 3:119–194.

Spinage, C. A. 1969. Naturalistic observations on the reproductive and maternal behaviour of the Uganda Defassa waterbuck *Kobus defassa ugandae* Neumann. Z. Tierpsychol. 26:39–47.

Sprunt, A. 1944. Remarkable aerial behavior of the purple martin. Auk 61:296–297.

Stanley, M. 1971. An ethogram of the hopping mouse, *Notomys alexis*. Z. Tierpsychol. 29:225–258.

Stearns, S. C. 1976. Life-history tactics: a review of the ideas. Q. Rev. Biol. 51:3–47.

Stearns, S. C. 1977. The evolution of life-history traits. Annu. Rev. Ecol. Syst. 8:145–171.

Stefansson, V. 1944. The friendly Arctic. Macmillan, N.Y.

Stegeman, L. C. 1937. Notes on young skunks in captivity. J. Mammal. 18:194–202.

Steinborn, W. 1973. Beobachtungen zum Verhalten des Alpensteinbocks, *Capra ibex ibex* Linne 1758. Säugetierk. Mitt. 21:37–65.

Steinemann, P. 1966. Künstliche Aufzucht eines Eisbären. Zool. Gart. 32:129–145.

Steiner, A. L. 1971. Play activity of Columbian ground squirrels. Z. Tierpsychol. 28:247–261.

Steiniger, B. 1976. Beiträge zum Verhalten und zur Soziologie des Bisams (*Ondatra zibethicus* L.). Z. Tierpsychol. 41:55–79.

Stephens, D. B. 1974. Studies on effect of social environment on behavior and growth-rates of artificially-reared British Friesian male calves. Anim. Prod. 18:23–34.

Stern, D. N. 1974a. The goal and structure of mother-infant play. J. Am. Acad. Child. Psychiatry 13:402–421.

Stern, D. N. 1974b. Mother and infant at play: the dyadic interaction involving facial, vocal and gaze behaviors. Pages 187–213 in M. Lewis & L. Rosenblum, eds., The effect of the infant on its care-giver. Wiley, N.Y.

Stevenson, M. F. 1973. Observations of maternal behaviour and infant development in the De Brazza monkey *Cercopithecus neglectus* in captivity. Int. Zoo Yb. 13:179–184.

Stevenson, M. F. 1976. Behavioural observations on groups of Callithricidae with an emphasis on playful behaviour. Jersey Wildlife Preservation Trust 13th Annual Report:47–52.

Stevenson, M. F. & T. B. Poole. 1976. An ethogram of the common marmoset (*Callithrix jacchus jacchus*): general behavioral repertoire. Anim. Behav. 24:428–451.

Stirling, I. 1970. Observations on the behavior of the New Zealand fur seal (*Arctocephalus forsteri*). J. Mammal. 51:766–778.

Stoinitzer, W. 1959. *Ursus arctos* (Ursidae)—Spiel der Jungtiere. Encyclopedia Cinematographica Film E205. Institut für den Wissenschaftlichen Film, Göttingen.

Stonehouse, B. & S. Stonehouse. 1963. The frigate bird *Fregata aquila* of Ascension Island. Ibis 103b:409–422.

Stoner, E. A. 1947. Anna hummingbirds at play. Condor 49:36.

Strayer, F. F., A. Bovenkirk & R. F. Koopman. 1975. Social affiliation and dominance in captive squirrel monkeys (*Saimiri sciureus*). J. Comp. Physiol. Psychol. 89:308–318.

Stringham, S. F. 1974. Mother-infant relations in moose. Nat. Can. 101:325–369.

Struhsaker, T. T. 1967a. Behavior of elk (*Cervus canadensis*) during the rut. Z. Tierpsychol. 24:80–114.

Struhsaker, T. T. 1967b. Behavior of vervet monkeys (*Cercopithecus aethiops*). Univ. Cal. Pub. Zool. 82.

Struhsaker, T. T. 1969. Correlates of ecology and social organization among African cercopithecines. Folia Primatol. 11:80–118.

Struhsaker, T. T. 1971. Social behaviour of mother and infant vervet monkeys (*Cercopithecus aethiops*). Anim. Behav. 19:233–250.

Struhsaker, T. T. 1975. The red colobus monkey. University of Chicago Press, Chicago.

Struhsaker, T. T. & J. S. Gartlan. 1970. Observations on the behavior and ecology of the patas monkey (*Erythrocebus patas*) in the Waza Reserve, Cameroon. J. Zool. Lond. 161:49–63.

Stumper, R. 1921. Études sur les fourmis. III. Recherches sur l'éthologie du *Formicoxenus nitidulus* Nyl. Bull. Soc. Entomol. Belg. 3:90–97.

Stynes, A. J., L. A. Rosenblum & I. C. Kaufman. 1968. The dominant male and behavior within heterospecific monkey groups. Folia Primatol. 9:123–134.

Sugiyama, Y. 1965a. Behavioral development and social structure in two troops of hanuman langurs (*Presbytis entellus*). Primates 6:213–247.

Sugiyama, Y. 1965b. On the social change of hanuman langurs (*Presbytis entellus*) in their natural condition. Primates 6:381–418.

Sugiyama, Y. 1969. Social behavior of chimpanzees in the Budongo Forest, Uganda. Primates 10:197–225.

Sugiyama, Y. 1971. Characteristics of the social life of bonnet macaques (*Macaca radiata*). Primates 12:247–266.

Sugiyama, Y. 1976a. Characteristics of the ecology of the Himalayan langurs. J. Human Evol. 5:249–277.

Sugiyama, Y. 1976b. Life-history of male Japanese monkeys. Adv. Stud. Behav. 7:255–284.

Sumner, E. L. 1931. Some observations on bird behavior. Condor 33:89–91.

Suomi, S. J. 1977. Development of attachment and other social behaviors in rhesus monkeys. Adv. Stud. Commun. Affect 3:197–224.

Suomi, S. J. & H. F. Harlow. 1977. Early separation and behavioral maturation. Pages 197–214 in A. Oliverio, ed., Genetics, environment and intelligence. Elsevier North-Holland, N.Y.

Susser, M. & W. B. Watson. 1969. Play and the sick child. Lancet 1:369.

Sussman, G. 1973. A computational model of skill acquisition. Technical Report 297, Artificial Intelligence Laboratory, Massachusetts Institute of Technology, Cambridge, Mass.

Sussman, G. 1975. A computer model of skill acquisition. American Elsevier, N.Y.

Sussman, R. W. 1977. Socialization, social structure, and ecology of two sympatric species of *Lemur*. Pages 515–528 in S. Chevalier-Skolnikoff & F. E. Poirier, eds., Primate bio-social development. Garland, N.Y.

Sutton-Smith, B. 1971. Conclusion. Pages 343–345 in R. E. Herron & B. Sutton-Smith, eds., Child's play. Wiley, N.Y.

Sutton-Smith, B. 1975. Play as adaptive potentiation. Sportswissenschaft 5:103–118.

Sutton-Smith, B., ed. 1979. Play and learning. Gardner Press, N.Y.

Sutton-Smith, B. & S. Sutton-Smith. 1974. How to play with your children (and when not to). Hawthorn, N.Y.

Svihla, A. 1931. Habits of the Louisiana mink (*Mustela vison vulgivagus*). J. Mammal. 12:366–368.

Svihla, R. D. 1930. A family of flying squirrels. J. Mammal. 11:211–213.

Swanberg, P. O. 1952. Observations on feeding, brooding, and bathing habits in a pair of kingfishers (*Alcedo atthis*). Vår. Fågelvärld 11:49–66.

Swynnerton, C. F. M. 1908. Further notes on the birds of Gazaland. Ibis 2:391–442.

Sylva, K. 1977. Play and learning. Pages 59–73 in B. Tizard & D. Harvey, eds., Biology of play. Heinemann, London & Lippincott, Philadelphia.

Sylva, K., J. S. Bruner & P. Genova. 1976. The role of play in the problem-solving of children 3-5 years old. Pages 244-260 in J. S. Bruner, A. Jolly & K. Sylva, eds., Play: its role in development and evolution. Basic Books, N.Y.

Symons, D. 1974. Aggressive play and communication in rhesus monkeys (*Macaca mulatta*). Am. Zool. 14:317-322.

Symons, D. 1978a. Play and aggression: a study of rhesus monkeys. Columbia University Press, N.Y.

Symons, D. 1978b. The question of function: dominance and play. Pages 193-230 in E. O. Smith, ed., Social play in primates. Academic Press, N.Y.

Symons, D. 1979. The evolution of human sexuality. Oxford University Press, N.Y.

Szederjei, A. & L. Fábián. 1975. Giraffengeburten im Zoo Budapest. Zool. Gart. 45:175-186.

Taber, F. W. 1945. Contribution on the life-history and ecology of the nine-banded armadillo. J. Mammal. 26:211-226.

Takeda, R. 1965. Development of vocal communication in man-raised Japanese monkeys. I. From birth until the sixth week. Primates 6:337-380.

Tattersall, I. 1977. Ecology and behavior of *Lemur fulvus mayottensis* (Primates, Lemuriformes). Anthropol. Pap. Am. Mus. Nat. Hist. 54:421-482.

Tattersall, I. & R. W. Sussman, eds, 1975. Lemur biology. Plenum Press, N.Y.

Tavolga, M. C. 1966. Behavior of the bottlenose dolphin (*Tursiops truncatus*): social interactions in a captive colony. Pages 718-730 in K. S. Norris, ed., Whales, dolphins and porpoises. University of Calif. Press, Berkeley & Los Angeles.

Tavolga, M. C. & F. S. Essapian. 1957. The behavior of the bottlenosed dolphin (*Tursiops trucatus*): mating, pregnancy, parturition, and mother-infant behavior. Zoologica 42:11-31.

Taylor, C. R. 1973. Energy cost of animal locomotion. Pages 23-42 in L. Bolis, K. Schmidt-Nielsen & S. H. P. Maddrell, eds., Comparative physiology. North-Holland Publishing Co., Amsterdam & London.

Taylor, C. R. 1978. Why change gaits? Recruitment of muscles and muscle fibers as a function of speed and gait. Am. Zool. 18:153-161.

Teitelbaum, P. 1977. Levels of integration of the operant. Pages 7-27 in W. K. Honig & J. E. R. Staddon, eds., Handbook of operant behavior. Prentice-Hall, Englewood Cliffs, N.J.

Tembrock, G. 1957. Zur Ethologie des Rotfuchses (*Vulpes vulpes* L.), unter besonderer Berücksichtigung der Fortpflanzung. Zool. Gart. 23:289-532.

Tembrock, G. 1958. Spielverhalten beim Rotfuchs. Zool. Beitr. (Berl.) 3:423-496.

Tembrock, G. 1960. Spielverhalten und vergleichende Ethologie. Beobachtungen zum Spiel von *Alopex lagopus*(L.) Z. Säugetierk. 25:1-14.

Tempel, E. 1971. Beobachtungen und Daten zur natürlichen und künstlichen Aufzucht eines Orang-Utans (*Pongo pygmaeus*) im Zoo Dresden. Zool. Gart. 40:232-241.

Tener, J. S. 1965. Muskoxen in Canada: a biological and taxonomic review. Monograph Series of the Canadian Wildlife Service, 2.

Thaler, E. 1977. Die postnatale Entwicklung eines Hybriden zwischen Alpenkrähen-♂ (*Pyrrhocorax pyrrhocorax*) und Alpendohlen-♀ (*P. graculus*). Zool. Gart. 47:241–260.

Tharp, G. D. & R. J. Buuck. 1974. Adrenal adaptation to chronic exercise. J. Appl. Physiol. 37:720–722.

Thoreau, H. D. 1949. Journal, June 25, 1858, Vol. 10. Riverside Press, Cambridge, Mass.

Thorington, R. W., Jr. 1968. Observations of squirrel monkeys in a Colombian forest. Pages 69–85 in L. A. Rosenblum & R. W. Cooper, eds., The squirrel monkey. Academic Press, N.Y.

Thornburn, A. 1921. British mammals, Vol. 2. Longmans, Green, London.

Thorpe, W. H. 1962. Biology and the nature of man. Oxford University Press, London.

Thorpe, W. H. 1963. Learning and instinct in animals, 2nd ed. Methuen, London.

Thorpe, W. H. 1966. Ritualization in ontogeny: I. Animal Play. Philos. Trans. R. Soc. Lond. Ser. B. Biol. Sci. 251:311–319.

Thorpe, W. H. 1972. The comparison of vocal communication in animals and man. Pages 27–47 in R. A. Hinde, ed., Non-verbal communication. Cambridge University Press, Cambridge, Mass.

Thorpe, W. H. 1974. Animal nature and human nature. Anchor, Doubleday, Garden City, N.Y.

Tileston, J. V. & R. R. Lechleitner. 1966. Some comparisons of the black-tailed and white-tailed prairie dogs in North-Central Colorado. Am. Midl. Nat. 75:292–316.

Tinbergen, N. 1951. The study of instinct. Clarendon Press, Oxford.

Tinbergen, N. 1963. On aims and methods of ethology. Z. Tierpsychol. 20:410–433.

Tinbergen, N. & M. Norton-Griffiths. 1964. Oystercatchers and mussels. Br. Birds 57:64–70.

Todd, F. S. 1974. Maturation and behaviour of the California condor *Gymnogyps californianus* at the Los Angeles Zoo. Int. Zoo Yb. 14:145–147.

Todd, F. S. & N. B. Gale. 1970. Further notes on the California condor *Gymnogyps californianus*, at Los Angeles Zoo. Int. Zoo Yb. 10:15–17.

Todd, N. B. 1963. The catnip response. Ph.D. diss., Harvard University.

Toepfer, I. 1971. Künstliche Aufzucht von Rehkitzen (*Capreolus capreolus* L.) Zool. Gart. 41:26–44.

Tomanek, R. J. 1970. Effects of age and exercise on the extent of the myocardial capillary bed. Anat. Rec. 167:55–62.

Tong, E. H. 1962. The breeding of the great Indian rhinoceros at Whipsnade Park. Int. Zoo Yb. 2:12–15.

Tonkin, B. A. & E. Kohler. 1978. Breeding the African golden cat *Felis* (*Profelis*) *aurata* in captivity. Int. Zoo Yb. 18:147–150.

Toweill, D. E. & D. B. Toweill. 1978. Growth and development of captive ringtails. Carnivore 1:46–53.

Townsend, M. T. & M. W. Smith. 1933. The white-tailed deer of the Adirondacks. Bull. N.Y. State Coll. Forestry Syracuse University 6:153–383.

Trapp, G. 1972. Some anatomical & behavioral adaptations of ringtails, *Bassariscus astutus*. J. Mammal. 53:549.

Treisman, M. 1977. The evolutionary restriction of aggression within a species: a game theory analysis. J. Math. Psychol. 16:167–203.

Trivers, R. 1971. The evolution of reciprocal altruism. Quart. Rev. Biol. 46:35–57.

Trivers, R. L. 1972. Parental investment and sexual selection. Pages 136–179 in B. Campbell, ed., Sexual selection and the descent of man, 1871–1971. Aldine, Chicago.

Trivers, R. L. 1974. Parent-offspring conflict. Am. Zool. 14:247–262.

Trivers, R. L. 1976. Introduction to R. Dawkins' *The Selfish Gene*. Oxford University Press, N.Y.

Trollope, J. 1971. Some aspects of behavior and reproduction in captive barn owls (*Tyto alba alba*). Avic. Mag. 77:117–125.

Trollope, J. & N. G. Blurton Jones. 1975. Aspects of reproduction and reproductive behaviour in *Macaca arctoides*. Primates 16:191–205.

Troughton, E. 1941. Furred mammals of Australia. Angus & Robertson, Sydney & London.

Trumler, E. 1959a. Beobachtungen an den Böhmzebras des Georg von Opel-Freigeheges für Tierforschung. 3. Beziehungen zum Artfremden. Säugetierk. Mitt. 7(Supp.):126–191.

Trumler, E. 1959b. Einige Beobachtungen beim Spiel jugendlicher Steppenelefanten, *Loxodonta africana knochenhaueri* Matschie, 1900. Säugetierk. Mitt. 7:52–58.

Turner, K. 1970. Breeding Tasmanian devils, *Sarcophilus harrisii*, at Westbury Zoo. Int. Zoo Yb. 10:65.

Tutin, C. E. G. & W. C. McGrew. 1973. Chimpanzee copulatory behavior. Folia Primatol. 19:237–256.

Tyler, S. J. 1972. The behaviour and social organization of the New Forest ponies. Anim. Behav. Monogr. 5:85–196.

Tyndale-Biscoe, H. 1973. Life of marsupials. American Elsevier, N.Y.

Ulmer, F. A. Jr. 1968. Breeding fishing cats, *Felis viverrina*, at Philadelphia Zoo. Int. Zoo Yb. 8:49–55.

Ulrich, S., V. Ziswiler & H. Bregulla. 1972. Biologie und Ethologie des Schmalbindenloris, *Trichoglossus haematodus massena* Bonaparte. Zool. Gart. 42:51–94.

van Hooff, J. A. R. A. M. 1967. The facial displays of the catarrhine monkeys and apes. Pages 7–68 in D. Morris, ed., Primate ethology. Aldine, Chicago.

van Hooff, J. A. R. A. M. 1970. A component analysis of the structure of the social behaviour of a semi-captive chimpanzee group. Experientia 26:549–550.

van Hooff, J. A. R. A. M. 1971. Aspecten van het sociale gedrag en de communicatie bij humane en hogere niet-humane primaten. Bronder-Offset n.v., Rotterdam.

van Hooff, J. A. R. A. M. 1972. A comparative approach to the phylogeny of laughter and smiling. Pages 209–241 in R. A. Hinde, ed., Nonverbal communication. University Press, Cambridge.

van Hooff, J. A. R. A. M. 1962. Facial expressions in higher primates. Symp. Zool. Soc. Lond. 8:97–125.

van Lawick, H. & J. van Lawick-Goodall. 1971. Innocent killers. Houghton Mifflin, Boston.

van Lawick-Goodall, J. 1967. Mother-offspring relationships in free-ranging chimpanzees. Pages 287–346 in D. Morris, ed., Primate ethology. Aldine, Chicago.

van Lawick-Goodall, J. 1968a. The behavior of free-living chimpanzees in the Gombe Stream Reserve. Anim. Behav. Monogr. 1:161–311.

van Lawick-Goodall, J. 1968b. A preliminary report on expressive movements and communication in the Gombe Stream chimpanzees. Pages 313–374 in P. C. Jay, ed., Primates: studies in adaptation and variability. Holt, Rinehart & Winston, N.Y.

van Lawick-Goodall, J. 1970. Tool-using in primates and other vertebrates. Adv. Study Behav. 3:195–249.

van Lawick-Goodall, J. 1971. In the shadow of man. Houghton Mifflin. Boston.

van Lawick-Goodall, J. 1973. Cultural elements in a chimpanzee community. Pages 144–184 in E. W. Menzel, ed., Precultural primate behavior. Karger, Basel.

Vanoli, T. 1967. Beobachtungen am Pudus, Mazama pudu (Molina, 1782). Säugetierk. Mitt. 15:155–163.

Van Valen, L. 1973. Festschrift. Science 180:488.

Varley, M. & D. Symmes. 1966. The hierarchy of dominance in a group of macaques. Behaviour 27:54–75.

Vaughn, R. 1974. Breeding the tayra Eira barbara at Antelope Zoo, Lincoln. Int. Zoo Yb. 14:120–122.

Vaz-Ferriera, R. 1975. Behavior of the Southern sea lion, Otaria flavescens (Shaw) in the Uruguayan islands. Rapp. P.-v. Réun. Cons. int. Explor. Mer 169:219–227.

Verbeek, N. A. M. 1972. Comparison of displays of the yellow-billed magpie (Pica nuttalli) and other corvids. J. Ornithol. 113:297–314.

Verheyen, R. 1955. Contribution à l'éthologie du waterbuck Kobus defassa ugandae Neumann et de l'antilope harnachée Tragelaphus scriptus (Pallas). Mammalia 19:309–319.

Verreaux, G. 1848. Observations sur l'ornithorhyque. Rev. Zool. 11:127–134.

Veselovský, Z. 1975. Notes on the breeding of cheetah (Acinonyx jubatus Schreber) at Prague Zoo. Zool. Gart. 45:28–44.

Vesey-Fitzgerald, B. 1965. Town fox, country fox. A. Deutsch, London.

Vick, L. G. & J. M. Conley. 1976. An ethogram for Lemur fulvus. Primates 17:125–144.

Vincent, L. E. & M. Bekoff. 1978. Quantitative analyses of the ontogeny of predatory behaviour in coyotes, Canis latrans. Anim. Behav. 26:225–231.

Vogel, C. 1962. Einige Gefangenschaftsbeobachtungen am weiblichen Fenek, *Fennecus zerda* (Zimm. 1780). Z. Säugetierk. 27:193–204.

Vogel, C. 1976. Ökologie, Lebensweise und Sozialverhalten der grauen Languren in verschiedenen Biotopen Indiens. Z. Tierpsychol. Beiheft 17.

Volkmar, F. R. & W. T. Greenough. 1972. Rearing complexity affects branching of dendrites in the visual cortex of the rat. Science 176:1445–1447.

Volmar, F. A. 1940. Das Bärenbuch. Paul Haupt, Bern.

von Frisch, O. & H. von Frisch. 1971. Beobachtungen bei der Handaufzucht und späteren Aussetzung einer Fuchsfähe (*Vulpes vulpes*). Z. Tierpsychol. 28:534–541.

Von Ketelhodt, H. F. 1966. Der Erdwolf, *Proteles cristatus* (Sparrman, 1783). Z. Säugetierk. 31:300–308.

von Uexküll, J. 1909. Umwelt und Innenwelt der Tiere. 2nd ed., 1921. Springer, Berlin.

Voss, G. 1963. Beobachtungen an einem jugen Wald-Caribou, *Rangifer caribou sylvestris* (Richardson). Z. Säugetierk. 28:184–186.

Voss, G. 1969. Breeding the pronghorn antelope and the saiga antelope, *Antilocapra americana* and *Saiga tatarica*, at Winnipeg Zoo. Int. Zoo Yb. 9:116–118.

Vosseler, J. 1928. Beobachtungen am Fleckenroller. Z. Säugetierk. 3:80–91.

Vowles, D. M. 1970. The psychobiology of aggression. Edinburgh University Press, Edinburgh.

Vygotsky, L. S. 1967. Play and its role in the mental development of the child. Sov. Psychol. 5:5–18.

Waddington, C. H. 1957. The strategy of the genes. Macmillan, N.Y.

Waddington, C. H. 1965. Introduction to the symposium. Pages 1–6 in H. G. Baker & G. L. Stebbins, eds., Genetics of colonizing species. Academic Press, N.Y.

Wade, M. J. 1978a. A critical review of the models of group selection. Q. Rev. Biol. 53:101–114.

Wade, M. J. 1978b. Kin selection: a classical approach and a general solution. Proc. Natl. Acad. Sci. U.S.A. 75:6154–6158.

Wagner, H. O. 1954. Versuch einer Analyse der Kolibribalz. Z. Tierpsychol. 11:182–212.

Wahlsten, D. 1974. A developmental time scale for postnatal changes in brain and behavior of B6D2F$_2$ mice. Brain Res. 72:251–264.

Walker, A. 1968. A note on hand-rearing a potto, *Perodicticus potto*. Int. Zoo Yb. 8:110–111.

Walker, E. P. 1975. Mammals of the world, 3rd ed., 2 vols. Johns Hopkins Univ. Press, Baltimore.

Wallace, A. R. 1962. The Malay archipelago. (Reprint of last revised ed., 1869). Dover Pubs., N.Y.

Wallace, B. 1973. Misinformation, fitness, and selection. Am. Nat. 107:1–7.

Wallace, H. F. 1913. The big game of central and western China, being an account of a journey from Shanghai to London overland across the Gobi desert. Duffield & Co., N.Y.

Wallmo, O. C. & S. Gallizioli. 1954. Status of the coati in Arizona. J. Mammal. 35:48–54.

Walther, F. 1958. Zum Kampf- und Paarungsverhalten einiger Antilopen. Z. Tierpsychol. 15:340–380.

Walther, F. 1961. Zum Kampfverhalten des Gerenuk (*Litocranius walleri*). Nat. Volk (Frankf.) 91:313–321.

Walther, F. 1962. Uber ein Spiel bei *Okapia johnstoni*. Z. Säugetierk. 27:245–251.

Walther, F. 1964. Summary zum Beitrag: Verhaltensstudien an der Gattung Tragelaphus De Blainville, 1816 in Gefangenschaft. Z. Tierpsychol. 21:642–646.

Walther, F. 1965a. Psychologische Beobachtungen zur Gesellschaftshaltung von Oryx-Antilopen (*Oryx gazella beisa* Rüpp). Zool. Gart. 31:1–58.

Walther, F. 1965b. Verhaltensstudien an der Grantgazelle (*Gazella granti* Brooke, 1872) im Ngorongoro-Krater. Z. Tierpsychol. 22:167–208.

Walther, F. 1968. Ethologische Beobachtungen bei der künstlichen Aufzucht eines Bleßbockkalbes (*Damaliscus dorcas philippsi* Harper, 1939). Zool. Gart. 36:191–215.

Walther, F. 1969. Flight behaviour and avoidance of predators in Thomson's gazelle. Behaviour 34:184–221.

Walther, F. R. 1972. Social grouping in Grant's gazelle (*Gazella granti* Brooke 1827) in the Serengeti National Park. Z. Tierpsychol. 31:348–403.

Walther, F. 1973. Round-the-clock activity of Thomson's gazelle (*Gazella thomsoni* Günther, 1884) in the Serengeti National Park. Z. Tierpsychol. 32:75–105.

Wandrey, R. 1975. Contribution to the study of the social behaviour of captive golden jackals (*Canis aureus* L.). Z. Tierpsychol. 39:365–402.

Warren, S. B. 1927. The beaver, its work and its ways. Williams & Wilkins, Baltimore.

Wasser, S. K. 1978. Structure and function of play in the tiger. Carnivore 1:27–40.

Watson, J. S. 1972. Smiling, cooing and "the game." Merrill-Palmer Q. Behav. Dev. 18:323–339.

Webb, C. S. 1955. A hare about the house. Hutchinson, London.

Weigel, R. M. 1979. The facial expressions of the brown capuchin monkey (*Cebus apella*). Behaviour 68:250–276.

Weiland, G. 1965. Zur Frage der Weidenhaltung von Kälbern. Arch. Tierzucht 8:199–226.

Weir, D. & N. Picozzi. 1975. Aspects of social behaviour in the buzzard. Br. Birds 68:125–141.

Weir, R. 1976. Playing with language. Pages 609–618 in J. S. Bruner, A. Jolly & K. Sylva, eds., Play: its role in development and evolution. Basic Books, N.Y.

Weisler, A. & R. B. McCall. 1976. Exploration and play: resumé and redirection. Am. Psychol. 31:492–508.

Weiss-Bürger, M. 1975. Lernspecifische Einflüsse im Erkundungs- und Spielverhalten von Iltisfrettchen (*Mustela putorius* x *M. furo*). Doktorarbeit, Philipps-Universität Marburg.

Welford, A. T. 1968. Fundamentals of skill. Methuen, London.

Welker, C. 1977. Zur Aktivitätsrhythmik von *Galago crassicaudatus* E. Geoffroy, 1812 (Prosimiae; Lorisiformes; Galagidae) in Gefangenschaft. Z. Säugetierk. 42:65–78.

Welker, W. I. 1959. Genesis of exploratory and play behavior in infant raccoons. Psychol. Rep. 5:764.

Welker, W. I. 1961. An analysis of exploratory and play behavior in animals. Pages 175–226 in D. W. Fiske & S. R. Maddi, eds., Functions of varied experience. Dorsey, Homewood, Ill.

Welker, W. I. 1971. Ontogeny of play and exploratory behaviors: a definition of problems and a search for new conceptual solutions. Pages 171–228 in H. Moltz, ed., The ontogeny of vertebrate behavior. Academic Press, N.Y.

Welles, R. E. & F. B. Welles. 1961. The bighorn of Death Valley. Fauna of the National Parks of the United States, Fauna Series No. 6. U.S. Government Printing Office, Washington, D.C.

Wemmer, C. M. 1977. Comparative ethology of the large-spotted genet (*Genetta tigrina*) and some related viverrids. Smithson. Contrib. Zool. 239:1–93.

Wemmer, C. & M. J. Fleming. 1974. Ontogeny of playful contact in a social mongoose, the meerkat, *Suricata suricatta*. Am. Zool. 14:415–426.

Wesley, F. 1967. Stereotypy and teat selection in pigs. Z. Säugetierk. 32:362–366.

West, M. 1974. Social play in the domestic cat. Am. Zool. 14:427–436.

West, M. J. 1977. Exploration and play with objects in domestic kittens. Dev. Psychobiol. 10:53–57.

Wheeler, R. 1943. Pacific gull at play? Emu 42:181.

White, B. L. 1970. Experience and the development of motor mechanisms in infancy. Pages 95–134 in K. J. Connolly, ed., Mechanisms of motor skill development. Academic Press, N.Y. & London.

White, L. 1977a. The nature of social play and its role in the development of the rhesus monkey. Ph.D. diss., Cambridge University.

White, L. 1977b. Play in animals. Pages 15–32 in B. Tizard & D. Harvey, eds., Biology of play. Heinemann, London & Lippincott, Philadelphia.

White, M. 1956. Toward reunion in philosophy. Harvard University Press, Cambridge, Mass.

Whiting, B. B. & C. P. Edwards. 1973. A cross-cultural analysis of sex differences in the behavior of children aged three through eleven. J. Soc. Psychol. 91:171–188.

Whiting, B. B. & J. W. M. Whiting. 1975. Children of six cultures: a psychocultural analysis. Harvard University Press, Cambridge, Mass.

Whitney, L. F. & A. B. Underwood. 1952. The raccoon. Practical Science Publishing Co., Orange, Conn.

Whittaker, R. H. & D. Goodman. 1979. Classifying species according to their demographic strategy. I. Population fluctuations and environmental heterogeneity. Am. Nat. 113:185–200.

Wilbur, H., D. W. Tinkle & J. P. Collins. 1974. Environmental certainty, trophic level, and resource availability in life history evolution. Am. Nat. 108:805–817.

Williams, E. & J. P. Scott. 1954. The development of social behavior patterns in the mouse, in relation to natural periods. Behaviour 6:35–65.

Williams, G. C. 1957. Pleiotropy, natural selection, and the evolution of senescence. Evolution 11:398–411.

Williams, G. C. 1966. Adaptation and natural selection. Princeton University Press, Princeton, N.J.

Willig, A. & S. Wendt. 1970. Aufzucht und Verhalten des Geoffroyi-Perückenäffchens, *Oedipomidas geoffroyi* Pucheran, 1845. Säugetierk. Mitt. 18:117–122.

Wilson, C. & E. Weston. 1946. The cats of Wildcat Hill. Duell, Stone & Pearce, N.Y.

Wilson, D. E. & J. F. Eisenberg. 1978. Relative brain size and feeding strategies in the Chiroptera. Evolution 32:740–751.

Wilson, D. S. 1975. The adequacy of body size as a niche difference. Am. Nat. 109:769–784.

Wilson, E. O. 1971a. Competitive and aggressive behavior. Pages 183–217 in J. F. Eisenberg & W. S. Dillon, eds., Man and beast: comparative social behavior. Smithsonian Institution Press, Washington.

Wilson, E. O. 1971b. The insect societies. The Belknap Press of Harvard University Press, Cambridge, Mass.

Wilson, E. O. 1975. Sociobiology. The Belknap Press of Harvard University Press, Cambridge, Mass.

Wilson, E. O. 1976. Some central problems of sociobiology. Soc. Sci. Inform. 14:5–18.

Wilson, E. O. 1977a. Animal and human sociobiology. Pages 273–281 in C. E. Goulden, ed., Changing scenes in natural sciences, 1776–1976. Special Publication 12, Academy of Natural Sciences, Philadelphia, Pa.

Wilson, E. O. 1977b. Biology and the social sciences. Daedalus 106(4):127–140.

Wilson, E. O. 1978. On human nature. Harvard University Press, Cambridge, Mass.

Wilson, E. O. & W. H. Bossert. 1971. A primer of population biology. Sinauer, Sunderland, Mass.

Wilson, J. 1860. Remarks on a tiger cat (*Felis pardalis minimus*). Proc. Acad. Nat. Sci. (Phila.) 1860:82–84.

Wilson, M. L. & J. G. Elicker. 1976. Establishment, maintenance, and behavior of free-ranging chimpanzees on Ossabaw Island, Georgia, U.S. Primates 17:451–473.

Wilson, S. 1973. The development of social behavior in the vole (*Microtus agrestis*). Zool. J. Linn. Soc. 52:45–62.

Wilson, S. 1974a. Juvenile play of the common seal *Phoca vitulina vitulina* with comparative notes on the grey seal *Halichoerus grypus*. Behaviour 48:37–60.

Wilson, S. 1974b. Mother-young interactions in the common seal, *Phoca vitulina vitulina*. Behaviour 48:23–36.

Wilson, S. & D. Kleiman. 1974. Eliciting play: a comparative study. Am. Zool. 14:341–370.

Wilson, V. & G. Child. 1966. Notes on development and behaviour of two captive leopards. Zool. Gart. 32:67–70.

Wilsson, L. 1968. My beaver colony. Trans. Joan Bulman. Doubleday & Co., Garden City, N.Y.

Wilsson, L. 1971. Observations and experiments on the ethology of the European beaver (*Castor fiber* L.). Viltrevy, Stockholm 8:115–226.

Winnicott, D. W. 1971. Playing and reality. Tavistock, London.

Winston, P. H., ed. 1975. The psychology of computer vision. McGraw-Hill, N.Y.

Winston, P. H. 1977. Artificial intelligence. Addison-Wesley, Reading, Mass.

Winter, P. 1968. Social communication in the squirrel monkey. Pages 235–253 in L. A. Rosenblum & R. W. Cooper, eds., The squirrel monkey. Academic Press, N.Y.

Winter, P. 1969. The variability of peep and twit calls in captive squirrel monkeys *Saimiri sciureus*. Folia Primatol. 10:204–215.

Wolfheim, J. H. 1977. Sex differences in behavior in a group of captive juvenile talapoin monkeys (*Miopithecus talapoin*). Behaviour 63:110–128.

Wolfheim, J. H. & T. E. Rowell. 1972. Communication among captive talapoin monkeys (*Miopithecus talapoin*). Folia Primatol. 18:224–255.

Woolpy, J. H. 1968. Socialization of wolves. Science and Psychoanalysis 12:82–94.

Wrogemann, N. 1975. Cheetah under the sun. McGraw-Hill, N.Y.

Wünschmann, A. 1966. Einige Gefangenschaftsbeobachtungen am Breitstirn-Wombat (*Lasiorhinus latifrons* Owen 1845). Z. Tierpsychol. 23:56–71.

Wüst, G. 1976. Geburt und perinatales Verhalten beim Steppenzebra, *Equus quagga*. Zool. Gart. 46:305–352.

Wüstehube, C. 1960. Beiträge zur Kenntnis besonders des Spiel- und Beuteverhaltens einheimischer Musteliden. Z. Tierpsychol. 17:579–613.

Yadav, R. N. 1967. Breeding the smooth-coated Indian otter *Lutra perspicillata* at Jaipur Zoo. Int. Zoo Yb. 7:130–131.

Yamada, M. 1963. A study of blood-relationship in the natural society of the Japanese macaque. Primates 4:43–65.

Yate, M. T. 1898. Polecats as pets. J. Bombay Nat. Hist. Soc. 11:737–739.

Yeaton, R. I. 1972. Social behavior and social organization in Richardson's ground squirrel (*Spermophilus richardsoni*) in Saskatchewan. J. Mammal. 53:139–147.

Yerkes, R. M. 1943. Chimpanzees, a laboratory colony. Yale University Press, New Haven.

Yerkes, R. M. & A. Petrunkevitch. 1925. Studies of chimpanzee vision by Ladygin-Kohts. J. Comp. Psychol. 5:99–108.

Yocom, C. F. 1967. Ecology of feral goats in Haleakala National Park, Maui, Hawaii. Am. Midl. Nat. 77:418–451.

Young, C. G. 1929. A contribution to the ornithology of the coastland of British Guiana. Ibis 5:1–38.

Zahavi, A. 1977. The testing of a bond. Anim. Behav. 25:246–247.

Zannier, F. 1965. Verhaltensuntersuchungen an der Zwergmanguste *Helogale undulata rufula* im Zoologischen Garten Frankfurt am Main. Z. Tierpsychol. 22:672–695.

Zeeb, K. 1977. Comparison du comportement inné des chevaux et des resultats du dressage au cirque. Ann. Med. Vet. 121:37–38.

Ziegler-Simon, J. 1957. Beobachtungen am Rüsseldikdik, *Rhynchotragus kirki* (Gthr.). Zool. Gart. 23:1–13.

Ziems, H. 1973. Erfahrungen und Beobachtungen bei der mutterlosen Aufzucht von Fischottern (*Lutra lutra* L. 1758). Zool. Gart. 43:305–318.

Zimen, E. 1972. Vergleichende Verhaltensbeobachtungen an Wölfen und Königspudeln. Piper, München.

Zimmermann, R. R., C. R. Geist & P. K. Ackles. 1975. Changes in the social behavior of rhesus monkeys during rehabilitation from prolonged protein-calorie malnutrition. Behav. Biol. 14:325–334.

Zippelius, H.-M. 1971. Soziale Hautpflege als Beschwichtigungsgebärde bei Säugetieren. Z. Säugetierk. 36:284–291.

Zippelius, H.-M. & F. Goethe. 1951. Ethologische Beobachtungen an Haselmäusen (*Muscardinus a. avellanarius* L.). Z. Tierpsychol. 8:348–367.

Zucker, E. L., G. Mitchell & T. Maple. 1978. Adult male-offspring play interactions within a captive group of orang-utans (*Pongo pygmaeus*). Primates 19:379–384.

Author Index

Page numbers in italics denote photographs or drawings by the author cited.

Abeelen, J.H.F. van, 231
Abelson, P.H., 312
Accordi, B., 178
Ackles, P.K., 370
Adamson, J., 238
Adamson, L., 327
Albert, I.B., 222
Albignac, R., 90, 91, 154, 156–58, 236, 237, 276, 288, 443
Alcock, J., 228, 229
Aldis, O., 39, 49, 51, 64, 92, 99, 114, 115, 116, 117, 118, 119, *120, 121,* 124, 229, 290, 291, 292-93, 310, 319, 326, 327, 372, 396, 410, 411, 412, 439, 473, 500
Aldous, S.E., 149, 234
Aldrich-Blake, F.P.G., 229
Alexander, B.K., 226
Alexander, R.D., 53, 74, 125, 252, 267, 330
Ali, S.A., 244
Allin, 500
Als, H., 327
Altman, J., 61, 62
Altmann, D., 233, 234, 244, 417
Altmann, J., 60, 101, 225, 227, 377, 403, 404, 446
Altmann, M., 191, 194, 241, 393, 404
Altmann, S.A., 47, 59, 62, 101, *102,* 223, 225, 226, 227, 319, 337, 339, 345, 378, 393, 394, 395, 399, 403, 413, 432, 446, 474
Anderson, C.O., 51
Anderson, D., 160
Anderson, M.A., 223
Anderson, V., 188, 241
Andersson, A.B., 222

Andrew, R.J., 41, 49, 423
Andrews, R.C., 70, 230, 243
Angot, M., 239
Angst, W., 226, 276, 362, 443
Angus, S., 49, 228
Anon., 55, 233, 234
Ansell, W.F.H., 235
Antonius, O., 240
Apfelbach, R., 158, 234, 237
Armitage, K.B., 128, 230, 260
Armstrong, J., 238
Arnold, G.W., 244
Arshavsky, I.A., 282, 300, 301, 307
Ashmole, N.P., 203, 207, 244
Aslin, H., 82, 222
Astrand, P.-O., 265, 281, 282, 283, 284, 296, 298
Audubon, J.J., 235
Austin, G.A., 450
Austin, H., 280, 319, 320, 321, 328, 331, 450
Autenrieth, R.E., 190, 241, 289, 361, 362
Autuori, M.P., 236
Avedon, E.M., 114, 123
Ayer, A.J., 38

Babault, G., 227, 237, 238
Bachman, J., 235
Backhaus, D., 242, 243
Baker, E.C.S., 244
Baker, R.P., 101, 226
Baker, R.R., 70, 268, 344, 352, 353, 472
Baldwin, E., 248
Baldwin, J.D., 14, 56, 59, 223, 224, 274, 275, 279, 292, 298, 306, 309, 315, 343, 349, 359, 370, 371, 377, 380, 393, 394, 399, 403, 416, 455

653

Sheldon, C., 234, 244
Shelley, F.W., 228
Shelley, O., 245
Shepherd, P., 211, 245
Sherwood, V., 122, 326, 362
Shettleworth, S.J., 264
Shillito, E.E., 86
Short, H.L., 87
Sibly, R., 406
Siddiqi, M.R., 410
Sidor, E.S., 493
Sielmann, H., 210, 246
Signoret, J.P., 301
Sikes, S.K., 178, 412, 490
Silverman, I.W., 124, 451
Simmons, K.E.L., 204
Simonds, P.E., 227, 416, 439
Simpson, A.E., 60
Simpson, C., 156, 237
Simpson, M.J.A., 41, 57, 60, 62, 94, 97,
    98, 106, 109, 226, 229, 251, 280,
    299, 310, 318, 323, 325, 350, 371
Sinclair, A.R.E., 59, 196, 243, 289
Singer, D.G., 123, 473
Singer, J.L., 123, 473
Siniff, D.B., 239
Skeldon, P.D., 236
Skidnevskii, R.N., 137
Skinner, B., 492
Skinner, M.P., 242
Skutch, A.F., 246
Sladen, W.J.L., 61
Slater, P.J.B., 264
Slatkin, M., 267
Slobodkin, L.B., 281
Slonim, A.D., 300, 361
Smilansky, S., 123
Smith, E.A., 174
Smith, E.O., 33, 42, 41, 58, 88, 226, 279,
    304, 361, 423, 451
Smith, G., 82, 222
Smith, G.A., 214
Smith, L.G., 228
Smith, M.W., 242
Smith, P.K., 58, 114, 116, 123, 124, 229,
    372
Smith, R.F., 237, 440
Smith, S.L., 129
Smith, S.M., 211, 328
Smith, W.J., 129, 229, 414, 475
Smythies, J.R., 38
Snow, C.J., 233
Snyder, P.A., 225
Soczka, L., 226, 343, 403
Sohn, J.J., 273
Sorenson, M.W., 89
Sosnovskii, J.P., 233
Southwick, C.H., 226, 227, 370, 410
Sparks, J., 227, 340, 393
Spector, S., 335
Spencer, C.C., 244, 394
Spencer, D.A., 232
Spencer-Booth, Y., 226, 394, 423

Spinage, C.A., 197, 243, 347
Spinelli, L., 237, 449
Spitzer, N.C., 247
Sprunt, A., 246
Sroges, R.W., 225
Stanley, M., 231
Stayman, K., 318
Stears, S.C., 13, 250, 255–56, 258, 359
Stefansson, V., 191-94, 233, 242, 447
Stegeman, L.C., 152, 236
Steinborn, W., 244, 275, 276
Steinemann, P., 234
Steiner, A.L., 129, 230, 288, 290, 393, 394,
    399, 417, 423
Steiniger, B., 231, 290
Stelzner, D., 289
Stephens, D.B., 242
Stern, D.N., 121-22, 229, 310, 326, 327
Stevenson, M.F., 92, 224, 225, 292
Stevenson-Hinde, J., 264
Stewart, J., 55, 133, 232, 371
Stingelin, W., 369-70
Stirling, I., 239
Stoinitzer, W., 234
Stokes, A.W., 230, 234
Stonehouse, B., 205, 244
Stonehouse, S., 205, 244
Stoner, E.A., 208, 245, 292
Strayer, F.F., 58, 224, 343
Stringham, S.F., 191, 241, 404
Stroman, H.R., 182
Struhsaker, T.T., 101, 194, 195, 225, 226,
    227, 241, 257, 274, 416, 417, 423,
    446, 447
Stuart, R.A., 268, 344, 353
Stumper, R., 247
Stynes, A.J., 227, 370, 380
Sugiyama, Y., 226, 227, 228, 229, 276,
    292, 304, 361
Sumner, E.L., 209, 245, 246
Susser, M., 642
Suomi, S.J., 41, 361, 429
Sussman, G., 57, 267, 319, 320, 321, 468,
    479
Sussman, R.W., 90, 91, 223
Sutton-Smith, B., 16, 114, 123, 124, 350,
    451, 503
Sutton-Smith, S., 350
Svihla, A., 148, 149, 230, 235
Svihla, R.D., 230
Swanberg, P.O., 208, 245
Swanson, H.H., 231, 371
Swynnerton, C.F.M., 208
Sylva, K., 16, 33, 123, 124, 308, 451, 468
Symmes, D., 394
Symons, D., 14, 16-17, 35, 44, 47, 52, 55,
    60, 63, 64, 72, 88, 94-98, 116, 124,
    159, 196, 226, 252, 268, 278, 279,
    280, 288, 290, 292, 295, 308, 309,
    311, 316, 317, 318, 323, 336, 338,
    339, 382, 393, 396, 399, 403, 410,
    411, 412, 413, 414, 416, 417, 419,
    423, 429, 430, 432, 433, 435, 443,

Topic and
Species Index

Page numbers in italics denote illustrations

668

hierarchical, 43
operational, 48
Degu. *See Octodon degus*
Demography. *See also* Survivorship
    curves and innovation, 455-56
    influence on play, 91, 191, 195, 197, 198,
    403-5
Deprivation of play, 306-7
    computer models, 380
    experimental, 35, 306
    and innovation, 455
    natural, 35, 302, 306-7, 317
    rebound from, 56, 315, 455
Description
    historical need in play research, 34
    molded by implicit theory, 71
    subjective, 36
Desert cavy. *See Microcavia australis*
Desmodus rotundus, 78, 87
Development of behavior
    evolutionary approach, 15, 248, 251,
    360
    play, 62
    regulation by play experience, 27,
    333-34
Developmental plasticity. *See* Behavioral
    flexibility
Developmental precursors, contrasted
    with play, 51
Developmental rate
    biological dependent variable, 203, 252,
    267, 335
    control and r3gulation by play, 333-34
    evolutionary significance, 251
    life-historical significance, 250
Developmental resilience. *See* Resilience
    in development, and cheating in
    play
Dhole. *See Cuon alpinus*
Diana monkey. *See Cercopithecus diana*
Diceros bicornis, 181, 240, 494
Dicrostonyx groenlandicus, 231
Dictyles tajacu, 240
Dolichotis patagonum, 135, 136, 232, 289
Dolichotis salinicola, 135, 232
Dingo. *See Canis dingo*
Dinomys branickii, 232
Dispersal, 268, 343, 352-53
    and cohesion, 342-43
    and demography, 405
    and ontogeny of play and dominance,
    438
    sex differences, 94, 353
    skills only beneficial in certain envi-
    ronments, 381
Diversity of play, 413
Diversive play, 9
Domestic cat. *See Felis catus*
Domestic cattle. *See Bos taurus*
Domestic dog. *See Canis familiaris*
Domestic fowl. *See Gallus gallus*
Domestic goat. *See Capra hircus*
Domestic guinea pig. *See Cavia porcellus*

Domestic horse. *See Equus caballus*
Domestic sheep. *See Ovis aries*
Domestic swine. *See Sus scrofa*
Dominance, 268
    and absence of play, 382
    less evident in play than in other be-
    haviors, 47, 396
    reversals, and social play relationships,
    432
Dorcas gazelle. *See Gazella dorcas*
Douc langur. *See Pygathrix nemaeus*
Drama, biology of, 3, 4
Dressage training, play movements in, 56
Drill. *See Mandrillus leucophaeus*
Dromedary. *See Camelus dromedarius*
Drug effects and use, relationships to
    human play, 490-91
Duality criterion for play, 319, 356
Dugong. *See Dugong dugon*
Dugong dugon, 179
Duino Elegy, Fourth (poem), 495
Dusicyon sechurae, 234
Dusky titi. *See Callicebus moloch*

Eagle owl. *See Bubo bubo*
East Caucasian tur. *See Capra cylindricornis*
Eastern chipmunk. *See Tamias striatus*
Eastern wood rat. *See Neotoma floridiana*
Ecological scaling of behavior, 260-61
    importance for the study of play, 17
    and tendency to play gregariously, 349
Economics of developmental program-
    ming. *See* Vitamin effect
Effects in play
    "dear enemy," 150
    McClintock, 470
    prestige, 437
    vitamin, 266, 281, 316
Effects of play. *See* Benefits of play; Costs
    of play
Eira barbara, 150, 151, 235
Elaborateness of play, in *Pan troglodytes*,
    107
Elephas maximus, 178, 179
Elk. *See Cervus canadensis*
Endrina. *See Indri indri*
Enhydra lutris, 152, 153, 235
Environmental enrichment
    and brain structure, 265, 284
    effects mediated by play experience,
    284, 351-52
Environmental enrichment/deprivation ef-
    fects
    Adaptation or damage?, 488
    Ecological significance?, 476-77
Environmental factors affecting play
    confinement, 302
    human observer, 50, 106, 127
    quatity and quality of sensory, cogni-
    tive, contingent-responsive stimu-
    lation, 284, 351-52, 478-79
    terrain, 177, 315
    weather, 302